STUDENT'S SOLUTIONS MANUAL PART ONE

WILLIAM ARDIS
Collin County Community College – Preston Ridge

UNIVERSITY CALCULUS
ELEMENTS WITH
EARLY TRANSCENDENTALS

Joel Hass
University of California, Davis

Maurice D. Weir
Naval Postgraduate School

George B. Thomas
Massachusetts Institute of Technology

PEARSON

Addison
Wesley

Boston San Francisco New York
London Toronto Sydney Tokyo Singapore Madrid
Mexico City Munich Paris Cape Town Hong Kong Montreal

Reproduced by Pearson Addison-Wesley from electronic files supplied by the author.

Copyright © 2009 Pearson Education, Inc.
Publishing as Pearson Addison-Wesley, 75 Arlington Street, Boston, MA 02116.

ISBN-13: 978-0-321-53608-2
ISBN-10: 0-321-53608-8

3 4 5 6 OPM 11 10 09 08

PREFACE TO THE STUDENT

The Student's Solutions Manual contains the solutions to all of the odd-numbered exercise in UNIVERSITY CALCULUS: ELEMENTS by Joel Hass, Maurice Weir and George Thomas, excluding the Computer Algebra System (CAS) exercises. We have worked each solution to ensure that it

- conforms exactly to the methods, procedures and steps presented in the text

- is mathematically correct

- includes all of the steps necessary so you can follow the logical argument and algebra

- includes a graph or figure whenever called for by the exercise, or if needed to help with the explanation

- is formatted in an appropriate style to aid in its understanding

How to use a solution's manual

- solve the assigned problem yourself

- if you get stuck along the way, refer to the solution in the manual as an aid but continue to solve the problem on your own

- if you cannot continue, reread the textbook section, or work through that section in the Student Study Guide, or consult your instructor

- if your answer is correct by your solution procedure seems to differ from the one in the manual, and you are unsure your method is correct, consult your instructor

- if your answer is incorrect and you cannot find your error, consult your instructor

Acknowledgments

Solutions Writers
      William Ardis, Collin County Community College-Preston Ridge Campus
      Joseph Borzellino, California Polytechnic State University
      Linda Buchanan, Howard College
      Duane Kouba, University of California-Davis
      Tim Mogill
      Patricia Nelson, University of Wisconsin-La Crosse

Accuracy Checkers
      Karl Kattchee, University of Wisconsin-La Crosse
      Debra McGivney
      Marie Vanisko, California State University, Stanislaus
      Tom Weigleitner, VISTA Information Technologies

Thanks to Elizabeth Bernardi, Rachel Reeve, Christine O'Brien, Sheila Spinney, Elka Block, and Joe Vetere for all their guidance and help at every step.

TABLE OF CONTENTS

CHAPTER 1 FUNCTIONS AND LIMITS

1.1 FUNCTIONS AND THEIR GRAPHS

1. domain $= (-\infty, \infty)$; range $= [1, \infty)$

3. domain $= (0, \infty)$; y in range $\Rightarrow$ $y = \frac{1}{\sqrt{t}}, t > 0$ $\Rightarrow$ $y^2 = \frac{1}{t}$ and $y > 0$ $\Rightarrow$ y can be any positive real number
   $\Rightarrow$ range $= (0, \infty)$.

5. (a) Not the graph of a function of x since it fails the vertical line test.
   (b) Is the graph of a function of x since any vertical line intersects the graph at most once.

7. base $= x$; (height)$^2 + \left(\frac{x}{2}\right)^2 = x^2$ $\Rightarrow$ height $= \frac{\sqrt{3}}{2}$ x; area is $a(x) = \frac{1}{2}$ (base)(height) $= \frac{1}{2}(x)\left(\frac{\sqrt{3}}{2}x\right) = \frac{\sqrt{3}}{4}x^2$;
   perimeter is $p(x) = x + x + x = 3x$.

9. Let D $=$ diagonal of a face of the cube and $\ell =$ the length of an edge. Then $\ell^2 + D^2 = d^2$ and
   $D^2 = 2\ell^2$ $\Rightarrow$ $3\ell^2 = d^2$ $\Rightarrow$ $\ell = \frac{d}{\sqrt{3}}$. The surface area is $6\ell^2 = \frac{6d^2}{3} = 2d^2$ and the volume is $\ell^3 = \left(\frac{d^2}{3}\right)^{3/2} = \frac{d^3}{3\sqrt{3}}$.

11. The domain is $(-\infty, \infty)$.

13. The domain is $(-\infty, \infty)$.

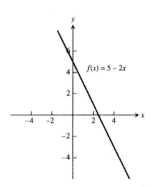

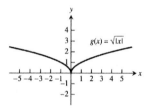

15. The domain is $(-\infty, 0) \cup (0, \infty)$.

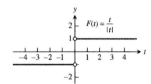

17. Neither graph passes the vertical line test

(a)

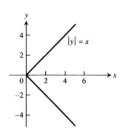

(b)

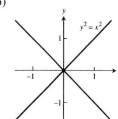

19.

x	0	1	2
y	0	1	0

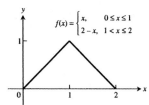

$$f(x) = \begin{cases} x, & 0 \le x \le 1 \\ 2-x, & 1 < x \le 2 \end{cases}$$

21. $G(x) = \begin{cases} 3-x, & x \le 1 \\ 2x, & x > 1 \end{cases}$

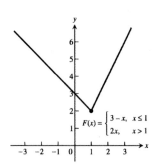

$$F(x) = \begin{cases} 3-x, & x \le 1 \\ 2x, & x > 1 \end{cases}$$

23. (a) Line through $(0, 0)$ and $(1, 1)$: $y = x$

Line through $(1, 1)$ and $(2, 0)$: $y = -x + 2$

$$f(x) = \begin{cases} x, & 0 \le x \le 1 \\ -x + 2, & 1 < x \le 2 \end{cases}$$

   (b) $f(x) = \begin{cases} 2, & 0 \le x < 1 \\ 0, & 1 \le x < 2 \\ 2, & 2 \le x < 3 \\ 0, & 3 \le x \le 4 \end{cases}$

25. (a) Line through $(-1, 1)$ and $(0, 0)$: $y = -x$

Line through $(0, 1)$ and $(1, 1)$: $y = 1$

Line through $(1, 1)$ and $(3, 0)$: $m = \frac{0-1}{3-1} = \frac{-1}{2} = -\frac{1}{2}$, so $y = -\frac{1}{2}(x - 1) + 1 = -\frac{1}{2}x + \frac{3}{2}$

$$f(x) = \begin{cases} -x & -1 \le x < 0 \\ 1 & 0 < x \le 1 \\ -\frac{1}{2}x + \frac{3}{2} & 1 < x < 3 \end{cases}$$

   (b) Line through $(-2, -1)$ and $(0, 0)$: $y = \frac{1}{2}x$

Line through $(0, 2)$ and $(1, 0)$: $y = -2x + 2$

Line through $(1, -1)$ and $(3, -1)$: $y = -1$

$$f(x) = \begin{cases} \frac{1}{2}x & -2 \le x \le 0 \\ -2x + 2 & 0 < x \le 1 \\ -1 & 1 < x \le 3 \end{cases}$$

27. (a) $\lfloor x \rfloor = 0$ for $x \in [0, 1)$ (b) $\lceil x \rceil = 0$ for $x \in (-1, 0]$

29. For any real number x, $n \le x \le n + 1$, where n is an integer. Now: $n \le x \le n + 1 \Rightarrow -(n + 1) \le -x \le -n$. By definition: $\lceil -x \rceil = -n$ and $\lfloor x \rfloor = n \Rightarrow -\lfloor x \rfloor = -n$. So $\lceil -x \rceil = -\lfloor x \rfloor$ for all $x \in \Re$.

31. Symmetric about the origin
    Dec: $-\infty < x < \infty$
    Inc: nowhere

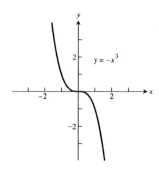

33. Symmetric about the origin
    Dec: nowhere
    Inc: $-\infty < x < 0$
    $\quad\quad 0 < x < \infty$

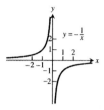

35. Symmetric about the y-axis
    Dec: $-\infty < x \le 0$
    Inc: $0 < x < \infty$

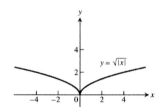

37. Symmetric about the origin
    Dec: nowhere
    Inc: $-\infty < x < \infty$

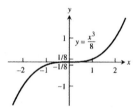

39. No symmetry
    Dec: $0 \le x < \infty$
    Inc: nowhere

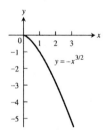

41. Symmetric about the y-axis
    Dec: $-\infty < x \le 0$
    Inc: $0 < x < \infty$

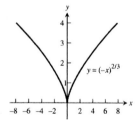

43. Since a horizontal line not through the origin is symmetric with respect to the y-axis, but not with respect to the origin, the function is even.

45. Since $f(x) = x^2 + 1 = (-x)^2 + 1 = -f(x)$. The function is even.

47. Since $g(x) = x^3 + x$, $g(-x) = -x^3 - x = -(x^3 + x) = -g(x)$. So the function is odd.

49. $g(x) = \frac{1}{x^2 - 1} = \frac{1}{(-x)^2 - 1} = g(-x)$. Thus the function is even.

51. $h(t) = \frac{1}{t-1}$; $h(-t) = \frac{1}{-t-1}$; $-h(t) = \frac{1}{1-t}$. Since $h(t) \ne -h(t)$ and $h(t) \ne h(-t)$, the function is neither even nor odd.

53. $h(t) = 2t + 1$, $h(-t) = -2t + 1$. So $h(t) \ne h(-t)$. $-h(t) = -2t - 1$, so $h(t) \ne -h(t)$. The function is neither even nor odd.

55. $f(t) = \sin 2t$ and $f(-t) = \sin 2(-t) = -\sin 2t$. So $f(t) = -f(-t)$ and the function is odd.

57. $g(t) = t\cos t$ and $g(-t) = (-t)\cos(-t) = -t\cos t$. So $g(t) = -g(-t)$ and the function is odd.

59. (a) From the graph, $\frac{x}{2} > 1 + \frac{4}{x} \Rightarrow x \in (-2, 0) \cup (4, \infty)$

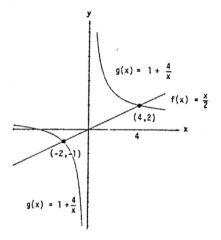

(b) $\frac{x}{2} > 1 + \frac{4}{x} \Rightarrow \frac{x}{2} - 1 - \frac{4}{x} > 0$

$x > 0$: $\frac{x}{2} - 1 - \frac{4}{x} > 0 \Rightarrow \frac{x^2 - 2x - 8}{2x} > 0 \Rightarrow \frac{(x-4)(x+2)}{2x} > 0$

$\Rightarrow x > 4$ since x is positive;

$x < 0$: $\frac{x}{2} - 1 - \frac{4}{x} > 0 \Rightarrow \frac{x^2 - 2x - 8}{2x} < 0 \Rightarrow \frac{(x-4)(x+2)}{2x} < 0$

$\Rightarrow x < -2$ since x is negative;

sign of $(x - 4)(x + 2)$

Solution interval: $(-2, 0) \cup (4, \infty)$

1.2 COMBINING FUNCTIONS; SHIFTING AND SCALING GRAPHS

1. D_f: $-\infty < x < \infty$, D_g: $x \geq 1 \Rightarrow D_{f+g} = D_{fg}$: $x \geq 1$. R_f: $-\infty < y < \infty$, R_g: $y \geq 0$, R_{f+g}: $y \geq 1$, R_{fg}: $y \geq 0$

3. D_f: $-\infty < x < \infty$, D_g: $-\infty < x < \infty$, $D_{f/g}$: $-\infty < x < \infty$, $D_{g/f}$: $-\infty < x < \infty$, R_f: $y = 2$, R_g: $y \geq 1$, $R_{f/g}$: $0 < y \leq 2$, $R_{g/f}$: $\frac{1}{2} \leq y < \infty$

5. (a) 2   (b) 22   (c) $x^2 + 2$
   (d) $(x+5)^2 - 3 = x^2 + 10x + 22$ (e) 5   (f) -2
   (g) $x + 10$ (h) $(x^2 - 3)^2 - 3 = x^4 - 6x^2 + 6$

7. (a) $\frac{4}{x^2} - 5$ (b) $\frac{4}{x^2} - 5$ (c) $\left(\frac{4}{x} - 5\right)^2$
   (d) $\frac{1}{(4x-5)^2}$ (e) $\frac{1}{4x^2 - 5}$ (f) $\frac{1}{(4x-5)^2}$

9. (a) $(f \circ g)(x)$ (b) $(j \circ g)(x)$ (c) $(g \circ g)(x)$
   (d) $(j \circ j)(x)$ (e) $(g \circ h \circ f)(x)$ (f) $(h \circ j \circ f)(x)$

11.

	$g(x)$	$f(x)$	$(f \circ g)(x)$
(a)	$x - 7$	$\sqrt{x}$	$\sqrt{x - 7}$
(b)	$x + 2$	$3x$	$3(x + 2) = 3x + 6$
(c)	x^2	$\sqrt{x - 5}$	$\sqrt{x^2 - 5}$
(d)	$\frac{x}{x-1}$	$\frac{x}{x-1}$	$\frac{\frac{x}{x-1}}{\frac{x}{x-1} - 1} = \frac{x}{x - (x-1)} = x$
(e)	$\frac{1}{x-1}$	$1 + \frac{1}{x}$	x
(f)	$\frac{1}{x}$	$\frac{1}{x}$	x

13. (a) Position 4   (b) Position 1   (c) Position 2   (d) Position 3

15.

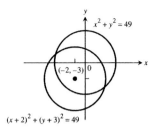
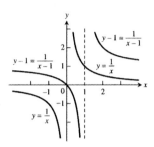

17.

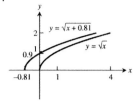
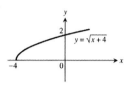

19.

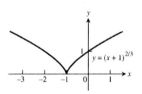
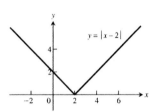

21.

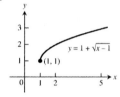

23.

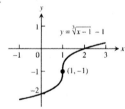

25.

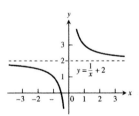

27.

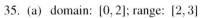

29.

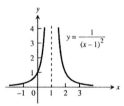

31.

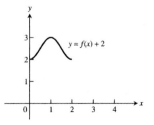

33.

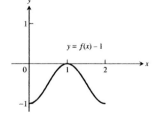

35. (a) domain: $[0, 2]$; range: $[2, 3]$ (b) domain: $[0, 2]$; range: $[-1, 0]$

(c) domain: $[0, 2]$; range: $[0, 2]$

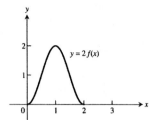

(d) domain: $[0, 2]$; range: $[-1, 0]$

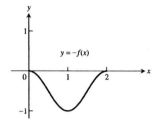

(e) domain: $[-2, 0]$; range: $[0, 1]$

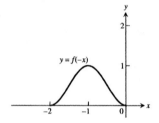

(f) domain: $[1, 3]$; range: $[0, 1]$

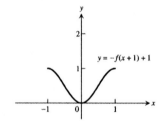

(g) domain: $[-2, 0]$; range: $[0, 1]$

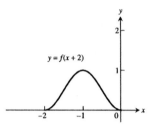

(h) domain: $[-1, 1]$; range: $[0, 1]$

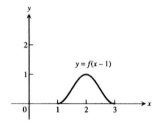

37. $y = 3x^2 - 3$

39. $y = \frac{1}{2}\left(1 + \frac{1}{x^2}\right) = \frac{1}{2} + \frac{1}{2x^2}$

41. $y = \sqrt{4x + 1}$

43. $y = \sqrt{4 - \left(\frac{x}{2}\right)^2} = \frac{1}{2}\sqrt{16 - x^2}$

45. $y = 1 - (3x)^3 = 1 - 27x^3$

47. Let $y = -\sqrt{2x + 1} = f(x)$ and let $g(x) = x^{1/2}$, $h(x) = \left(x + \frac{1}{2}\right)^{1/2}$, $i(x) = \sqrt{2}\left(x + \frac{1}{2}\right)^{1/2}$, and

$j(x) = -\left[\sqrt{2}\left(x + \frac{1}{2}\right)^{1/2}\right] = f(x)$. The graph of $h(x)$ is the graph of $g(x)$ shifted left $\frac{1}{2}$ unit; the graph of $i(x)$ is the graph of $h(x)$ stretched vertically by a factor of $\sqrt{2}$; and the graph of $j(x) = f(x)$ is the graph of $i(x)$ reflected across the x-axis.

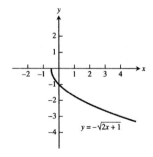

49. $y = f(x) = x^3$. Shift $f(x)$ one unit right followed by a shift two units up to get $g(x) = (x - 1)^3 + 2$.

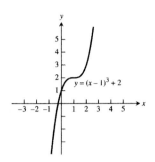

51. Compress the graph of $f(x) = \frac{1}{x}$ horizontally by a factor of 2 to get $g(x) = \frac{1}{2x}$. Then shift $g(x)$ vertically down 1 unit to get $h(x) = \frac{1}{2x} - 1$.

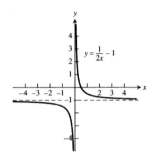

53. Reflect the graph of $y = f(x) = \sqrt[3]{x}$ across the x-axis to get $g(x) = -\sqrt[3]{x}$.

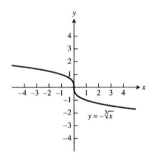

55.

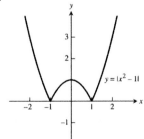

57. (a) $(fg)(-x) = f(-x)g(-x) = f(x)(-g(x)) = -(fg)(x)$, odd

(b) $\left(\frac{f}{g}\right)(-x) = \frac{f(-x)}{g(-x)} = \frac{f(x)}{-g(x)} = -\left(\frac{f}{g}\right)(x)$, odd

(c) $\left(\frac{g}{f}\right)(-x) = \frac{g(-x)}{f(-x)} = \frac{-g(x)}{f(x)} = -\left(\frac{g}{f}\right)(x)$, odd

(d) $f^2(-x) = f(-x)f(-x) = f(x)f(x) = f^2(x)$, even

(e) $g^2(-x) = (g(-x))^2 = (-g(x))^2 = g^2(x)$, even

(f) $(f \circ g)(-x) = f(g(-x)) = f(-g(x)) = f(g(x)) = (f \circ g)(x)$, even

(g) $(g \circ f)(-x) = g(f(-x)) = g(f(x)) = (g \circ f)(x)$, even

(h) $(f \circ f)(-x) = f(f(-x)) = f(f(x)) = (f \circ f)(x)$, even

(i) $(g \circ g)(-x) = g(g(-x)) = g(-g(x)) = -g(g(x)) = -(g \circ g)(x)$, odd

59. (a)

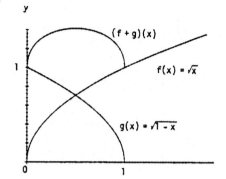

(b)

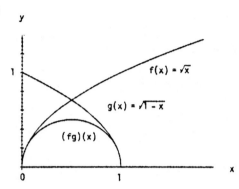

(c)

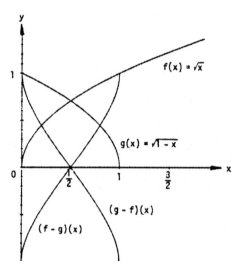

(d)

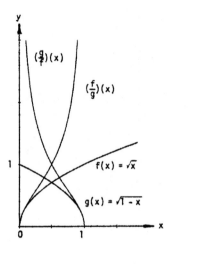

1.3 RATES OF CHANGE AND TANGENTS TO CURVES

1. (a) $\frac{\Delta f}{\Delta x} = \frac{f(3) - f(2)}{3 - 2} = \frac{28 - 9}{1} = 19$

(b) $\frac{\Delta f}{\Delta x} = \frac{f(1) - f(-1)}{1 - (-1)} = \frac{2 - 0}{2} = 1$

3. (a) $\frac{\Delta h}{\Delta t} = \frac{h\left(\frac{3\pi}{4}\right) - h\left(\frac{\pi}{4}\right)}{\frac{3\pi}{4} - \frac{\pi}{4}} = \frac{-1 - 1}{\frac{\pi}{2}} = -\frac{4}{\pi}$

(b) $\frac{\Delta h}{\Delta t} = \frac{h\left(\frac{\pi}{2}\right) - h\left(\frac{\pi}{6}\right)}{\frac{\pi}{2} - \frac{\pi}{6}} = \frac{0 - \sqrt{3}}{\frac{\pi}{3}} = \frac{-3\sqrt{3}}{\pi}$

5. $\frac{\Delta R}{\Delta \theta} = \frac{R(2) - R(0)}{2 - 0} = \frac{\sqrt{8+1} - \sqrt{1}}{2} = \frac{3 - 1}{2} = 1$

7. (a)

Q	Slope of PQ $= \frac{\Delta p}{\Delta t}$
$Q_1(10, 225)$	$\frac{650 - 225}{20 - 10} = 42.5$ m/sec
$Q_2(14, 375)$	$\frac{650 - 375}{20 - 14} = 45.83$ m/sec
$Q_3(16.5, 475)$	$\frac{650 - 475}{20 - 16.5} = 50.00$ m/sec
$Q_4(18, 550)$	$\frac{650 - 550}{20 - 18} = 50.00$ m/sec

(b) At $t = 20$, the sportscar was traveling approximately 50 m/sec or 180 km/h.

9. (a) $\frac{\Delta y}{\Delta x} = \frac{\left((2+h)^2 - 3\right) - (2^2 - 3)}{h} = \frac{4 + 4h + h^2 - 3 - 1}{h} = \frac{4h + h^2}{h} = 4 + h$. As $h \to 0$, $4 + h \to 4 \Rightarrow$ at $P(2, 1)$ the slope is 4.

(b) $y - 1 = 4(x - 2) \Rightarrow y - 1 = 4x - 8 \Rightarrow y = 4x - 7$

11. (a) $\frac{\Delta y}{\Delta x} = \frac{\left((2+h)^2 - 2(2+h) - 3\right) - \left(2^2 - 2(2) - 3\right)}{h} = \frac{4 + 4h + h^2 - 4 - 2h - 3 - (-3)}{h} = \frac{2h + h^2}{h} = 2 + h.$ As $h \to 0$, $2 + h \to 2 \Rightarrow$ at
 P$(2, -3)$ the slope is 2.

 (b) $y - (-3) = 2(x - 2) \Rightarrow y + 3 = 2x - 4 \Rightarrow y = 2x - 7.$

13. (a) $\frac{\Delta y}{\Delta x} = \frac{(2+h)^3 - 2^3}{h} = \frac{8 + 12h + 4h^2 + h^3 - 8}{h} = \frac{12h + 4h^2 + h^3}{h} = 12 + 4h + h^2.$ As $h \to 0$, $12 + 4h + h^2 \to 12$, $\Rightarrow$ at
 P$(2, 8)$ the slope is 12.

 (b) $y - 8 = 12(x - 2) \Rightarrow y - 8 = 12x - 24 \Rightarrow y = 12x - 16.$

15. (a) $\frac{\Delta y}{\Delta x} = \frac{(1+h)^3 - 12(1+h) - (1^3 - 12(1))}{h} = \frac{1 + 3h + 3h^2 + h^3 - 12 - 12h - (-11)}{h} = \frac{-9h + 3h^2 + h^3}{h} = -9 + 3h + h^2.$ As $h \to 0$,
 $-9 + 3h + h^2 \to -9 \Rightarrow$ at P$(1, -11)$ the slope is -9.

 (b) $y - (-11) = (-9)(x - 1) \Rightarrow y + 11 = -9x + 9 \Rightarrow y = -9x - 2.$

17. (a)

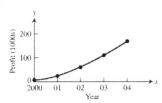

 (b) $\frac{\Delta p}{\Delta t} = \frac{174 - 62}{2004 - 2002} = \frac{112}{2} = 56$ thousand dollars per year

 (c) The average rate of change from 2001 to 2002 is $\frac{\Delta p}{\Delta t} = \frac{62 - 27}{20022 - 2001} = 35$ thousand dollars per year.
  The average rate of change from 2002 to 2003 is $\frac{\Delta p}{\Delta t} = \frac{111 - 62}{2003 - 2002} = 49$ thousand dollars per year.
  So, the rate at which profits were changing in 2002 is approximatley $\frac{1}{2}(35 + 49) = 42$ thousand dollars per year.

1.4 LIMIT OF A FUNCTION AND LIMIT LAWS

1. (a) Does not exist. As x approaches 1 from the right, g(x) approaches 0. As x approaches 1 from the left, g(x)
  approaches 1. There is no single number L that all the values g(x) get arbitrarily close to as $x \to 1$.

 (b) 1

 (c) 0

3. (a) True   (b) True   (c) False

 (d) False   (e) False   (f) True

5. $\lim\limits_{x \to 0} \frac{x}{|x|}$ does not exist because $\frac{x}{|x|} = \frac{x}{x} = 1$ if $x > 0$ and $\frac{x}{|x|} = \frac{x}{-x} = -1$ if $x < 0$. As x approaches 0 from the left,
 $\frac{x}{|x|}$ approaches -1. As x approaches 0 from the right, $\frac{x}{|x|}$ approaches 1. There is no single number L that all
 the function values get arbitrarily close to as $x \to 0$.

7. Nothing can be said about f(x) because the existence of a limit as $x \to x_0$ does not depend on how the function
 is defined at x_0. In order for a limit to exist, f(x) must be arbitrarily close to a single real number L when
 x is close enough to x_0. That is, the existence of a limit depends on the values of f(x) for x <u>near</u> x_0, not on the
 definition of f(x) at x_0 itself.

9. No, the definition does not require that f be defined at $x = 1$ in order for a limiting value to exist there. If f(1)
 is defined, it can be any real number, so we can conclude nothing about f(1) from $\lim\limits_{x \to 1} f(x) = 5.$

11. $\lim\limits_{x \to -7} (2x + 5) = 2(-7) + 5 = -14 + 5 = -9$

13. $\lim\limits_{x \to 2} (-x^2 + 5x - 2) = -(2)^2 + 5(2) - 2 = -4 + 10 - 2 = 4$

15. $\lim\limits_{t \to 6} 8(t-5)(t-7) = 8(6-5)(6-7) = -8$

17. $\lim\limits_{x \to 2} \frac{x+3}{x+6} = \frac{2+3}{2+6} = \frac{5}{8}$

19. $\lim\limits_{y \to -5} \frac{y^2}{5-y} = \frac{(-5)^2}{5-(-5)} = \frac{25}{10} = \frac{5}{2}$

21. $\lim\limits_{x \to -1} 3(2x-1)^2 = 3(2(-1)-1)^2 = 3(-3)^2 = 27$

23. $\lim\limits_{y \to -3} (5-y)^{4/3} = [5-(-3)]^{4/3} = (8)^{4/3} = \left((8)^{1/3}\right)^4 = 2^4 = 16$

25. $\lim\limits_{h \to 0} \frac{3}{\sqrt{3h+1}+1} = \frac{3}{\sqrt{3(0)+1}+1} = \frac{3}{\sqrt{1}+1} = \frac{3}{2}$

27. $\lim\limits_{h \to 0} \frac{\sqrt{3h+1}-1}{h} = \lim\limits_{h \to 0} \frac{\sqrt{3h+1}-1}{h} \cdot \frac{\sqrt{3h+1}+1}{\sqrt{3h+1}+1} = \lim\limits_{h \to 0} \frac{(3h+1)-1}{h\left(\sqrt{3h+1}+1\right)} = \lim\limits_{h \to 0} \frac{3h}{h\left(\sqrt{3h+1}+1\right)} = \lim\limits_{h \to 0} \frac{3}{\sqrt{3h+1}+1}$
$= \frac{3}{\sqrt{1}+1} = \frac{3}{2}$

29. $\lim\limits_{x \to 5} \frac{x-5}{x^2-25} = \lim\limits_{x \to 5} \frac{x-5}{(x+5)(x-5)} = \lim\limits_{x \to 5} \frac{1}{x+5} = \frac{1}{5+5} = \frac{1}{10}$

31. $\lim\limits_{x \to -5} \frac{x^2+3x-10}{x+5} = \lim\limits_{x \to -5} \frac{(x+5)(x-2)}{x+5} = \lim\limits_{x \to -5} (x-2) = -5 - 2 = -7$

33. $\lim\limits_{t \to 1} \frac{t^2+t-2}{t^2-1} = \lim\limits_{t \to 1} \frac{(t+2)(t-1)}{(t-1)(t+1)} = \lim\limits_{t \to 1} \frac{t+2}{t+1} = \frac{1+2}{1+1} = \frac{3}{2}$

35. $\lim\limits_{x \to -2} \frac{-2x-4}{x^3+2x^2} = \lim\limits_{x \to -2} \frac{-2(x+2)}{x^2(x+2)} = \lim\limits_{x \to -2} \frac{-2}{x^2} = \frac{-2}{4} = -\frac{1}{2}$

37. $\lim\limits_{u \to 1} \frac{u^4-1}{u^3-1} = \lim\limits_{u \to 1} \frac{(u^2+1)(u+1)(u-1)}{(u^2+u+1)(u-1)} = \lim\limits_{u \to 1} \frac{(u^2+1)(u+1)}{u^2+u+1} = \frac{(1+1)(1+1)}{1+1+1} = \frac{4}{3}$

39. $\lim\limits_{x \to 9} \frac{\sqrt{x}-3}{x-9} = \lim\limits_{x \to 9} \frac{\sqrt{x}-3}{(\sqrt{x}-3)(\sqrt{x}+3)} = \lim\limits_{x \to 9} \frac{1}{\sqrt{x}+3} = \frac{1}{\sqrt{9}+3} = \frac{1}{6}$

41. $\lim\limits_{x \to 1} \frac{x-1}{\sqrt{x+3}-2} = \lim\limits_{x \to 1} \frac{(x-1)\left(\sqrt{x+3}+2\right)}{\left(\sqrt{x+3}-2\right)\left(\sqrt{x+3}+2\right)} = \lim\limits_{x \to 1} \frac{(x-1)\left(\sqrt{x+3}+2\right)}{(x+3)-4} = \lim\limits_{x \to 1} \left(\sqrt{x+3}+2\right)$
$= \sqrt{4} + 2 = 4$

43. $\lim\limits_{x \to 2} \frac{\sqrt{x^2+12}-4}{x-2} = \lim\limits_{x \to 2} \frac{\left(\sqrt{x^2+12}-4\right)\left(\sqrt{x^2+12}+4\right)}{(x-2)\left(\sqrt{x^2+12}+4\right)} = \lim\limits_{x \to 2} \frac{(x^2+12)-16}{(x-2)\left(\sqrt{x^2+12}+4\right)}$
$= \lim\limits_{x \to 2} \frac{(x-2)(x+2)}{(x-2)\left(\sqrt{x^2+12}+4\right)} = \lim\limits_{x \to 2} \frac{x+2}{\sqrt{x^2+12}+4} = \frac{4}{\sqrt{16}+4} = \frac{1}{2}$

45. $\lim\limits_{x \to -3} \frac{2-\sqrt{x^2-5}}{x+3} = \lim\limits_{x \to -3} \frac{\left(2-\sqrt{x^2-5}\right)\left(2+\sqrt{x^2-5}\right)}{(x+3)\left(2+\sqrt{x^2-5}\right)} = \lim\limits_{x \to -3} \frac{4-(x^2-5)}{(x+3)\left(2+\sqrt{x^2-5}\right)}$
$= \lim\limits_{x \to -3} \frac{9-x^2}{(x+3)\left(2+\sqrt{x^2-5}\right)} = \lim\limits_{x \to -3} \frac{(3-x)(3+x)}{(x+3)\left(2+\sqrt{x^2-5}\right)} = \lim\limits_{x \to -3} \frac{3-x}{2+\sqrt{x^2-5}} = \frac{6}{2+\sqrt{4}} = \frac{3}{2}$

47. $\lim\limits_{x \to 0} (2\sin x - 1) = 2\sin 0 - 1 = 0 - 1 = -1$

49. $\lim\limits_{x \to 0} \sec x = \lim\limits_{x \to 0} \frac{1}{\cos x} = \frac{1}{\cos 0} = \frac{1}{1} = 1$

51. $\lim\limits_{x \to 0} \frac{1+x+\sin x}{3\cos x} = \frac{1+0+\sin 0}{3\cos 0} = \frac{1+0+0}{3} = \frac{1}{3}$

53. $\lim\limits_{x \to 0} \sqrt{x+1}\, \cos^{1/3} x = \lim\limits_{x \to 0} \sqrt{x+1} \cdot \sqrt[3]{\cos x} = \sqrt{0+1} \cdot \sqrt[3]{\cos 0} = \sqrt{1} \cdot \sqrt[3]{1} = 1 \cdot 1 = 1$

55. (a) $\lim\limits_{x \to c} f(x)\, g(x) = \left[\lim\limits_{x \to c} f(x)\right]\left[\lim\limits_{x \to c} g(x)\right] = (5)(-2) = -10$

   (b) $\lim\limits_{x \to c} 2f(x)\, g(x) = 2\left[\lim\limits_{x \to c} f(x)\right]\left[\lim\limits_{x \to c} g(x)\right] = 2(5)(-2) = -20$

   (c) $\lim\limits_{x \to c} [f(x) + 3g(x)] = \lim\limits_{x \to c} f(x) + 3\lim\limits_{x \to c} g(x) = 5 + 3(-2) = -1$

   (d) $\lim\limits_{x \to c} \frac{f(x)}{f(x) - g(x)} = \frac{\lim\limits_{x \to c} f(x)}{\lim\limits_{x \to c} f(x) - \lim\limits_{x \to c} g(x)} = \frac{5}{5 - (-2)} = \frac{5}{7}$

57. $\lim\limits_{h \to 0} \frac{(1+h)^2 - 1^2}{h} = \lim\limits_{h \to 0} \frac{1 + 2h + h^2 - 1}{h} = \lim\limits_{h \to 0} \frac{h(2+h)}{h} = \lim\limits_{h \to 0} (2+h) = 2$

59. $\lim\limits_{h \to 0} \frac{[3(2+h) - 4] - [3(2) - 4]}{h} = \lim\limits_{h \to 0} \frac{3h}{h} = 3$

61. $\lim\limits_{h \to 0} \frac{\sqrt{7+h} - \sqrt{7}}{h} = \lim\limits_{h \to 0} \frac{\left(\sqrt{7+h} - \sqrt{7}\right)\left(\sqrt{7+h} + \sqrt{7}\right)}{h\left(\sqrt{7+h} + \sqrt{7}\right)} = \lim\limits_{h \to 0} \frac{(7+h) - 7}{h\left(\sqrt{7+h} + \sqrt{7}\right)}$

   $= \lim\limits_{h \to 0} \frac{h}{h\left(\sqrt{7+h} + \sqrt{7}\right)} = \lim\limits_{h \to 0} \frac{1}{\sqrt{7+h} + \sqrt{7}} = \frac{1}{2\sqrt{7}}$

63. $\lim\limits_{x \to 0} \sqrt{5 - 2x^2} = \sqrt{5 - 2(0)^2} = \sqrt{5}$ and $\lim\limits_{x \to 0} \sqrt{5 - x^2} = \sqrt{5 - (0)^2} = \sqrt{5}$; by the sandwich theorem,

   $\lim\limits_{x \to 0} f(x) = \sqrt{5}$

65. (a) $\lim\limits_{x \to 0} \left(1 - \frac{x^2}{6}\right) = 1 - \frac{0}{6} = 1$ and $\lim\limits_{x \to 0} 1 = 1$; by the sandwich theorem, $\lim\limits_{x \to 0} \frac{x \sin x}{2 - 2\cos x} = 1$

   (b) For $x \neq 0$, $y = (x \sin x)/(2 - 2 \cos x)$
      lies between the other two graphs in the
      figure, and the graphs converge as $x \to 0$.

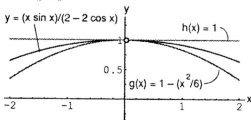

67. (a) $f(x) = (x^2 - 9)/(x + 3)$

x	-3.1	-3.01	-3.001	-3.0001	-3.00001	-3.000001
f(x)	-6.1	-6.01	-6.001	-6.0001	-6.00001	-6.000001

x	-2.9	-2.99	-2.999	-2.9999	-2.99999	-2.999999
f(x)	-5.9	-5.99	-5.999	-5.9999	-5.99999	-5.999999

   The estimate is $\lim\limits_{x \to -3} f(x) = -6$

   (b)

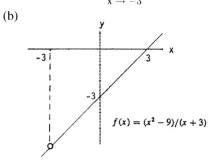

(c) $f(x) = \frac{x^2-9}{x+3} = \frac{(x+3)(x-3)}{x+3} = x - 3$ if $x \neq -3$, and $\lim_{x \to -3} (x - 3) = -3 - 3 = -6$.

69. (a) $f(x) = (x^2 - 1)/(|x| - 1)$

x	−1.1	−1.01	−1.001	−1.0001	−1.00001	−1.000001
f(x)	2.1	2.01	2.001	2.0001	2.00001	2.000001

x	−.9	−.99	−.999	−.9999	−.99999	−.999999
f(x)	1.9	1.99	1.999	1.9999	1.99999	1.999999

(b)

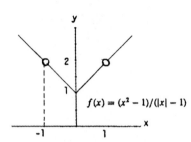

(c) $f(x) = \frac{x^2-1}{|x|-1} = \begin{cases} \frac{(x+1)(x-1)}{x-1} = x + 1, & x \geq 0 \text{ and } x \neq 1 \\ \frac{(x+1)(x-1)}{-(x+1)} = 1 - x, & x < 0 \text{ and } x \neq -1 \end{cases}$, and $\lim_{x \to -1} (1 - x) = 1 - (-1) = 2$.

71. (a) $g(\theta) = (\sin \theta)/\theta$

θ	.1	.01	.001	.0001	.00001	.000001
$g(\theta)$	.998334	.999983	.999999	.999999	.999999	.999999

θ	−.1	−.01	−.001	−.0001	−.00001	−.000001
$g(\theta)$	.998334	.999983	.999999	.999999	.999999	.999999

$\lim_{\theta \to 0} g(\theta) = 1$

(b)

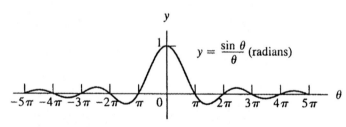

NOT TO SCALE

73. $\lim_{x \to c} f(x)$ exists at those points c where $\lim_{x \to c} x^4 = \lim_{x \to c} x^2$. Thus, $c^4 = c^2 \Rightarrow c^2(1 - c^2) = 0$

$\Rightarrow c = 0, 1, \text{ or } -1$. Moreover, $\lim_{x \to 0} f(x) = \lim_{x \to 0} x^2 = 0$ and $\lim_{x \to -1} f(x) = \lim_{x \to 1} f(x) = 1$.

75. (a) $0 = 3 \cdot 0 = \left[\lim_{x \to 2} \frac{f(x)-5}{x-2}\right]\left[\lim_{x \to 2}(x-2)\right] = \lim_{x \to 2}\left[\left(\frac{f(x)-5}{x-2}\right)(x-2)\right] = \lim_{x \to 2}[f(x) - 5] = \lim_{x \to 2} f(x) - 5$

$\Rightarrow \lim_{x \to 2} f(x) = 5$.

(b) $0 = 4 \cdot 0 = \left[\lim_{x \to 2} \frac{f(x)-5}{x-2}\right]\left[\lim_{x \to 2}(x-2)\right] \Rightarrow \lim_{x \to 2} f(x) = 5$ as in part (a).

77. (a) $\lim\limits_{x \to 0} x \sin \frac{1}{x} = 0$

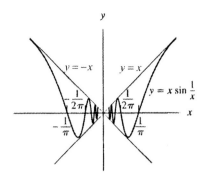

(b) $-1 \le \sin \frac{1}{x} \le 1$ for $x \ne 0$:

$x > 0 \Rightarrow -x \le x \sin \frac{1}{x} \le x \Rightarrow \lim\limits_{x \to 0} x \sin \frac{1}{x} = 0$ by the sandwich theorem;

$x < 0 \Rightarrow -x \ge x \sin \frac{1}{x} \ge x \Rightarrow \lim\limits_{x \to 0} x \sin \frac{1}{x} = 0$ by the sandwich theorem.

1.5 PRECISE DEFINITION OF A LIMIT

1.

   Step 1: $|x - 5| < \delta \Rightarrow -\delta < x - 5 < \delta \Rightarrow -\delta + 5 < x < \delta + 5$

   Step 2: $\delta + 5 = 7 \Rightarrow \delta = 2$, or $-\delta + 5 = 1 \Rightarrow \delta = 4$.

   The value of δ which assures $|x - 5| < \delta \Rightarrow 1 < x < 7$ is the smaller value, $\delta = 2$.

3.

   Step 1: $|x - (-3)| < \delta \Rightarrow -\delta < x + 3 < \delta \Rightarrow -\delta - 3 < x < \delta - 3$

   Step 2: $-\delta - 3 = -\frac{7}{2} \Rightarrow \delta = \frac{1}{2}$, or $\delta - 3 = -\frac{1}{2} \Rightarrow \delta = \frac{5}{2}$.

   The value of δ which assures $|x - (-3)| < \delta \Rightarrow -\frac{7}{2} < x < -\frac{1}{2}$ is the smaller value, $\delta = \frac{1}{2}$.

5.

   Step 1: $\left|x - \frac{1}{2}\right| < \delta \Rightarrow -\delta < x - \frac{1}{2} < \delta \Rightarrow -\delta + \frac{1}{2} < x < \delta + \frac{1}{2}$

   Step 2: $-\delta + \frac{1}{2} = \frac{4}{9} \Rightarrow \delta = \frac{1}{18}$, or $\delta + \frac{1}{2} = \frac{4}{7} \Rightarrow \delta = \frac{1}{14}$.

   The value of δ which assures $\left|x - \frac{1}{2}\right| < \delta \Rightarrow \frac{4}{9} < x < \frac{4}{7}$ is the smaller value, $\delta = \frac{1}{18}$.

7. Step 1: $|x - 5| < \delta \Rightarrow -\delta < x - 5 < \delta \Rightarrow -\delta + 5 < x < \delta + 5$

   Step 2: From the graph, $-\delta + 5 = 4.9 \Rightarrow \delta = 0.1$, or $\delta + 5 = 5.1 \Rightarrow \delta = 0.1$; thus $\delta = 0.1$ in either case.

9. Step 1: $|x - 2| < \delta \Rightarrow -\delta < x - 2 < \delta \Rightarrow -\delta + 2 < x < \delta + 2$

   Step 2: From the graph, $-\delta + 2 = \sqrt{3} \Rightarrow \delta = 2 - \sqrt{3} \approx 0.2679$, or $\delta + 2 = \sqrt{5} \Rightarrow \delta = \sqrt{5} - 2 \approx 0.2361$; thus $\delta = \sqrt{5} - 2$.

11. Step 1: $|(x + 1) - 5| < 0.01 \Rightarrow |x - 4| < 0.01 \Rightarrow -0.01 < x - 4 < 0.01 \Rightarrow 3.99 < x < 4.01$

    Step 2: $|x - 4| < \delta \Rightarrow -\delta < x - 4 < \delta \Rightarrow -\delta + 4 < x < \delta + 4 \Rightarrow \delta = 0.01$.

13. Step 1: $\left|\sqrt{x + 1} - 1\right| < 0.1 \Rightarrow -0.1 < \sqrt{x + 1} - 1 < 0.1 \Rightarrow 0.9 < \sqrt{x + 1} < 1.1 \Rightarrow 0.81 < x + 1 < 1.21$

    $\Rightarrow -0.19 < x < 0.21$

    Step 2: $|x - 0| < \delta \Rightarrow -\delta < x < \delta$. Then, $-\delta = -0.19 \Rightarrow \delta = 0.19$ or $\delta = 0.21$; thus, $\delta = 0.19$.

15. Step 1: $\left|\sqrt{19-x}-3\right| < 1 \Rightarrow -1 < \sqrt{19-x}-3 < 1 \Rightarrow 2 < \sqrt{19-x} < 4 \Rightarrow 4 < 19-x < 16$

$\Rightarrow -4 > x - 19 > -16 \Rightarrow 15 > x > 3$ or $3 < x < 15$

    Step 2: $|x-10| < \delta \Rightarrow -\delta < x - 10 < \delta \Rightarrow -\delta + 10 < x < \delta + 10.$

Then $-\delta + 10 = 3 \Rightarrow \delta = 7$, or $\delta + 10 = 15 \Rightarrow \delta = 5$; thus $\delta = 5$.

17. Step 1: $\left|\frac{1}{x} - \frac{1}{4}\right| < 0.05 \Rightarrow -0.05 < \frac{1}{x} - \frac{1}{4} < 0.05 \Rightarrow 0.2 < \frac{1}{x} < 0.3 \Rightarrow \frac{10}{2} > x > \frac{10}{3}$ or $\frac{10}{3} < x < 5.$

    Step 2: $|x-4| < \delta \Rightarrow -\delta < x - 4 < \delta \Rightarrow -\delta + 4 < x < \delta + 4.$

Then $-\delta + 4 = \frac{10}{3}$ or $\delta = \frac{2}{3}$, or $\delta + 4 = 5$ or $\delta = 1$; thus $\delta = \frac{2}{3}$.

19. Step 1: $|x^2 - 4| < 0.5 \Rightarrow -0.5 < x^2 - 4 < 0.5 \Rightarrow 3.5 < x^2 < 4.5 \Rightarrow \sqrt{3.5} < |x| < \sqrt{4.5} \Rightarrow -\sqrt{4.5} < x < -\sqrt{3.5},$

for x near -2.

    Step 2: $|x - (-2)| < \delta \Rightarrow -\delta < x + 2 < \delta \Rightarrow -\delta - 2 < x < \delta - 2.$

Then $-\delta - 2 = -\sqrt{4.5} \Rightarrow \delta = \sqrt{4.5} - 2 \approx 0.1213$, or $\delta - 2 = -\sqrt{3.5} \Rightarrow \delta = 2 - \sqrt{3.5} \approx 0.1292$;

thus $\delta = \sqrt{4.5} - 2 \approx 0.12.$

21. Step 1: $|(x^2 - 5) - 11| < 1 \Rightarrow |x^2 - 16| < 1 \Rightarrow -1 < x^2 - 16 < 1 \Rightarrow 15 < x^2 < 17 \Rightarrow \sqrt{15} < x < \sqrt{17}.$

    Step 2: $|x - 4| < \delta \Rightarrow -\delta < x - 4 < \delta \Rightarrow -\delta + 4 < x < \delta + 4.$

Then $-\delta + 4 = \sqrt{15} \Rightarrow \delta = 4 - \sqrt{15} \approx 0.1270$, or $\delta + 4 = \sqrt{17} \Rightarrow \delta = \sqrt{17} - 4 \approx 0.1231$;

thus $\delta = \sqrt{17} - 4 \approx 0.12.$

23. $\lim\limits_{x \to 3} (3 - 2x) = 3 - 2(3) = -3$

    Step 1: $|(3 - 2x) - (-3)| < 0.02 \Rightarrow -0.02 < 6 - 2x < 0.02 \Rightarrow -6.02 < -2x < -5.98 \Rightarrow 3.01 > x > 2.99$ or

$2.99 < x < 3.01.$

    Step 2: $0 < |x - 3| < \delta \Rightarrow -\delta < x - 3 < \delta \Rightarrow -\delta + 3 < x < \delta + 3.$

Then $-\delta + 3 = 2.99 \Rightarrow \delta = 0.01$, or $\delta + 3 = 3.01 \Rightarrow \delta = 0.01$; thus $\delta = 0.01.$

25. $\lim\limits_{x \to 2} \frac{x^2 - 4}{x - 2} = \lim\limits_{x \to 2} \frac{(x+2)(x-2)}{(x-2)} = \lim\limits_{x \to 2} (x + 2) = 2 + 2 = 4, x \neq 2$

    Step 1: $\left|\left(\frac{x^2 - 4}{x - 2}\right) - 4\right| < 0.05 \Rightarrow -0.05 < \frac{(x+2)(x-2)}{(x-2)} - 4 < 0.05 \Rightarrow 3.95 < x + 2 < 4.05, x \neq 2$

$\Rightarrow 1.95 < x < 2.05, x \neq 2.$

    Step 2: $|x - 2| < \delta \Rightarrow -\delta < x - 2 < \delta \Rightarrow -\delta + 2 < x < \delta + 2.$

Then $-\delta + 2 = 1.95 \Rightarrow \delta = 0.05$, or $\delta + 2 = 2.05 \Rightarrow \delta = 0.05$; thus $\delta = 0.05.$

27. Step 1: $|(9 - x) - 5| < \epsilon \Rightarrow -\epsilon < 4 - x < \epsilon \Rightarrow -\epsilon - 4 < -x < \epsilon - 4 \Rightarrow \epsilon + 4 > x > 4 - \epsilon \Rightarrow 4 - \epsilon < x < 4 + \epsilon.$

    Step 2: $|x - 4| < \delta \Rightarrow -\delta < x - 4 < \delta \Rightarrow -\delta + 4 < x < \delta + 4.$

Then $-\delta + 4 = -\epsilon + 4 \Rightarrow \delta = \epsilon$, or $\delta + 4 = \epsilon + 4 \Rightarrow \delta = \epsilon$. Thus choose $\delta = \epsilon.$

29. Step 1: $\left|\sqrt{x-5} - 2\right| < \epsilon \Rightarrow -\epsilon < \sqrt{x-5} - 2 < \epsilon \Rightarrow 2 - \epsilon < \sqrt{x-5} < 2 + \epsilon \Rightarrow (2 - \epsilon)^2 < x - 5 < (2 + \epsilon)^2$

$\Rightarrow (2 - \epsilon)^2 + 5 < x < (2 + \epsilon)^2 + 5.$

    Step 2: $|x - 9| < \delta \Rightarrow -\delta < x - 9 < \delta \Rightarrow -\delta + 9 < x < \delta + 9.$

Then $-\delta + 9 = \epsilon^2 - 4\epsilon + 9 \Rightarrow \delta = 4\epsilon - \epsilon^2$, or $\delta + 9 = \epsilon^2 + 4\epsilon + 9 \Rightarrow \delta = 4\epsilon + \epsilon^2$. Thus choose

the smaller distance, $\delta = 4\epsilon - \epsilon^2.$

31. Step 1:   For $x \neq 1$, $|x^2 - 1| < \epsilon \Rightarrow -\epsilon < x^2 - 1 < \epsilon \Rightarrow 1 - \epsilon < x^2 < 1 + \epsilon \Rightarrow \sqrt{1 - \epsilon} < |x| < \sqrt{1 + \epsilon}$

$\Rightarrow \sqrt{1 - \epsilon} < x < \sqrt{1 + \epsilon}$ near $x = 1.$

Step 2: $|x - 1| < \delta \Rightarrow -\delta < x - 1 < \delta \Rightarrow -\delta + 1 < x < \delta + 1.$

Then $-\delta + 1 = \sqrt{1 - \epsilon} \Rightarrow \delta = 1 - \sqrt{1 - \epsilon}$, or $\delta + 1 = \sqrt{1 + \epsilon} \Rightarrow \delta = \sqrt{1 + \epsilon} - 1.$ Choose

$\delta = \min\left\{1 - \sqrt{1 - \epsilon}, \sqrt{1 + \epsilon} - 1\right\}$, that is, the smaller of the two distances.

33. Step 1: $\left|\frac{1}{x} - 1\right| < \epsilon \Rightarrow -\epsilon < \frac{1}{x} - 1 < \epsilon \Rightarrow 1 - \epsilon < \frac{1}{x} < 1 + \epsilon \Rightarrow \frac{1}{1+\epsilon} < x < \frac{1}{1-\epsilon}.$

Step 2: $|x - 1| < \delta \Rightarrow -\delta < x - 1 < \delta \Rightarrow 1 - \delta < x < 1 + \delta.$

Then $1 - \delta = \frac{1}{1+\epsilon} \Rightarrow \delta = 1 - \frac{1}{1+\epsilon} = \frac{\epsilon}{1+\epsilon}$, or $1 + \delta = \frac{1}{1-\epsilon} \Rightarrow \delta = \frac{1}{1-\epsilon} - 1 = \frac{\epsilon}{1-\epsilon}.$

Choose $\delta = \frac{\epsilon}{1+\epsilon}$, the smaller of the two distances.

35. Step 1: $\left|\left(\frac{x^2 - 9}{x + 3}\right) - (-6)\right| < \epsilon \Rightarrow -\epsilon < (x - 3) + 6 < \epsilon, x \neq -3 \Rightarrow -\epsilon < x + 3 < \epsilon \Rightarrow -\epsilon - 3 < x < \epsilon - 3.$

Step 2: $|x - (-3)| < \delta \Rightarrow -\delta < x + 3 < \delta \Rightarrow -\delta - 3 < x < \delta - 3.$

Then $-\delta - 3 = -\epsilon - 3 \Rightarrow \delta = \epsilon$, or $\delta - 3 = \epsilon - 3 \Rightarrow \delta = \epsilon.$ Choose $\delta = \epsilon.$

37. By the figure, $-x \leq x \sin \frac{1}{x} \leq x$ for all $x > 0$ and $-x \geq x \sin \frac{1}{x} \geq x$ for $x < 0$. Since $\lim_{x \to 0} (-x) = \lim_{x \to 0} x = 0$,

then by the sandwich theorem, in either case, $\lim_{x \to 0} x \sin \frac{1}{x} = 0.$

39. As x approaches the value 0, the values of g(x) approach k. Thus for every number $\epsilon > 0$, there exists a $\delta > 0$
such that $0 < |x - 0| < \delta \Rightarrow |g(x) - k| < \epsilon.$

41. Let $f(x) = x^2$. The function values do get closer to -1 as x approaches 0, but $\lim_{x \to 0} f(x) = 0$, not -1. The
function $f(x) = x^2$ never gets <u>arbitrarily close</u> to -1 for x near 0.

43. $|A - 9| \leq 0.01 \Rightarrow -0.01 \leq \pi\left(\frac{x}{2}\right)^2 - 9 \leq 0.01 \Rightarrow 8.99 \leq \frac{\pi x^2}{4} \leq 9.01 \Rightarrow \frac{4}{\pi}(8.99) \leq x^2 \leq \frac{4}{\pi}(9.01)$

$\Rightarrow 2\sqrt{\frac{8.99}{\pi}} \leq x \leq 2\sqrt{\frac{9.01}{\pi}}$ or $3.384 \leq x \leq 3.387.$ To be safe, the left endpoint was rounded up and the right
endpoint was rounded down.

45. (a) $-\delta < x - 1 < 0 \Rightarrow 1 - \delta < x < 1 \Rightarrow f(x) = x.$ Then $|f(x) - 2| = |x - 2| = 2 - x > 2 - 1 = 1.$ That is,
$|f(x) - 2| \geq 1 \geq \frac{1}{2}$ no matter how small δ is taken when $1 - \delta < x < 1 \Rightarrow \lim_{x \to 1} f(x) \neq 2.$

(b) $0 < x - 1 < \delta \Rightarrow 1 < x < 1 + \delta \Rightarrow f(x) = x + 1.$ Then $|f(x) - 1| = |(x + 1) - 1| = |x| = x > 1.$ That is,
$|f(x) - 1| \geq 1$ no matter how small δ is taken when $1 < x < 1 + \delta \Rightarrow \lim_{x \to 1} f(x) \neq 1.$

(c) $-\delta < x - 1 < 0 \Rightarrow 1 - \delta < x < 1 \Rightarrow f(x) = x.$ Then $|f(x) - 1.5| = |x - 1.5| = 1.5 - x > 1.5 - 1 = 0.5.$
Also, $0 < x - 1 < \delta \Rightarrow 1 < x < 1 + \delta \Rightarrow f(x) = x + 1.$ Then $|f(x) - 1.5| = |(x + 1) - 1.5| = |x - 0.5|$
$= x - 0.5 > 1 - 0.5 = 0.5.$ Thus, no matter how small δ is taken, there exists a value of x such that
$-\delta < x - 1 < \delta$ but $|f(x) - 1.5| \geq \frac{1}{2} \Rightarrow \lim_{x \to 1} f(x) \neq 1.5.$

1.6 ONE-SIDED LIMITS

1. (a) True (b) True (c) False (d) True
   (e) True (f) True (g) False (h) False
   (i) False (j) False (k) True (l) False

3. (a) $\lim_{x \to 2^+} f(x) = \frac{2}{2} + 1 = 2$, $\lim_{x \to 2^-} f(x) = 3 - 2 = 1$

   (b) No, $\lim_{x \to 2} f(x)$ does not exist because $\lim_{x \to 2^+} f(x) \neq \lim_{x \to 2^-} f(x)$

(c) $\lim\limits_{x \to 4^-} f(x) = \frac{4}{2} + 1 = 3$, $\lim\limits_{x \to 4^+} f(x) = \frac{4}{2} + 1 = 3$

(d) Yes, $\lim\limits_{x \to 4} f(x) = 3$ because $3 = \lim\limits_{x \to 4^-} f(x) = \lim\limits_{x \to 4^+} f(x)$

5. (a)

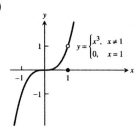

$y = \begin{cases} x^3, & x \neq 1 \\ 0, & x = 1 \end{cases}$

(b) $\lim\limits_{x \to 1^-} f(x) = 1 = \lim\limits_{x \to 1^+} f(x)$

(c) Yes, $\lim\limits_{x \to 1} f(x) = 1$ since the right-hand and left-hand limits exist and equal 1

7. (a) domain: $0 \leq x \leq 2$

range: $0 < y \leq 1$ and $y = 2$

(b) $\lim\limits_{x \to c} f(x)$ exists for c belonging to $(0, 1) \cup (1, 2)$

(c) $x = 2$

(d) $x = 0$

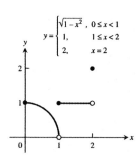

$y = \begin{cases} \sqrt{1 - x^2}, & 0 \leq x < 1 \\ 1, & 1 \leq x < 2 \\ 2, & x = 2 \end{cases}$

9. $\lim\limits_{x \to -0.5^-} \sqrt{\frac{x+2}{x-1}} = \sqrt{\frac{-0.5+2}{-0.5+1}} = \sqrt{\frac{3/2}{1/2}} = \sqrt{3}$

11. $\lim\limits_{x \to -2^+} \left(\frac{x}{x+1}\right)\left(\frac{2x+5}{x^2+x}\right) = \left(\frac{-2}{-2+1}\right)\left(\frac{2(-2)+5}{(-2)^2+(-2)}\right) = (2)\left(\frac{1}{2}\right) = 1$

13. $\lim\limits_{h \to 0^+} \frac{\sqrt{h^2+4h+5} - \sqrt{5}}{h} = \lim\limits_{h \to 0^+} \left(\frac{\sqrt{h^2+4h+5} - \sqrt{5}}{h}\right)\left(\frac{\sqrt{h^2+4h+5} + \sqrt{5}}{\sqrt{h^2+4h+5} + \sqrt{5}}\right)$

$= \lim\limits_{h \to 0^+} \frac{(h^2+4h+5) - 5}{h\left(\sqrt{h^2+4h+5} + \sqrt{5}\right)} = \lim\limits_{h \to 0^+} \frac{h(h+4)}{h\left(\sqrt{h^2+4h+5} + \sqrt{5}\right)} = \frac{0+4}{\sqrt{5}+\sqrt{5}} = \frac{2}{\sqrt{5}}$

15. (a) $\lim\limits_{x \to -2^+} (x + 3)\frac{|x+2|}{x+2} = \lim\limits_{x \to -2^+} (x + 3)\frac{(x+2)}{(x+2)}$ $(|x + 2| = x + 2 \text{ for } x > -2)$

$= \lim\limits_{x \to -2^+} (x + 3) = (-2) + 3 = 1$

(b) $\lim\limits_{x \to -2^-} (x + 3)\frac{|x+2|}{x+2} = \lim\limits_{x \to -2^-} (x + 3)\left[\frac{-(x+2)}{(x+2)}\right]$ $(|x + 2| = -(x + 2) \text{ for } x < -2)$

$= \lim\limits_{x \to -2^-} (x + 3)(-1) = -(-2 + 3) = -1$

17. (a) $\lim\limits_{\theta \to 3^+} \frac{|\theta|}{\theta} = \frac{3}{3} = 1$

(b) $\lim\limits_{\theta \to 3^-} \frac{|\theta|}{\theta} = \frac{2}{3}$

19. $\lim\limits_{\theta \to 0} \frac{\sin \sqrt{2\theta}}{\sqrt{2\theta}} = \lim\limits_{x \to 0} \frac{\sin x}{x} = 1$ (where $x = \sqrt{2\theta}$)

21. $\lim\limits_{y \to 0} \frac{\sin 3y}{4y} = \frac{1}{4} \lim\limits_{y \to 0} \frac{3 \sin 3y}{3y} = \frac{3}{4} \lim\limits_{y \to 0} \frac{\sin 3y}{3y} = \frac{3}{4} \lim\limits_{\theta \to 0} \frac{\sin \theta}{\theta} = \frac{3}{4}$ (where $\theta = 3y$)

23. $\lim\limits_{x \to 0} \frac{\tan 2x}{x} = \lim\limits_{x \to 0} \frac{\left(\frac{\sin 2x}{\cos 2x}\right)}{x} = \lim\limits_{x \to 0} \frac{\sin 2x}{x \cos 2x} = \left(\lim\limits_{x \to 0} \frac{1}{\cos 2x}\right)\left(\lim\limits_{x \to 0} \frac{2 \sin 2x}{2x}\right) = 1 \cdot 2 = 2$

25. $\lim\limits_{x \to 0} \frac{x \csc 2x}{\cos 5x} = \lim\limits_{x \to 0} \left(\frac{x}{\sin 2x} \cdot \frac{1}{\cos 5x}\right) = \left(\frac{1}{2} \lim\limits_{x \to 0} \frac{2x}{\sin 2x}\right)\left(\lim\limits_{x \to 0} \frac{1}{\cos 5x}\right) = \left(\frac{1}{2} \cdot 1\right)(1) = \frac{1}{2}$

27. $\lim\limits_{x \to 0} \frac{x + x\cos x}{\sin x \cos x} = \lim\limits_{x \to 0} \left(\frac{x}{\sin x \cos x} + \frac{x\cos x}{\sin x \cos x} \right) = \lim\limits_{x \to 0} \left(\frac{x}{\sin x} \cdot \frac{1}{\cos x} \right) + \lim\limits_{x \to 0} \frac{x}{\sin x}$

$= \lim\limits_{x \to 0} \left(\frac{1}{\frac{\sin x}{x}} \right) \cdot \lim\limits_{x \to 0} \left(\frac{1}{\cos x} \right) + \lim\limits_{x \to 0} \left(\frac{1}{\frac{\sin x}{x}} \right) = (1)(1) + 1 = 2$

29. $\lim\limits_{t \to 0} \frac{\sin(1 - \cos t)}{1 - \cos t} = \lim\limits_{\theta \to 0} \frac{\sin \theta}{\theta} = 1$ since $\theta = 1 - \cos t \to 0$ as $t \to 0$

31. $\lim\limits_{\theta \to 0} \frac{\sin \theta}{\sin 2\theta} = \lim\limits_{\theta \to 0} \left(\frac{\sin \theta}{\sin 2\theta} \cdot \frac{2\theta}{2\theta} \right) = \frac{1}{2} \lim\limits_{\theta \to 0} \left(\frac{\sin \theta}{\theta} \cdot \frac{2\theta}{\sin 2\theta} \right) = \frac{1}{2} \cdot 1 \cdot 1 = \frac{1}{2}$

33. $\lim\limits_{x \to 0} \frac{\tan 3x}{\sin 8x} = \lim\limits_{x \to 0} \left(\frac{\sin 3x}{\cos 3x} \cdot \frac{1}{\sin 8x} \right) = \lim\limits_{x \to 0} \left(\frac{\sin 3x}{\cos 3x} \cdot \frac{1}{\sin 8x} \cdot \frac{8x}{3x} \cdot \frac{3}{8} \right)$

$= \frac{3}{8} \lim\limits_{x \to 0} \left(\frac{1}{\cos 3x} \right) \left(\frac{\sin 3x}{3x} \right) \left(\frac{8x}{\sin 8x} \right) = \frac{3}{8} \cdot 1 \cdot 1 \cdot 1 = \frac{3}{8}$

35. Yes. If $\lim\limits_{x \to a^+} f(x) = L = \lim\limits_{x \to a^-} f(x)$, then $\lim\limits_{x \to a} f(x) = L$. If $\lim\limits_{x \to a^+} f(x) \neq \lim\limits_{x \to a^-} f(x)$, then $\lim\limits_{x \to a} f(x)$ does not exist.

37. If f is an odd function of x, then $f(-x) = -f(x)$. Given $\lim\limits_{x \to 0^+} f(x) = 3$, then $\lim\limits_{x \to 0^-} f(x) = -3$.

39. $I = (5, 5 + \delta) \Rightarrow 5 < x < 5 + \delta$. Also, $\sqrt{x - 5} < \epsilon \Rightarrow x - 5 < \epsilon^2 \Rightarrow x < 5 + \epsilon^2$. Choose $\delta = \epsilon^2$

$\Rightarrow \lim\limits_{x \to 5^+} \sqrt{x - 5} = 0$.

41. As $x \to 0^-$ the number x is always negative. Thus, $\left| \frac{x}{|x|} - (-1) \right| < \epsilon \Rightarrow \left| \frac{x}{-x} + 1 \right| < \epsilon \Rightarrow 0 < \epsilon$ which is always

true independent of the value of x. Hence we can choose any $\delta > 0$ with $-\delta < x < 0 \Rightarrow \lim\limits_{x \to 0^-} \frac{x}{|x|} = -1$.

43. (a) $\lim\limits_{x \to 400^+} \lfloor x \rfloor = 400$. Just observe that if $400 < x < 401$, then $\lfloor x \rfloor = 400$. Thus if we choose $\delta = 1$, we have for any

number $\epsilon > 0$ that $400 < x < 400 + \delta \Rightarrow |\lfloor x \rfloor - 400| = |400 - 400| = 0 < \epsilon$.

(b) $\lim\limits_{x \to 400^-} \lfloor x \rfloor = 399$. Just observe that if $399 < x < 400$ then $\lfloor x \rfloor = 399$. Thus if we choose $\delta = 1$, we have for any

number $\epsilon > 0$ that $400 - \delta < x < 400 \Rightarrow |\lfloor x \rfloor - 399| = |399 - 399| = 0 < \epsilon$.

(c) Since $\lim\limits_{x \to 400^+} \lfloor x \rfloor \neq \lim\limits_{x \to 400^-} \lfloor x \rfloor$ we conclude that $\lim\limits_{x \to 400} \lfloor x \rfloor$ does not exist.

1.7 CONTINUITY

1. No, discontinuous at $x = 2$, not defined at $x = 2$

3. Continuous on $[-1, 3]$

5. (a) Yes

(b) Yes, $\lim\limits_{x \to -1^+} f(x) = 0$

(c) Yes

(d) Yes

7. (a) No

(b) No

9. Nonremovable discontinuity at $x = 1$ because $\lim\limits_{x \to 1} f(x)$ fails to exist ($\lim\limits_{x \to 1^-} f(x) = 1$ and $\lim\limits_{x \to 1^+} f(x) = 0$).

Removable discontinuity at $x = 0$ by assigning the number $\lim\limits_{x \to 0} f(x) = 0$ to be the value of $f(0)$ rather than $f(0) = 1$.

11. Discontinuous only when $x - 2 = 0 \Rightarrow x = 2$

13. Discontinuous only when $x^2 - 4x + 3 = 0 \Rightarrow (x - 3)(x - 1) = 0 \Rightarrow x = 3$ or $x = 1$

15. Continuous everywhere. ($|x - 1| + \sin x$ defined for all x; limits exist and are equal to function values.)

17. Discontinuous only at $x = 0$

19. Discontinuous when 2x is an integer multiple of π, i.e., $2x = n\pi$, n an integer $\Rightarrow x = \frac{n\pi}{2}$, n an integer, but continuous at all other x.

21. Discontinuous at odd integer multiples of $\frac{\pi}{2}$, i.e., $x = (2n - 1)\frac{\pi}{2}$, n an integer, but continuous at all other x.

23. Discontinuous when $2x + 3 < 0$ or $x < -\frac{3}{2} \Rightarrow$ continuous on the interval $\left[-\frac{3}{2}, \infty\right)$.

25. Continuous everywhere: $(2x - 1)^{1/3}$ is defined for all x; limits exist and are equal to function values.

27. $\lim\limits_{x \to \pi} \sin(x - \sin x) = \sin(\pi - \sin \pi) = \sin(\pi - 0) = \sin \pi = 0$, and function continuous at $x = \pi$.

29. $\lim\limits_{y \to 1} \sec(y \sec^2 y - \tan^2 y - 1) = \lim\limits_{y \to 1} \sec(y \sec^2 y - \sec^2 y) = \lim\limits_{y \to 1} \sec((y - 1)\sec^2 y) = \sec((1 - 1)\sec^2 1)$
    $= \sec 0 = 1$, and function continuous at $y = 1$.

31. As defined, $\lim\limits_{x \to 3^-} f(x) = (3)^2 - 1 = 8$ and $\lim\limits_{x \to 3^+} (2a)(3) = 6a$. For f(x) to be continuous we must have $6a = 8 \Rightarrow a = \frac{4}{3}$.

33. Let $f(x) = x^3 - 15x + 1$, which is continuous on $[-4, 4]$. Then $f(-4) = -3$, $f(-1) = 15$, $f(1) = -13$, and $f(4) = 5$. By the Intermediate Value Theorem, $f(x) = 0$ for some x in each of the intervals $-4 < x < -1$, $-1 < x < 1$, and $1 < x < 4$. That is, $x^3 - 15x + 1 = 0$ has three solutions in $[-4, 4]$. Since a polynomial of degree 3 can have at most 3 solutions, these are the only solutions.

35. Answers may vary. Note that f is continuous for every value of x.
    (a) $f(0) = 10$, $f(1) = 1^3 - 8(1) + 10 = 3$. Since $3 < \pi < 10$, by the Intermediate Value Theorem, there exists a c so that $0 < c < 1$ and $f(c) = \pi$.
    (b) $f(0) = 10$, $f(-4) = (-4)^3 - 8(-4) + 10 = -22$. Since $-22 < -\sqrt{3} < 10$, by the Intermediate Value Theorem, there exists a c so that $-4 < c < 0$ and $f(c) = -\sqrt{3}$.
    (c) $f(0) = 10$, $f(1000) = (1000)^3 - 8(1000) + 10 = 999,992,010$. Since $10 < 5,000,000 < 999,992,010$, by the Intermediate Value Theorem, there exists a c so that $0 < c < 1000$ and $f(c) = 5,000,000$.

37. Answers may vary. For example, $f(x) = \frac{\sin(x - 2)}{x - 2}$ is discontinuous at $x = 2$ because it is not defined there. However, the discontinuity can be removed because f has a limit (namely 1) as $x \to 2$.

39. (a) Suppose x_0 is rational $\Rightarrow f(x_0) = 1$. Choose $\epsilon = \frac{1}{2}$. For any $\delta > 0$ there is an irrational number x (actually infinitely many) in the interval $(x_0 - \delta, x_0 + \delta) \Rightarrow f(x) = 0$. Then $0 < |x - x_0| < \delta$ but $|f(x) - f(x_0)|$
    $= 1 > \frac{1}{2} = \epsilon$, so $\lim\limits_{x \to x_0} f(x)$ fails to exist $\Rightarrow$ f is discontinuous at x_0 rational.
    On the other hand, x_0 irrational $\Rightarrow f(x_0) = 0$ and there is a rational number x in $(x_0 - \delta, x_0 + \delta) \Rightarrow f(x)$
    $= 1$. Again $\lim\limits_{x \to x_0} f(x)$ fails to exist $\Rightarrow$ f is discontinuous at x_0 irrational. That is, f is discontinuous at every point.

(b) f is neither right-continuous nor left-continuous at any point x_0 because in every interval $(x_0 - \delta, x_0)$ or $(x_0, x_0 + \delta)$ there exist both rational and irrational real numbers. Thus neither limits $\lim_{x \to x_0^-} f(x)$ and $\lim_{x \to x_0^+} f(x)$ exist by the same arguments used in part (a).

41. No. For instance, if $f(x) = 0$, $g(x) = \lceil x \rceil$, then $h(x) = 0(\lceil x \rceil) = 0$ is continuous at $x = 0$ and $g(x)$ is not.

43. Yes, because of the Intermediate Value Theorem. If $f(a)$ and $f(b)$ did have different signs then f would have to equal zero at some point between a and b since f is continuous on $[a, b]$.

45. If $f(0) = 0$ or $f(1) = 1$, we are done (i.e., $c = 0$ or $c = 1$ in those cases). Then let $f(0) = a > 0$ and $f(1) = b < 1$ because $0 \le f(x) \le 1$. Define $g(x) = f(x) - x \Rightarrow$ g is continuous on $[0, 1]$. Moreover, $g(0) = f(0) - 0 = a > 0$ and $g(1) = f(1) - 1 = b - 1 < 0 \Rightarrow$ by the Intermediate Value Theorem there is a number c in $(0, 1)$ such that $g(c) = 0 \Rightarrow f(c) - c = 0$ or $f(c) = c$.

47. By Exercises 40 in Section 1.5, we have $\lim_{x \to c} f(x) = L \Leftrightarrow \lim_{h \to 0} f(c + h) = L$.
Thus, $f(x)$ is continuous at $x = c \Leftrightarrow \lim_{x \to c} f(x) = f(c) \Leftrightarrow \lim_{h \to 0} f(c + h) = f(c)$.

49. $x \approx 1.8794, -1.5321, -0.3473$ 51. $x \approx 1.7549$

53. $x \approx 3.5156$ 55. $x \approx 0.7391$

1.8 LIMITS INVOLVING INFINITY

Note: In these exercises we use the result $\lim_{x \to \pm\infty} \frac{1}{x^{m/n}} = 0$ whenever $\frac{m}{n} > 0$. This result follows immediately from Theorem 8 and the power rule in Theorem 1: $\lim_{x \to \pm\infty} \left(\frac{1}{x^{m/n}}\right) = \lim_{x \to \pm\infty} \left(\frac{1}{x}\right)^{m/n} = \left(\lim_{x \to \pm\infty} \frac{1}{x}\right)^{m/n} = 0^{m/n} = 0$.

1. (a) -3 (b) -3

3. (a) $\frac{1}{2}$ (b) $\frac{1}{2}$

5. (a) $-\frac{5}{3}$ (b) $-\frac{5}{3}$

7. $-\frac{1}{x} \le \frac{\sin 2x}{x} \le \frac{1}{x} \Rightarrow \lim_{x \to \infty} \frac{\sin 2x}{x} = 0$ by the Sandwich Theorem

9. $\lim_{t \to \infty} \frac{2 - t + \sin t}{t + \cos t} = \lim_{t \to \infty} \frac{\frac{2}{t} - 1 + \left(\frac{\sin t}{t}\right)}{1 + \left(\frac{\cos t}{t}\right)} = \frac{0 - 1 + 0}{1 + 0} = -1$

11. (a) $\lim_{x \to \infty} \frac{2x + 3}{5x + 7} = \lim_{x \to \infty} \frac{2 + \frac{3}{x}}{5 + \frac{7}{x}} = \frac{2}{5}$ (b) $\frac{2}{5}$ (same process as part (a))

13. (a) $\lim_{x \to \infty} \frac{x + 1}{x^2 + 3} = \lim_{x \to \infty} \frac{\frac{1}{x} + \frac{1}{x^2}}{1 + \frac{3}{x^2}} = 0$ (b) 0 (same process as part (a))

15. (a) $\lim_{x \to \infty} \frac{7x^3}{x^3 - 3x^2 + 6x} = \lim_{x \to \infty} \frac{7}{1 - \frac{3}{x} + \frac{6}{x^2}} = 7$ (b) 7 (same process as part (a))

17. (a) $\displaystyle \lim_{x \to \infty} \frac{10x^5 + x^4 + 31}{x^6} = \lim_{x \to \infty} \frac{\frac{10}{x} + \frac{1}{x^2} + \frac{31}{x^6}}{1} = 0$

(b) 0 (same process as part (a))

19. (a) $\displaystyle \lim_{x \to \infty} \frac{-2x^3 - 2x + 3}{3x^3 + 3x^2 - 5x} = \lim_{x \to \infty} \frac{-2 - \frac{2}{x^2} + \frac{3}{x^3}}{3 + \frac{3}{x} - \frac{5}{x^2}} = -\frac{2}{3}$

(b) $-\frac{2}{3}$ (same process as part (a))

21. $\displaystyle \lim_{x \to \infty} \frac{2\sqrt{x} + x^{-1}}{3x - 7} = \lim_{x \to \infty} \frac{\left(\frac{2}{x^{1/2}}\right) + \left(\frac{1}{x^2}\right)}{3 - \frac{7}{x}} = 0$

23. $\displaystyle \lim_{x \to -\infty} \frac{\sqrt[3]{x} - \sqrt[5]{x}}{\sqrt[3]{x} + \sqrt[5]{x}} = \lim_{x \to -\infty} \frac{1 - x^{(1/5) - (1/3)}}{1 + x^{(1/5) - (1/3)}} = \lim_{x \to -\infty} \frac{1 - \left(\frac{1}{x^{2/15}}\right)}{1 + \left(\frac{1}{x^{2/15}}\right)} = 1$

25. $\displaystyle \lim_{x \to \infty} \frac{2x^{5/3} - x^{1/3} + 7}{x^{8/5} + 3x + \sqrt{x}} = \lim_{x \to \infty} \frac{2x^{1/15} - \frac{1}{x^{19/15}} + \frac{7}{x^{8/5}}}{1 + \frac{3}{x^{3/5}} + \frac{1}{x^{11/10}}} = \infty$

27. $\displaystyle \lim_{x \to 0^+} \frac{1}{3x} = \infty$ $\qquad \left(\dfrac{positive}{positive}\right)$

29. $\displaystyle \lim_{x \to 2^-} \frac{3}{x-2} = -\infty$ $\qquad \left(\dfrac{positive}{negative}\right)$

31. $\displaystyle \lim_{x \to -8^+} \frac{2x}{x+8} = -\infty$ $\qquad \left(\dfrac{negative}{positive}\right)$

33. $\displaystyle \lim_{x \to 7} \frac{4}{(x-7)^2} = \infty$ $\qquad \left(\dfrac{positive}{positive}\right)$

35. (a) $\displaystyle \lim_{x \to 0^+} \frac{2}{3x^{1/3}} = \infty$

(b) $\displaystyle \lim_{x \to 0^-} \frac{2}{3x^{1/3}} = -\infty$

37. $\displaystyle \lim_{x \to 0} \frac{4}{x^{2/5}} = \lim_{x \to 0} \frac{4}{\left(x^{1/5}\right)^2} = \infty$

39. $\displaystyle \lim_{x \to \left(\frac{\pi}{2}\right)^-} \tan x = \infty$

41. $\displaystyle \lim_{\theta \to 0^-} (1 + \csc \theta) = -\infty$

43. (a) $\displaystyle \lim_{x \to 2^+} \frac{1}{x^2 - 4} = \lim_{x \to 2^+} \frac{1}{(x+2)(x-2)} = \infty$ $\qquad \left(\dfrac{1}{positive \cdot positive}\right)$

(b) $\displaystyle \lim_{x \to 2^-} \frac{1}{x^2 - 4} = \lim_{x \to 2^-} \frac{1}{(x+2)(x-2)} = -\infty$ $\qquad \left(\dfrac{1}{positive \cdot negative}\right)$

(c) $\displaystyle \lim_{x \to -2^+} \frac{1}{x^2 - 4} = \lim_{x \to -2^+} \frac{1}{(x+2)(x-2)} = -\infty$ $\qquad \left(\dfrac{1}{positive \cdot negative}\right)$

(d) $\displaystyle \lim_{x \to -2^-} \frac{1}{x^2 - 4} = \lim_{x \to -2^-} \frac{1}{(x+2)(x-2)} = \infty$ $\qquad \left(\dfrac{1}{negative \cdot negative}\right)$

45. (a) $\displaystyle \lim_{x \to 0^+} \frac{x^2}{2} - \frac{1}{x} = 0 + \lim_{x \to 0^+} \frac{1}{-x} = -\infty$ $\qquad \left(\dfrac{1}{negative}\right)$

(b) $\displaystyle \lim_{x \to 0^-} \frac{x^2}{2} - \frac{1}{x} = 0 + \lim_{x \to 0^-} \frac{1}{-x} = \infty$ $\qquad \left(\dfrac{1}{positive}\right)$

(c) $\displaystyle \lim_{x \to \sqrt[3]{2}} \frac{x^2}{2} - \frac{1}{x} = \frac{2^{2/3}}{2} - \frac{1}{2^{1/3}} = 2^{-1/3} - 2^{-1/3} = 0$

(d) $\displaystyle \lim_{x \to -1} \frac{x^2}{2} - \frac{1}{x} = \frac{1}{2} - \left(\frac{1}{-1}\right) = \frac{3}{2}$

47. (a) $\displaystyle \lim_{x \to 0^+} \frac{x^2 - 3x + 2}{x^3 - 2x^2} = \lim_{x \to 0^+} \frac{(x-2)(x-1)}{x^2(x-2)} = -\infty$ $\qquad \left(\dfrac{negative \cdot negative}{positive \cdot negative}\right)$

(b) $\displaystyle \lim_{x \to 2^+} \frac{x^2 - 3x + 2}{x^3 - 2x^2} = \lim_{x \to 2^+} \frac{(x-2)(x-1)}{x^2(x-2)} = \lim_{x \to 2^+} \frac{x-1}{x^2} = \frac{1}{4}, x \neq 2$

(c) $\displaystyle \lim_{x \to 2^-} \frac{x^2 - 3x + 2}{x^3 - 2x^2} = \lim_{x \to 2^-} \frac{(x-2)(x-1)}{x^2(x-2)} = \lim_{x \to 2^-} \frac{x-1}{x^2} = \frac{1}{4}, x \neq 2$

(d) $\displaystyle \lim_{x \to 2} \frac{x^2 - 3x + 2}{x^3 - 2x^2} = \lim_{x \to 2} \frac{(x-2)(x-1)}{x^2(x-2)} = \lim_{x \to 2} \frac{x-1}{x^2} = \frac{1}{4}, x \neq 2$

(e) $\lim\limits_{x \to 0} \frac{x^2 - 3x + 2}{x^3 - 2x^2} = \lim\limits_{x \to 0} \frac{(x-2)(x-1)}{x^2(x-2)} = -\infty$ $\left(\dfrac{\text{negative·negative}}{\text{positive·negative}}\right)$

49. (a) $\lim\limits_{t \to 0^+} \left[2 - \frac{3}{t^{1/3}}\right] = -\infty$ (b) $\lim\limits_{t \to 0^-} \left[2 - \frac{3}{t^{1/3}}\right] = \infty$

51. (a) $\lim\limits_{x \to 0^+} \left[\frac{1}{x^{2/3}} + \frac{2}{(x-1)^{2/3}}\right] = \infty$ (b) $\lim\limits_{x \to 0^-} \left[\frac{1}{x^{2/3}} + \frac{2}{(x-1)^{2/3}}\right] = \infty$

    (c) $\lim\limits_{x \to 1^+} \left[\frac{1}{x^{2/3}} + \frac{2}{(x-1)^{2/3}}\right] = \infty$ (d) $\lim\limits_{x \to 1^-} \left[\frac{1}{x^{2/3}} + \frac{2}{(x-1)^{2/3}}\right] = \infty$

53. $y = \frac{1}{x-1}$ 55. $y = \frac{1}{2x+4}$

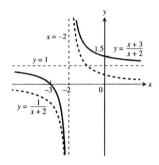   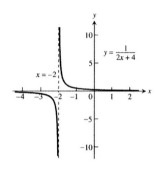

57. $y = \frac{x+3}{x+2} = 1 + \frac{1}{x+2}$

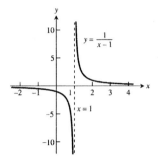

59. Here is one possibility.   61. Here is one possibility.

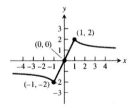   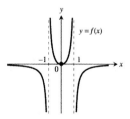

63. Here is one possibility.   65. Here is one possibility.

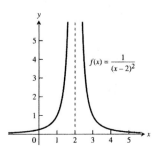   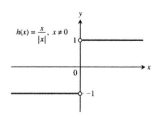

67. Yes. If $\lim\limits_{x \to \infty} \frac{f(x)}{g(x)} = 2$ then the ratio of the polynomials' leading coefficients is 2, so $\lim\limits_{x \to -\infty} \frac{f(x)}{g(x)} = 2$ as well.

69. At most 1 horizontal asymptote: If $\lim\limits_{x \to \infty} \frac{f(x)}{g(x)} = L$, then the ratio of the polynomials' leading coefficients is L, so $\lim\limits_{x \to -\infty} \frac{f(x)}{g(x)} = L$ as well.

71. For any $\epsilon > 0$, take N = 1. Then for all x > N we have that $|f(x) - k| = |k - k| = 0 < \epsilon$.

73. For every real number $-B < 0$, we must find a $\delta > 0$ such that for all x, $0 < |x - 0| < \delta \Rightarrow \frac{-1}{x^2} < -B$. Now,
$-\frac{1}{x^2} < -B < 0 \Leftrightarrow \frac{1}{x^2} > B > 0 \Leftrightarrow x^2 < \frac{1}{B} \Leftrightarrow |x| < \frac{1}{\sqrt{B}}$. Choose $\delta = \frac{1}{\sqrt{B}}$, then $0 < |x| < \delta \Rightarrow |x| < \frac{1}{\sqrt{B}}$
$\Rightarrow \frac{-1}{x^2} < -B$ so that $\lim\limits_{x \to 0} -\frac{1}{x^2} = -\infty$.

75. For every real number $-B < 0$, we must find a $\delta > 0$ such that for all x, $0 < |x - 3| < \delta \Rightarrow \frac{-2}{(x-3)^2} < -B$.
Now, $\frac{-2}{(x-3)^2} < -B < 0 \Leftrightarrow \frac{2}{(x-3)^2} > B > 0 \Leftrightarrow \frac{(x-3)^2}{2} < \frac{1}{B} \Leftrightarrow (x-3)^2 < \frac{2}{B} \Leftrightarrow 0 < |x - 3| < \sqrt{\frac{2}{B}}$. Choose
$\delta = \sqrt{\frac{2}{B}}$, then $0 < |x - 3| < \delta \Rightarrow \frac{-2}{(x-3)^2} < -B < 0$ so that $\lim\limits_{x \to 3} \frac{-2}{(x-3)^2} = -\infty$.

77. (a) We say that f(x) approaches infinity as x approaches x_0 from the left, and write $\lim\limits_{x \to x_0^-} f(x) = \infty$, if
for every positive number B, there exists a corresponding number $\delta > 0$ such that for all x,
$x_0 - \delta < x < x_0 \Rightarrow f(x) > B$.

(b) We say that f(x) approaches minus infinity as x approaches x_0 from the right, and write $\lim\limits_{x \to x_0^+} f(x) = -\infty$,
if for every positive number B (or negative number $-B$) there exists a corresponding number $\delta > 0$ such
that for all x, $x_0 < x < x_0 + \delta \Rightarrow f(x) < -B$.

(c) We say that f(x) approaches minus infinity as x approaches x_0 from the left, and write $\lim\limits_{x \to x_0^-} f(x) = -\infty$,
if for every positive number B (or negative number $-B$) there exists a corresponding number $\delta > 0$ such
that for all x, $x_0 - \delta < x < x_0 \Rightarrow f(x) < -B$.

79. For $B > 0$, $\frac{1}{x} < -B < 0 \Leftrightarrow -\frac{1}{x} > B > 0 \Leftrightarrow -x < \frac{1}{B} \Leftrightarrow -\frac{1}{B} < x$. Choose $\delta = \frac{1}{B}$. Then $-\delta < x < 0$
$\Rightarrow -\frac{1}{B} < x \Rightarrow \frac{1}{x} < -B$ so that $\lim\limits_{x \to 0^-} \frac{1}{x} = -\infty$.

81. For $B > 0$, $\frac{1}{x-2} > B \Leftrightarrow 0 < x - 2 < \frac{1}{B}$. Choose $\delta = \frac{1}{B}$. Then $2 < x < 2 + \delta \Rightarrow 0 < x - 2 < \delta \Rightarrow 0 < x - 2 < \frac{1}{B}$
$\Rightarrow \frac{1}{x-2} > B > 0$ so that $\lim\limits_{x \to 2^+} \frac{1}{x-2} = \infty$.

83. $y = \frac{x^2}{x-1} = x + 1 + \frac{1}{x-1}$ 85. $y = \frac{x^2-4}{x-1} = x + 1 - \frac{3}{x-1}$

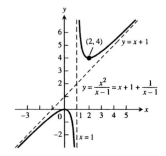

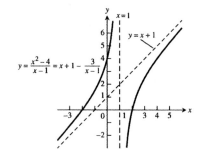

87. $y = \frac{x^2-1}{x} = x - \frac{1}{x}$

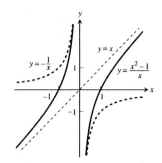

89. $y = \frac{x}{\sqrt{4-x^2}}$

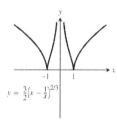

91. $y = x^{2/3} + \frac{1}{x^{1/3}}$

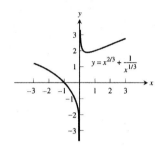

93. (a) $y \to \infty$ (see accompanying graph)
    (b) $y \to \infty$ (see accompanying graph)
    (c) cusps at $x = \pm 1$ (see accompanying graph)

CHAPTER 1 PRACTICE AND ADDITIONAL EXERCISES

1. (a) The function is defined for all values of x, so the domain is $(-\infty, \infty)$.
   (b) Since $|x|$ attains all nonnegative values, the range is $[-2, \infty)$.

3. (a) Since the square root requires $16 - x^2 \geq 0$, the domain is $[-4, 4]$.
   (b) For values of x in the domain, $0 \leq 16 - x^2 \leq 16$, so $0 \leq \sqrt{16-x^2} \leq 4$. The range is $[0, 4]$.

5. (a) The function is defined for all values of x, so the domain is $(-\infty, \infty)$.
   (b) The sine function attains values from -1 to 1, so $-2 \leq 2\sin(3x + \pi) \leq 2$ and hence $-3 \leq 2\sin(3x + \pi) - 1 \leq 1$. The range is $[-3, 1]$.

7. (a) The function is defined for $-4 \leq x \leq 4$, so the domain is $[-4, 4]$.
   (b) The function is equivalent to $y = \sqrt{|x|}$, $-4 \leq x \leq 4$, which attains values from 0 to 2 for x in the domain. The range is $[0, 2]$.

9. First piece: Line through $(0, 1)$ and $(1, 0)$. $m = \frac{0-1}{1-0} = \frac{-1}{1} = -1 \Rightarrow y = -x + 1 = 1 - x$

Second piece: Line through $(1, 1)$ and $(2, 0)$. $m = \frac{0-1}{2-1} = \frac{-1}{1} = -1 \Rightarrow y = -(x - 1) + 1 = -x + 2 = 2 - x$

$$f(x) = \begin{cases} 1 - x, & 0 \le x < 1 \\ 2 - x, & 1 \le x \le 2 \end{cases}$$

11. (a) $(f \circ g)(-1) = f(g(-1)) = f\left(\frac{1}{\sqrt{-1+2}}\right) = f(1) = \frac{1}{1} = 1$

(b) $(g \circ f)(2) = g(f(2)) = g\left(\frac{1}{2}\right) = \frac{1}{\sqrt{\frac{1}{2}+2}} = \frac{1}{\sqrt{2.5}}$ or $\sqrt{\frac{2}{5}}$

(c) $(f \circ f)(x) = f(f(x)) = f\left(\frac{1}{x}\right) = \frac{1}{1/x} = x, \, x \ne 0$

(d) $(g \circ g)(x) = g(g(x)) = g\left(\frac{1}{\sqrt{x+2}}\right) = \frac{1}{\sqrt{\frac{1}{\sqrt{x+2}}+2}} = \frac{\sqrt[4]{x+2}}{\sqrt{1+2\sqrt{x+2}}}$

13. (a) $(f \circ g)(x) = f(g(x)) = f\left(\sqrt{x+2}\right) = 2 - \left(\sqrt{x+2}\right)^2 = -x, \, x \ge -2.$

$(g \circ f)(x) = f(g(x)) = g(2 - x^2) = \sqrt{(2 - x^2) + 2} = \sqrt{4 - x^2}$

(b) Domain of $f \circ g$: $[-2, \infty)$.

Domain of $g \circ f$: $[-2, 2]$.

(c) Range of $f \circ g$: $(-\infty, 2]$.

Range of $g \circ f$: $[0, 2]$.

15. $y(-x) = (-x)^2 + 1 = x^2 + 1 = y(x)$. Even.

17. $y(-x) = \frac{(-x)^4 + 1}{(-x)^3 - 2(-x)} = \frac{x^4 + 1}{-x^3 + 2x} = -\frac{x^4 + 1}{x^3 - 2x} = -y(x)$. Odd.

19. $y(-x) = -x + \cos(-x) = -x + \cos x$. Neither even nor odd.

21. At $x = -1$: $\lim_{x \to -1^-} f(x) = \lim_{x \to -1^+} f(x) = 1$

$\Rightarrow \lim_{x \to -1} f(x) = 1 = f(-1)$

$\Rightarrow f$ is continuous at $x = -1$.

At $x = 0$: $\lim_{x \to 0^-} f(x) = \lim_{x \to 0^+} f(x) = 0 \Rightarrow \lim_{x \to 0} f(x) = 0.$

But $f(0) = 1 \ne \lim_{x \to 0} f(x)$

$\Rightarrow f$ is discontinuous at $x = 0$.

If we define $f(0) = 0$, then the discontinuity at $x = 0$ is removable.

At $x = 1$: $\lim_{x \to 1^-} f(x) = -1$ and $\lim_{x \to 1^+} f(x) = 1$

$\Rightarrow \lim_{x \to 1} f(x)$ does not exist

$\Rightarrow f$ is discontinuous at $x = 1$.

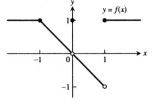

23. (a) $\lim_{t \to t_0} (3f(t)) = 3 \lim_{t \to t_0} f(t) = 3(-7) = -21$

(b) $\lim_{t \to t_0} (f(t))^2 = \left(\lim_{t \to t_0} f(t)\right)^2 = (-7)^2 = 49$

(c) $\lim_{t \to t_0} (f(t) \cdot g(t)) = \lim_{t \to t_0} f(t) \cdot \lim_{t \to t_0} g(t) = (-7)(0) = 0$

(d) $\lim_{t \to t_0} \frac{f(t)}{g(t) - 7} = \frac{\lim_{t \to t_0} f(t)}{\lim_{t \to t_0} (g(t) - 7)} = \frac{\lim_{t \to t_0} f(t)}{\lim_{t \to t_0} g(t) - \lim_{t \to t_0} 7} = \frac{-7}{0 - 7} = 1$

(e) $\lim_{t \to t_0} \cos(g(t)) = \cos\left(\lim_{t \to t_0} g(t)\right) = \cos 0 = 1$

(f) $\lim\limits_{t \to t_0} |f(t)| = \left| \lim\limits_{t \to t_0} f(t) \right| = |-7| = 7$

(g) $\lim\limits_{t \to t_0} (f(t) + g(t)) = \lim\limits_{t \to t_0} f(t) + \lim\limits_{t \to t_0} g(t) = -7 + 0 = -7$

(h) $\lim\limits_{t \to t_0} \left(\frac{1}{f(t)} \right) = \frac{1}{\lim\limits_{t \to t_0} f(t)} = \frac{1}{-7} = -\frac{1}{7}$

25. Since $\lim\limits_{x \to 0} x = 0$ we must have that $\lim\limits_{x \to 0} (4 - g(x)) = 0$. Otherwise, if $\lim\limits_{x \to 0} (4 - g(x))$ is a finite positive

number, we would have $\lim\limits_{x \to 0^-} \left[\frac{4 - g(x)}{x} \right] = -\infty$ and $\lim\limits_{x \to 0^+} \left[\frac{4 - g(x)}{x} \right] = \infty$ so the limit could not equal 1 as

$x \to 0$. Similar reasoning holds if $\lim\limits_{x \to 0} (4 - g(x))$ is a finite negative number. We conclude that $\lim\limits_{x \to 0} g(x) = 4$.

27. (a) $\lim\limits_{x \to c} f(x) = \lim\limits_{x \to c} x^{1/3} = c^{1/3} = f(c)$ for every real number $c \Rightarrow f$ is continuous on $(-\infty, \infty)$.

(b) $\lim\limits_{x \to c} g(x) = \lim\limits_{x \to c} x^{3/4} = c^{3/4} = g(c)$ for every nonnegative real number $c \Rightarrow g$ is continuous on $[0, \infty)$.

(c) $\lim\limits_{x \to c} h(x) = \lim\limits_{x \to c} x^{-2/3} = \frac{1}{c^{2/3}} = h(c)$ for every nonzero real number $c \Rightarrow h$ is continuous on $(-\infty, 0)$ and $(0, \infty)$.

(d) $\lim\limits_{x \to c} k(x) = \lim\limits_{x \to c} x^{-1/6} = \frac{1}{c^{1/6}} = k(c)$ for every positive real number $c \Rightarrow k$ is continuous on $(0, \infty)$

29. (a) $\lim\limits_{x \to 0} \frac{x^2 - 4x + 4}{x^3 + 5x^2 - 14x} = \lim\limits_{x \to 0} \frac{(x - 2)(x - 2)}{x(x + 7)(x - 2)} = \lim\limits_{x \to 0} \frac{x - 2}{x(x + 7)}$, $x \neq 2$; the limit does not exist because

$\lim\limits_{x \to 0^-} \frac{x - 2}{x(x + 7)} = \infty$ and $\lim\limits_{x \to 0^+} \frac{x - 2}{x(x + 7)} = -\infty$

(b) $\lim\limits_{x \to 2} \frac{x^2 - 4x + 4}{x^3 + 5x^2 - 14x} = \lim\limits_{x \to 2} \frac{(x - 2)(x - 2)}{x(x + 7)(x - 2)} = \lim\limits_{x \to 2} \frac{x - 2}{x(x + 7)}$, $x \neq 2$, and $\lim\limits_{x \to 2} \frac{x - 2}{x(x + 7)} = \frac{0}{2(9)} = 0$

31. $\lim\limits_{x \to 1} \frac{1 - \sqrt{x}}{1 - x} = \lim\limits_{x \to 1} \frac{1 - \sqrt{x}}{(1 - \sqrt{x})(1 + \sqrt{x})} = \lim\limits_{x \to 1} \frac{1}{1 + \sqrt{x}} = \frac{1}{2}$

33. $\lim\limits_{h \to 0} \frac{(x + h)^2 - x^2}{h} = \lim\limits_{h \to 0} \frac{(x^2 + 2hx + h^2) - x^2}{h} = \lim\limits_{h \to 0} (2x + h) = 2x$

35. $\lim\limits_{x \to 0} \frac{\frac{1}{2 + x} - \frac{1}{2}}{x} = \lim\limits_{x \to 0} \frac{2 - (2 + x)}{2x(2 + x)} = \lim\limits_{x \to 0} \frac{-1}{4 + 2x} = -\frac{1}{4}$

37. $\lim\limits_{x \to 0} \frac{\tan 2x}{\tan \pi x} = \lim\limits_{x \to 0} \frac{\sin 2x}{\cos 2x} \cdot \frac{\cos \pi x}{\sin \pi x} = \lim\limits_{x \to 0} \left(\frac{\sin 2x}{2x} \right) \left(\frac{\cos \pi x}{\cos 2x} \right) \left(\frac{\pi x}{\sin \pi x} \right) \left(\frac{2x}{\pi x} \right) = 1 \cdot 1 \cdot 1 \cdot \frac{2}{\pi} = \frac{2}{\pi}$

39. $\lim\limits_{x \to \pi} \sin \left(\frac{x}{2} + \sin x \right) = \sin \left(\frac{\pi}{2} + \sin \pi \right) = \sin \left(\frac{\pi}{2} \right) = 1$

41. $\lim\limits_{x \to 0^+} [4 g(x)]^{1/3} = 2 \Rightarrow \left[\lim\limits_{x \to 0^+} 4 g(x) \right]^{1/3} = 2 \Rightarrow \lim\limits_{x \to 0^+} 4 g(x) = 8$, since $2^3 = 8$. Then $\lim\limits_{x \to 0^+} g(x) = 2$.

43. $\lim\limits_{x \to 1} \frac{3x^2 + 1}{g(x)} = \infty \Rightarrow \lim\limits_{x \to 1} g(x) = 0$ since $\lim\limits_{x \to 1} (3x^2 + 1) = 4$

45. $\lim\limits_{x \to \infty} \frac{2x + 3}{5x + 7} = \lim\limits_{x \to \infty} \frac{2 + \frac{3}{x}}{5 + \frac{7}{x}} = \frac{2 + 0}{5 + 0} = \frac{2}{5}$

47. $\lim\limits_{x \to -\infty} \frac{x^2 - 4x + 8}{3x^3} = \lim\limits_{x \to -\infty} \left(\frac{1}{3x} - \frac{4}{3x^2} + \frac{8}{3x^3} \right) = 0 - 0 + 0 = 0$

49. $\lim\limits_{x \to -\infty} \frac{x^2 - 7x}{x + 1} = \lim\limits_{x \to -\infty} \frac{x - 7}{1 + \frac{1}{x}} = -\infty$

51. $\lim\limits_{x \to \infty} \frac{\sin x}{\lfloor x \rfloor} \leq \lim\limits_{x \to \infty} \frac{1}{\lfloor x \rfloor} = 0$ since int $x \to \infty$ as $x \to \infty \Rightarrow \lim\limits_{x \to \infty} \frac{\sin x}{\lfloor x \rfloor} = 0$.

53. $\displaystyle\lim_{x \to \infty} \frac{x + \sin x + 2\sqrt{x}}{x + \sin x} = \lim_{x \to \infty} \frac{1 + \frac{\sin x}{x} + \frac{2}{\sqrt{x}}}{1 + \frac{\sin x}{x}} = \frac{1 + 0 + 0}{1 + 0} = 1$

55. (a) $f(-1) = -1$ and $f(2) = 5 \Rightarrow f$ has a root between -1 and 2 by the Intermediate Value Theorem.

    (b), (c) root is 1.32471795724

57. Show $\displaystyle\lim_{x \to 1} f(x) = \lim_{x \to 1} (x^2 - 7) = -6 = f(1)$.

    Step 1: $|(x^2 - 7) + 6| < \epsilon \Rightarrow -\epsilon < x^2 - 1 < \epsilon \Rightarrow 1 - \epsilon < x^2 < 1 + \epsilon \Rightarrow \sqrt{1 - \epsilon} < x < \sqrt{1 + \epsilon}$.

    Step 2: $|x - 1| < \delta \Rightarrow -\delta < x - 1 < \delta \Rightarrow -\delta + 1 < x < \delta + 1$.

    Then $-\delta + 1 = \sqrt{1 - \epsilon}$ or $\delta + 1 = \sqrt{1 + \epsilon}$. Choose $\delta = \min\left\{1 - \sqrt{1 - \epsilon}, \sqrt{1 + \epsilon} - 1\right\}$, then

    $0 < |x - 1| < \delta \Rightarrow |(x^2 - 7) - 6| < \epsilon$ and $\displaystyle\lim_{x \to 1} f(x) = -6$. By the continuity test, $f(x)$ is continuous at $x = 1$.

59. Show $\displaystyle\lim_{x \to 2} h(x) = \lim_{x \to 2} \sqrt{2x - 3} = 1 = h(2)$.

    Step 1: $\left|\sqrt{2x - 3} - 1\right| < \epsilon \Rightarrow -\epsilon < \sqrt{2x - 3} - 1 < \epsilon \Rightarrow 1 - \epsilon < \sqrt{2x - 3} < 1 + \epsilon \Rightarrow \frac{(1 - \epsilon)^2 + 3}{2} < x < \frac{(1 + \epsilon)^2 + 3}{2}$.

    Step 2: $|x - 2| < \delta \Rightarrow -\delta < x - 2 < \delta$ or $-\delta + 2 < x < \delta + 2$.

    Then $-\delta + 2 = \frac{(1 - \epsilon)^2 + 3}{2} \Rightarrow \delta = 2 - \frac{(1 - \epsilon)^2 + 3}{2} = \frac{1 - (1 - \epsilon)^2}{2} = \epsilon - \frac{\epsilon^2}{2}$, or $\delta + 2 = \frac{(1 + \epsilon)^2 + 3}{2}$

    $\Rightarrow \delta = \frac{(1 + \epsilon)^2 + 3}{2} - 2 = \frac{(1 + \epsilon)^2 - 1}{2} = \epsilon + \frac{\epsilon^2}{2}$. Choose $\delta = \epsilon - \frac{\epsilon^2}{2}$, the smaller of the two values. Then,

    $0 < |x - 2| < \delta \Rightarrow \left|\sqrt{2x - 3} - 1\right| < \epsilon$, so $\displaystyle\lim_{x \to 2} \sqrt{2x - 3} = 1$. By the continuity test, $h(x)$ is continuous at $x = 2$.

61. (a) Let $\epsilon > 0$ be given. If x is rational, then $f(x) = x \Rightarrow |f(x) - 0| = |x - 0| < \epsilon \Leftrightarrow |x - 0| < \epsilon$; i.e., choose

    $\delta = \epsilon$. Then $|x - 0| < \delta \Rightarrow |f(x) - 0| < \epsilon$ for x rational. If x is irrational, then $f(x) = 0 \Rightarrow |f(x) - 0| < \epsilon$

    $\Leftrightarrow 0 < \epsilon$ which is true no matter how close irrational x is to 0, so again we can choose $\delta = \epsilon$. In either case,

    given $\epsilon > 0$ there is a $\delta = \epsilon > 0$ such that $0 < |x - 0| < \delta \Rightarrow |f(x) - 0| < \epsilon$. Therefore, f is continuous at

    $x = 0$.

    (b) Choose $x = c > 0$. Then within any interval $(c - \delta, c + \delta)$ there are both rational and irrational numbers.

    If c is rational, pick $\epsilon = \frac{c}{2}$. No matter how small we choose $\delta > 0$ there is an irrational number x in

    $(c - \delta, c + \delta) \Rightarrow |f(x) - f(c)| = |0 - c| = c > \frac{c}{2} = \epsilon$. That is, f is not continuous at any rational $c > 0$. On

    the other hand, suppose c is irrational $\Rightarrow f(c) = 0$. Again pick $\epsilon = \frac{c}{2}$. No matter how small we choose $\delta > 0$

    there is a rational number x in $(c - \delta, c + \delta)$ with $|x - c| < \frac{c}{2} = \epsilon \Leftrightarrow \frac{c}{2} < x < \frac{3c}{2}$. Then $|f(x) - f(c)| = |x - 0|$

    $= |x| > \frac{c}{2} = \epsilon \Rightarrow f$ is not continuous at any irrational $c > 0$.

    If $x = c < 0$, repeat the argument picking $\epsilon = \frac{|c|}{2} = \frac{-c}{2}$. Therefore f fails to be continuous at any

    nonzero value $x = c$.

63. (a) The function f is bounded on D if $f(x) \geq M$ and $f(x) \leq N$ for all x in D. This means $M \leq f(x) \leq N$ for all x

    in D. Choose B to be $\max\{|M|, |N|\}$. Then $|f(x)| \leq B$. On the other hand, if $|f(x)| \leq B$, then

    $-B \leq f(x) \leq B \Rightarrow f(x) \geq -B$ and $f(x) \leq B \Rightarrow f(x)$ is bounded on D with $N = B$ an upper bound and

    $M = -B$ a lower bound.

    (b) Assume $f(x) \leq N$ for all x and that $L > N$. Let $\epsilon = \frac{L - N}{2}$. Since $\displaystyle\lim_{x \to x_0} f(x) = L$ there is a $\delta > 0$ such that

    $0 < |x - x_0| < \delta \Rightarrow |f(x) - L| < \epsilon \Leftrightarrow L - \epsilon < f(x) < L + \epsilon \Leftrightarrow L - \frac{L - N}{2} < f(x) < L + \frac{L - N}{2}$

    $\Leftrightarrow \frac{L + N}{2} < f(x) < \frac{3L - N}{2}$. But $L > N \Rightarrow \frac{L + N}{2} > N \Rightarrow N < f(x)$ contrary to the boundedness assumption

    $f(x) \leq N$. This contradiction proves $L \leq N$.

    (c) Assume $M \leq f(x)$ for all x and that $L < M$. Let $\epsilon = \frac{M - L}{2}$. As in part (b), $0 < |x - x_0| < \delta$

    $\Rightarrow L - \frac{M - L}{2} < f(x) < L + \frac{M - L}{2} \Leftrightarrow \frac{3L - M}{2} < f(x) < \frac{M + L}{2} < M$, a contradiction.

CHAPTER 2 DIFFERENTIATION

2.1 TANGENTS AND DERIVATIVES AT A POINT

1. P_1: $m_1 = 1$, P_2: $m_2 = 5$

3. P_1: $m_1 = \frac{5}{2}$, P_2: $m_2 = -\frac{1}{2}$

5. $m = \lim\limits_{h \to 0} \frac{[3-(-1+h)^2]-(3-(-1)^2)}{h}$

$= \lim\limits_{h \to 0} \frac{3-(1-2h+h^2)-2}{h} = \lim\limits_{h \to 0} \frac{h(2-h)}{h} = 2$;

at $(-1, 2)$: $y = 2 + 2(x-(-1)) \Rightarrow y = 2x + 4$,

tangent line

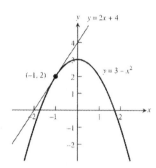

7. $m = \lim\limits_{h \to 0} \frac{2\sqrt{1+h}-2\sqrt{1}}{h} = \lim\limits_{h \to 0} \frac{2\sqrt{1+h}-2}{h} \cdot \frac{2\sqrt{1+h}+2}{2\sqrt{1+h}+2}$

$= \lim\limits_{h \to 0} \frac{4(1+h)-4}{2h\left(\sqrt{1+h}+1\right)} = \lim\limits_{h \to 0} \frac{2}{\sqrt{1+h}+1} = 1$;

at $(1, 2)$: $y = 2 + 1(x - 1) \Rightarrow y = x + 1$, tangent line

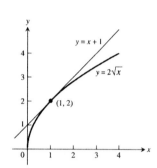

9. $m = \lim\limits_{h \to 0} \frac{(-2+h)^3+1-((-2)^3+1)}{h} = \lim\limits_{h \to 0} \frac{-8+12h-6h^2+h^3+8}{h}$

$= \lim\limits_{h \to 0} (12 - 6h + h^2) = 12$;

at $(-2, -7)$: $y = -7 + 12(x-(-2)) \Rightarrow y = 12x + 17$,

tangent line

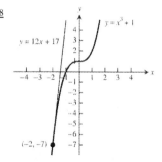

11. $m = \lim\limits_{h \to 0} \frac{[(2+h)^2+1]-5}{h} = \lim\limits_{h \to 0} \frac{(5+4h+h^2)-5}{h} = \lim\limits_{h \to 0} \frac{h(4+h)}{h} = 4$;

at $(2, 5)$: $y - 5 = 4(x - 2)$, tangent line

13. $m = \lim\limits_{h \to 0} \frac{\frac{3+h}{(3+h)-2}-3}{h} = \lim\limits_{h \to 0} \frac{(3+h)-3(h+1)}{h(h+1)} = \lim\limits_{h \to 0} \frac{-2h}{h(h+1)} = -2$;

at $(3, 3)$: $y - 3 = -2(x - 3)$, tangent line

15. $m = \lim\limits_{h \to 0} \frac{((2+h)^3-(2+h))-6}{h} = \lim\limits_{h \to 0} \frac{(6+11h+6h^2+h^3)-6}{h} = \lim\limits_{h \to 0} \frac{h(11+6h+h^2)}{h} = 11$;

at $(2, 6)$: $y - 6 = 11(t - 2)$, tangent line

17. $m = \lim\limits_{h \to 0} \frac{\sqrt{4+h}-2}{h} = \lim\limits_{h \to 0} \frac{\sqrt{4+h}-2}{h} \cdot \frac{\sqrt{4+h}+2}{\sqrt{4+h}+2} = \lim\limits_{h \to 0} \frac{(4+h)-4}{h\left(\sqrt{4+h}+2\right)} = \lim\limits_{h \to 0} \frac{h}{h\left(\sqrt{4+h}+2\right)} = \frac{1}{\sqrt{4}+2}$

   $= \frac{1}{4}$; at $(4,2)$: $y - 2 = \frac{1}{4}(x-4)$, tangent line

19. At $x = -1$, $y = 5 \Rightarrow m = \lim\limits_{h \to 0} \frac{5(-1+h)^2 - 5}{h} = \lim\limits_{h \to 0} \frac{5(1-2h+h^2)-5}{h} = \lim\limits_{h \to 0} \frac{5h(-2+h)}{h} = -10$, slope

21. At $x = 3$, $y = \frac{1}{2} \Rightarrow m = \lim\limits_{h \to 0} \frac{\frac{1}{(3+h)-1} - \frac{1}{2}}{h} = \lim\limits_{h \to 0} \frac{2-(2+h)}{2h(2+h)} = \lim\limits_{h \to 0} \frac{-h}{2h(2+h)} = -\frac{1}{4}$, slope

23. At a horizontal tangent the slope $m = 0 \Rightarrow 0 = m = \lim\limits_{h \to 0} \frac{[(x+h)^2 + 4(x+h) - 1] - (x^2 + 4x - 1)}{h}$

   $= \lim\limits_{h \to 0} \frac{(x^2 + 2xh + h^2 + 4x + 4h - 1) - (x^2 + 4x - 1)}{h} = \lim\limits_{h \to 0} \frac{(2xh + h^2 + 4h)}{h} = \lim\limits_{h \to 0} (2x + h + 4) = 2x + 4$;

   $2x + 4 = 0 \Rightarrow x = -2$. Then $f(-2) = 4 - 8 - 1 = -5 \Rightarrow (-2, -5)$ is the point on the graph where there is a horizontal tangent.

25. $-1 = m = \lim\limits_{h \to 0} \frac{\frac{1}{(x+h)-1} - \frac{1}{x-1}}{h} = \lim\limits_{h \to 0} \frac{(x-1)-(x+h-1)}{h(x-1)(x+h-1)} = \lim\limits_{h \to 0} \frac{-h}{h(x-1)(x+h-1)} = -\frac{1}{(x-1)^2}$

   $\Rightarrow (x-1)^2 = 1 \Rightarrow x^2 - 2x = 0 \Rightarrow x(x-2) = 0 \Rightarrow x = 0$ or $x = 2$. If $x = 0$, then $y = -1$ and $m = -1$

   $\Rightarrow y = -1 - (x - 0) = -(x+1)$. If $x = 2$, then $y = 1$ and $m = -1 \Rightarrow y = 1 - (x-2) = -(x-3)$.

27. $\lim\limits_{h \to 0} \frac{f(2+h) - f(2)}{h} = \lim\limits_{h \to 0} \frac{(100 - 4.9(2+h)^2) - (100 - 4.9(2)^2)}{h} = \lim\limits_{h \to 0} \frac{-4.9(4 + 4h + h^2) + 4.9(4)}{h}$

   $= \lim\limits_{h \to 0} (-19.6 - 4.9h) = -19.6$. The minus sign indicates the object is falling <u>downward</u> at a speed of 19.6 m/sec.

29. $\lim\limits_{h \to 0} \frac{f(3+h) - f(3)}{h} = \lim\limits_{h \to 0} \frac{\pi(3+h)^2 - \pi(3)^2}{h} = \lim\limits_{h \to 0} \frac{\pi[9 + 6h + h^2 - 9]}{h} = \lim\limits_{h \to 0} \pi(6+h) = 6\pi$

31. $\lim\limits_{h \to 0} \frac{(m(x_0 + h) + b) - (mx_0 + b)}{h} = \lim\limits_{h \to 0} \frac{mx_0 + mh + b - mx_0 - b}{h} = \lim\limits_{h \to 0} \frac{mh}{h} = \lim\limits_{h \to 0} m = m$

   $y - (mx_0 + b) = m(x - x_0) \to y - mx_0 - b = mx - mx_0 \to y = mx + b$

33. Slope at origin $= \lim\limits_{h \to 0} \frac{f(0+h) - f(0)}{h} = \lim\limits_{h \to 0} \frac{h^2 \sin\left(\frac{1}{h}\right)}{h} = \lim\limits_{h \to 0} h \sin\left(\frac{1}{h}\right) = 0 \Rightarrow$ yes, $f(x)$ does have a tangent at the origin with slope 0.

35. $\lim\limits_{h \to 0^-} \frac{f(0+h) - f(0)}{h} = \lim\limits_{h \to 0^-} \frac{-1-0}{h} = \infty$, and $\lim\limits_{h \to 0^+} \frac{f(0+h) - f(0)}{h} = \lim\limits_{h \to 0^+} \frac{1-0}{h} = \infty$. Therefore,

   $\lim\limits_{h \to 0} \frac{f(0+h) - f(0)}{h} = \infty \Rightarrow$ yes, the graph of f has a vertical tangent at the origin.

37. (a) The graph appears to have a cusp at $x = 0$.

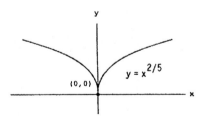

   (b) $\lim\limits_{h \to 0^-} \frac{f(0+h) - f(0)}{h} = \lim\limits_{h \to 0^-} \frac{h^{2/5} - 0}{h} = \lim\limits_{h \to 0^-} \frac{1}{h^{3/5}} = -\infty$ and $\lim\limits_{h \to 0^+} \frac{1}{h^{3/5}} = \infty \Rightarrow$ limit does not exist

   $\Rightarrow$ the graph of $y = x^{2/5}$ does not have a vertical tangent at $x = 0$.

39. (a) The graph appears to have a vertical tangent at x = 0.

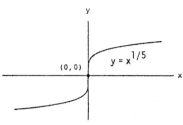

   (b) $\lim\limits_{h \to 0} \frac{f(0+h)-f(0)}{h} = \lim\limits_{h \to 0} \frac{h^{1/5}-0}{h} = \lim\limits_{h \to 0} \frac{1}{h^{4/5}} = \infty \Rightarrow y = x^{1/5}$ has a vertical tangent at x = 0.

41. (a) The graph appears to have a cusp at x = 0.

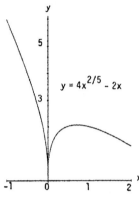

   (b) $\lim\limits_{h \to 0^-} \frac{f(0+h)-f(0)}{h} = \lim\limits_{h \to 0^-} \frac{4h^{2/5}-2h}{h} = \lim\limits_{h \to 0^-} \frac{4}{h^{3/5}} - 2 = -\infty$ and $\lim\limits_{h \to 0^+} \frac{4}{h^{3/5}} - 2 = \infty$
   $\Rightarrow$ limit does not exist $\Rightarrow$ the graph of $y = 4x^{2/5} - 2x$ does not have a vertical tangent at x = 0.

43. (a) The graph appears to have a vertical tangent at x = 1
   and a cusp at x = 0.

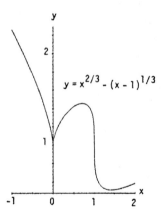

   (b) x = 1: $\lim\limits_{h \to 0} \frac{(1+h)^{2/3}-(1+h-1)^{1/3}-1}{h} = \lim\limits_{h \to 0} \frac{(1+h)^{2/3}-h^{1/3}-1}{h} = -\infty$
   $\Rightarrow y = x^{2/3} - (x-1)^{1/3}$ has a vertical tangent at x = 1;
   x = 0: $\lim\limits_{h \to 0} \frac{f(0+h)-f(0)}{h} = \lim\limits_{h \to 0} \frac{h^{2/3}-(h-1)^{1/3}-(-1)^{1/3}}{h} = \lim\limits_{h \to 0} \left[\frac{1}{h^{1/3}} - \frac{(h-1)^{1/3}}{h} + \frac{1}{h} \right]$
   does not exist $\Rightarrow y = x^{2/3} - (x-1)^{1/3}$ does not have a vertical tangent at x = 0.

45. (a) The graph appears to have a vertical tangent at x = 0.

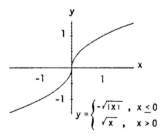

(b) $\lim\limits_{h \to 0^+} \frac{f(0+h) - f(0)}{h} = \lim\limits_{x \to 0^+} \frac{\sqrt{h} - 0}{h} = \lim\limits_{h \to 0} \frac{1}{\sqrt{h}} = \infty;$

$\lim\limits_{h \to 0^-} \frac{f(0+h) - f(0)}{h} = \lim\limits_{h \to 0^-} \frac{-\sqrt{|h|} - 0}{h} = \lim\limits_{h \to 0^-} \frac{-\sqrt{|h|}}{-|h|} = \lim\limits_{h \to 0^-} \frac{1}{\sqrt{|h|}} = \infty$

$\Rightarrow$ y has a vertical tangent at x = 0.

2.2 THE DERIVATIVE AS A FUNCTION

1. Step 1: $f(x) = 4 - x^2$ and $f(x + h) = 4 - (x + h)^2$

Step 2: $\frac{f(x+h) - f(x)}{h} = \frac{[4 - (x+h)^2] - (4 - x^2)}{h} = \frac{(4 - x^2 - 2xh - h^2) - 4 + x^2}{h} = \frac{-2xh - h^2}{h} = \frac{h(-2x - h)}{h}$

$= -2x - h$

Step 3: $f'(x) = \lim\limits_{h \to 0} (-2x - h) = -2x;\ f'(-3) = 6,\ f'(0) = 0,\ f'(1) = -2$

3. Step 1: $g(t) = \frac{1}{t^2}$ and $g(t + h) = \frac{1}{(t+h)^2}$

Step 2: $\frac{g(t+h) - g(t)}{h} = \frac{\frac{1}{(t+h)^2} - \frac{1}{t^2}}{h} = \frac{\left(\frac{t^2 - (t+h)^2}{(t+h)^2 \cdot t^2}\right)}{h} = \frac{t^2 - (t^2 + 2th + h^2)}{(t+h)^2 \cdot t^2 \cdot h} = \frac{-2th - h^2}{(t+h)^2 \, t^2 h}$

$= \frac{h(-2t - h)}{(t+h)^2 \, t^2 h} = \frac{-2t - h}{(t+h)^2 \, t^2}$

Step 3: $g'(t) = \lim\limits_{h \to 0} \frac{-2t - h}{(t+h)^2 \, t^2} = \frac{-2t}{t^2 \cdot t^2} = \frac{-2}{t^3};\ g'(-1) = 2,\ g'(2) = -\frac{1}{4},\ g'\left(\sqrt{3}\right) = -\frac{2}{3\sqrt{3}}$

5. Step 1: $p(\theta) = \sqrt{3\theta}$ and $p(\theta + h) = \sqrt{3(\theta + h)}$

Step 2: $\frac{p(\theta+h) - p(\theta)}{h} = \frac{\sqrt{3(\theta+h)} - \sqrt{3\theta}}{h} = \frac{\left(\sqrt{3\theta + 3h} - \sqrt{3\theta}\right)}{h} \cdot \frac{\left(\sqrt{3\theta + 3h} + \sqrt{3\theta}\right)}{\left(\sqrt{3\theta + 3h} + \sqrt{3\theta}\right)} = \frac{(3\theta + 3h) - 3\theta}{h\left(\sqrt{3\theta + 3h} + \sqrt{3\theta}\right)}$

$= \frac{3h}{h\left(\sqrt{3\theta + 3h} + \sqrt{3\theta}\right)} = \frac{3}{\sqrt{3\theta + 3h} + \sqrt{3\theta}}$

Step 3: $p'(\theta) = \lim\limits_{h \to 0} \frac{3}{\sqrt{3\theta + 3h} + \sqrt{3\theta}} = \frac{3}{\sqrt{3\theta} + \sqrt{3\theta}} = \frac{3}{2\sqrt{3\theta}};\ p'(1) = \frac{3}{2\sqrt{3}},\ p'(3) = \frac{1}{2},\ p'\left(\frac{2}{3}\right) = \frac{3}{2\sqrt{2}}$

7. $y = f(x) = 2x^3$ and $f(x + h) = 2(x + h)^3 \Rightarrow \frac{dy}{dx} = \lim\limits_{h \to 0} \frac{2(x+h)^3 - 2x^3}{h} = \lim\limits_{h \to 0} \frac{2(x^3 + 3x^2 h + 3xh^2 + h^3) - 2x^3}{h}$

$= \lim\limits_{h \to 0} \frac{6x^2 h + 6xh^2 + 2h^3}{h} = \lim\limits_{h \to 0} \frac{h(6x^2 + 6xh + 2h^2)}{h} = \lim\limits_{h \to 0} (6x^2 + 6xh + 2h^2) = 6x^2$

9. $s = r(t) = \frac{t}{2t+1}$ and $r(t + h) = \frac{t+h}{2(t+h)+1} \Rightarrow \frac{ds}{dt} = \lim\limits_{h \to 0} \frac{\left(\frac{t+h}{2(t+h)+1}\right) - \left(\frac{t}{2t+1}\right)}{h}$

$= \lim\limits_{h \to 0} \frac{\left(\frac{(t+h)(2t+1) - t(2t+2h+1)}{(2t+2h+1)(2t+1)}\right)}{h} = \lim\limits_{h \to 0} \frac{(t+h)(2t+1) - t(2t+2h+1)}{(2t+2h+1)(2t+1)h} = \lim\limits_{h \to 0} \frac{2t^2 + t + 2ht + h - 2t^2 - 2ht - t}{(2t+2h+1)(2t+1)h}$

$= \lim\limits_{h \to 0} \frac{h}{(2t+2h+1)(2t+1)h} = \lim\limits_{h \to 0} \frac{1}{(2t+2h+1)(2t+1)} = \frac{1}{(2t+1)(2t+1)} = \frac{1}{(2t+1)^2}$

11. $p = f(q) = \frac{1}{\sqrt{q+1}}$ and $f(q + h) = \frac{1}{\sqrt{(q+h)+1}} \Rightarrow \frac{dp}{dq} = \lim\limits_{h \to 0} \frac{\left(\frac{1}{\sqrt{(q+h)+1}}\right) - \left(\frac{1}{\sqrt{q+1}}\right)}{h}$

$= \lim\limits_{h \to 0} \frac{\left(\frac{\sqrt{q+1} - \sqrt{q+h+1}}{\sqrt{q+h+1}\sqrt{q+1}}\right)}{h} = \lim\limits_{h \to 0} \frac{\sqrt{q+1} - \sqrt{q+h+1}}{h\sqrt{q+h+1}\sqrt{q+1}} = \lim\limits_{h \to 0} \frac{(\sqrt{q+1} - \sqrt{q+h+1})}{h\sqrt{q+h+1}\sqrt{q+1}} \cdot \frac{(\sqrt{q+1} + \sqrt{q+h+1})}{(\sqrt{q+1} + \sqrt{q+h+1})}$

$= \lim\limits_{h \to 0} \frac{(q+1) - (q+h+1)}{h\sqrt{q+h+1}\sqrt{q+1}\left(\sqrt{q+1} + \sqrt{q+h+1}\right)} = \lim\limits_{h \to 0} \frac{-h}{h\sqrt{q+h+1}\sqrt{q+1}\left(\sqrt{q+1} + \sqrt{q+h+1}\right)}$

$= \lim\limits_{h \to 0} \frac{-1}{\sqrt{q+h+1}\sqrt{q+1}\left(\sqrt{q+1} + \sqrt{q+h+1}\right)} = \frac{-1}{\sqrt{q+1}\sqrt{q+1}\left(\sqrt{q+1} + \sqrt{q+1}\right)} = \frac{-1}{2(q+1)\sqrt{q+1}}$

13. $f(x) = x + \frac{9}{x}$ and $f(x + h) = (x + h) + \frac{9}{(x+h)} \Rightarrow \frac{f(x+h) - f(x)}{h} = \frac{\left[(x+h) + \frac{9}{(x+h)}\right] - \left[x + \frac{9}{x}\right]}{h}$

$= \frac{x(x+h)^2 + 9x - x^2(x+h) - 9(x+h)}{x(x+h)h} = \frac{x^3 + 2x^2 h + xh^2 + 9x - x^3 - x^2 h - 9x - 9h}{x(x+h)h} = \frac{x^2 h + xh^2 - 9h}{x(x+h)h}$

$= \frac{h(x^2 + xh - 9)}{x(x+h)h} = \frac{x^2 + xh - 9}{x(x+h)};\ f'(x) = \lim\limits_{h \to 0} \frac{x^2 + xh - 9}{x(x+h)} = \frac{x^2 - 9}{x^2} = 1 - \frac{9}{x^2};\ m = f'(-3) = 0$

15. $\dfrac{ds}{dt} = \lim\limits_{h \to 0} \dfrac{[(t+h)^3 - (t+h)^2] - (t^3 - t^2)}{h} = \lim\limits_{h \to 0} \dfrac{(t^3 + 3t^2h + 3th^2 + h^3) - (t^2 + 2th + h^2) - t^3 + t^2}{h}$

$= \lim\limits_{h \to 0} \dfrac{3t^2h + 3th^2 + h^3 - 2th - h^2}{h} = \lim\limits_{h \to 0} \dfrac{h(3t^2 + 3th + h^2 - 2t - h)}{h} = \lim\limits_{h \to 0} (3t^2 + 3th + h^2 - 2t - h)$

$= 3t^2 - 2t; \; m = \dfrac{ds}{dt}\Big|_{t=-1} = 5$

17. $f(x) = \dfrac{8}{\sqrt{x}-2}$ and $f(x+h) = \dfrac{8}{\sqrt{(x+h)}-2}$ $\Rightarrow \dfrac{f(x+h)-f(x)}{h} = \dfrac{\frac{8}{\sqrt{(x+h)}-2} - \frac{8}{\sqrt{x}-2}}{h}$

$= \dfrac{8\left(\sqrt{x-2} - \sqrt{x+h-2}\right)}{h\sqrt{x+h-2}\sqrt{x-2}} \cdot \dfrac{\left(\sqrt{x-2} + \sqrt{x+h-2}\right)}{\left(\sqrt{x-2} + \sqrt{x+h-2}\right)} = \dfrac{8[(x-2) - (x+h-2)]}{h\sqrt{x+h-2}\sqrt{x-2}\left(\sqrt{x-2} + \sqrt{x+h-2}\right)}$

$= \dfrac{-8h}{h\sqrt{x+h-2}\sqrt{x-2}\left(\sqrt{x-2} + \sqrt{x+h-2}\right)} \Rightarrow f'(x) = \lim\limits_{h \to 0} \dfrac{-8}{\sqrt{x+h-2}\sqrt{x-2}\left(\sqrt{x-2} + \sqrt{x+h-2}\right)}$

$= \dfrac{-8}{\sqrt{x-2}\sqrt{x-2}\left(\sqrt{x-2} + \sqrt{x-2}\right)} = \dfrac{-4}{(x-2)\sqrt{x-2}}; \; m = f'(6) = \dfrac{-4}{4\sqrt{4}} = -\dfrac{1}{2} \Rightarrow$ the equation of the tangent

line at $(6, 4)$ is $y - 4 = -\dfrac{1}{2}(x - 6) \Rightarrow y = -\dfrac{1}{2}x + 3 + 4 \Rightarrow y = -\dfrac{1}{2}x + 7.$

19. $s = f(t) = 1 - 3t^2$ and $f(t+h) = 1 - 3(t+h)^2 = 1 - 3t^2 - 6th - 3h^2 \Rightarrow \dfrac{ds}{dt} = \lim\limits_{h \to 0} \dfrac{f(t+h)-f(t)}{h}$

$= \lim\limits_{h \to 0} \dfrac{(1 - 3t^2 - 6th - 3h^2) - (1 - 3t^2)}{h} = \lim\limits_{h \to 0} (-6t - 3h) = -6t \Rightarrow \dfrac{ds}{dt}\Big|_{t=-1} = 6$

21. $r = f(\theta) = \dfrac{2}{\sqrt{4-\theta}}$ and $f(\theta+h) = \dfrac{2}{\sqrt{4-(\theta+h)}} \Rightarrow \dfrac{dr}{d\theta} = \lim\limits_{h \to 0} \dfrac{f(\theta+h)-f(\theta)}{h} = \lim\limits_{h \to 0} \dfrac{\frac{2}{\sqrt{4-\theta-h}} - \frac{2}{\sqrt{4-\theta}}}{h}$

$= \lim\limits_{h \to 0} \dfrac{2\sqrt{4-\theta} - 2\sqrt{4-\theta-h}}{h\sqrt{4-\theta}\sqrt{4-\theta-h}} = \lim\limits_{h \to 0} \dfrac{2\sqrt{4-\theta} - 2\sqrt{4-\theta-h}}{h\sqrt{4-\theta}\sqrt{4-\theta-h}} \cdot \dfrac{\left(2\sqrt{4-\theta} + 2\sqrt{4-\theta-h}\right)}{\left(2\sqrt{4-\theta} + 2\sqrt{4-\theta-h}\right)}$

$= \lim\limits_{h \to 0} \dfrac{4(4-\theta) - 4(4-\theta-h)}{2h\sqrt{4-\theta}\sqrt{4-\theta-h}\left(\sqrt{4-\theta} + \sqrt{4-\theta-h}\right)} = \lim\limits_{h \to 0} \dfrac{2}{\sqrt{4-\theta}\sqrt{4-\theta-h}\left(\sqrt{4-\theta} + \sqrt{4-\theta-h}\right)}$

$= \dfrac{2}{(4-\theta)\left(2\sqrt{4-\theta}\right)} = \dfrac{1}{(4-\theta)\sqrt{4-\theta}} \Rightarrow \dfrac{dr}{d\theta}\Big|_{\theta=0} = \dfrac{1}{8}$

23. $f'(x) = \lim\limits_{z \to x} \dfrac{f(z)-f(x)}{z-x} = \lim\limits_{z \to x} \dfrac{\frac{1}{z+2} - \frac{1}{x+2}}{z-x} = \lim\limits_{z \to x} \dfrac{(x+2)-(z+2)}{(z-x)(z+2)(x+2)} = \lim\limits_{z \to x} \dfrac{x-z}{(z-x)(z+2)(x+2)} = \lim\limits_{z \to x} \dfrac{-1}{(z+2)(x+2)} = \dfrac{-1}{(x+2)^2}$

25. $g'(x) = \lim\limits_{z \to x} \dfrac{g(z)-g(x)}{z-x} = \lim\limits_{z \to x} \dfrac{\frac{z}{z-1} - \frac{x}{x-1}}{z-x} = \lim\limits_{z \to x} \dfrac{z(x-1) - x(z-1)}{(z-x)(z-1)(x-1)} = \lim\limits_{z \to x} \dfrac{-z+x}{(z-x)(z-1)(x-1)} = \lim\limits_{z \to x} \dfrac{-1}{(z-1)(x-1)} = \dfrac{-1}{(x-1)^2}$

27. Note that as x increases, the slope of the tangent line to the curve is first negative, then zero (when $x = 0$), then positive $\Rightarrow$ the slope is always increasing which matches (b).

29. $f_3(x)$ is an oscillating function like the cosine. Everywhere that the graph of f_3 has a horizontal tangent we expect f_3' to be zero, and (d) matches this condition.

31. (a) f' is not defined at $x = 0, 1, 4$. At these points, the left-hand and right-hand derivatives do not agree.

For example, $\lim\limits_{x \to 0^-} \dfrac{f(x)-f(0)}{x-0} =$ slope of line joining $(-4, 0)$ and $(0, 2) = \dfrac{1}{2}$ but $\lim\limits_{x \to 0^+} \dfrac{f(x)-f(0)}{x-0} =$ slope of

line joining $(0, 2)$ and $(1, -2) = -4$. Since these values are not equal, $f'(0) = \lim\limits_{x \to 0} \dfrac{f(x)-f(0)}{x-0}$ does not exist.

(b)

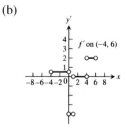

33. Left-hand derivative: For $h < 0$, $f(0 + h) = f(h) = h^2$ (using $y = x^2$ curve) $\Rightarrow \lim\limits_{h \to 0^-} \frac{f(0 + h) - f(0)}{h}$

$= \lim\limits_{h \to 0^-} \frac{h^2 - 0}{h} = \lim\limits_{h \to 0^-} h = 0$;

Right-hand derivative: For $h > 0$, $f(0 + h) = f(h) = h$ (using $y = x$ curve) $\Rightarrow \lim\limits_{h \to 0^+} \frac{f(0 + h) - f(0)}{h}$

$= \lim\limits_{h \to 0^+} \frac{h - 0}{h} = \lim\limits_{h \to 0^+} 1 = 1$;

Then $\lim\limits_{h \to 0^-} \frac{f(0 + h) - f(0)}{h} \neq \lim\limits_{h \to 0^+} \frac{f(0 + h) - f(0)}{h} \Rightarrow$ the derivative $f'(0)$ does not exist.

35. Left-hand derivative: When $h < 0$, $1 + h < 1 \Rightarrow f(1 + h) = \sqrt{1 + h} \Rightarrow \lim\limits_{h \to 0^-} \frac{f(1 + h) - f(1)}{h}$

$= \lim\limits_{h \to 0^-} \frac{\sqrt{1 + h} - 1}{h} = \lim\limits_{h \to 0^-} \frac{\left(\sqrt{1 + h} - 1\right)}{h} \cdot \frac{\left(\sqrt{1 + h} + 1\right)}{\left(\sqrt{1 + h} + 1\right)} = \lim\limits_{h \to 0^-} \frac{(1 + h) - 1}{h\left(\sqrt{1 + h} + 1\right)} = \lim\limits_{h \to 0^-} \frac{1}{\sqrt{1 + h} + 1} = \frac{1}{2}$;

Right-hand derivative: When $h > 0$, $1 + h > 1 \Rightarrow f(1 + h) = 2(1 + h) - 1 = 2h + 1 \Rightarrow \lim\limits_{h \to 0^+} \frac{f(1 + h) - f(1)}{h}$

$= \lim\limits_{h \to 0^+} \frac{(2h + 1) - 1}{h} = \lim\limits_{h \to 0^+} 2 = 2$;

Then $\lim\limits_{h \to 0^-} \frac{f(1 + h) - f(1)}{h} \neq \lim\limits_{h \to 0^+} \frac{f(1 + h) - f(1)}{h} \Rightarrow$ the derivative $f'(1)$ does not exist.

37. (a) The function is differentiable on its domain $-3 \leq x \leq 2$ (it is smooth)
    (b) none
    (c) none

39. (a) The function is differentiable on $-3 \leq x < 0$ and $0 < x \leq 3$
    (b) none
    (c) The function is neither continuous nor differentiable at $x = 0$ since $\lim\limits_{x \to 0^-} f(x) \neq \lim\limits_{x \to 0^+} f(x)$

41. (a) f is differentiable on $-1 \leq x < 0$ and $0 < x \leq 2$
    (b) f is continuous but not differentiable at $x = 0$: $\lim\limits_{x \to 0} f(x) = 0$ exists but there is a cusp at $x = 0$, so
    $f'(0) = \lim\limits_{h \to 0} \frac{f(0 + h) - f(0)}{h}$ does not exist
    (c) none

43. $y' = \lim\limits_{h \to 0} \frac{(2(x + h)^2 - 13(x + h) + 5) - (2x^2 - 13x + 5)}{h} = \lim\limits_{h \to 0} \frac{2x^2 + 4xh + 2h^2 - 13x - 13h + 5 - 2x^2 + 13x - 5}{h}$

$= \lim\limits_{h \to 0} \frac{4xh + 2h^2 - 13h}{h} = \lim\limits_{h \to 0} (4x + 2h - 13) = 4x - 13$, slope at x. The slope is -1 when $4x - 13 = -1$

$\Rightarrow 4x = 12 \Rightarrow x = 3 \Rightarrow y = 2 \cdot 3^2 - 13 \cdot 3 + 5 = -16$. Thus the tangent line is $y + 16 = (-1)(x - 3)$

$\Rightarrow y = -x - 13$ and the point of tangency is $(3, -16)$.

45. (a) Suppose $|f(x)| \leq x^2$ for $-1 \leq x \leq 1$. Then $|f(0)| \leq 0^2 \Rightarrow f(0) = 0$. Then $f'(0) = \lim\limits_{h \to 0} \frac{f(0 + h) - f(0)}{h}$

$= \lim\limits_{h \to 0} \frac{f(h) - 0}{h} = \lim\limits_{h \to 0} \frac{f(h)}{h}$. For $|h| \leq 1$, $-h^2 \leq f(h) \leq h^2 \Rightarrow -h \leq \frac{f(h)}{h} \leq h \Rightarrow f'(0) = \lim\limits_{h \to 0} \frac{f(h)}{h} = 0$

by the Sandwich Theorem for limits.

(b) Note that for $x \neq 0$, $|f(x)| = \left|x^2 \sin \frac{1}{x}\right| = |x^2| |\sin x| \leq |x^2| \cdot 1 = x^2$ (since $-1 \leq \sin x \leq 1$). By part (a),

f is differentiable at $x = 0$ and $f'(0) = 0$.

2.3 DIFFERENTIATION RULES

1. $y = -x^2 + 3 \Rightarrow \frac{dy}{dx} = \frac{d}{dx}\left(-x^2\right) + \frac{d}{dx}(3) = -2x + 0 = -2x \Rightarrow \frac{d^2y}{dx^2} = -2$

3. $s = 5t^3 - 3t^5 \Rightarrow \frac{ds}{dt} = \frac{d}{dt}\left(5t^3\right) - \frac{d}{dt}\left(3t^5\right) = 15t^2 - 15t^4 \Rightarrow \frac{d^2s}{dt^2} = \frac{d}{dt}\left(15t^2\right) - \frac{d}{dt}\left(15t^4\right) = 30t - 60t^3$

5. $y = \frac{4}{3} x^3 - x \Rightarrow \frac{dy}{dx} = 4x^2 - 1 \Rightarrow \frac{d^2y}{dx^2} = 8x$

7. $w = 3z^{-2} - z^{-1} \Rightarrow \frac{dw}{dz} = -6z^{-3} + z^{-2} = \frac{-6}{z^3} + \frac{1}{z^2} \Rightarrow \frac{d^2w}{dz^2} = 18z^{-4} - 2z^{-3} = \frac{18}{z^4} - \frac{2}{z^3}$

9. $y = 6x^2 - 10x - 5x^{-2} \Rightarrow \frac{dy}{dx} = 12x - 10 + 10x^{-3} = 12x - 10 + \frac{10}{x^3} \Rightarrow \frac{d^2y}{dx^2} = 12 - 0 - 30x^{-4} = 12 - \frac{30}{x^4}$

11. $r = \frac{1}{3} s^{-2} - \frac{5}{2} s^{-1} \Rightarrow \frac{dr}{ds} = -\frac{2}{3} s^{-3} + \frac{5}{2} s^{-2} = \frac{-2}{3s^3} + \frac{5}{2s^2} \Rightarrow \frac{d^2r}{ds^2} = 2s^{-4} - 5s^{-3} = \frac{2}{s^4} - \frac{5}{s^3}$

13. (a) $y = (3 - x^2)(x^3 - x + 1) \Rightarrow y' = (3 - x^2) \cdot \frac{d}{dx} (x^3 - x + 1) + (x^3 - x + 1) \cdot \frac{d}{dx} (3 - x^2)$

$= (3 - x^2)(3x^2 - 1) + (x^3 - x + 1)(-2x) = -5x^4 + 12x^2 - 2x - 3$

(b) $y = -x^5 + 4x^3 - x^2 - 3x + 3 \Rightarrow y' = -5x^4 + 12x^2 - 2x - 3$

15. (a) $y = (x^2 + 1)\left(x + 5 + \frac{1}{x}\right) \Rightarrow y' = (x^2 + 1) \cdot \frac{d}{dx}\left(x + 5 + \frac{1}{x}\right) + \left(x + 5 + \frac{1}{x}\right) \cdot \frac{d}{dx} (x^2 + 1)$

$= (x^2 + 1)(1 - x^{-2}) + (x + 5 + x^{-1})(2x) = (x^2 - 1 + 1 - x^{-2}) + (2x^2 + 10x + 2) = 3x^2 + 10x + 2 - \frac{1}{x^2}$

(b) $y = x^3 + 5x^2 + 2x + 5 + \frac{1}{x} \Rightarrow y' = 3x^2 + 10x + 2 - \frac{1}{x^2}$

17. $y = \frac{2x + 5}{3x - 2}$; use the quotient rule: $u = 2x + 5$ and $v = 3x - 2 \Rightarrow u' = 2$ and $v' = 3 \Rightarrow y' = \frac{vu' - uv'}{v^2}$

$= \frac{(3x - 2)(2) - (2x + 5)(3)}{(3x - 2)^2} = \frac{6x - 4 - 6x - 15}{(3x - 2)^2} = \frac{-19}{(3x - 2)^2}$

19. $g(x) = \frac{x^2 - 4}{x + 0.5}$; use the quotient rule: $u = x^2 - 4$ and $v = x + 0.5 \Rightarrow u' = 2x$ and $v' = 1 \Rightarrow g'(x) = \frac{vu' - uv'}{v^2}$

$= \frac{(x + 0.5)(2x) - (x^2 - 4)(1)}{(x + 0.5)^2} = \frac{2x^2 + x - x^2 + 4}{(x + 0.5)^2} = \frac{x^2 + x + 4}{(x + 0.5)^2}$

21. $v = (1 - t)(1 + t^2)^{-1} = \frac{1 - t}{1 + t^2} \Rightarrow \frac{dv}{dt} = \frac{(1 + t^2)(-1) - (1 - t)(2t)}{(1 + t^2)^2} = \frac{-1 - t^2 - 2t + 2t^2}{(1 + t^2)^2} = \frac{t^2 - 2t - 1}{(1 + t^2)^2}$

23. $f(s) = \frac{\sqrt{s} - 1}{\sqrt{s} + 1} \Rightarrow f'(s) = \frac{(\sqrt{s} + 1)\left(\frac{1}{2\sqrt{s}}\right) - (\sqrt{s} - 1)\left(\frac{1}{2\sqrt{s}}\right)}{(\sqrt{s} + 1)^2} = \frac{(\sqrt{s} + 1) - (\sqrt{s} - 1)}{2\sqrt{s}(\sqrt{s} + 1)^2} = \frac{1}{\sqrt{s}(\sqrt{s} + 1)^2}$

NOTE: $\frac{d}{ds}\left(\sqrt{s}\right) = \frac{1}{2\sqrt{s}}$ from Example 2 in Section 2.1

25. $v = \frac{1 + x - 4\sqrt{x}}{x} \Rightarrow v' = \frac{x\left(1 - \frac{2}{\sqrt{x}}\right) - (1 + x - 4\sqrt{x})}{x^2} = \frac{2\sqrt{x} - 1}{x^2}$

27. $y = \frac{1}{(x^2 - 1)(x^2 + x + 1)}$; use the quotient rule: $u = 1$ and $v = (x^2 - 1)(x^2 + x + 1) \Rightarrow u' = 0$ and

$v' = (x^2 - 1)(2x + 1) + (x^2 + x + 1)(2x) = 2x^3 + x^2 - 2x - 1 + 2x^3 + 2x^2 + 2x = 4x^3 + 3x^2 - 1$

$\Rightarrow \frac{dy}{dx} = \frac{vu' - uv'}{v^2} = \frac{0 - 1(4x^3 + 3x^2 - 1)}{(x^2 - 1)^2(x^2 + x + 1)^2} = \frac{-4x^3 - 3x^2 + 1}{(x^2 - 1)^2(x^2 + x + 1)^2}$

29. $y = \frac{1}{2} x^4 - \frac{3}{2} x^2 - x \Rightarrow y' = 2x^3 - 3x - 1 \Rightarrow y'' = 6x^2 - 3 \Rightarrow y''' = 12x \Rightarrow y^{(4)} = 12 \Rightarrow y^{(n)} = 0$ for all $n \geq 5$

31. $y = \frac{x^3 + 7}{x} = x^2 + 7x^{-1} \Rightarrow \frac{dy}{dx} = 2x - 7x^{-2} = 2x - \frac{7}{x^2} \Rightarrow \frac{d^2y}{dx^2} = 2 + 14x^{-3} = 2 + \frac{14}{x^3}$

33. $r = \frac{(\theta - 1)(\theta^2 + \theta + 1)}{\theta^3} = \frac{\theta^3 - 1}{\theta^3} = 1 - \frac{1}{\theta^3} = 1 - \theta^{-3} \Rightarrow \frac{dr}{d\theta} = 0 + 3\theta^{-4} = 3\theta^{-4} = \frac{3}{\theta^4} \Rightarrow \frac{d^2r}{d\theta^2} = -12\theta^{-5} = \frac{-12}{\theta^5}$

35. $w = \left(\frac{1 + 3z}{3z}\right)(3 - z) = \left(\frac{1}{3} z^{-1} + 1\right)(3 - z) = z^{-1} - \frac{1}{3} + 3 - z = z^{-1} + \frac{8}{3} - z \Rightarrow \frac{dw}{dz} = -z^{-2} + 0 - 1 = -z^{-2} - 1$

$= \frac{-1}{z^2} - 1 \Rightarrow \frac{d^2w}{dz^2} = 2z^{-3} - 0 = 2z^{-3} = \frac{2}{z^3}$

37. $p = \left(\frac{q^2+3}{12q}\right)\left(\frac{q^4-1}{q^3}\right) = \frac{q^6-q^2+3q^4-3}{12q^4} = \frac{1}{12}q^2 - \frac{1}{12}q^{-2} + \frac{1}{4} - \frac{1}{4}q^{-4} \Rightarrow \frac{dp}{dq} = \frac{1}{6}q + \frac{1}{6}q^{-3} + q^{-5} = \frac{1}{6}q + \frac{1}{6q^3} + \frac{1}{q^5}$

$\Rightarrow \frac{d^2p}{dq^2} = \frac{1}{6} - \frac{1}{2}q^{-4} - 5q^{-6} = \frac{1}{6} - \frac{1}{2q^4} - \frac{5}{q^6}$

39. $u(0) = 5, u'(0) = -3, v(0) = -1, v'(0) = 2$

(a) $\frac{d}{dx}(uv) = uv' + vu' \Rightarrow \frac{d}{dx}(uv)\big|_{x=0} = u(0)v'(0) + v(0)u'(0) = 5 \cdot 2 + (-1)(-3) = 13$

(b) $\frac{d}{dx}\left(\frac{u}{v}\right) = \frac{vu'-uv'}{v^2} \Rightarrow \frac{d}{dx}\left(\frac{u}{v}\right)\big|_{x=0} = \frac{v(0)u'(0)-u(0)v'(0)}{(v(0))^2} = \frac{(-1)(-3)-(5)(2)}{(-1)^2} = -7$

(c) $\frac{d}{dx}\left(\frac{v}{u}\right) = \frac{uv'-vu'}{u^2} \Rightarrow \frac{d}{dx}\left(\frac{v}{u}\right)\big|_{x=0} = \frac{u(0)v'(0)-v(0)u'(0)}{(u(0))^2} = \frac{(5)(2)-(-1)(-3)}{(5)^2} = \frac{7}{25}$

(d) $\frac{d}{dx}(7v - 2u) = 7v' - 2u' \Rightarrow \frac{d}{dx}(7v - 2u)\big|_{x=0} = 7v'(0) - 2u'(0) = 7 \cdot 2 - 2(-3) = 20$

41. $y = x^3 - 4x + 1$. Note that $(2, 1)$ is on the curve: $1 = 2^3 - 4(2) + 1$

(a) Slope of the tangent at (x, y) is $y' = 3x^2 - 4 \Rightarrow$ slope of the tangent at $(2, 1)$ is $y'(2) = 3(2)^2 - 4 = 8$. Thus the slope of the line perpendicular to the tangent at $(2, 1)$ is $-\frac{1}{8} \Rightarrow$ the equation of the line perpendicular to the tangent line at $(2, 1)$ is $y - 1 = -\frac{1}{8}(x - 2)$ or $y = -\frac{x}{8} + \frac{5}{4}$.

(b) The slope of the curve at x is $m = 3x^2 - 4$ and the smallest value for m is -4 when $x = 0$ and $y = 1$.

(c) We want the slope of the curve to be $8 \Rightarrow y' = 8 \Rightarrow 3x^2 - 4 = 8 \Rightarrow 3x^2 = 12 \Rightarrow x^2 = 4 \Rightarrow x = \pm 2$. When $x = 2, y = 1$ and the tangent line has equation $y - 1 = 8(x - 2)$ or $y = 8x - 15$; when $x = -2$, $y = (-2)^3 - 4(-2) + 1 = 1$, and the tangent line has equation $y - 1 = 8(x + 2)$ or $y = 8x + 17$.

43. $y = \frac{4x}{x^2+1} \Rightarrow \frac{dy}{dx} = \frac{(x^2+1)(4)-(4x)(2x)}{(x^2+1)^2} = \frac{4x^2+4-8x^2}{(x^2+1)^2} = \frac{4(-x^2+1)}{(x^2+1)^2}$. When $x = 0, y = 0$ and $y' = \frac{4(0+1)}{1}$

$= 4$, so the tangent to the curve at $(0, 0)$ is the line $y = 4x$. When $x = 1, y = 2 \Rightarrow y' = 0$, so the tangent to the curve at $(1, 2)$ is the line $y = 2$.

45. $y = ax^2 + bx + c$ passes through $(0, 0) \Rightarrow 0 = a(0) + b(0) + c \Rightarrow c = 0; y = ax^2 + bx$ passes through $(1, 2)$ $\Rightarrow 2 = a + b; y' = 2ax + b$ and since the curve is tangent to $y = x$ at the origin, its slope is 1 at $x = 0 \Rightarrow y' = 1$ when $x = 0 \Rightarrow 1 = 2a(0) + b \Rightarrow b = 1$. Then $a + b = 2 \Rightarrow a = 1$. In summary $a = b = 1$ and $c = 0$ so the curve is $y = x^2 + x$.

47. (a) $y = x^3 - x \Rightarrow y' = 3x^2 - 1$. When $x = -1, y = 0$ and $y' = 2 \Rightarrow$ the tangent line to the curve at $(-1, 0)$ is $y = 2(x + 1)$ or $y = 2x + 2$.

(b)

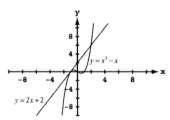

(c) $\left.\begin{array}{l} y = x^3 - x \\ y = 2x + 2 \end{array}\right\} \Rightarrow x^3 - x = 2x + 2 \Rightarrow x^3 - 3x - 2 = (x - 2)(x + 1)^2 = 0 \Rightarrow x = 2$ or $x = -1$. Since $y = 2(2) + 2 = 6$; the other intersection point is $(2, 6)$

49. $P(x) = a_nx^n + a_{n-1}x^{n-1} + \cdots + a_2x^2 + a_1x + a_0 \Rightarrow P'(x) = na_nx^{n-1} + (n-1)a_{n-1}x^{n-2} + \cdots + 2a_2x + a_1$

51. Let c be a constant $\Rightarrow \frac{dc}{dx} = 0 \Rightarrow \frac{d}{dx}(u \cdot c) = u \cdot \frac{dc}{dx} + c \cdot \frac{du}{dx} = u \cdot 0 + c\frac{du}{dx} = c\frac{du}{dx}$. Thus when one of the functions is a constant, the Product Rule is just the Constant Multiple Rule $\Rightarrow$ the Constant Multiple Rule is a special case of the Product Rule.

53. $P = \frac{nRT}{V-nb} - \frac{an^2}{V^2}$. We are holding T constant, and a, b, n, R are also constant so their derivatives are zero

$\Rightarrow \frac{dP}{dV} = \frac{(V-nb)\cdot 0 - (nRT)(1)}{(V-nb)^2} - \frac{V^2(0) - (an^2)(2V)}{(V^2)^2} = \frac{-nRT}{(V-nb)^2} + \frac{2an^2}{V^3}$

2.4 THE DERIVATIVE AS A RATE OF CHANGE

1. $s = t^2 - 3t + 2, 0 \le t \le 2$
   (a) displacement $= \Delta s = s(2) - s(0) = 0m - 2m = -2$ m, $v_{av} = \frac{\Delta s}{\Delta t} = \frac{-2}{2} = -1$ m/sec
   (b) $v = \frac{ds}{dt} = 2t - 3 \Rightarrow |v(0)| = |-3| = 3$ m/sec and $|v(2)| = 1$ m/sec;
   $a = \frac{d^2s}{dt^2} = 2 \Rightarrow a(0) = 2$ m/sec^2 and $a(2) = 2$ m/sec^2
   (c) $v = 0 \Rightarrow 2t - 3 = 0 \Rightarrow t = \frac{3}{2}$. v is negative in the interval $0 < t < \frac{3}{2}$ and v is positive when $\frac{3}{2} < t < 2 \Rightarrow$ the body changes direction at $t = \frac{3}{2}$.

3. $s = -t^3 + 3t^2 - 3t, 0 \le t \le 3$
   (a) displacement $= \Delta s = s(3) - s(0) = -9$ m, $v_{av} = \frac{\Delta s}{\Delta t} = \frac{-9}{3} = -3$ m/sec
   (b) $v = \frac{ds}{dt} = -3t^2 + 6t - 3 \Rightarrow |v(0)| = |-3| = 3$ m/sec and $|v(3)| = |-12| = 12$ m/sec; $a = \frac{d^2s}{dt^2} = -6t + 6$
   $\Rightarrow a(0) = 6$ m/sec^2 and $a(3) = -12$ m/sec^2
   (c) $v = 0 \Rightarrow -3t^2 + 6t - 3 = 0 \Rightarrow t^2 - 2t + 1 = 0 \Rightarrow (t-1)^2 = 0 \Rightarrow t = 1$. For all other values of t in the interval the velocity v is negative (the graph of $v = -3t^2 + 6t - 3$ is a parabola with vertex at $t = 1$ which opens downward $\Rightarrow$ the body never changes direction).

5. $s = \frac{25}{t^2} - \frac{5}{t}, 1 \le t \le 5$
   (a) $\Delta s = s(5) - s(1) = -20$ m, $v_{av} = \frac{-20}{4} = -5$ m/sec
   (b) $v = \frac{-50}{t^3} + \frac{5}{t^2} \Rightarrow |v(1)| = 45$ m/sec and $|v(5)| = \frac{1}{5}$ m/sec; $a = \frac{150}{t^4} - \frac{10}{t^3} \Rightarrow a(1) = 140$ m/sec^2 and $a(5) = \frac{4}{25}$ m/sec^2
   (c) $v = 0 \Rightarrow \frac{-50 + 5t}{t^3} = 0 \Rightarrow -50 + 5t = 0 \Rightarrow t = 10 \Rightarrow$ the body does not change direction in the interval

7. $s = t^3 - 6t^2 + 9t$ and let the positive direction be to the right on the s-axis.
   (a) $v = 3t^2 - 12t + 9$ so that $v = 0 \Rightarrow t^2 - 4t + 3 = (t-3)(t-1) = 0 \Rightarrow t = 1$ or 3; $a = 6t - 12 \Rightarrow a(1)$
   $= -6$ m/sec^2 and $a(3) = 6$ m/sec^2. Thus the body is motionless but being accelerated left when $t = 1$, and motionless but being accelerated right when $t = 3$.
   (b) $a = 0 \Rightarrow 6t - 12 = 0 \Rightarrow t = 2$ with speed $|v(2)| = |12 - 24 + 9| = 3$ m/sec
   (c) The body moves to the right or forward on $0 \le t < 1$, and to the left or backward on $1 < t < 2$. The positions are $s(0) = 0, s(1) = 4$ and $s(2) = 2 \Rightarrow$ total distance $= |s(1) - s(0)| + |s(2) - s(1)| = |4| + |-2| = 6$ m.

9. $s_m = 1.86t^2 \Rightarrow v_m = 3.72t$ and solving $3.72t = 27.8 \Rightarrow t \approx 7.5$ sec on Mars; $s_j = 11.44t^2 \Rightarrow v_j = 22.88t$ and solving $22.88t = 27.8 \Rightarrow t \approx 1.2$ sec on Jupiter.

11. (a) $s = 179 - 16t^2 \Rightarrow v = -32t \Rightarrow$ speed $= |v| = 32t$ ft/sec and $a = -32$ ft/sec^2
    (b) $s = 0 \Rightarrow 179 - 16t^2 = 0 \Rightarrow t = \sqrt{\frac{179}{16}} \approx 3.3$ sec
    (c) When $t = \sqrt{\frac{179}{16}}$, $v = -32\sqrt{\frac{179}{16}} = -8\sqrt{179} \approx -107.0$ ft/sec

13. (a) at 2 and 7 seconds   (b) between 3 and 6 seconds: $3 \le t \le 6$

(c)

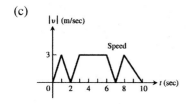

(d)

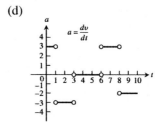

15. (a) 190 ft/sec   (b) 2 sec
    (c) at 8 sec, 0 ft/sec   (d) 10.8 sec, 90 ft/sec
    (e) From t = 8 until t = 10.8 sec, a total of 2.8 sec
    (f) Greatest acceleration happens 2 sec after launch
    (g) From t = 2 to t = 10.8 sec; during this period, $a = \frac{v(10.8) - v(2)}{10.8 - 2} \approx -32$ ft/sec^2

17. $s = 490t^2 \Rightarrow v = 980t \Rightarrow a = 980$
    (a) Solving $160 = 490t^2 \Rightarrow t = \frac{4}{7}$ sec. The average velocity was $\frac{s(4/7) - s(0)}{4/7} = 280$ cm/sec.
    (b) At the 160 cm mark the balls are falling at $v(4/7) = 560$ cm/sec. The acceleration at the 160 cm mark
         was 980 cm/sec^2.
    (c) The light was flashing at a rate of $\frac{17}{4/7} = 29.75$ flashes per second.

19. (a) $c(100) = 11,000 \Rightarrow c_{av} = \frac{11,000}{100} = \$110$
    (b) $c(x) = 2000 + 100x - .1x^2 \Rightarrow c'(x) = 100 - .2x$. Marginal cost $= c'(x) \Rightarrow$ the marginal cost of producing 100
         machines is $c'(100) = \$80$
    (c) The cost of producing the 101st machine is $c(101) - c(100) = 100 - \frac{201}{10} = \$79.90$

21. $b(t) = 10^6 + 10^4 t - 10^3 t^2 \Rightarrow b'(t) = 10^4 - (2)(10^3 t) = 10^3(10 - 2t)$
    (a) $b'(0) = 10^4$ bacteria/hr   (b) $b'(5) = 0$ bacteria/hr
    (c) $b'(10) = -10^4$ bacteria/hr

23. 200 km/hr $= 55 \frac{5}{9}$ m/sec $= \frac{500}{9}$ m/sec, and $D = \frac{10}{9} t^2 \Rightarrow V = \frac{20}{9} t$. Thus $V = \frac{500}{9} \Rightarrow \frac{20}{9} t = \frac{500}{9} \Rightarrow t = 25$ sec. When
     $t = 25$, $D = \frac{10}{9}(25)^2 = \frac{6250}{9}$ m

25.

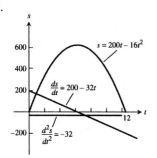

    (a) $v = 0$ when $t = 6.25$ sec
    (b) $v > 0$ when $0 \le t < 6.25 \Rightarrow$ body moves up; $v < 0$ when $6.25 < t \le 12.5 \Rightarrow$ body moves down
    (c) body changes direction at $t = 6.25$ sec
    (d) body speeds up on $(6.25, 12.5]$ and slows down on $[0, 6.25)$
    (e) The body is moving fastest at the endpoints $t = 0$ and $t = 12.5$ when it is traveling 200 ft/sec. It's moving slowest at
         $t = 6.25$ when the speed is 0.
    (f) When $t = 6.25$ the body is $s = 625$ m from the origin and farthest away.

27.

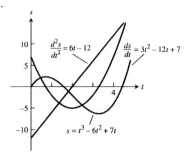

(a) $v = 0$ when $t = \frac{6 \pm \sqrt{15}}{3}$ sec

(b) $v < 0$ when $\frac{6 - \sqrt{15}}{3} < t < \frac{6 + \sqrt{15}}{3} \Rightarrow$ body moves left; $v > 0$ when $0 \leq t < \frac{6 - \sqrt{15}}{3}$ or $\frac{6 + \sqrt{15}}{3} < t \leq 4$
    $\Rightarrow$ body moves right

(c) body changes direction at $t = \frac{6 \pm \sqrt{15}}{3}$ sec

(d) body speeds up on $\left(\frac{6 - \sqrt{15}}{3}, 2 \right) \cup \left(\frac{6 + \sqrt{15}}{3}, 4 \right]$ and slows down on $\left[0, \frac{6 - \sqrt{15}}{3} \right) \cup \left(2, \frac{6 + \sqrt{15}}{3} \right)$.

(e) The body is moving fastest at $t = 0$ and $t = 4$ when it is moving 7 units/sec and slowest at $t = \frac{6 \pm \sqrt{15}}{3}$ sec

(f) When $t = \frac{6 + \sqrt{15}}{3}$ the body is at position $s \approx -6.303$ units and farthest from the origin.

2.5 DERIVATIVES OF TRIGONOMETRIC FUNCTIONS

1. $y = -10x + 3 \cos x \Rightarrow \frac{dy}{dx} = -10 + 3 \frac{d}{dx} (\cos x) = -10 - 3 \sin x$

3. $y = \csc x - 4\sqrt{x} + 7 \Rightarrow \frac{dy}{dx} = -\csc x \cot x - \frac{4}{2\sqrt{x}} + 0 = -\csc x \cot x - \frac{2}{\sqrt{x}}$

5. $y = (\sec x + \tan x)(\sec x - \tan x) \Rightarrow \frac{dy}{dx} = (\sec x + \tan x) \frac{d}{dx} (\sec x - \tan x) + (\sec x - \tan x) \frac{d}{dx} (\sec x + \tan x)$
    $= (\sec x + \tan x)(\sec x \tan x - \sec^2 x) + (\sec x - \tan x)(\sec x \tan x + \sec^2 x)$
    $= (\sec^2 x \tan x + \sec x \tan^2 x - \sec^3 x - \sec^2 x \tan x) + (\sec^2 x \tan x - \sec x \tan^2 x + \sec^3 x - \tan x \sec^2 x) = 0.$
    $\left(\text{Note also that } y = \sec^2 x - \tan^2 x = (\tan^2 x + 1) - \tan^2 x = 1 \Rightarrow \frac{dy}{dx} = 0. \right)$

7. $y = \frac{\cot x}{1 + \cot x} \Rightarrow \frac{dy}{dx} = \frac{(1 + \cot x) \frac{d}{dx} (\cot x) - (\cot x) \frac{d}{dx} (1 + \cot x)}{(1 + \cot x)^2} = \frac{(1 + \cot x)(-\csc^2 x) - (\cot x)(-\csc^2 x)}{(1 + \cot x)^2}$
    $= \frac{-\csc^2 x - \csc^2 x \cot x + \csc^2 x \cot x}{(1 + \cot x)^2} = \frac{-\csc^2 x}{(1 + \cot x)^2}$

9. $y = \frac{4}{\cos x} + \frac{1}{\tan x} = 4 \sec x + \cot x \Rightarrow \frac{dy}{dx} = 4 \sec x \tan x - \csc^2 x$

11. $y = x^2 \sin x + 2x \cos x - 2 \sin x \Rightarrow \frac{dy}{dx} = \left(x^2 \cos x + (\sin x)(2x) \right) + \left((2x)(-\sin x) + (\cos x)(2) \right) - 2 \cos x$
     $= x^2 \cos x + 2x \sin x - 2x \sin x + 2 \cos x - 2 \cos x = x^2 \cos x$

13. $s = \tan t - e^{-t} \Rightarrow \frac{ds}{dt} = \sec^2 t + e^{-t}$

15. $s = \frac{1 + \csc t}{1 - \csc t} \Rightarrow \frac{ds}{dt} = \frac{(1 - \csc t)(-\csc t \cot t) - (1 + \csc t)(\csc t \cot t)}{(1 - \csc t)^2} = \frac{-\csc t \cot t + \csc^2 t \cot t - \csc t \cot t - \csc^2 t \cot t}{(1 - \csc t)^2} = \frac{-2 \csc t \cot t}{(1 - \csc t)^2}$

17. $r = 4 - \theta^2 \sin \theta \Rightarrow \frac{dr}{d\theta} = - \left(\theta^2 \frac{d}{d\theta} (\sin \theta) + (\sin \theta)(2\theta) \right) = - \left(\theta^2 \cos \theta + 2\theta \sin \theta \right) = -\theta(\theta \cos \theta + 2 \sin \theta)$

19. $r = \sec \theta \csc \theta \Rightarrow \frac{dr}{d\theta} = (\sec \theta)(-\csc \theta \cot \theta) + (\csc \theta)(\sec \theta \tan \theta)$
     $= \left(\frac{-1}{\cos \theta} \right) \left(\frac{1}{\sin \theta} \right) \left(\frac{\cos \theta}{\sin \theta} \right) + \left(\frac{1}{\sin \theta} \right) \left(\frac{1}{\cos \theta} \right) \left(\frac{\sin \theta}{\cos \theta} \right) = \frac{-1}{\sin^2 \theta} + \frac{1}{\cos^2 \theta} = \sec^2 \theta - \csc^2 \theta$

21. $p = 5 + \frac{1}{\cot q} = 5 + \tan q \Rightarrow \frac{dp}{dq} = \sec^2 q$

23. $p = \frac{\sin q + \cos q}{\cos q} \Rightarrow \frac{dp}{dq} = \frac{(\cos q)(\cos q - \sin q) - (\sin q + \cos q)(-\sin q)}{\cos^2 q} = \frac{\cos^2 q - \cos q \sin q + \sin^2 q + \cos q \sin q}{\cos^2 q} = \frac{1}{\cos^2 q} = \sec^2 q$

25. (a) $y = \csc x \Rightarrow y' = -\csc x \cot x \Rightarrow y'' = -\left((\csc x)(-\csc^2 x) + (\cot x)(-\csc x \cot x)\right) = \csc^3 x + \csc x \cot^2 x$

      $= (\csc x)(\csc^2 x + \cot^2 x) = (\csc x)(\csc^2 x + \csc^2 x - 1) = 2 \csc^3 x - \csc x$

   (b) $y = \sec x \Rightarrow y' = \sec x \tan x \Rightarrow y'' = (\sec x)(\sec^2 x) + (\tan x)(\sec x \tan x) = \sec^3 x + \sec x \tan^2 x$

      $= (\sec x)(\sec^2 x + \tan^2 x) = (\sec x)(\sec^2 x + \sec^2 x - 1) = 2 \sec^3 x - \sec x$

27. $y = \sin x \Rightarrow y' = \cos x \Rightarrow$ slope of tangent at
    $x = -\pi$ is $y'(-\pi) = \cos(-\pi) = -1$; slope of
    tangent at $x = 0$ is $y'(0) = \cos(0) = 1$; and
    slope of tangent at $x = \frac{3\pi}{2}$ is $y'\left(\frac{3\pi}{2}\right) = \cos \frac{3\pi}{2}$
    $= 0$. The tangent at $(-\pi, 0)$ is $y - 0 = -1(x + \pi)$,
    or $y = -x - \pi$; the tangent at $(0, 0)$ is
    $y - 0 = 1(x - 0)$, or $y = x$; and the tangent at
    $\left(\frac{3\pi}{2}, -1\right)$ is $y = -1$.

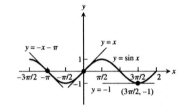

29. $y = \sec x \Rightarrow y' = \sec x \tan x \Rightarrow$ slope of tangent at
    $x = -\frac{\pi}{3}$ is $\sec\left(-\frac{\pi}{3}\right) \tan\left(-\frac{\pi}{3}\right) = -2\sqrt{3}$; slope of tangent
    at $x = \frac{\pi}{4}$ is $\sec\left(\frac{\pi}{4}\right) \tan\left(\frac{\pi}{4}\right) = \sqrt{2}$. The tangent at the point
    $\left(-\frac{\pi}{3}, \sec\left(-\frac{\pi}{3}\right)\right) = \left(-\frac{\pi}{3}, 2\right)$ is $y - 2 = -2\sqrt{3}\left(x + \frac{\pi}{3}\right)$;
    the tangent at the point $\left(\frac{\pi}{4}, \sec\left(\frac{\pi}{4}\right)\right) = \left(\frac{\pi}{4}, \sqrt{2}\right)$ is $y - \sqrt{2}$
    $= \sqrt{2}\left(x - \frac{\pi}{4}\right)$.

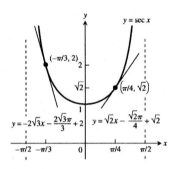

31. Yes, $y = x + \sin x \Rightarrow y' = 1 + \cos x$; horizontal tangent occurs where $1 + \cos x = 0 \Rightarrow \cos x = -1 \Rightarrow x = \pi$

33. No, $y = x - \cot x \Rightarrow y' = 1 + \csc^2 x$; horizontal tangent occurs where $1 + \csc^2 x = 0 \Rightarrow \csc^2 x = -1$. But there are no
    x-values for which $\csc^2 x = -1$.

35. We want all points on the curve where the tangent
    line has slope 2. Thus, $y = \tan x \Rightarrow y' = \sec^2 x$ so
    that $y' = 2 \Rightarrow \sec^2 x = 2 \Rightarrow \sec x = \pm\sqrt{2}$
    $\Rightarrow x = \pm\frac{\pi}{4}$. Then the tangent line at $\left(\frac{\pi}{4}, 1\right)$ has
    equation $y - 1 = 2\left(x - \frac{\pi}{4}\right)$; the tangent line at
    $\left(-\frac{\pi}{4}, -1\right)$ has equation $y + 1 = 2\left(x + \frac{\pi}{4}\right)$.

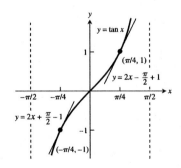

37. $\lim\limits_{x \to 2} \sin\left(\frac{1}{x} - \frac{1}{2}\right) = \sin\left(\frac{1}{2} - \frac{1}{2}\right) = \sin 0 = 0$

39. $\lim\limits_{x \to 0} \sec\left[e^x + \pi \tan\left(\frac{\pi}{4 \sec x}\right) - 1\right] = \sec\left[1 + \pi \tan\left(\frac{\pi}{4 \sec 0}\right) - 1\right] = = \sec\left[\pi \tan\left(\frac{\pi}{4}\right)\right] = \sec \pi = -1$

41. $\lim\limits_{t \to 0} \tan\left(1 - \frac{\sin t}{t}\right) = \tan\left(1 - \lim\limits_{t \to 0} \frac{\sin t}{t}\right) = \tan(1 - 1) = 0$

43. $s = 2 - 2 \sin t \Rightarrow v = \frac{ds}{dt} = -2 \cos t \Rightarrow a = \frac{dv}{dt} = 2 \sin t \Rightarrow j = \frac{da}{dt} = 2 \cos t$. Therefore, velocity $= v\left(\frac{\pi}{4}\right)$

   $= -\sqrt{2}$ m/sec; speed $= \left|v\left(\frac{\pi}{4}\right)\right| = \sqrt{2}$ m/sec; acceleration $= a\left(\frac{\pi}{4}\right) = \sqrt{2}$ m/sec^2; jerk $= j\left(\frac{\pi}{4}\right) = \sqrt{2}$ m/sec^3.

45. $\lim\limits_{x \to 0} f(x) = \lim\limits_{x \to 0} \frac{\sin^2 3x}{x^2} = \lim\limits_{x \to 0} 9\left(\frac{\sin 3x}{3x}\right)\left(\frac{\sin 3x}{3x}\right) = 9$ so that f is continuous at $x = 0 \Rightarrow \lim\limits_{x \to 0} f(x) = f(0) \Rightarrow 9 = c$.

47. $\frac{d^{999}}{dx^{999}}(\cos x) = \sin x$ because $\frac{d^4}{dx^4}(\cos x) = \cos x \Rightarrow$ the derivative of cos x any number of times that is a multiple of 4 is

   cos x. Thus, dividing 999 by 4 gives $999 = 249 \cdot 4 + 3 \Rightarrow \frac{d^{999}}{dx^{999}}(\cos x) = \frac{d^3}{dx^3}\left[\frac{d^{249 \cdot 4}}{dx^{249 \cdot 4}}(\cos x)\right] = \frac{d^3}{dx^3}(\cos x) = \sin x$.

49. (a)

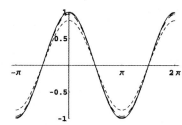

   The dashed curves of $y = \frac{\sin(x+h) - \sin(x-h)}{2h}$ are closer to the black curve $y = \cos x$ than the corresponding dashed curves in Exercise 51 illustrating that the centered difference quotient is a better approximation of the derivative of this function.

   (b)

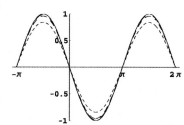

   The dashed curves of $y = \frac{\cos(x+h) - \cos(x-h)}{2h}$ are closer to the black curve $y = -\sin x$ than the corresponding dashed curves in Exercise 52 illustrating that the centered difference quotient is a better approximation of the derivative of this function.

51. $y = \tan x \Rightarrow y' = \sec^2 x$, so the smallest value
    $y' = \sec^2 x$ takes on is $y' = 1$ when $x = 0$;
    y' has no maximum value since $\sec^2 x$ has no
    largest value on $\left(-\frac{\pi}{2}, \frac{\pi}{2}\right)$; y' is never negative
    since $\sec^2 x \geq 1$.

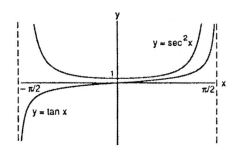

2.6 EXPONENTIAL FUNCTIONS

1.

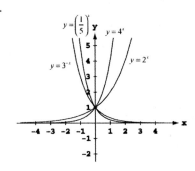

3.

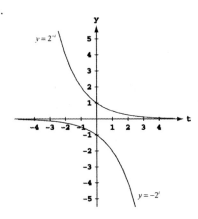

5.

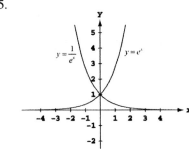

7.

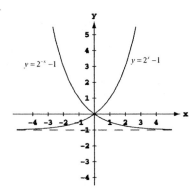

9.

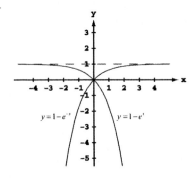

11. $16^2 \cdot 16^{-1.75} = 16^{2+(-1.75)} = 16^{0.25} = 16^{1/4} = 2$

13. $\frac{4^{4.2}}{4^{3.7}} = 4^{4.2-3.7} = 4^{0.5} = 4^{1/2} = 2$

15. $\left(25^{1/8}\right)^4 = 25^{4/8} = 25^{1/2} = 5$

17. $2^{\sqrt{3}} \cdot 7^{\sqrt{3}} = (2 \cdot 7)^{\sqrt{3}} = 14^{\sqrt{3}}$

19. $\left(\frac{2}{\sqrt{2}}\right)^4 = \frac{2^4}{\left(2^{1/2}\right)^4} = \frac{16}{2^2} = 4$

21. Domain: $(-\infty, \infty)$; y in range $\Rightarrow y = \frac{1}{2+e^x}$. As x increases, e^x becomes infinitely large and y becomes a smaller and smaller positive real number. As x decreases, e^x becomes a smaller and smaller positive real number, $y < \frac{1}{2}$, and y gets arbitrarily close to $\frac{1}{2} \Rightarrow$ Range: $\left(0, \frac{1}{2}\right)$.

23. Domain: $(-\infty, \infty)$; y in range $\Rightarrow y = \sqrt{1 + 3^{-t}}$. Since the values of 3^{-t} are $(0, \infty) \Rightarrow$ Range: $(1, \infty)$.

25. $y = x^3 e^x \Rightarrow y' = x^3 \cdot e^x + 3x^2 \cdot e^x = (x^3 + 3x^2)e^x$

27. $y = x^{9/4} \Rightarrow y' = \frac{9}{4} x^{5/4}$

29. $s = 2t^{3/2} + 3e^2 \Rightarrow s' = 3t^{1/2} + 0 \Rightarrow s' = 3t^{1/2}$

31. $y = \sqrt[7]{x^2} - x^e = x^{2/7} - x^e \Rightarrow y' = \frac{2}{7} x^{-5/7} - e\, x^{e-1} \Rightarrow y' = \frac{2}{7x^{5/7}} - e\, x^{e-1}$

33. $r = \frac{e^s}{s} \Rightarrow r' = \frac{s \cdot e^s - e^s(1)}{s^2} \Rightarrow r' = \frac{e^s(s-1)}{s^2}$

35. $y = \frac{1}{2} x^4 - \frac{3}{2} x^2 - x \Rightarrow y' = 2x^3 - 3x - 1 \Rightarrow y'' = 6x^2 - 3 \Rightarrow y''' = 12x \Rightarrow y^{(4)} = 12 \Rightarrow y^{(n)} = 0$ for all $n \geq 5$

37. $y = \frac{x^3+7}{x} = x^2 + 7x^{-1} \Rightarrow \frac{dy}{dx} = 2x - 7x^{-2} = 2x - \frac{7}{x^2} \Rightarrow \frac{d^2y}{dx^2} = 2 + 14x^{-3} = 2 + \frac{14}{x^3}$

39. $r = \frac{(\theta-1)(\theta^2+\theta+1)}{\theta^3} = \frac{\theta^3-1}{\theta^3} = 1 - \frac{1}{\theta^3} = 1 - \theta^{-3} \Rightarrow \frac{dr}{d\theta} = 0 + 3\theta^{-4} = 3\theta^{-4} = \frac{3}{\theta^4} \Rightarrow \frac{d^2r}{d\theta^2} = -12\theta^{-5} = \frac{-12}{\theta^5}$

41. $w = 3z^2 e^z \Rightarrow w' = 3z^2 \cdot e^z + 6z \cdot e^z = (3z^2 + 6z)e^z$ and $w'' = (3z^2 + 6z)e^z + (6z + 6)e^z = (3z^2 + 12z + 6)e^z$

43. $p = \left(\frac{q^2+3}{12q}\right)\left(\frac{q^4-1}{q^3}\right) = \frac{q^6 - q^2 + 3q^4 - 3}{12q^4} = \frac{1}{12}q^2 - \frac{1}{12}q^{-2} + \frac{1}{4} - \frac{1}{4}q^{-4} \Rightarrow \frac{dp}{dq} = \frac{1}{6}q + \frac{1}{6}q^{-3} + q^{-5} = \frac{1}{6}q + \frac{1}{6q^3} + \frac{1}{q^5}$

$\Rightarrow \frac{d^2p}{dq^2} = \frac{1}{6} - \frac{1}{2}q^{-4} - 5q^{-6} = \frac{1}{6} - \frac{1}{2q^4} - \frac{5}{q^6}$

45. $\frac{dy}{dx} = e^{-x}(2x) + (x^2 - 3)e^{-x}(-1) = e^{-x}(2x - (x^2 - 3)) = e^{-x}(3 + 2x - x^2)$

$\Rightarrow \frac{d^2y}{dx^2} = e^{-x}(2 - 2x) + (3 + 2x - x^2)e^{-x}(-1) = e^{-x}(2 - 2x - 3 - 2x + x^2) = e^{-x}(x^2 - 4x - 1)$

47.

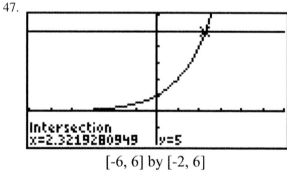

[-6, 6] by [-2, 6]

$x \approx 2.3219$

49. Let t be the number of years. Solving $500,000(1.0375)^t = 1,000,000$ graphically, we find that $t \approx 18.828$. The population will reach 1 million in about 19 years.

51. Let A be the amount of the initial investment, and let t be the number of years. We wish to solve $A(1.0625)^t = 2A$, which is equivalent to $1.0625^t = 2$. Solving graphically, we find that $t \approx 11.433$. It will take about 11.433 years. (If the interest is credited at the end of each year, it will take 12 years.)

53. $-\frac{1}{e^x} \leq \frac{\sin x}{e^x} \leq \frac{1}{e^x} \Rightarrow \lim_{x \to \infty} e^{-x}\sin x = 0$ by the Sandwich Theorem.

55. $\lim_{x \to -\infty} \frac{e^x - e^{-x}}{e^x + e^{-x}} = \lim_{x \to -\infty} \frac{e^x - \frac{1}{e^x}}{e^x + \frac{1}{e^x}} = \lim_{x \to -\infty} \frac{\frac{e^{2x}-1}{e^x}}{\frac{e^{2x}+1}{e^x}} = \lim_{x \to -\infty} \frac{e^{2x} - 1}{e^{2x} + 1} = \frac{0-1}{0+1} = -1$

57. From the graph at the right, can see that the exponential
    function $y = 2^x$ grows more rapidly than $y = x^2$ if $x > 4$.

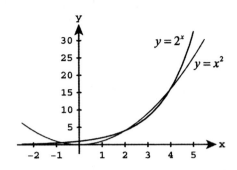

2.7 THE CHAIN RULE

1. $f(u) = 6u - 9 \Rightarrow f'(u) = 6 \Rightarrow f'(g(x)) = 6$; $g(x) = \frac{1}{2}x^4 \Rightarrow g'(x) = 2x^3$; therefore $\frac{dy}{dx} = f'(g(x))g'(x) = 6 \cdot 2x^3 = 12x^3$

3. $f(u) = \sin u \Rightarrow f'(u) = \cos u \Rightarrow f'(g(x)) = \cos(3x + 1)$; $g(x) = 3x + 1 \Rightarrow g'(x) = 3$; therefore $\frac{dy}{dx} = f'(g(x))g'(x)$
   $= (\cos(3x + 1))(3) = 3\cos(3x + 1)$

5. $f(u) = \cos u \Rightarrow f'(u) = -\sin u \Rightarrow f'(g(x)) = -\sin(\sin x)$; $g(x) = \sin x \Rightarrow g'(x) = \cos x$; therefore
   $\frac{dy}{dx} = f'(g(x))g'(x) = -(\sin(\sin x))\cos x$

7. $f(u) = \tan u \Rightarrow f'(u) = \sec^2 u \Rightarrow f'(g(x)) = \sec^2(10x - 5)$; $g(x) = 10x - 5 \Rightarrow g'(x) = 10$; therefore
   $\frac{dy}{dx} = f'(g(x))g'(x) = (\sec^2(10x - 5))(10) = 10\sec^2(10x - 5)$

9. With $u = (2x + 1)$, $y = u^5$: $\frac{dy}{dx} = \frac{dy}{du}\frac{du}{dx} = 5u^4 \cdot 2 = 10(2x + 1)^4$

11. With $u = \left(1 - \frac{x}{7}\right)$, $y = u^{-7}$: $\frac{dy}{dx} = \frac{dy}{du}\frac{du}{dx} = -7u^{-8} \cdot \left(-\frac{1}{7}\right) = \left(1 - \frac{x}{7}\right)^{-8}$

13. With $u = (1 - 6x)$, $y = u^{2/3} \Rightarrow \frac{dy}{dx} = \frac{dy}{du}\frac{du}{dx} = \frac{2}{3}u^{-1/3}(-6) = -4(1 - 6x)^{-1/3}$

15. With $u = \tan x$, $y = \sec u$: $\frac{dy}{dx} = \frac{dy}{du}\frac{du}{dx} = (\sec u \tan u)(\sec^2 x) = (\sec(\tan x)\tan(\tan x))\sec^2 x$

17. With $u = \sin x$, $y = u^3$: $\frac{dy}{dx} = \frac{dy}{du}\frac{du}{dx} = 3u^2 \cos x = 3(\sin^2 x)(\cos x)$

19. With $u = -5x$, $y = e^u$: $\frac{dy}{dx} = \frac{dy}{du}\frac{du}{dx} = e^u(-5) = -5e^{-5x}$

21. With $u = 5 - 7x$, $y = e^u$: $\frac{dy}{dx} = \frac{dy}{du}\frac{du}{dx} = e^u(-7) = -7e^{5-7x}$

23. $p = \sqrt{3 - t} = (3 - t)^{1/2} \Rightarrow \frac{dp}{dt} = \frac{1}{2}(3 - t)^{-1/2} \cdot \frac{d}{dt}(3 - t) = -\frac{1}{2}(3 - t)^{-1/2} = \frac{-1}{2\sqrt{3 - t}}$

25. $s = \frac{4}{3\pi}\sin 3t + \frac{4}{5\pi}\cos 5t \Rightarrow \frac{ds}{dt} = \frac{4}{3\pi}\cos 3t \cdot \frac{d}{dt}(3t) + \frac{4}{5\pi}(-\sin 5t) \cdot \frac{d}{dt}(5t) = \frac{4}{\pi}\cos 3t - \frac{4}{\pi}\sin 5t = \frac{4}{\pi}(\cos 3t - \sin 5t)$

27. $r = (\csc\theta + \cot\theta)^{-1} \Rightarrow \frac{dr}{d\theta} = -(\csc\theta + \cot\theta)^{-2}\frac{d}{d\theta}(\csc\theta + \cot\theta) = \frac{\csc\theta\cot\theta + \csc^2\theta}{(\csc\theta + \cot\theta)^2} = \frac{\csc\theta(\cot\theta + \csc\theta)}{(\csc\theta + \cot\theta)^2} = \frac{\csc\theta}{\csc\theta + \cot\theta}$

29. $y = x^2\sin^4 x + x\cos^{-2} x \Rightarrow \frac{dy}{dx} = x^2\frac{d}{dx}(\sin^4 x) + \sin^4 x \cdot \frac{d}{dx}(x^2) + x\frac{d}{dx}(\cos^{-2} x) + \cos^{-2} x \cdot \frac{d}{dx}(x)$
    $= x^2\left(4\sin^3 x\frac{d}{dx}(\sin x)\right) + 2x\sin^4 x + x\left(-2\cos^{-3} x \cdot \frac{d}{dx}(\cos x)\right) + \cos^{-2} x$

$$= x^2 \left(4 \sin^3 x \cos x\right) + 2x \sin^4 x + x\left((-2 \cos^{-3} x)(-\sin x)\right) + \cos^{-2} x$$
$$= 4x^2 \sin^3 x \cos x + 2x \sin^4 x + 2x \sin x \cos^{-3} x + \cos^{-2} x$$

31. $y = \frac{1}{21}(3x - 2)^7 + \left(4 - \frac{1}{2x^2}\right)^{-1} \Rightarrow \frac{dy}{dx} = \frac{7}{21}(3x - 2)^6 \cdot \frac{d}{dx}(3x - 2) + (-1)\left(4 - \frac{1}{2x^2}\right)^{-2} \cdot \frac{d}{dx}\left(4 - \frac{1}{2x^2}\right)$

$= \frac{7}{21}(3x - 2)^6 \cdot 3 + (-1)\left(4 - \frac{1}{2x^2}\right)^{-2}\left(\frac{1}{x^3}\right) = (3x - 2)^6 - \frac{1}{x^3\left(4 - \frac{1}{2x^2}\right)^2}$

33. $y = (4x + 3)^4(x + 1)^{-3} \Rightarrow \frac{dy}{dx} = (4x + 3)^4(-3)(x + 1)^{-4} \cdot \frac{d}{dx}(x + 1) + (x + 1)^{-3}(4)(4x + 3)^3 \cdot \frac{d}{dx}(4x + 3)$

$= (4x + 3)^4(-3)(x + 1)^{-4}(1) + (x + 1)^{-3}(4)(4x + 3)^3(4) = -3(4x + 3)^4(x + 1)^{-4} + 16(4x + 3)^3(x + 1)^{-3}$

$= \frac{(4x + 3)^3}{(x + 1)^4}\left[-3(4x + 3) + 16(x + 1)\right] = \frac{(4x + 3)^3(4x + 7)}{(x + 1)^4}$

35. $y = xe^{-x} + e^{3x} \Rightarrow y' = x \cdot e^{-x}(-1) + (1) \cdot e^{-x} + 3e^{3x} = (1 - x)e^{-x} + 3e^{3x}$

37. $y = (x^2 - 2x + 2)e^{5x/2} \Rightarrow y' = (x^2 - 2x + 2) \cdot e^{5x/2}\left(\frac{5}{2}\right) + (2x - 2) \cdot e^{5x/2} = \left(\frac{5}{2}x^2 - 3x + 3\right)e^{5x/2}$

39. $h(x) = x \tan\left(2\sqrt{x}\right) + 7 \Rightarrow h'(x) = x \frac{d}{dx}\left(\tan\left(2x^{1/2}\right)\right) + \tan\left(2x^{1/2}\right) \cdot \frac{d}{dx}(x) + 0$

$= x \sec^2\left(2x^{1/2}\right) \cdot \frac{d}{dx}\left(2x^{1/2}\right) + \tan\left(2x^{1/2}\right) = x \sec^2\left(2\sqrt{x}\right) \cdot \frac{1}{\sqrt{x}} + \tan\left(2\sqrt{x}\right) = \sqrt{x} \sec^2\left(2\sqrt{x}\right) + \tan\left(2\sqrt{x}\right)$

41. $f(\theta) = \left(\frac{\sin \theta}{1 + \cos \theta}\right)^2 \Rightarrow f'(\theta) = 2\left(\frac{\sin \theta}{1 + \cos \theta}\right) \cdot \frac{d}{d\theta}\left(\frac{\sin \theta}{1 + \cos \theta}\right) = \frac{2 \sin \theta}{1 + \cos \theta} \cdot \frac{(1 + \cos \theta)(\cos \theta) - (\sin \theta)(-\sin \theta)}{(1 + \cos \theta)^2}$

$= \frac{(2 \sin \theta)(\cos \theta + \cos^2 \theta + \sin^2 \theta)}{(1 + \cos \theta)^3} = \frac{(2 \sin \theta)(\cos \theta + 1)}{(1 + \cos \theta)^3} = \frac{2 \sin \theta}{(1 + \cos \theta)^2}$

43. $h(\theta) = \sqrt[3]{1 + \cos(2\theta)} = (1 + \cos 2\theta)^{1/3} \Rightarrow h'(\theta) = \frac{1}{3}(1 + \cos 2\theta)^{-2/3} \cdot (-\sin 2\theta) \cdot 2 = -\frac{2}{3}(\sin 2\theta)(1 + \cos 2\theta)^{-2/3}$

45. $q = \sin\left(\frac{t}{\sqrt{t + 1}}\right) \Rightarrow \frac{dq}{dt} = \cos\left(\frac{t}{\sqrt{t + 1}}\right) \cdot \frac{d}{dt}\left(\frac{t}{\sqrt{t + 1}}\right) = \cos\left(\frac{t}{\sqrt{t + 1}}\right) \cdot \frac{\sqrt{t + 1}(1) - t \cdot \frac{d}{dt}\left(\sqrt{t + 1}\right)}{\left(\sqrt{t + 1}\right)^2}$

$= \cos\left(\frac{t}{\sqrt{t + 1}}\right) \cdot \frac{\sqrt{t + 1} - \frac{t}{2\sqrt{t + 1}}}{t + 1} = \cos\left(\frac{t}{\sqrt{t + 1}}\right)\left(\frac{2(t + 1) - t}{2(t + 1)^{3/2}}\right) = \left(\frac{t + 2}{2(t + 1)^{3/2}}\right)\cos\left(\frac{t}{\sqrt{t + 1}}\right)$

47. $y = \cos\left(e^{-\theta^2}\right) \Rightarrow \frac{dy}{d\theta} = -\sin\left(e^{-\theta^2}\right)\frac{d}{d\theta}\left(e^{-\theta^2}\right) = \left(-\sin\left(e^{-\theta^2}\right)\right)\left(e^{-\theta^2}\right)\frac{d}{d\theta}\left(-\theta^2\right) = 2\theta e^{-\theta^2}\sin\left(e^{-\theta^2}\right)$

49. $y = \sin^2(\pi t - 2) \Rightarrow \frac{dy}{dt} = 2 \sin(\pi t - 2) \cdot \frac{d}{dt}\sin(\pi t - 2) = 2 \sin(\pi t - 2) \cdot \cos(\pi t - 2) \cdot \frac{d}{dt}(\pi t - 2)$

$= 2\pi \sin(\pi t - 2)\cos(\pi t - 2)$

51. $y = (1 + \cos 2t)^{-4} \Rightarrow \frac{dy}{dt} = -4(1 + \cos 2t)^{-5} \cdot \frac{d}{dt}(1 + \cos 2t) = -4(1 + \cos 2t)^{-5}(-\sin 2t) \cdot \frac{d}{dt}(2t) = \frac{8 \sin 2t}{(1 + \cos 2t)^5}$

53. $y = e^{\cos^2(\pi t - 1)} \Rightarrow \frac{dy}{dt} = e^{\cos^2(\pi t - 1)} \cdot 2\cos(\pi t - 1) \cdot (-\sin(\pi t - 1)) \cdot \pi = -2\pi \sin(\pi t - 1)\cos(\pi t - 1)e^{\cos^2(\pi t - 1)}$

55. $y = \sin(\cos(2t - 5)) \Rightarrow \frac{dy}{dt} = \cos(\cos(2t - 5)) \cdot \frac{d}{dt}\cos(2t - 5) = \cos(\cos(2t - 5)) \cdot (-\sin(2t - 5)) \cdot \frac{d}{dt}(2t - 5)$

$= -2 \cos(\cos(2t - 5))(\sin(2t - 5))$

57. $y = \left[1 + \tan^4\left(\frac{t}{12}\right)\right]^3 \Rightarrow \frac{dy}{dt} = 3\left[1 + \tan^4\left(\frac{t}{12}\right)\right]^2 \cdot \frac{d}{dt}\left[1 + \tan^4\left(\frac{t}{12}\right)\right] = 3\left[1 + \tan^4\left(\frac{t}{12}\right)\right]^2\left[4 \tan^3\left(\frac{t}{12}\right) \cdot \frac{d}{dt}\tan\left(\frac{t}{12}\right)\right]$

$= 12\left[1 + \tan^4\left(\frac{t}{12}\right)\right]^2\left[\tan^3\left(\frac{t}{12}\right)\sec^2\left(\frac{t}{12}\right) \cdot \frac{1}{12}\right] = \left[1 + \tan^4\left(\frac{t}{12}\right)\right]^2\left[\tan^3\left(\frac{t}{12}\right)\sec^2\left(\frac{t}{12}\right)\right]$

59. $y = (1 + \cos(t^2))^{1/2} \Rightarrow \frac{dy}{dt} = \frac{1}{2}(1 + \cos(t^2))^{-1/2} \cdot \frac{d}{dt}(1 + \cos(t^2)) = \frac{1}{2}(1 + \cos(t^2))^{-1/2}\left(-\sin(t^2) \cdot \frac{d}{dt}(t^2)\right)$

$= -\frac{1}{2}(1 + \cos(t^2))^{-1/2}(\sin(t^2)) \cdot 2t = -\frac{t \sin(t^2)}{\sqrt{1 + \cos(t^2)}}$

61. $y = \left(1 + \frac{1}{x}\right)^3 \Rightarrow y' = 3\left(1 + \frac{1}{x}\right)^2\left(-\frac{1}{x^2}\right) = -\frac{3}{x^2}\left(1 + \frac{1}{x}\right)^2 \Rightarrow y'' = \left(-\frac{3}{x^2}\right) \cdot \frac{d}{dx}\left(1 + \frac{1}{x}\right)^2 - \left(1 + \frac{1}{x}\right)^2 \cdot \frac{d}{dx}\left(\frac{3}{x^2}\right)$

$= \left(-\frac{3}{x^2}\right)\left(2\left(1 + \frac{1}{x}\right)\left(-\frac{1}{x^2}\right)\right) + \left(\frac{6}{x^3}\right)\left(1 + \frac{1}{x}\right)^2 = \frac{6}{x^4}\left(1 + \frac{1}{x}\right) + \frac{6}{x^3}\left(1 + \frac{1}{x}\right)^2 = \frac{6}{x^3}\left(1 + \frac{1}{x}\right)\left(\frac{1}{x} + 1 + \frac{1}{x}\right)$

$= \frac{6}{x^3}\left(1 + \frac{1}{x}\right)\left(1 + \frac{2}{x}\right)$

63. $y = \frac{1}{9}\cot(3x - 1) \Rightarrow y' = -\frac{1}{9}\csc^2(3x - 1)(3) = -\frac{1}{3}\csc^2(3x - 1) \Rightarrow y'' = \left(-\frac{2}{3}\right)\left(\csc(3x - 1) \cdot \frac{d}{dx}\csc(3x - 1)\right)$

$= -\frac{2}{3}\csc(3x - 1)(-\csc(3x - 1)\cot(3x - 1) \cdot \frac{d}{dx}(3x - 1)) = 2\csc^2(3x - 1)\cot(3x - 1)$

65. $y = e^{x^2} + 5x \Rightarrow y' = 2xe^{x^2} + 5 \Rightarrow y'' = 2x \cdot e^{x^2}(2x) + 2e^{x^2} = (4x^2 + 2)e^{x^2}$

67. $g(x) = \sqrt{x} \Rightarrow g'(x) = \frac{1}{2\sqrt{x}} \Rightarrow g(1) = 1$ and $g'(1) = \frac{1}{2}$; $f(u) = u^5 + 1 \Rightarrow f'(u) = 5u^4 \Rightarrow f'(g(1)) = f'(1) = 5$;

therefore, $(f \circ g)'(1) = f'(g(1)) \cdot g'(1) = 5 \cdot \frac{1}{2} = \frac{5}{2}$

69. $g(x) = 5\sqrt{x} \Rightarrow g'(x) = \frac{5}{2\sqrt{x}} \Rightarrow g(1) = 5$ and $g'(1) = \frac{5}{2}$; $f(u) = \cot\left(\frac{\pi u}{10}\right) \Rightarrow f'(u) = -\csc^2\left(\frac{\pi u}{10}\right)\left(\frac{\pi}{10}\right)$

$= \frac{-\pi}{10}\csc^2\left(\frac{\pi u}{10}\right) \Rightarrow f'(g(1)) = f'(5) = -\frac{\pi}{10}\csc^2\left(\frac{\pi}{2}\right) = -\frac{\pi}{10}$; therefore, $(f \circ g)'(1) = f'(g(1))g'(1) = -\frac{\pi}{10} \cdot \frac{5}{2} = -\frac{\pi}{4}$

71. $g(x) = 10x^2 + x + 1 \Rightarrow g'(x) = 20x + 1 \Rightarrow g(0) = 1$ and $g'(0) = 1$; $f(u) = \frac{2u}{u^2+1} \Rightarrow f'(u) = \frac{(u^2+1)(2) - (2u)(2u)}{(u^2+1)^2}$

$= \frac{-2u^2+2}{(u^2+1)^2} \Rightarrow f'(g(0)) = f'(1) = 0$; therefore, $(f \circ g)'(0) = f'(g(0))g'(0) = 0 \cdot 1 = 0$

73. (a) $y = 2f(x) \Rightarrow \frac{dy}{dx} = 2f'(x) \Rightarrow \frac{dy}{dx}\Big|_{x=2} = 2f'(2) = 2\left(\frac{1}{3}\right) = \frac{2}{3}$

(b) $y = f(x) + g(x) \Rightarrow \frac{dy}{dx} = f'(x) + g'(x) \Rightarrow \frac{dy}{dx}\Big|_{x=3} = f'(3) + g'(3) = 2\pi + 5$

(c) $y = f(x) \cdot g(x) \Rightarrow \frac{dy}{dx} = f(x)g'(x) + g(x)f'(x) \Rightarrow \frac{dy}{dx}\Big|_{x=3} = f(3)g'(3) + g(3)f'(3) = 3 \cdot 5 + (-4)(2\pi) = 15 - 8\pi$

(d) $y = \frac{f(x)}{g(x)} \Rightarrow \frac{dy}{dx} = \frac{g(x)f'(x) - f(x)g'(x)}{[g(x)]^2} \Rightarrow \frac{dy}{dx}\Big|_{x=2} = \frac{g(2)f'(2) - f(2)g'(2)}{[g(2)]^2} = \frac{(2)\left(\frac{1}{3}\right) - (8)(-3)}{2^2} = \frac{37}{6}$

(e) $y = f(g(x)) \Rightarrow \frac{dy}{dx} = f'(g(x))g'(x) \Rightarrow \frac{dy}{dx}\Big|_{x=2} = f'(g(2))g'(2) = f'(2)(-3) = \frac{1}{3}(-3) = -1$

(f) $y = (f(x))^{1/2} \Rightarrow \frac{dy}{dx} = \frac{1}{2}(f(x))^{-1/2} \cdot f'(x) = \frac{f'(x)}{2\sqrt{f(x)}} \Rightarrow \frac{dy}{dx}\Big|_{x=2} = \frac{f'(2)}{2\sqrt{f(2)}} = \frac{\left(\frac{1}{3}\right)}{2\sqrt{8}} = \frac{1}{6\sqrt{8}} = \frac{1}{12\sqrt{2}} = \frac{\sqrt{2}}{24}$

(g) $y = (g(x))^{-2} \Rightarrow \frac{dy}{dx} = -2(g(x))^{-3} \cdot g'(x) \Rightarrow \frac{dy}{dx}\Big|_{x=3} = -2(g(3))^{-3}g'(3) = -2(-4)^{-3} \cdot 5 = \frac{5}{32}$

(h) $y = \left((f(x))^2 + (g(x))^2\right)^{1/2} \Rightarrow \frac{dy}{dx} = \frac{1}{2}\left((f(x))^2 + (g(x))^2\right)^{-1/2}\left(2f(x) \cdot f'(x) + 2g(x) \cdot g'(x)\right)$

$\Rightarrow \frac{dy}{dx}\Big|_{x=2} = \frac{1}{2}\left((f(2))^2 + (g(2))^2\right)^{-1/2}\left(2f(2)f'(2) + 2g(2)g'(2)\right) = \frac{1}{2}\left(8^2 + 2^2\right)^{-1/2}\left(2 \cdot 8 \cdot \frac{1}{3} + 2 \cdot 2 \cdot (-3)\right) = -\frac{5}{3\sqrt{17}}$

75. $\frac{ds}{dt} = \frac{ds}{d\theta} \cdot \frac{d\theta}{dt}$: $s = \cos\theta \Rightarrow \frac{ds}{d\theta} = -\sin\theta \Rightarrow \frac{ds}{d\theta}\Big|_{\theta=\frac{3\pi}{2}} = -\sin\left(\frac{3\pi}{2}\right) = 1$ so that $\frac{ds}{dt} = \frac{ds}{d\theta} \cdot \frac{d\theta}{dt} = 1 \cdot 5 = 5$

77. With $y = x$, we should get $\frac{dy}{dx} = 1$ for both (a) and (b):

(a) $y = \frac{u}{5} + 7 \Rightarrow \frac{dy}{du} = \frac{1}{5}$; $u = 5x - 35 \Rightarrow \frac{du}{dx} = 5$; therefore, $\frac{dy}{dx} = \frac{dy}{du} \cdot \frac{du}{dx} = \frac{1}{5} \cdot 5 = 1$, as expected

(b) $y = 1 + \frac{1}{u} \Rightarrow \frac{dy}{du} = -\frac{1}{u^2}$; $u = (x - 1)^{-1} \Rightarrow \frac{du}{dx} = -(x - 1)^{-2}(1) = \frac{-1}{(x-1)^2}$; therefore $\frac{dy}{dx} = \frac{dy}{du} \cdot \frac{du}{dx}$

$= \frac{-1}{u^2} \cdot \frac{-1}{(x-1)^2} = \frac{-1}{((x-1)^{-1})^2} \cdot \frac{-1}{(x-1)^2} = (x - 1)^2 \cdot \frac{1}{(x-1)^2} = 1$, again as expected

79. $y = 2\tan\left(\frac{\pi x}{4}\right) \Rightarrow \frac{dy}{dx} = \left(2\sec^2\frac{\pi x}{4}\right)\left(\frac{\pi}{4}\right) = \frac{\pi}{2}\sec^2\frac{\pi x}{4}$

(a) $\frac{dy}{dx}\Big|_{x=1} = \frac{\pi}{2}\sec^2\left(\frac{\pi}{4}\right) = \pi \Rightarrow$ slope of tangent is 2; thus, $y(1) = 2\tan\left(\frac{\pi}{4}\right) = 2$ and $y'(1) = \pi \Rightarrow$ tangent line is

given by $y - 2 = \pi(x - 1) \Rightarrow y = \pi x + 2 - \pi$

(b) $y' = \frac{\pi}{2}\sec^2\left(\frac{\pi x}{4}\right)$ and the smallest value the secant function can have in $-2 < x < 2$ is $1 \Rightarrow$ the minimum

value of y' is $\frac{\pi}{2}$ and that occurs when $\frac{\pi}{2} = \frac{\pi}{2}\sec^2\left(\frac{\pi x}{4}\right) \Rightarrow 1 = \sec^2\left(\frac{\pi x}{4}\right) \Rightarrow \pm 1 = \sec\left(\frac{\pi x}{4}\right) \Rightarrow x = 0$.

81. $s = (1 + 4t)^{1/2} \Rightarrow v = \frac{ds}{dt} = \frac{1}{2}(1 + 4t)^{-1/2}(4) = 2(1 + 4t)^{-1/2} \Rightarrow v(6) = 2(1 + 4 \cdot 6)^{-1/2} = \frac{2}{5}$ m/sec;

   $v = 2(1 + 4t)^{-1/2} \Rightarrow a = \frac{dv}{dt} = -\frac{1}{2} \cdot 2(1 + 4t)^{-3/2}(4) = -4(1 + 4t)^{-3/2} \Rightarrow a(6) = -4(1 + 4 \cdot 6)^{-3/2} = -\frac{4}{125}$ m

83. v proportional to $\frac{1}{\sqrt{s}} \Rightarrow v = \frac{k}{\sqrt{s}}$ for some constant k $\Rightarrow \frac{dv}{ds} = -\frac{k}{2s^{3/2}}$. Thus, $a = \frac{dv}{dt} = \frac{dv}{ds} \cdot \frac{ds}{dt} = \frac{dv}{ds} \cdot v$

   $= -\frac{k}{2s^{3/2}} \cdot \frac{k}{\sqrt{s}} = -\frac{k^2}{2}\left(\frac{1}{s^2}\right) \Rightarrow$ acceleration is a constant times $\frac{1}{s^2}$ so a is inversely proportional to s^2.

85. No. The chain rule says that when g is differentiable at 0 and f is differentiable at g(0), then f o g is differentiable at 0. But the chain rule says nothing about what happens when g is not differentiable at 0 so there is no contradiction.

87. From the power rule, with $y = x^{1/4}$, we get $\frac{dy}{dx} = \frac{1}{4}x^{-3/4}$. From the chain rule, $y = \sqrt{\sqrt{x}}$

   $\Rightarrow \frac{dy}{dx} = \frac{1}{2\sqrt{\sqrt{x}}} \cdot \frac{d}{dx}\left(\sqrt{x}\right) = \frac{1}{2\sqrt{\sqrt{x}}} \cdot \frac{1}{2\sqrt{x}} = \frac{1}{4}x^{-3/4}$, in agreement.

2.8 IMPLICIT DIFFERENTIATION

1. $x^2y + xy^2 = 6$:

   Step 1: $\left(x^2\frac{dy}{dx} + y \cdot 2x\right) + \left(x \cdot 2y\frac{dy}{dx} + y^2 \cdot 1\right) = 0$

   Step 2: $x^2\frac{dy}{dx} + 2xy\frac{dy}{dx} = -2xy - y^2$

   Step 3: $\frac{dy}{dx}\left(x^2 + 2xy\right) = -2xy - y^2$

   Step 4: $\frac{dy}{dx} = \frac{-2xy - y^2}{x^2 + 2xy}$

3. $2xy + y^2 = x + y$:

   Step 1: $\left(2x\frac{dy}{dx} + 2y\right) + 2y\frac{dy}{dx} = 1 + \frac{dy}{dx}$

   Step 2: $2x\frac{dy}{dx} + 2y\frac{dy}{dx} - \frac{dy}{dx} = 1 - 2y$

   Step 3: $\frac{dy}{dx}(2x + 2y - 1) = 1 - 2y$

   Step 4: $\frac{dy}{dx} = \frac{1 - 2y}{2x + 2y - 1}$

5. $x^2(x - y)^2 = x^2 - y^2$:

   Step 1: $x^2\left[2(x - y)\left(1 - \frac{dy}{dx}\right)\right] + (x - y)^2(2x) = 2x - 2y\frac{dy}{dx}$

   Step 2: $-2x^2(x - y)\frac{dy}{dx} + 2y\frac{dy}{dx} = 2x - 2x^2(x - y) - 2x(x - y)^2$

   Step 3: $\frac{dy}{dx}\left[-2x^2(x - y) + 2y\right] = 2x\left[1 - x(x - y) - (x - y)^2\right]$

   Step 4: $\frac{dy}{dx} = \frac{2x[1 - x(x - y) - (x - y)^2]}{-2x^2(x - y) + 2y} = \frac{x[1 - x(x - y) - (x - y)^2]}{y - x^2(x - y)} = \frac{x(1 - x^2 + xy - x^2 + 2xy - y^2)}{x^2y - x^3 + y}$

   $= \frac{x - 2x^3 + 3x^2y - xy^2}{x^2y - x^3 + y}$

7. $y^2 = \frac{x - 1}{x + 1} \Rightarrow 2y\frac{dy}{dx} = \frac{(x + 1) - (x - 1)}{(x + 1)^2} = \frac{2}{(x + 1)^2} \Rightarrow \frac{dy}{dx} = \frac{1}{y(x + 1)^2}$

9. $x = \tan y \Rightarrow 1 = (\sec^2 y)\frac{dy}{dx} \Rightarrow \frac{dy}{dx} = \frac{1}{\sec^2 y} = \cos^2 y$

11. $e^{2x} = \sin(x + 3y) \Rightarrow 2e^{2x} = (1 + 3y')\cos(x + 3y) \Rightarrow 1 + 3y' = \frac{2e^{2x}}{\cos(x + 3y)} \Rightarrow 3y' = \frac{2e^{2x}}{\cos(x + 3y)} - 1$

   $\Rightarrow y' = \frac{2e^{2x} - \cos(x + 3y)}{3\cos(x + 3y)}$

13. $y \sin\left(\frac{1}{y}\right) = 1 - xy \Rightarrow y\left[\cos\left(\frac{1}{y}\right) \cdot (-1)\frac{1}{y^2} \cdot \frac{dy}{dx}\right] + \sin\left(\frac{1}{y}\right) \cdot \frac{dy}{dx} = -x\frac{dy}{dx} - y \Rightarrow$

$\frac{dy}{dx}\left[-\frac{1}{y}\cos\left(\frac{1}{y}\right) + \sin\left(\frac{1}{y}\right) + x\right] = -y \Rightarrow \frac{dy}{dx} = \frac{-y}{-\frac{1}{y}\cos\left(\frac{1}{y}\right) + \sin\left(\frac{1}{y}\right) + x} = \frac{-y^2}{y\sin\left(\frac{1}{y}\right) - \cos\left(\frac{1}{y}\right) + xy}$

15. $\theta^{1/2} + r^{1/2} = 1 \Rightarrow \frac{1}{2}\theta^{-1/2} + \frac{1}{2}r^{-1/2} \cdot \frac{dr}{d\theta} = 0 \Rightarrow \frac{dr}{d\theta}\left[\frac{1}{2\sqrt{r}}\right] = \frac{-1}{2\sqrt{\theta}} \Rightarrow \frac{dr}{d\theta} = -\frac{2\sqrt{r}}{2\sqrt{\theta}} = -\frac{\sqrt{r}}{\sqrt{\theta}}$

17. $\sin(r\theta) = \frac{1}{2} \Rightarrow [\cos(r\theta)]\left(r + \theta\frac{dr}{d\theta}\right) = 0 \Rightarrow \frac{dr}{d\theta}[\theta\cos(r\theta)] = -r\cos(r\theta) \Rightarrow \frac{dr}{d\theta} = \frac{-r\cos(r\theta)}{\theta\cos(r\theta)} = -\frac{r}{\theta}$,

$\cos(r\theta) \neq 0$

19. $x^2 + y^2 = 1 \Rightarrow 2x + 2yy' = 0 \Rightarrow 2yy' = -2x \Rightarrow \frac{dy}{dx} = y' = -\frac{x}{y}$; now to find $\frac{d^2y}{dx^2}$, $\frac{d}{dx}(y') = \frac{d}{dx}\left(-\frac{x}{y}\right)$

$\Rightarrow y'' = \frac{y(-1) + xy'}{y^2} = \frac{-y + x\left(-\frac{x}{y}\right)}{y^2}$ since $y' = -\frac{x}{y} \Rightarrow \frac{d^2y}{dx^2} = y'' = \frac{-y^2 - x^2}{y^3} = \frac{-y^2 - (1 - y^2)}{y^3} = \frac{-1}{y^3}$

21. $y^2 = e^{x^2} + 2x \Rightarrow 2yy' = 2xe^{x^2} + 2 \Rightarrow \frac{dy}{dx} = \frac{xe^{x^2} + 1}{y} \Rightarrow \frac{d^2y}{dx^2} = \frac{y\left(2x^2e^{x^2} + e^{x^2}\right) - \left(xe^{x^2} + 1\right)y'}{y^2}$

$= \frac{y\left(2x^2e^{x^2} + e^{x^2}\right) - \left(xe^{x^2} + 1\right) \cdot \frac{xe^{x^2} + 1}{y}}{y^2} = \frac{y^2\left(2x^2e^{x^2} + e^{x^2}\right) - \left(x^2e^{2x^2} + 2xe^{x^2} + 1\right)}{y^3}$

$= \frac{\left(2x^2y^2 + y^2 - 2x\right)e^{x^2} - x^2e^{2x^2} - 1}{y^3}$

23. $2\sqrt{y} = x - y \Rightarrow y^{-1/2}y' = 1 - y' \Rightarrow y'\left(y^{-1/2} + 1\right) = 1 \Rightarrow \frac{dy}{dx} = y' = \frac{1}{y^{-1/2} + 1} = \frac{\sqrt{y}}{\sqrt{y} + 1}$; we can

differentiate the equation $y'\left(y^{-1/2} + 1\right) = 1$ again to find y'': $y'\left(-\frac{1}{2}y^{-3/2}y'\right) + \left(y^{-1/2} + 1\right)y'' = 0$

$\Rightarrow \left(y^{-1/2} + 1\right)y'' = \frac{1}{2}[y']^2y^{-3/2} \Rightarrow \frac{d^2y}{dx^2} = y'' = \frac{\frac{1}{2}\left(\frac{1}{y^{-1/2} + 1}\right)^2 y^{-3/2}}{\left(y^{-1/2} + 1\right)} = \frac{1}{2y^{3/2}\left(y^{-1/2} + 1\right)^3} = \frac{1}{2\left(1 + \sqrt{y}\right)^3}$

25. $x^3 + y^3 = 16 \Rightarrow 3x^2 + 3y^2y' = 0 \Rightarrow 3y^2y' = -3x^2 \Rightarrow y' = -\frac{x^2}{y^2}$; we differentiate $y^2y' = -x^2$ to find y'':

$y^2y'' + y'[2y \cdot y'] = -2x \Rightarrow y^2y'' = -2x - 2y[y']^2 \Rightarrow y'' = \frac{-2x - 2y\left(-\frac{x^2}{y^2}\right)^2}{y^2} = \frac{-2x - \frac{2x^4}{y^3}}{y^2}$

$= \frac{-2xy^3 - 2x^4}{y^5} \Rightarrow \frac{d^2y}{dx^2}\Big|_{(2,2)} = \frac{-32 - 32}{32} = -2$

27. $y^2 + x^2 = y^4 - 2x$ at $(-2, 1)$ and $(-2, -1) \Rightarrow 2y\frac{dy}{dx} + 2x = 4y^3\frac{dy}{dx} - 2 \Rightarrow 2y\frac{dy}{dx} - 4y^3\frac{dy}{dx} = -2 - 2x$

$\Rightarrow \frac{dy}{dx}(2y - 4y^3) = -2 - 2x \Rightarrow \frac{dy}{dx} = \frac{x + 1}{2y^3 - y} \Rightarrow \frac{dy}{dx}\Big|_{(-2,1)} = -1$ and $\frac{dy}{dx}\Big|_{(-2,-1)} = 1$

29. $x^2 + xy - y^2 = 1 \Rightarrow 2x + y + xy' - 2yy' = 0 \Rightarrow (x - 2y)y' = -2x - y \Rightarrow y' = \frac{2x + y}{2y - x}$;

(a) the slope of the tangent line $m = y'|_{(2,3)} = \frac{7}{4} \Rightarrow$ the tangent line is $y - 3 = \frac{7}{4}(x - 2) \Rightarrow y = \frac{7}{4}x - \frac{1}{2}$

(b) the normal line is $y - 3 = -\frac{4}{7}(x - 2) \Rightarrow y = -\frac{4}{7}x + \frac{29}{7}$

31. $x^2y^2 = 9 \Rightarrow 2xy^2 + 2x^2yy' = 0 \Rightarrow x^2yy' = -xy^2 \Rightarrow y' = -\frac{y}{x}$;

(a) the slope of the tangent line $m = y'|_{(-1,3)} = -\frac{y}{x}\Big|_{(-1,3)} = 3 \Rightarrow$ the tangent line is $y - 3 = 3(x + 1) \Rightarrow y = 3x + 6$

(b) the normal line is $y - 3 = -\frac{1}{3}(x + 1) \Rightarrow y = -\frac{1}{3}x + \frac{8}{3}$

33. $6x^2 + 3xy + 2y^2 + 17y - 6 = 0 \Rightarrow 12x + 3y + 3xy' + 4yy' + 17y' = 0 \Rightarrow y'(3x + 4y + 17) = -12x - 3y$

$\Rightarrow y' = \frac{-12x - 3y}{3x + 4y + 17}$;

(a) the slope of the tangent line m $= y'|_{(-1,0)} = \frac{-12x - 3y}{3x + 4y + 17}\Big|_{(-1,0)} = \frac{6}{7} \Rightarrow$ the tangent line is $y - 0 = \frac{6}{7}(x + 1)$

$\Rightarrow y = \frac{6}{7}x + \frac{6}{7}$

(b) the normal line is $y - 0 = -\frac{7}{6}(x + 1) \Rightarrow y = -\frac{7}{6}x - \frac{7}{6}$

35. $2xy + \pi \sin y = 2\pi \Rightarrow 2xy' + 2y + \pi(\cos y)y' = 0 \Rightarrow y'(2x + \pi \cos y) = -2y \Rightarrow y' = \frac{-2y}{2x + \pi \cos y}$;

(a) the slope of the tangent line m $= y'|_{(1, \frac{\pi}{2})} = \frac{-2y}{2x + \pi \cos y}\Big|_{(1, \frac{\pi}{2})} = -\frac{\pi}{2} \Rightarrow$ the tangent line is

$y - \frac{\pi}{2} = -\frac{\pi}{2}(x - 1) \Rightarrow y = -\frac{\pi}{2}x + \pi$

(b) the normal line is $y - \frac{\pi}{2} = \frac{2}{\pi}(x - 1) \Rightarrow y = \frac{2}{\pi}x - \frac{2}{\pi} + \frac{\pi}{2}$

37. $y = 2\sin(\pi x - y) \Rightarrow y' = 2[\cos(\pi x - y)] \cdot (\pi - y') \Rightarrow y'[1 + 2\cos(\pi x - y)] = 2\pi \cos(\pi x - y)$

$\Rightarrow y' = \frac{2\pi \cos(\pi x - y)}{1 + 2\cos(\pi x - y)}$;

(a) the slope of the tangent line m $= y'|_{(1,0)} = \frac{2\pi \cos(\pi x - y)}{1 + 2\cos(\pi x - y)}\Big|_{(1,0)} = 2\pi \Rightarrow$ the tangent line is

$y - 0 = 2\pi(x - 1) \Rightarrow y = 2\pi x - 2\pi$

(b) the normal line is $y - 0 = -\frac{1}{2\pi}(x - 1) \Rightarrow y = -\frac{x}{2\pi} + \frac{1}{2\pi}$

39. Solving $x^2 + xy + y^2 = 7$ and $y = 0 \Rightarrow x^2 = 7 \Rightarrow x = \pm\sqrt{7} \Rightarrow \left(-\sqrt{7}, 0\right)$ and $\left(\sqrt{7}, 0\right)$ are the points where the

curve crosses the x-axis. Now $x^2 + xy + y^2 = 7 \Rightarrow 2x + y + xy' + 2yy' = 0 \Rightarrow (x + 2y)y' = -2x - y$

$\Rightarrow y' = -\frac{2x + y}{x + 2y} \Rightarrow m = -\frac{2x + y}{x + 2y} \Rightarrow$ the slope at $\left(-\sqrt{7}, 0\right)$ is $m = -\frac{-2\sqrt{7}}{-\sqrt{7}} = -2$ and the slope at $\left(\sqrt{7}, 0\right)$ is

$m = -\frac{2\sqrt{7}}{\sqrt{7}} = -2$. Since the slope is -2 in each case, the corresponding tangents must be parallel.

41. $y^4 = y^2 - x^2 \Rightarrow 4y^3y' = 2yy' - 2x \Rightarrow 2(2y^3 - y)y' = -2x \Rightarrow y' = \frac{x}{y - 2y^3}$; the slope of the tangent line at

$\left(\frac{\sqrt{3}}{4}, \frac{\sqrt{3}}{2}\right)$ is $\frac{x}{y - 2y^3}\Big|_{\left(\frac{\sqrt{3}}{4}, \frac{\sqrt{3}}{2}\right)} = \frac{\frac{\sqrt{3}}{4}}{\frac{\sqrt{3}}{2} - \frac{6\sqrt{3}}{8}} = \frac{\frac{1}{4}}{\frac{1}{2} - \frac{3}{4}} = \frac{1}{2 - 3} = -1$; the slope of the tangent line at $\left(\frac{\sqrt{3}}{4}, \frac{1}{2}\right)$

is $\frac{x}{y - 2y^3}\Big|_{\left(\frac{\sqrt{3}}{4}, \frac{1}{2}\right)} = \frac{\frac{\sqrt{3}}{4}}{\frac{1}{2} - \frac{2}{8}} = \frac{2\sqrt{3}}{4 - 2} = \sqrt{3}$

43. $x^2 + 2xy - 3y^2 = 0 \Rightarrow 2x + 2xy' + 2y - 6yy' = 0 \Rightarrow y'(2x - 6y) = -2x - 2y \Rightarrow y' = \frac{x + y}{3y - x} \Rightarrow$ the slope of the tangent

line m $= y'|_{(1,1)} = \frac{x + y}{3y - x}\Big|_{(1,1)} = 1 \Rightarrow$ the equation of the normal line at $(1, 1)$ is $y - 1 = -1(x - 1) \Rightarrow y = -x + 2$. To find

where the normal line intersects the curve we substitute into its equation: $x^2 + 2x(2 - x) - 3(2 - x)^2 = 0$

$\Rightarrow x^2 + 4x - 2x^2 - 3(4 - 4x + x^2) = 0 \Rightarrow -4x^2 + 16x - 12 = 0 \Rightarrow x^2 - 4x + 3 = 0 \Rightarrow (x - 3)(x - 1) = 0$

$\Rightarrow x = 3$ and $y = -x + 2 = -1$. Therefore, the normal to the curve at $(1, 1)$ intersects the curve at the point $(3, -1)$.

Note that it also intersects the curve at $(1, 1)$.

45. $xy^3 + x^2y = 6 \Rightarrow x\left(3y^2 \frac{dy}{dx}\right) + y^3 + x^2 \frac{dy}{dx} + 2xy = 0 \Rightarrow \frac{dy}{dx}(3xy^2 + x^2) = -y^3 - 2xy \Rightarrow \frac{dy}{dx} = \frac{-y^3 - 2xy}{3xy^2 + x^2}$

$= -\frac{y^3 + 2xy}{3xy^2 + x^2}$; also, $xy^3 + x^2y = 6 \Rightarrow x(3y^2) + y^3 \frac{dx}{dy} + x^2 + y\left(2x \frac{dx}{dy}\right) = 0 \Rightarrow \frac{dx}{dy}(y^3 + 2xy) = -3xy^2 - x^2$

$\Rightarrow \frac{dx}{dy} = -\frac{3xy^2 + x^2}{y^3 + 2xy}$; thus $\frac{dx}{dy}$ appears to equal $\frac{1}{\frac{dy}{dx}}$. The two different treatments view the graphs as functions

symmetric across the line $y = x$, so their slopes are reciprocals of one another at the corresponding points

(a, b) and (b, a).

2.9 INVERSE FUNCTIONS AND THEIR DERIVATIVES

1. Yes one-to-one, the graph passes the horizontal test.

3. Not one-to-one since (for example) the horizontal line $y = 2$ intersects the graph twice.

5. Domain: $-1 \leq x \leq 1$, Range: $-\frac{\pi}{2} \leq y \leq \frac{\pi}{2}$

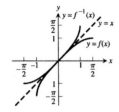

7. Step 1: $y = x^2 + 1 \Rightarrow x^2 = y - 1 \Rightarrow x = \sqrt{y - 1}$
    Step 2: $y = \sqrt{x - 1} = f^{-1}(x)$

9. Step 1: $y = x^3 - 1 \Rightarrow x^3 = y + 1 \Rightarrow x = (y + 1)^{1/3}$
    Step 2: $y = \sqrt[3]{x + 1} = f^{-1}(x)$

11. Step 1: $y = (x + 1)^2 \Rightarrow \sqrt{y} = x + 1$, since $x \geq -1 \Rightarrow x = \sqrt{y} - 1$
    Step 2: $y = \sqrt{x} - 1 = f^{-1}(x)$

13. Step 1: $y = x^5 \Rightarrow x = y^{1/5}$
    Step 2: $y = \sqrt[5]{x} = f^{-1}(x)$;
    Domain and Range of f^{-1}: all reals;
    $f\left(f^{-1}(x)\right) = \left(x^{1/5}\right)^5 = x$ and $f^{-1}(f(x)) = \left(x^5\right)^{1/5} = x$

15. Step 1: $y = \frac{1}{x^2} \Rightarrow x^2 = \frac{1}{y} \Rightarrow x = \frac{1}{\sqrt{y}}$
    Step 2: $y = \frac{1}{\sqrt{x}} = f^{-1}(x)$
    Domain of f^{-1}: $x > 0$, Range of f^{-1}: $y > 0$;
    $f\left(f^{-1}(x)\right) = \frac{1}{\left(\frac{1}{\sqrt{x}}\right)^2} = \frac{1}{\left(\frac{1}{x}\right)} = x$ and $f^{-1}(f(x)) = \frac{1}{\sqrt{\frac{1}{x^2}}} = \frac{1}{\left(\frac{1}{x}\right)} = x$ since $x > 0$

17. (a) $y = 2x + 3 \Rightarrow 2x = y - 3$
        $\Rightarrow x = \frac{y}{2} - \frac{3}{2} \Rightarrow f^{-1}(x) = \frac{x}{2} - \frac{3}{2}$

    (b)

    (c) $\frac{df}{dx}\Big|_{x=-1} = 2$, $\frac{df^{-1}}{dx}\Big|_{x=1} = \frac{1}{2}$

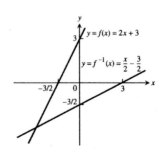

19. (a) $y = 5 - 4x \Rightarrow 4x = 5 - y$

$\Rightarrow x = \frac{5}{4} - \frac{y}{4} \Rightarrow f^{-1}(x) = \frac{5}{4} - \frac{x}{4}$

(b)

(c) $\frac{df}{dx}\big|_{x=1/2} = -4$, $\frac{df^{-1}}{dx}\big|_{x=3} = -\frac{1}{4}$

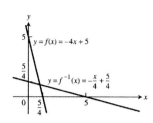

21. (a) $f(g(x)) = \left(\sqrt[3]{x}\right)^3 = x$, $g(f(x)) = \sqrt[3]{x^3} = x$

(b)

(c) $f'(x) = 3x^2 \Rightarrow f'(1) = 3, f'(-1) = 3$;

$g'(x) = \frac{1}{3}x^{-2/3} \Rightarrow g'(1) = \frac{1}{3}, g'(-1) = \frac{1}{3}$

(d) The line $y = 0$ is tangent to $f(x) = x^3$ at $(0,0)$;

the line $x = 0$ is tangent to $g(x) = \sqrt[3]{x}$ at $(0,0)$

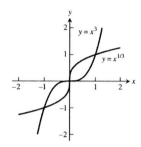

23. $\frac{df}{dx} = 3x^2 - 6x \Rightarrow \frac{df^{-1}}{dx}\big|_{x=f(3)} = \frac{1}{\frac{df}{dx}}\big|_{x=3} = \frac{1}{9}$

25. (a) $y = mx \Rightarrow x = \frac{1}{m}y \Rightarrow f^{-1}(x) = \frac{1}{m}x$

(b) The graph of $y = f^{-1}(x)$ is a line through the origin with slope $\frac{1}{m}$.

27. Let $x_1 \neq x_2$ be two numbers in the domain of an increasing function f. Then, either $x_1 < x_2$ or $x_1 > x_2$ which implies $f(x_1) < f(x_2)$ or $f(x_1) > f(x_2)$, since $f(x)$ is increasing. In either case, $f(x_1) \neq f(x_2)$ and f is one-to-one. Similar arguments hold if f is decreasing.

29. $f(x)$ is increasing since $x_2 > x_1 \Rightarrow 27x_2^3 > 27x_1^3$; $y = 27x^3 \Rightarrow x = \frac{1}{3}y^{1/3} \Rightarrow f^{-1}(x) = \frac{1}{3}x^{1/3}$;

$\frac{df}{dx} = 81x^2 \Rightarrow \frac{df^{-1}}{dx} = \frac{1}{81x^2}\big|_{\frac{1}{3}x^{1/3}} = \frac{1}{9x^{2/3}} = \frac{1}{9}x^{-2/3}$

31. $f(x)$ is decreasing since $x_2 > x_1 \Rightarrow (1 - x_2)^3 < (1 - x_1)^3$; $y = (1 - x)^3 \Rightarrow x = 1 - y^{1/3} \Rightarrow f^{-1}(x) = 1 - x^{1/3}$;

$\frac{df}{dx} = -3(1 - x)^2 \Rightarrow \frac{df^{-1}}{dx} = \frac{1}{-3(1-x)^2}\big|_{1-x^{1/3}} = \frac{-1}{3x^{2/3}} = -\frac{1}{3}x^{-2/3}$

33. The function $g(x)$ is also one-to-one. The reasoning: $f(x)$ is one-to-one means that if $x_1 \neq x_2$ then $f(x_1) \neq f(x_2)$, so $-f(x_1) \neq -f(x_2)$ and therefore $g(x_1) \neq g(x_2)$. Therefore $g(x)$ is one-to-one as well.

35. The composite is one-to-one also. The reasoning: If $x_1 \neq x_2$ then $g(x_1) \neq g(x_2)$ because g is one-to-one. Since $g(x_1) \neq g(x_2)$, we also have $f(g(x_1)) \neq f(g(x_2))$ because f is one-to-one. Thus, f ∘ g is one-to-one because $x_1 \neq x_2 \Rightarrow f(g(x_1)) \neq f(g(x_2))$.

37. $(g \circ f)(x) = x \Rightarrow g(f(x)) = x \Rightarrow g'(f(x))f'(x) = 1$

2.10 LOGARITHMIC FUNCTIONS

1. (a) $\ln 0.75 = \ln \frac{3}{4} = \ln 3 - \ln 4 = \ln 3 - \ln 2^2 = \ln 3 - 2 \ln 2$

(b) $\ln \frac{4}{9} = \ln 4 - \ln 9 = \ln 2^2 - \ln 3^2 = 2 \ln 2 - 2 \ln 3$

(c) $\ln \frac{1}{2} = \ln 1 - \ln 2 = -\ln 2$

(d) $\ln \sqrt[3]{9} = \frac{1}{3} \ln 9 = \frac{1}{3} \ln 3^2 = \frac{2}{3} \ln 3$

(e) $\ln 3\sqrt{2} = \ln 3 + \ln 2^{1/2} = \ln 3 + \frac{1}{2}\ln 2$

(f) $\ln \sqrt{13.5} = \frac{1}{2}\ln 13.5 = \frac{1}{2}\ln \frac{27}{2} = \frac{1}{2}\left(\ln 3^3 - \ln 2\right) = \frac{1}{2}\left(3\ln 3 - \ln 2\right)$

3. (a) $\ln \sin \theta - \ln\left(\frac{\sin \theta}{5}\right) = \ln\left(\frac{\sin \theta}{\left(\frac{\sin \theta}{5}\right)}\right) = \ln 5$ (b) $\ln\left(3x^2 - 9x\right) + \ln\left(\frac{1}{3x}\right) = \ln\left(\frac{3x^2 - 9x}{3x}\right) = \ln\left(x - 3\right)$

    (c) $\frac{1}{2}\ln\left(4t^4\right) - \ln 2 = \ln \sqrt{4t^4} - \ln 2 = \ln 2t^2 - \ln 2 = \ln\left(\frac{2t^2}{2}\right) = \ln\left(t^2\right)$

5. (a) $e^{\ln 7.2} = 7.2$ (b) $e^{-\ln x^2} = \frac{1}{e^{\ln x^2}} = \frac{1}{x^2}$ (c) $e^{\ln x - \ln y} = e^{\ln(x/y)} = \frac{x}{y}$

7. (a) $2\ln\sqrt{e} = 2\ln e^{1/2} = (2)\left(\frac{1}{2}\right)\ln e = 1$ (b) $\ln\left(\ln e^e\right) = \ln\left(e\ln e\right) = \ln e = 1$

    (c) $\ln e^{(-x^2 - y^2)} = \left(-x^2 - y^2\right)\ln e = -x^2 - y^2$

9. $\ln y = 2t + 4 \Rightarrow e^{\ln y} = e^{2t+4} \Rightarrow y = e^{2t+4}$

11. $\ln(y - 40) = 5t \Rightarrow e^{\ln(y-40)} = e^{5t} \Rightarrow y - 40 = e^{5t} \Rightarrow y = e^{5t} + 40$

13. $\ln(y - 1) - \ln 2 = x + \ln x \Rightarrow \ln(y - 1) - \ln 2 - \ln x = x \Rightarrow \ln\left(\frac{y-1}{2x}\right) = x \Rightarrow e^{\ln\left(\frac{y-1}{2x}\right)} = e^x \Rightarrow \frac{y-1}{2x} = e^x$

    $\Rightarrow y - 1 = 2xe^x \Rightarrow y = 2xe^x + 1$

15. (a) $e^{2k} = 4 \Rightarrow \ln e^{2k} = \ln 4 \Rightarrow 2k\ln e = \ln 2^2 \Rightarrow 2k = 2\ln 2 \Rightarrow k = \ln 2$

    (b) $100e^{10k} = 200 \Rightarrow e^{10k} = 2 \Rightarrow \ln e^{10k} = \ln 2 \Rightarrow 10k\ln e = \ln 2 \Rightarrow 10k = \ln 2 \Rightarrow k = \frac{\ln 2}{10}$

    (c) $e^{k/1000} = a \Rightarrow \ln e^{k/1000} = \ln a \Rightarrow \frac{k}{1000}\ln e = \ln a \Rightarrow \frac{k}{1000} = \ln a \Rightarrow k = 1000\ln a$

17. (a) $e^{-0.3t} = 27 \Rightarrow \ln e^{-0.3t} = \ln 3^3 \Rightarrow (-0.3t)\ln e = 3\ln 3 \Rightarrow -0.3t = 3\ln 3 \Rightarrow t = -10\ln 3$

    (b) $e^{kt} = \frac{1}{2} \Rightarrow \ln e^{kt} = \ln 2^{-1} = kt\ln e = -\ln 2 \Rightarrow t = -\frac{\ln 2}{k}$

    (c) $e^{(\ln 0.2)t} = 0.4 \Rightarrow \left(e^{\ln 0.2}\right)^t = 0.4 \Rightarrow 0.2^t = 0.4 \Rightarrow \ln 0.2^t = \ln 0.4 \Rightarrow t\ln 0.2 = \ln 0.4 \Rightarrow t = \frac{\ln 0.4}{\ln 0.2}$

19. $e^{\sqrt{t}} = x^2 \Rightarrow \ln e^{\sqrt{t}} = \ln x^2 \Rightarrow \sqrt{t} = 2\ln x \Rightarrow t = 4(\ln x)^2$

21. (a) $5^{\log_5 7} = 7$ (b) $8^{\log_8 \sqrt{2}} = \sqrt{2}$ (c) $1.3^{\log_{1.3} 75} = 75$

    (d) $\log_4 16 = \log_4 4^2 = 2\log_4 4 = 2 \cdot 1 = 2$ (e) $\log_3 \sqrt{3} = \log_3 3^{1/2} = \frac{1}{2}\log_3 3 = \frac{1}{2}\cdot 1 = \frac{1}{2} = 0.5$

    (f) $\log_4\left(\frac{1}{4}\right) = \log_4 4^{-1} = -1\log_4 4 = -1 \cdot 1 = -1$

23. (a) Let $z = \log_4 x \Rightarrow 4^z = x \Rightarrow 2^{2z} = x \Rightarrow \left(2^z\right)^2 = x \Rightarrow 2^z = \sqrt{x}$

    (b) Let $z = \log_3 x \Rightarrow 3^z = x \Rightarrow \left(3^z\right)^2 = x^2 \Rightarrow 3^{2z} = x^2 \Rightarrow 9^z = x^2$

    (c) $\log_2\left(e^{(\ln 2)\sin x}\right) = \log_2 2^{\sin x} = \sin x$

25. (a) $\frac{\log_2 x}{\log_3 x} = \frac{\ln x}{\ln 2} \div \frac{\ln x}{\ln 3} = \frac{\ln x}{\ln 2}\cdot \frac{\ln 3}{\ln x} = \frac{\ln 3}{\ln 2}$ (b) $\frac{\log_2 x}{\log_8 x} = \frac{\ln x}{\ln 2} \div \frac{\ln x}{\ln 8} = \frac{\ln x}{\ln 2}\cdot \frac{\ln 8}{\ln x} = \frac{3\ln 2}{\ln 2} = 3$

    (c) $\frac{\log_x a}{\log_{x^2} a} = \frac{\ln a}{\ln x} \div \frac{\ln a}{\ln x^2} = \frac{\ln a}{\ln x}\cdot \frac{\ln x^2}{\ln a} = \frac{2\ln x}{\ln x} = 2$

27. $y = \ln 3x \Rightarrow y' = \left(\frac{1}{3x}\right)(3) = \frac{1}{x}$ 29. $y = \ln\left(t^2\right) \Rightarrow \frac{dy}{dt} = \left(\frac{1}{t^2}\right)(2t) = \frac{2}{t}$

31. $y = \ln\frac{3}{x} = \ln 3x^{-1} \Rightarrow \frac{dy}{dx} = \left(\frac{1}{3x^{-1}}\right)\left(-3x^{-2}\right) = -\frac{1}{x}$

33. $y = \ln(\theta + 1) \Rightarrow \frac{dy}{d\theta} = \left(\frac{1}{\theta + 1}\right)(1) = \frac{1}{\theta + 1}$

35. $y = \ln x^3 \Rightarrow \frac{dy}{dx} = \left(\frac{1}{x^3}\right)(3x^2) = \frac{3}{x}$

37. $y = t(\ln t)^2 \Rightarrow \frac{dy}{dt} = (\ln t)^2 + 2t(\ln t) \cdot \frac{d}{dt}(\ln t) = (\ln t)^2 + \frac{2t \ln t}{t} = (\ln t)^2 + 2 \ln t$

39. $y = \frac{x^4}{4} \ln x - \frac{x^4}{16} \Rightarrow \frac{dy}{dx} = x^3 \ln x + \frac{x^4}{4} \cdot \frac{1}{x} - \frac{4x^3}{16} = x^3 \ln x$

41. $y = \frac{\ln t}{t} \Rightarrow \frac{dy}{dt} = \frac{t\left(\frac{1}{t}\right) - (\ln t)(1)}{t^2} = \frac{1 - \ln t}{t^2}$

43. $y = \frac{\ln x}{1 + \ln x} \Rightarrow y' = \frac{(1 + \ln x)\left(\frac{1}{x}\right) - (\ln x)\left(\frac{1}{x}\right)}{(1 + \ln x)^2} = \frac{\frac{1}{x} + \frac{\ln x}{x} - \frac{\ln x}{x}}{(1 + \ln x)^2} = \frac{1}{x(1 + \ln x)^2}$

45. $y = \ln(\ln x) \Rightarrow y' = \left(\frac{1}{\ln x}\right)\left(\frac{1}{x}\right) = \frac{1}{x \ln x}$

47. $y = \theta[\sin(\ln \theta) + \cos(\ln \theta)] \Rightarrow \frac{dy}{d\theta} = [\sin(\ln \theta) + \cos(\ln \theta)] + \theta\left[\cos(\ln \theta) \cdot \frac{1}{\theta} - \sin(\ln \theta) \cdot \frac{1}{\theta}\right]$
$= \sin(\ln \theta) + \cos(\ln \theta) + \cos(\ln \theta) - \sin(\ln \theta) = 2 \cos(\ln \theta)$

49. $y = \ln \frac{1}{x\sqrt{x + 1}} = -\ln x - \frac{1}{2} \ln(x + 1) \Rightarrow y' = -\frac{1}{x} - \frac{1}{2}\left(\frac{1}{x + 1}\right) = -\frac{2(x + 1) + x}{2x(x + 1)} = -\frac{3x + 2}{2x(x + 1)}$

51. $y = \frac{1 + \ln t}{1 - \ln t} \Rightarrow \frac{dy}{dt} = \frac{(1 - \ln t)\left(\frac{1}{t}\right) - (1 + \ln t)\left(\frac{-1}{t}\right)}{(1 - \ln t)^2} = \frac{\frac{1}{t} - \frac{\ln t}{t} + \frac{1}{t} + \frac{\ln t}{t}}{(1 - \ln t)^2} = \frac{2}{t(1 - \ln t)^2}$

53. $y = \ln(\sec(\ln \theta)) \Rightarrow \frac{dy}{d\theta} = \frac{1}{\sec(\ln \theta)} \cdot \frac{d}{d\theta}(\sec(\ln \theta)) = \frac{\sec(\ln \theta)\tan(\ln \theta)}{\sec(\ln \theta)} \cdot \frac{d}{d\theta}(\ln \theta) = \frac{\tan(\ln \theta)}{\theta}$

55. $y = \ln\left(\frac{(x^2 + 1)^5}{\sqrt{1 - x}}\right) = 5 \ln(x^2 + 1) - \frac{1}{2} \ln(1 - x) \Rightarrow y' = \frac{5 \cdot 2x}{x^2 + 1} - \frac{1}{2}\left(\frac{1}{1 - x}\right)(-1) = \frac{10x}{x^2 + 1} + \frac{1}{2(1 - x)}$

57. $y = \sqrt{x(x + 1)} = (x(x + 1))^{1/2} \Rightarrow \ln y = \frac{1}{2} \ln(x(x + 1)) \Rightarrow 2 \ln y = \ln(x) + \ln(x + 1) \Rightarrow \frac{2y'}{y} = \frac{1}{x} + \frac{1}{x + 1}$
$\Rightarrow y' = \left(\frac{1}{2}\right)\sqrt{x(x + 1)}\left(\frac{1}{x} + \frac{1}{x + 1}\right) = \frac{\sqrt{x(x + 1)}(2x + 1)}{2x(x + 1)} = \frac{2x + 1}{2\sqrt{x(x + 1)}}$

59. $y = \sqrt{\frac{t}{t + 1}} = \left(\frac{t}{t + 1}\right)^{1/2} \Rightarrow \ln y = \frac{1}{2}[\ln t - \ln(t + 1)] \Rightarrow \frac{1}{y}\frac{dy}{dt} = \frac{1}{2}\left(\frac{1}{t} - \frac{1}{t + 1}\right)$
$\Rightarrow \frac{dy}{dt} = \frac{1}{2}\sqrt{\frac{t}{t + 1}}\left(\frac{1}{t} - \frac{1}{t + 1}\right) = \frac{1}{2}\sqrt{\frac{t}{t + 1}}\left[\frac{1}{t(t + 1)}\right] = \frac{1}{2\sqrt{t}(t + 1)^{3/2}}$

61. $y = \sqrt{\theta + 3}(\sin \theta) = (\theta + 3)^{1/2} \sin \theta \Rightarrow \ln y = \frac{1}{2} \ln(\theta + 3) + \ln(\sin \theta) \Rightarrow \frac{1}{y}\frac{dy}{d\theta} = \frac{1}{2(\theta + 3)} + \frac{\cos \theta}{\sin \theta}$
$\Rightarrow \frac{dy}{d\theta} = \sqrt{\theta + 3}(\sin \theta)\left[\frac{1}{2(\theta + 3)} + \cot \theta\right]$

63. $y = t(t + 1)(t + 2) \Rightarrow \ln y = \ln t + \ln(t + 1) + \ln(t + 2) \Rightarrow \frac{1}{y}\frac{dy}{dt} = \frac{1}{t} + \frac{1}{t + 1} + \frac{1}{t + 2}$
$\Rightarrow \frac{dy}{dt} = t(t + 1)(t + 2)\left(\frac{1}{t} + \frac{1}{t + 1} + \frac{1}{t + 2}\right) = t(t + 1)(t + 2)\left[\frac{(t + 1)(t + 2) + t(t + 2) + t(t + 1)}{t(t + 1)(t + 2)}\right] = 3t^2 + 6t + 2$

65. $y = \frac{\theta + 5}{\theta \cos \theta} \Rightarrow \ln y = \ln(\theta + 5) - \ln \theta - \ln(\cos \theta) \Rightarrow \frac{1}{y}\frac{dy}{d\theta} = \frac{1}{\theta + 5} - \frac{1}{\theta} + \frac{\sin \theta}{\cos \theta}$
$\Rightarrow \frac{dy}{d\theta} = \left(\frac{\theta + 5}{\theta \cos \theta}\right)\left(\frac{1}{\theta + 5} - \frac{1}{\theta} + \tan \theta\right)$

67. $y = \frac{x\sqrt{x^2 + 1}}{(x + 1)^{2/3}} \Rightarrow \ln y = \ln x + \frac{1}{2} \ln(x^2 + 1) - \frac{2}{3} \ln(x + 1) \Rightarrow \frac{y'}{y} = \frac{1}{x} + \frac{x}{x^2 + 1} - \frac{2}{3(x + 1)}$
$\Rightarrow y' = \frac{x\sqrt{x^2 + 1}}{(x + 1)^{2/3}}\left[\frac{1}{x} + \frac{x}{x^2 + 1} - \frac{2}{3(x + 1)}\right]$

69. $y = \sqrt[3]{\frac{x(x-2)}{x^2+1}} \Rightarrow \ln y = \frac{1}{3}\left[\ln x + \ln(x-2) - \ln(x^2+1)\right] \Rightarrow \frac{y'}{y} = \frac{1}{3}\left(\frac{1}{x} + \frac{1}{x-2} - \frac{2x}{x^2+1}\right)$

$\Rightarrow y' = \frac{1}{3}\sqrt[3]{\frac{x(x-2)}{x^2+1}}\left(\frac{1}{x} + \frac{1}{x-2} - \frac{2x}{x^2+1}\right)$

71. $y = \ln(\cos^2\theta) \Rightarrow \frac{dy}{d\theta} = \frac{1}{\cos^2\theta} \cdot 2\cos\theta \cdot (-\sin\theta) = -2\tan\theta$

73. $y = \ln(3te^{-t}) = \ln 3 + \ln t + \ln e^{-t} = \ln 3 + \ln t - t \Rightarrow \frac{dy}{dt} = \frac{1}{t} - 1 = \frac{1-t}{t}$

75. $y = \ln\frac{e^\theta}{1+e^\theta} = \ln e^\theta - \ln\left(1+e^\theta\right) = \theta - \ln\left(1+e^\theta\right) \Rightarrow \frac{dy}{d\theta} = 1 - \left(\frac{1}{1+e^\theta}\right)\frac{d}{d\theta}\left(1+e^\theta\right) = 1 - \frac{e^\theta}{1+e^\theta} = \frac{1}{1+e^\theta}$

77. $y = e^{(\cos t + \ln t)} = e^{\cos t}e^{\ln t} = te^{\cos t} \Rightarrow \frac{dy}{dt} = e^{\cos t} + te^{\cos t}\frac{d}{dt}(\cos t) = (1 - t\sin t)e^{\cos t}$

79. $\ln y = e^y\sin x \Rightarrow \left(\frac{1}{y}\right)y' = (y'e^y)(\sin x) + e^y\cos x \Rightarrow y'\left(\frac{1}{y} - e^y\sin x\right) = e^y\cos x$

$\Rightarrow y'\left(\frac{1-ye^y\sin x}{y}\right) = e^y\cos x \Rightarrow y' = \frac{ye^y\cos x}{1 - ye^y\sin x}$

81. $x^y = y^x \Rightarrow \ln x^y = \ln y^x \Rightarrow y\ln x = x\ln y \Rightarrow y\cdot\frac{1}{x} + y'\cdot\ln x = x\cdot\frac{1}{y}\cdot y' + (1)\cdot\ln y \Rightarrow \ln x\cdot y' - \frac{x}{y}\cdot y' = \ln y - \frac{y}{x}$

$\Rightarrow y' = \frac{\ln y - \frac{y}{x}}{\ln x - \frac{x}{y}} = \frac{xy\ln y - y^2}{xy\ln x - x^2} = \frac{y}{x}\left(\frac{x\ln y - y}{y\ln x - x}\right)$

83. $y = 2^x \Rightarrow y' = 2^x\ln 2$

85. $y = 5^{\sqrt{s}} \Rightarrow \frac{dy}{ds} = 5^{\sqrt{s}}(\ln 5)\left(\frac{1}{2}s^{-1/2}\right) = \left(\frac{\ln 5}{2\sqrt{s}}\right)5^{\sqrt{s}}$

87. $y = x^\pi \Rightarrow y' = \pi x^{(\pi-1)}$

89. $y = \log_2 5\theta = \frac{\ln 5\theta}{\ln 2} \Rightarrow \frac{dy}{d\theta} = \left(\frac{1}{\ln 2}\right)\left(\frac{1}{5\theta}\right)(5) = \frac{1}{\theta\ln 2}$

91. $y = \log_4 x + \log_4 x^2 = \frac{\ln x}{\ln 4} + \frac{\ln x^2}{\ln 4} = \frac{\ln x}{\ln 4} + 2\frac{\ln x}{\ln 4} = 3\frac{\ln x}{\ln 4} \Rightarrow y' = \frac{3}{x\ln 4}$

93. $y = \log_2 r\cdot\log_4 r = \left(\frac{\ln r}{\ln 2}\right)\left(\frac{\ln r}{\ln 4}\right) = \frac{\ln^2 r}{(\ln 2)(\ln 4)} \Rightarrow \frac{dy}{dr} = \left[\frac{1}{(\ln 2)(\ln 4)}\right](2\ln r)\left(\frac{1}{r}\right) = \frac{2\ln r}{r(\ln 2)(\ln 4)}$

95. $y = \log_3\left(\left(\frac{x+1}{x-1}\right)^{\ln 3}\right) = \frac{\ln\left(\frac{x+1}{x-1}\right)^{\ln 3}}{\ln 3} = \frac{(\ln 3)\ln\left(\frac{x+1}{x-1}\right)}{\ln 3} = \ln\left(\frac{x+1}{x-1}\right) = \ln(x+1) - \ln(x-1)$

$\Rightarrow \frac{dy}{dx} = \frac{1}{x+1} - \frac{1}{x-1} = \frac{-2}{(x+1)(x-1)}$

97. $y = \theta\sin(\log_7\theta) = \theta\sin\left(\frac{\ln\theta}{\ln 7}\right) \Rightarrow \frac{dy}{d\theta} = \sin\left(\frac{\ln\theta}{\ln 7}\right) + \theta\left[\cos\left(\frac{\ln\theta}{\ln 7}\right)\right]\left(\frac{1}{\theta\ln 7}\right) = \sin(\log_7\theta) + \frac{1}{\ln 7}\cos(\log_7\theta)$

99. $y = \log_5 e^x = \frac{\ln e^x}{\ln 5} = \frac{x}{\ln 5} \Rightarrow y' = \frac{1}{\ln 5}$

101. $y = 3^{\log_2 t} = 3^{(\ln t)/(\ln 2)} \Rightarrow \frac{dy}{dt} = \left[3^{(\ln t)/(\ln 2)}(\ln 3)\right]\left(\frac{1}{t\ln 2}\right) = \frac{1}{t}(\log_2 3)\,3^{\log_2 t}$

103. $y = \log_2\left(8t^{\ln 2}\right) = \frac{\ln 8 + \ln\left(t^{\ln 2}\right)}{\ln 2} = \frac{3\ln 2 + (\ln 2)(\ln t)}{\ln 2} = 3 + \ln t \Rightarrow \frac{dy}{dt} = \frac{1}{t}$

105. $y = (x+1)^x \Rightarrow \ln y = \ln(x+1)^x = x\ln(x+1) \Rightarrow \frac{y'}{y} = \ln(x+1) + x\cdot\frac{1}{(x+1)} \Rightarrow y' = (x+1)^x\left[\frac{x}{x+1} + \ln(x+1)\right]$

107. $y = \left(\sqrt{t}\right)^t = \left(t^{1/2}\right)^t = t^{t/2} \Rightarrow \ln y = \ln t^{t/2} = \left(\frac{t}{2}\right) \ln t \Rightarrow \frac{1}{y} \frac{dy}{dt} = \left(\frac{1}{2}\right)(\ln t) + \left(\frac{t}{2}\right)\left(\frac{1}{t}\right) = \frac{\ln t}{2} + \frac{1}{2}$
$\Rightarrow \frac{dy}{dt} = \left(\sqrt{t}\right)^t \left(\frac{\ln t}{2} + \frac{1}{2}\right)$

109. $y = (\sin x)^x \Rightarrow \ln y = \ln (\sin x)^x = x \ln (\sin x) \Rightarrow \frac{y'}{y} = \ln (\sin x) + x \left(\frac{\cos x}{\sin x}\right) \Rightarrow y' = (\sin x)^x [\ln (\sin x) + x \cot x]$

111. $y = x^{\ln x}, x > 0 \Rightarrow \ln y = (\ln x)^2 \Rightarrow \frac{y'}{y} = 2(\ln x)\left(\frac{1}{x}\right) \Rightarrow y' = \left(x^{\ln x}\right)\left(\frac{\ln x^2}{x}\right)$

113. (a) Begin with $y = \ln x$ and reduce the y-value by $3 \Rightarrow y = \ln x - 3$.
    (b) Begin with $y = \ln x$ and replace x with $x - 1 \Rightarrow y = \ln(x - 1)$.
    (c) Begin with $y = \ln x$, replace x with $x + 1$, and increase the y-value by $3 \Rightarrow y = \ln(x + 1) + 3$.
    (d) Begin with $y = \ln x$, reduce the y-value by 4, and replace x with $x - 2 \Rightarrow y = \ln(x - 2) - 4$.
    (e) Begin with $y = \ln x$ and replace x with $-x \Rightarrow y = \ln(-x)$.
    (f) Begin with $y = \ln x$ and switch x and $y \Rightarrow x = \ln y$ or $y = e^x$.

115. $y = A \sin(\ln x) + B \cos(\ln x) \Rightarrow y' = A \cos(\ln x) \cdot \frac{1}{x} - B \sin(\ln x) \cdot \frac{1}{x} = (A \cos(\ln x) - B \sin(\ln x)) \cdot \frac{1}{x}$
$\Rightarrow y'' = (A \cos(\ln x) - B \sin(\ln x)) \cdot \frac{-1}{x^2} + \left(-A \sin(\ln x) \cdot \frac{1}{x} - B \cos(\ln x) \cdot \frac{1}{x}\right) \cdot \frac{1}{x}$
$= (-A(\cos(\ln x) + \sin(\ln x)) + B(\sin(\ln x) - \cos(\ln x))) \cdot \frac{1}{x^2}$
$\Rightarrow x^2 y'' + x y' + y = (-A(\cos(\ln x) + \sin(\ln x)) + B(\sin(\ln x) - \cos(\ln x))) + (A \cos(\ln x) - B \sin(\ln x))$
$+ (A \sin(\ln x) + B \cos(\ln x)) = 0$

117. (a) Amount $= 8\left(\frac{1}{2}\right)^{t/12}$
    (b) $8\left(\frac{1}{2}\right)^{t/12} = 1 \rightarrow \left(\frac{1}{2}\right)^{t/12} = \frac{1}{8} \rightarrow \left(\frac{1}{2}\right)^{t/12} = \left(\frac{1}{2}\right)^3 \rightarrow \frac{t}{12} = 3 \rightarrow t = 36$
    There will be 1 gram remaining after 36 hours.

2.11 INVERSE TRIGONOMETRIC FUNCTIONS

1. (a) $\frac{\pi}{4}$ (b) $-\frac{\pi}{3}$ (c) $\frac{\pi}{6}$ 3. (a) $-\frac{\pi}{6}$ (b) $\frac{\pi}{4}$ (c) $-\frac{\pi}{3}$

5. (a) $\frac{\pi}{3}$ (b) $\frac{3\pi}{4}$ (c) $\frac{\pi}{6}$ 7. (a) $\frac{3\pi}{4}$ (b) $\frac{\pi}{6}$ (c) $\frac{2\pi}{3}$

9. (a) $\frac{\pi}{4}$ (b) $-\frac{\pi}{3}$ (c) $\frac{\pi}{6}$ 11. (a) $\frac{3\pi}{4}$ (b) $\frac{\pi}{6}$ (c) $\frac{2\pi}{3}$

13. $\lim\limits_{x \to 1^-} \sin^{-1} x = \frac{\pi}{2}$ 15. $\lim\limits_{x \to \infty} \tan^{-1} x = \frac{\pi}{2}$

17. $\lim\limits_{x \to \infty} \sec^{-1} x = \frac{\pi}{2}$ 19. $\lim\limits_{x \to \infty} \csc^{-1} x = \lim\limits_{x \to \infty} \sin^{-1}\left(\frac{1}{x}\right) = 0$

21. $y = \cos^{-1}\left(x^2\right) \Rightarrow \frac{dy}{dx} = -\frac{2x}{\sqrt{1 - (x^2)^2}} = \frac{-2x}{\sqrt{1 - x^4}}$ 23. $y = \sin^{-1}\sqrt{2t} \Rightarrow \frac{dy}{dt} = \frac{\sqrt{2}}{\sqrt{1 - \left(\sqrt{2t}\right)^2}} = \frac{\sqrt{2}}{\sqrt{1 - 2t^2}}$

25. $y = \sec^{-1}(2s + 1) \Rightarrow \frac{dy}{ds} = \frac{2}{|2s + 1|\sqrt{(2s + 1)^2 - 1}} = \frac{2}{|2s + 1|\sqrt{4s^2 + 4s}} = \frac{1}{|2s + 1|\sqrt{s^2 + s}}$

27. $y = \csc^{-1}\left(x^2 + 1\right) \Rightarrow \frac{dy}{dx} = -\frac{2x}{|x^2 + 1|\sqrt{(x^2 + 1)^2 - 1}} = \frac{-2x}{(x^2 + 1)\sqrt{x^4 + 2x^2}}$

29. $y = \sec^{-1}\left(\frac{1}{t}\right) = \cos^{-1} t \Rightarrow \frac{dy}{dt} = \frac{-1}{\sqrt{1 - t^2}}$

31. $y = \cot^{-1}\sqrt{t} = \cot^{-1} t^{1/2} \Rightarrow \frac{dy}{dt} = -\frac{\left(\frac{1}{2}\right)t^{-1/2}}{1+(t^{1/2})^2} = \frac{-1}{2\sqrt{t}(1+t)}$

33. $y = \ln(\tan^{-1} x) \Rightarrow \frac{dy}{dx} = \frac{\left(\frac{1}{1+x^2}\right)}{\tan^{-1} x} = \frac{1}{(\tan^{-1} x)(1+x^2)}$

35. $y = \csc^{-1}(e^t) \Rightarrow \frac{dy}{dt} = -\frac{e^t}{|e^t|\sqrt{(e^t)^2-1}} = \frac{-1}{\sqrt{e^{2t}-1}}$

37. $y = s\sqrt{1-s^2} + \cos^{-1} s = s(1-s^2)^{1/2} + \cos^{-1} s \Rightarrow \frac{dy}{ds} = (1-s^2)^{1/2} + s\left(\frac{1}{2}\right)(1-s^2)^{-1/2}(-2s) - \frac{1}{\sqrt{1-s^2}}$

    $= \sqrt{1-s^2} - \frac{s^2}{\sqrt{1-s^2}} - \frac{1}{\sqrt{1-s^2}} = \sqrt{1-s^2} - \frac{s^2+1}{\sqrt{1-s^2}} = \frac{1-s^2-s^2-1}{\sqrt{1-s^2}} = \frac{-2s^2}{\sqrt{1-s^2}}$

39. $y = \tan^{-1}\sqrt{x^2-1} + \csc^{-1} x = \tan^{-1}(x^2-1)^{1/2} + \csc^{-1} x \Rightarrow \frac{dy}{dx} = \frac{\left(\frac{1}{2}\right)(x^2-1)^{-1/2}(2x)}{1+\left[(x^2-1)^{1/2}\right]^2} - \frac{1}{|x|\sqrt{x^2-1}}$

    $= \frac{1}{x\sqrt{x^2-1}} - \frac{1}{|x|\sqrt{x^2-1}} = 0$, for $x > 1$

41. $y = x\sin^{-1} x + \sqrt{1-x^2} = x\sin^{-1} x + (1-x^2)^{1/2} \Rightarrow \frac{dy}{dx} = \sin^{-1} x + x\left(\frac{1}{\sqrt{1-x^2}}\right) + \left(\frac{1}{2}\right)(1-x^2)^{-1/2}(-2x)$

    $= \sin^{-1} x + \frac{x}{\sqrt{1-x^2}} - \frac{x}{\sqrt{1-x^2}} = \sin^{-1} x$

43. The angle α is the large angle between the wall and the right end of the blackboard minus the small angle between the left end of the blackboard and the wall $\Rightarrow \alpha = \cot^{-1}\left(\frac{x}{15}\right) - \cot^{-1}\left(\frac{x}{3}\right)$.

45. If $x = 1$: $\sin^{-1}(1) + \cos^{-1}(1) = \frac{\pi}{2} + 0 = \frac{\pi}{2}$.
    If $x = 0$: $\sin^{-1}(0) + \cos^{-1}(0) = 0 + \frac{\pi}{2} = \frac{\pi}{2}$.
    If $x = -1$: $\sin^{-1}(-1) + \cos^{-1}(-1) = -\frac{\pi}{2} + \pi = \frac{\pi}{2}$.
    The identity $\sin^{-1}(x) + \cos^{-1}(x) = \frac{\pi}{2}$ has been established for x in $(0, 1)$, by Figure 1.6.7. So now if x is in $(-1, 0)$, note that $-x$ is in $(0, 1)$, and we have that
    $\sin^{-1}(x) + \cos^{-1}(x) = -\sin^{-1}(-x) + \cos^{-1}(x)$ since $\sin^{-1}$ is odd
    $\qquad\qquad\qquad\qquad\quad = -\sin^{-1}(-x) + \pi - \cos^{-1}(-x)$ by Eq. 3, Section 1.6
    $\qquad\qquad\qquad\qquad\quad = -(\sin^{-1}(-x) + \cos^{-1}(-x)) + \pi$
    $\qquad\qquad\qquad\qquad\quad = -\frac{\pi}{2} + \pi$
    $\qquad\qquad\qquad\qquad\quad = \frac{\pi}{2}$
    This establishes the identity for all x in $[-1, 1]$.

47. (a) Defined; there is an angle whose tangent is 2.
    (b) Not defined; there is no angle whose cosine is 2.
    (c) Not defined; there is no angle whose sine is $\sqrt{2}$.

49. $\csc^{-1} u = \frac{\pi}{2} - \sec^{-1} u \Rightarrow \frac{d}{dx}(\csc^{-1} u) = \frac{d}{dx}\left(\frac{\pi}{2} - \sec^{-1} u\right) = 0 - \frac{\frac{du}{dx}}{|u|\sqrt{u^2-1}} = -\frac{\frac{du}{dx}}{|u|\sqrt{u^2-1}}$, $|u| > 1$

51. $f(x) = \sec x \Rightarrow f'(x) = \sec x \tan x \Rightarrow \left.\frac{df^{-1}}{dx}\right|_{x=b} = \frac{1}{\left.\frac{df}{dx}\right|_{x=f^{-1}(b)}} = \frac{1}{\sec(\sec^{-1} b)\tan(\sec^{-1} b)} = \frac{1}{b\left(\pm\sqrt{b^2-1}\right)}$.

    Since the slope of $\sec^{-1} x$ is always positive, we the right sign by writing $\frac{d}{dx}\sec^{-1} x = \frac{1}{|x|\sqrt{x^2-1}}$.

53. (a) Domain: all real numbers except those having
the form $\frac{\pi}{2} + k\pi$ where k is an integer.
Range: $-\frac{\pi}{2} < y < \frac{\pi}{2}$

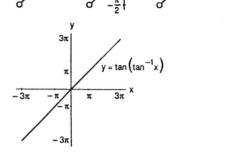

(b) Domain: $-\infty < x < \infty$; Range: $-\infty < y < \infty$
The graph of $y = \tan^{-1}(\tan x)$ is periodic, the
graph of $y = \tan(\tan^{-1} x) = x$ for $-\infty \leq x < \infty$.

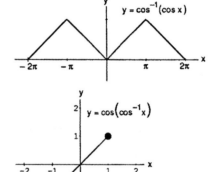

55. (a) Domain: $-\infty < x < \infty$; Range: $0 \leq y \leq \pi$

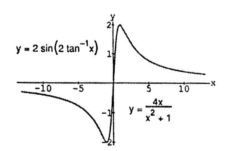

(b) Domain: $-1 \leq x \leq 1$; Range: $-1 \leq y \leq 1$
The graph of $y = \cos^{-1}(\cos x)$ is periodic; the
graph of $y = \cos(\cos^{-1} x) = x$ for $-1 \leq x \leq 1$.

57. The graphs are identical for $y = 2\sin(2\tan^{-1} x)$
$$= 4[\sin(\tan^{-1} x)][\cos(\tan^{-1} x)] = 4\left(\frac{x}{\sqrt{x^2+1}}\right)\left(\frac{1}{\sqrt{x^2+1}}\right)$$
$$= \frac{4x}{x^2+1} \text{ from the triangle}$$

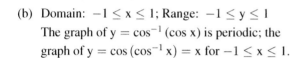

2.12 RELATED RATES

1. $A = \pi r^2 \Rightarrow \frac{dA}{dt} = 2\pi r \frac{dr}{dt}$

3. (a) $V = \pi r^2 h \Rightarrow \frac{dV}{dt} = \pi r^2 \frac{dh}{dt}$ (b) $V = \pi r^2 h \Rightarrow \frac{dV}{dt} = 2\pi r h \frac{dr}{dt}$
    (c) $V = \pi r^2 h \Rightarrow \frac{dV}{dt} = \pi r^2 \frac{dh}{dt} + 2\pi r h \frac{dr}{dt}$

5. (a) $\frac{dV}{dt} = 1$ volt/sec   (b) $\frac{dI}{dt} = -\frac{1}{3}$ amp/sec
    (c) $\frac{dV}{dt} = R\left(\frac{dI}{dt}\right) + I\left(\frac{dR}{dt}\right) \Rightarrow \frac{dR}{dt} = \frac{1}{I}\left(\frac{dV}{dt} - R\frac{dI}{dt}\right) \Rightarrow \frac{dR}{dt} = \frac{1}{I}\left(\frac{dV}{dt} - \frac{V}{I}\frac{dI}{dt}\right)$
    (d) $\frac{dR}{dt} = \frac{1}{2}\left[1 - \frac{12}{2}\left(-\frac{1}{3}\right)\right] = \left(\frac{1}{2}\right)(3) = \frac{3}{2}$ ohms/sec, R is increasing

7. (a) $s = \sqrt{x^2 + y^2} = (x^2 + y^2)^{1/2} \Rightarrow \frac{ds}{dt} = \frac{x}{\sqrt{x^2+y^2}} \frac{dx}{dt}$

   (b) $s = \sqrt{x^2 + y^2} = (x^2 + y^2)^{1/2} \Rightarrow \frac{ds}{dt} = \frac{x}{\sqrt{x^2+y^2}} \frac{dx}{dt} + \frac{y}{\sqrt{x^2+y^2}} \frac{dy}{dt}$

   (c) $s = \sqrt{x^2 + y^2} \Rightarrow s^2 = x^2 + y^2 \Rightarrow 2s \frac{ds}{dt} = 2x \frac{dx}{dt} + 2y \frac{dy}{dt} \Rightarrow 2s \cdot 0 = 2x \frac{dx}{dt} + 2y \frac{dy}{dt} \Rightarrow \frac{dx}{dt} = -\frac{y}{x} \frac{dy}{dt}$

9. (a) $A = \frac{1}{2} ab \sin \theta \Rightarrow \frac{dA}{dt} = \frac{1}{2} ab \cos \theta \frac{d\theta}{dt}$ (b) $A = \frac{1}{2} ab \sin \theta \Rightarrow \frac{dA}{dt} = \frac{1}{2} ab \cos \theta \frac{d\theta}{dt} + \frac{1}{2} b \sin \theta \frac{da}{dt}$

   (c) $A = \frac{1}{2} ab \sin \theta \Rightarrow \frac{dA}{dt} = \frac{1}{2} ab \cos \theta \frac{d\theta}{dt} + \frac{1}{2} b \sin \theta \frac{da}{dt} + \frac{1}{2} a \sin \theta \frac{db}{dt}$

11. Given $\frac{d\ell}{dt} = -2$ cm/sec, $\frac{dw}{dt} = 2$ cm/sec, $\ell = 12$ cm and $w = 5$ cm.

   (a) $A = \ell w \Rightarrow \frac{dA}{dt} = \ell \frac{dw}{dt} + w \frac{d\ell}{dt} \Rightarrow \frac{dA}{dt} = 12(2) + 5(-2) = 14$ cm^2/sec, increasing

   (b) $P = 2\ell + 2w \Rightarrow \frac{dP}{dt} = 2 \frac{d\ell}{dt} + 2 \frac{dw}{dt} = 2(-2) + 2(2) = 0$ cm/sec, constant

   (c) $D = \sqrt{w^2 + \ell^2} = (w^2 + \ell^2)^{1/2} \Rightarrow \frac{dD}{dt} = \frac{1}{2}(w^2 + \ell^2)^{-1/2}(2w \frac{dw}{dt} + 2\ell \frac{d\ell}{dt}) \Rightarrow \frac{dD}{dt} = \frac{w \frac{dw}{dt} + \ell \frac{d\ell}{dt}}{\sqrt{w^2 + \ell^2}}$

   $= \frac{(5)(2) + (12)(-2)}{\sqrt{25 + 144}} = -\frac{14}{13}$ cm/sec, decreasing

13. Given: $\frac{dx}{dt} = 5$ ft/sec, the ladder is 13 ft long, and $x = 12$, $y = 5$ at the instant of time

   (a) Since $x^2 + y^2 = 169 \Rightarrow \frac{dy}{dt} = -\frac{x}{y} \frac{dx}{dt} = -\left(\frac{12}{5}\right)(5) = -12$ ft/sec, the ladder is sliding down the wall

   (b) The area of the triangle formed by the ladder and walls is $A = \frac{1}{2} xy \Rightarrow \frac{dA}{dt} = \left(\frac{1}{2}\right)\left(x \frac{dy}{dt} + y \frac{dx}{dt}\right)$. The area

   is changing at $\frac{1}{2}[12(-12) + 5(5)] = -\frac{119}{2} = -59.5$ ft^2/sec.

   (c) $\cos \theta = \frac{x}{13} \Rightarrow -\sin \theta \frac{d\theta}{dt} = \frac{1}{13} \cdot \frac{dx}{dt} \Rightarrow \frac{d\theta}{dt} = -\frac{1}{13 \sin \theta} \cdot \frac{dx}{dt} = -\left(\frac{1}{5}\right)(5) = -1$ rad/sec

15. Let s represent the distance between the girl and the kite and x represents the horizontal distance between the girl and kite

   $\Rightarrow s^2 = (300)^2 + x^2 \Rightarrow \frac{ds}{dt} = \frac{x}{s} \frac{dx}{dt} = \frac{400(25)}{500} = 20$ ft/sec.

17. $V = \frac{1}{3} \pi r^2 h$, $h = \frac{3}{8}(2r) = \frac{3r}{4} \Rightarrow r = \frac{4h}{3} \Rightarrow V = \frac{1}{3} \pi \left(\frac{4h}{3}\right)^2 h = \frac{16\pi h^3}{27} \Rightarrow \frac{dV}{dt} = \frac{16\pi h^2}{9} \frac{dh}{dt}$

   (a) $\frac{dh}{dt}\big|_{h=4} = \left(\frac{9}{16\pi 4^2}\right)(10) = \frac{90}{256\pi} \approx 0.1119$ m/sec $= 11.19$ cm/sec

   (b) $r = \frac{4h}{3} \Rightarrow \frac{dr}{dt} = \frac{4}{3} \frac{dh}{dt} = \frac{4}{3}\left(\frac{90}{256\pi}\right) = \frac{15}{32\pi} \approx 0.1492$ m/sec $= 14.92$ cm/sec

19. (a) $V = \frac{\pi}{3} y^2 (3R - y) \Rightarrow \frac{dV}{dt} = \frac{\pi}{3}[2y(3R - y) + y^2(-1)] \frac{dy}{dt} \Rightarrow \frac{dy}{dt} = \left[\frac{\pi}{3}(6Ry - 3y^2)\right]^{-1} \frac{dV}{dt} \Rightarrow$ at $R = 13$ and

   $y = 8$ we have $\frac{dy}{dt} = \frac{1}{144\pi}(-6) = \frac{-1}{24\pi}$ m/min

   (b) The hemisphere is on the circle $r^2 + (13 - y)^2 = 169 \Rightarrow r = \sqrt{26y - y^2}$ m

   (c) $r = (26y - y^2)^{1/2} \Rightarrow \frac{dr}{dt} = \frac{1}{2}(26y - y^2)^{-1/2}(26 - 2y) \frac{dy}{dt} \Rightarrow \frac{dr}{dt} = \frac{13 - y}{\sqrt{26y - y^2}} \frac{dy}{dt} \Rightarrow \frac{dr}{dt}\big|_{y=8} = \frac{13 - 8}{\sqrt{26 \cdot 8 - 64}}\left(\frac{-1}{24\pi}\right)$

   $= \frac{-5}{288\pi}$ m/min

21. If $V = \frac{4}{3} \pi r^3$, $r = 5$, and $\frac{dV}{dt} = 100\pi$ ft^3/min, then $\frac{dV}{dt} = 4\pi r^2 \frac{dr}{dt} \Rightarrow \frac{dr}{dt} = 1$ ft/min. Then $S = 4\pi r^2 \Rightarrow \frac{dS}{dt}$

   $= 8\pi r \frac{dr}{dt} = 8\pi(5)(1) = 40\pi$ ft^2/min, the rate at which the surface area is increasing.

23. Let s represent the distance between the bicycle and balloon, h the height of the balloon and x the horizontal

   distance between the balloon and the bicycle. The relationship between the variables is $s^2 = h^2 + x^2$

   $\Rightarrow \frac{ds}{dt} = \frac{1}{s}\left(h \frac{dh}{dt} + x \frac{dx}{dt}\right) \Rightarrow \frac{ds}{dt} = \frac{1}{85}[68(1) + 51(17)] = 11$ ft/sec.

25. Let $P(x, y)$ represent a point on the curve $y = x^2$ and θ the angle of inclination of a line containing P and the

   origin. Consequently, $\tan \theta = \frac{y}{x} \Rightarrow \tan \theta = \frac{x^2}{x} = x \Rightarrow \sec^2 \theta \frac{d\theta}{dt} = \frac{dx}{dt} \Rightarrow \frac{d\theta}{dt} = \cos^2 \theta \frac{dx}{dt}$. Since $\frac{dx}{dt} = 10$ m/sec

   and $\cos^2 \theta\big|_{x=3} = \frac{x^2}{y^2 + x^2} = \frac{3^2}{9^2 + 3^2} = \frac{1}{10}$, we have $\frac{d\theta}{dt}\big|_{x=3} = 1$ rad/sec.

27. The distance from the origin is s $= \sqrt{x^2 + y^2}$ and we wish to find $\left.\frac{ds}{dt}\right|_{(5,12)}$

$= \frac{1}{2}\left(x^2 + y^2\right)^{-1/2}\left(2x\,\frac{dx}{dt} + 2y\,\frac{dy}{dt}\right)\Big|_{(5,12)} = \frac{(5)(-1) + (12)(-5)}{\sqrt{25 + 144}} = -5$ m/sec

29. Let s $= 16t^2$ represent the distance the ball has fallen, h the distance between the ball and the ground, and I the distance between the shadow and the point directly beneath the ball. Accordingly, s $+$ h $= 50$ and since the triangle LOQ and triangle PRQ are similar we have

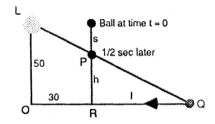

Ball at time t = 0
1/2 sec later

I $= \frac{30h}{50 - h}$ $\Rightarrow$ h $= 50 - 16t^2$ and I $= \frac{30\left(50 - 16t^2\right)}{50 - \left(50 - 16t^2\right)}$

$= \frac{1500}{16t^2} - 30$ $\Rightarrow$ $\frac{dI}{dt} = -\frac{1500}{8t^3}$ $\Rightarrow$ $\left.\frac{dI}{dt}\right|_{t=\frac{1}{2}} = -1500$ ft/sec.

31. The volume of the ice is V $= \frac{4}{3}\pi r^3 - \frac{4}{3}\pi 4^3$ $\Rightarrow$ $\frac{dV}{dt} = 4\pi r^2\,\frac{dr}{dt}$ $\Rightarrow$ $\left.\frac{dr}{dt}\right|_{r=6} = \frac{-5}{72\pi}$ in./min when $\frac{dV}{dt} = -10$ in³/min, the thickness of the ice is decreasing at $\frac{5}{72\pi}$ in/min. The surface area is S $= 4\pi r^2$ $\Rightarrow$ $\frac{dS}{dt} = 8\pi r\,\frac{dr}{dt}$ $\Rightarrow$ $\left.\frac{dS}{dt}\right|_{r=6} = 48\pi\left(\frac{-5}{72\pi}\right)$ $= -\frac{10}{3}$ in²/min, the outer surface area of the ice is decreasing at $\frac{10}{3}$ in²/min.

33. When x represents the length of the shadow, then $\tan\theta = \frac{80}{x}$ $\Rightarrow$ $\sec^2\theta\,\frac{d\theta}{dt} = -\frac{80}{x^2}\,\frac{dx}{dt}$ $\Rightarrow$ $\frac{dx}{dt} = \frac{-x^2\sec^2\theta}{80}\,\frac{d\theta}{dt}$. We are given that $\frac{d\theta}{dt} = 0.27° = \frac{3\pi}{2000}$ rad/min. At x $= 60$, $\cos\theta = \frac{3}{5}$ $\Rightarrow$

$\left|\frac{dx}{dt}\right| = \left|\frac{-x^2\sec^2\theta}{80}\,\frac{d\theta}{dt}\right|_{\left(\frac{d\theta}{dt} = \frac{3\pi}{2000}\text{ and }\sec\theta = \frac{5}{3}\right)} = \frac{3\pi}{16}$ ft/min ≈ 0.589 ft/min ≈ 7.1 in./min.

2.13 LINEARIZATION AND DIFFERENTIALS

1. f(x) $= x^3 - 2x + 3$ $\Rightarrow$ f'(x) $= 3x^2 - 2$ $\Rightarrow$ L(x) $=$ f'(2)(x $- 2) + $ f(2) $= 10$(x $- 2) + 7$ $\Rightarrow$ L(x) $= 10$x $- 13$ at x $= 2$

3. f(x) $= x + \frac{1}{x}$ $\Rightarrow$ f'(x) $= 1 - x^{-2}$ $\Rightarrow$ L(x) $=$ f(1) $+$ f'(1)(x $- 1) = 2 + 0$(x $- 1) = 2$

5. f(x) $= \tan x$ $\Rightarrow$ f'(x) $= \sec^2 x$ $\Rightarrow$ L(x) $=$ f(π) $+$ f'(π)(x $- \pi) = 0 + 1$(x $- \pi) = x - \pi$

7. f(x) $= x^2 + 2x$ $\Rightarrow$ f'(x) $= 2x + 2$ $\Rightarrow$ L(x) $=$ f'(0)(x $- 0) + $ f(0) $= 2$(x $- 0) + 0$ $\Rightarrow$ L(x) $= 2$x at x $= 0$

9. f(x) $= 2x^2 + 4x - 3$ $\Rightarrow$ f'(x) $= 4x + 4$ $\Rightarrow$ L(x) $=$ f'(-1)(x $+ 1) + $ f(-1) $= 0$(x $+ 1) + (-5)$ $\Rightarrow$ L(x) $= -5$ at x $= -1$

11. f(x) $= \sqrt[3]{x} = x^{1/3}$ $\Rightarrow$ f'(x) $= \left(\frac{1}{3}\right)x^{-2/3}$ $\Rightarrow$ L(x) $=$ f'(8)(x $- 8) + $ f(8) $= \frac{1}{12}$(x $- 8) + 2$ $\Rightarrow$ L(x) $= \frac{1}{12}$x $+ \frac{4}{3}$ at x $= 8$

13. f(x) $= e^{-x}$ $\Rightarrow$ f'(x) $= -e^{-x}$ $\Rightarrow$ L(x) $=$ f(0) $+$ f'(0)(x $- 0) = -x + 1$

15. f'(x) $= k(1 + x)^{k-1}$. We have f(0) $= 1$ and f'(0) $= k$. L(x) $=$ f(0) $+$ f'(0)(x $- 0) = 1 + k$(x $- 0) = 1 + kx$

17. (a) $(1.0002)^{50} = (1 + 0.0002)^{50} \approx 1 + 50(0.0002) = 1 + .01 = 1.01$
    (b) $\sqrt[3]{1.009} = (1 + 0.009)^{1/3} \approx 1 + \left(\frac{1}{3}\right)(0.009) = 1 + 0.003 = 1.003$

19. y $= x^3 - 3\sqrt{x} = x^3 - 3x^{1/2}$ $\Rightarrow$ dy $= \left(3x^2 - \frac{3}{2}x^{-1/2}\right)$ dx $\Rightarrow$ dy $= \left(3x^2 - \frac{3}{2\sqrt{x}}\right)$ dx

21. $2y^{3/2} + xy - x = 0$ $\Rightarrow$ $3y^{1/2}$ dy $+$ y dx $+$ x dy $-$ dx $= 0$ $\Rightarrow$ $\left(3y^{1/2} + x\right)$ dy $= (1 - y)$ dx $\Rightarrow$ dy $= \frac{1 - y}{3\sqrt{y} + x}$ dx

23. $y = \sin\left(5\sqrt{x}\right) = \sin\left(5x^{1/2}\right) \Rightarrow dy = \left(\cos\left(5x^{1/2}\right)\right)\left(\frac{5}{2}x^{-1/2}\right) dx \Rightarrow dy = \frac{5\cos\left(5\sqrt{x}\right)}{2\sqrt{x}} dx$

25. $y = e^{\sqrt{x}} \Rightarrow dy = \frac{e^{\sqrt{x}}}{2\sqrt{x}} dx$

27. $y = \ln(1 + x^2) \Rightarrow dy = \frac{2x}{1+x^2} dx$

29. $f(x) = x^2 + 2x, \ x_0 = 1, \ dx = 0.1 \Rightarrow f'(x) = 2x + 2$
    (a) $\Delta f = f(x_0 + dx) - f(x_0) = f(1.1) - f(1) = 3.41 - 3 = 0.41$
    (b) $df = f'(x_0)\,dx = [2(1) + 2](0.1) = 0.4$
    (c) $|\Delta f - df| = |0.41 - 0.4| = 0.01$

31. $f(x) = x^{-1}, \ x_0 = 0.5, \ dx = 0.1 \Rightarrow f'(x) = -x^{-2}$
    (a) $\Delta f = f(x_0 + dx) - f(x_0) = f(.6) - f(.5) = -\frac{1}{3}$
    (b) $df = f'(x_0)\,dx = (-4)\left(\frac{1}{10}\right) = -\frac{2}{5}$
    (c) $|\Delta f - df| = \left|-\frac{1}{3} + \frac{2}{5}\right| = \frac{1}{15}$

33. $V = \frac{4}{3}\pi r^3 \Rightarrow dV = 4\pi r_0^2\,dr$

35. $V = \pi r^2 h,$ height constant $\Rightarrow dV = 2\pi r_0 h\,dr$

37. Given $r = 2$ m, $dr = .02$ m
    (a) $A = \pi r^2 \Rightarrow dA = 2\pi r\,dr = 2\pi(2)(.02) = .08\pi$ m^2
    (b) $\left(\frac{.08\pi}{4\pi}\right)(100\%) = 2\%$

39. The volume of a cylinder is $V = \pi r^2 h$. When h is held fixed, we have $\frac{dV}{dr} = 2\pi rh$, and so $dV = 2\pi rh\,dr$. For $h = 30$ in., $r = 6$ in., and $dr = 0.5$ in., the volume of the material in the shell is approximately $dV = 2\pi rh\,dr = 2\pi(6)(30)(0.5)$ $= 180\pi \approx 565.5\,\text{in}^3$.

41. $V = \pi h^3 \Rightarrow dV = 3\pi h^2\,dh;$ recall that $\Delta V \approx dV$. Then $|\Delta V| \leq (1\%)(V) = \frac{(1)\,(\pi h^3)}{100} \Rightarrow |dV| \leq \frac{(1)\,(\pi h^3)}{100}$
    $\Rightarrow |3\pi h^2\,dh| \leq \frac{(1)\,(\pi h^3)}{100} \Rightarrow |dh| \leq \frac{1}{300}\,h = \left(\frac{1}{3}\%\right)h$. Therefore the greatest tolerated error in the measurement of h is $\frac{1}{3}\%$.

43. $V = \pi r^2 h,\ h$ is constant $\Rightarrow dV = 2\pi rh\,dr;$ recall that $\Delta V \approx dV$. We want $|\Delta V| \leq \frac{1}{1000} V \Rightarrow |dV| \leq \frac{\pi r^2 h}{1000}$
    $\Rightarrow |2\pi rh\,dr| \leq \frac{\pi r^2 h}{1000} \Rightarrow |dr| \leq \frac{r}{2000} = (.05\%)r \Rightarrow$ a .05% variation in the radius can be tolerated.

45. The error in measurement $dx = (1\%)(10) = 0.1$ cm; $V = x^3 \Rightarrow dV = 3x^2\,dx = 3(10)^2(0.1) = 30\,\text{cm}^3 \Rightarrow$ the percentage error in the volume calculation is $\left(\frac{30}{1000}\right)(100\%) = 3\%$

47. Given $D = 100$ cm, $dD = 1$ cm, $V = \frac{4}{3}\pi\left(\frac{D}{2}\right)^3 = \frac{\pi D^3}{6} \Rightarrow dV = \frac{\pi}{2}D^2\,dD = \frac{\pi}{2}(100)^2(1) = \frac{10^4\pi}{2}$. Then $\frac{dV}{V}(100\%)$
    $= \left[\frac{\frac{10^4\pi}{2}}{\frac{10^6\pi}{6}}\right](10^2\%) = \left[\frac{\frac{10^6\pi}{2}}{\frac{10^6\pi}{6}}\right]\% = 3\%$

49. $E(x) = f(x) - g(x) \Rightarrow E(x) = f(x) - m(x - a) - c$. Then $E(a) = 0 \Rightarrow f(a) - m(a - a) - c = 0 \Rightarrow c = f(a)$. Next we calculate m: $\lim\limits_{x \to a} \frac{E(x)}{x - a} = 0 \Rightarrow \lim\limits_{x \to a} \frac{f(x) - m(x - a) - c}{x - a} = 0 \Rightarrow \lim\limits_{x \to a}\left[\frac{f(x) - f(a)}{x - a} - m\right] = 0$ (since $c = f(a)$)
    $\Rightarrow f'(a) - m = 0 \Rightarrow m = f'(a)$. Therefore, $g(x) = m(x - a) + c = f'(a)(x - a) + f(a)$ is the linear approximation, as claimed.

51. (a) $f(x) = 2^x \Rightarrow f'(x) = 2^x \ln 2$; $L(x) = (2^0 \ln 2) x + 2^0 = x \ln 2 + 1 \approx 0.69x + 1$

(b)

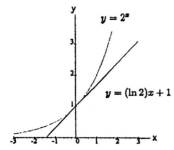   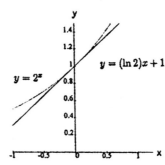

CHAPTER 2 PRACTICE AND ADDITIONAL EXERCISES

1. $y = x^5 - 0.125x^2 + 0.25x \Rightarrow \frac{dy}{dx} = 5x^4 - 0.25x + 0.25$

3. $y = x^3 - 3(x^2 + \pi^2) \Rightarrow \frac{dy}{dx} = 3x^2 - 3(2x + 0) = 3x^2 - 6x = 3x(x - 2)$

5. $y = (x + 1)^2 (x^2 + 2x) \Rightarrow \frac{dy}{dx} = (x + 1)^2(2x + 2) + (x^2 + 2x)(2(x + 1)) = 2(x + 1) [(x + 1)^2 + x(x + 2)]$
    $= 2(x + 1)(2x^2 + 4x + 1)$

7. $y = (\theta^2 + \sec\theta + 1)^3 \Rightarrow \frac{dy}{d\theta} = 3(\theta^2 + \sec\theta + 1)^2(2\theta + \sec\theta\tan\theta)$

9. $s = \frac{\sqrt{t}}{1 + \sqrt{t}} \Rightarrow \frac{ds}{dt} = \frac{(1 + \sqrt{t}) \cdot \frac{1}{2\sqrt{t}} - \sqrt{t}\left(\frac{1}{2\sqrt{t}}\right)}{(1 + \sqrt{t})^2} = \frac{(1 + \sqrt{t}) - \sqrt{t}}{2\sqrt{t}(1 + \sqrt{t})^2} = \frac{1}{2\sqrt{t}(1 + \sqrt{t})^2}$

11. $y = 2\tan^2 x - \sec^2 x \Rightarrow \frac{dy}{dx} = (4\tan x)(\sec^2 x) - (2\sec x)(\sec x \tan x) = 2\sec^2 x \tan x$

13. $s = \cos^4(1 - 2t) \Rightarrow \frac{ds}{dt} = 4\cos^3(1 - 2t)(-\sin(1 - 2t))(-2) = 8\cos^3(1 - 2t)\sin(1 - 2t)$

15. $s = (\sec t + \tan t)^5 \Rightarrow \frac{ds}{dt} = 5(\sec t + \tan t)^4(\sec t \tan t + \sec^2 t) = 5(\sec t)(\sec t + \tan t)^5$

17. $r = \sqrt{2\theta\sin\theta} = (2\theta\sin\theta)^{1/2} \Rightarrow \frac{dr}{d\theta} = \frac{1}{2}(2\theta\sin\theta)^{-1/2}(2\theta\cos\theta + 2\sin\theta) = \frac{\theta\cos\theta + \sin\theta}{\sqrt{2\theta\sin\theta}}$

19. $r = \sin\sqrt{2\theta} = \sin(2\theta)^{1/2} \Rightarrow \frac{dr}{d\theta} = \cos(2\theta)^{1/2}\left(\frac{1}{2}(2\theta)^{-1/2}(2)\right) = \frac{\cos\sqrt{2\theta}}{\sqrt{2\theta}}$

21. $y = \frac{1}{2}x^2 \csc\frac{2}{x} \Rightarrow \frac{dy}{dx} = \frac{1}{2}x^2\left(-\csc\frac{2}{x}\cot\frac{2}{x}\right)\left(\frac{-2}{x^2}\right) + \left(\csc\frac{2}{x}\right)\left(\frac{1}{2}\cdot 2x\right) = \csc\frac{2}{x}\cot\frac{2}{x} + x\csc\frac{2}{x}$

23. $y = x^{-1/2}\sec(2x)^2 \Rightarrow \frac{dy}{dx} = x^{-1/2}\sec(2x)^2\tan(2x)^2(2(2x)\cdot 2) + \sec(2x)^2\left(-\frac{1}{2}x^{-3/2}\right)$
    $= 8x^{1/2}\sec(2x)^2\tan(2x)^2 - \frac{1}{2}x^{-3/2}\sec(2x)^2 = \frac{1}{2}x^{1/2}\sec(2x)^2[16\tan(2x)^2 - x^{-2}]$ or $\frac{1}{2x^{3/2}}\sec(2x)^2[16x^2\tan(2x)^2 - 1]$

25. $y = 5\cot x^2 \Rightarrow \frac{dy}{dx} = 5(-\csc^2 x^2)(2x) = -10x\csc^2(x^2)$

27. $y = x^2\sin^2(2x^2) \Rightarrow \frac{dy}{dx} = x^2(2\sin(2x^2))(\cos(2x^2))(4x) + \sin^2(2x^2)(2x) = 8x^3\sin(2x^2)\cos(2x^2) + 2x\sin^2(2x^2)$

29. $s = \left(\frac{4t}{t+1}\right)^{-2} \Rightarrow \frac{ds}{dt} = -2\left(\frac{4t}{t+1}\right)^{-3}\left(\frac{(t+1)(4) - (4t)(1)}{(t+1)^2}\right) = -2\left(\frac{4t}{t+1}\right)^{-3}\frac{4}{(t+1)^2} = -\frac{(t+1)}{8t^3}$

31. $y = \left(\frac{\sqrt{x}}{x+1}\right)^2 \Rightarrow \frac{dy}{dx} = 2\left(\frac{\sqrt{x}}{x+1}\right) \cdot \frac{(x+1)\left(\frac{1}{2\sqrt{x}}\right) - (\sqrt{x})(1)}{(x+1)^2} = \frac{(x+1) - 2x}{(x+1)^3} = \frac{1-x}{(x+1)^3}$

33. $y = \sqrt{\frac{x^2+x}{x^2}} = \left(1 + \frac{1}{x}\right)^{1/2} \Rightarrow \frac{dy}{dx} = \frac{1}{2}\left(1 + \frac{1}{x}\right)^{-1/2}\left(-\frac{1}{x^2}\right) = -\frac{1}{2x^2\sqrt{1 + \frac{1}{x}}}$

35. $r = \left(\frac{\sin\theta}{\cos\theta - 1}\right)^2 \Rightarrow \frac{dr}{d\theta} = 2\left(\frac{\sin\theta}{\cos\theta - 1}\right)\left[\frac{(\cos\theta - 1)(\cos\theta) - (\sin\theta)(-\sin\theta)}{(\cos\theta - 1)^2}\right]$

$= 2\left(\frac{\sin\theta}{\cos\theta - 1}\right)\left(\frac{\cos^2\theta - \cos\theta + \sin^2\theta}{(\cos\theta - 1)^2}\right) = \frac{(2\sin\theta)(1 - \cos\theta)}{(\cos\theta - 1)^3} = \frac{-2\sin\theta}{(\cos\theta - 1)^2}$

37. $y = (2x + 1)\sqrt{2x + 1} = (2x + 1)^{3/2} \Rightarrow \frac{dy}{dx} = \frac{3}{2}(2x + 1)^{1/2}(2) = 3\sqrt{2x + 1}$

39. $y = 3(5x^2 + \sin 2x)^{-3/2} \Rightarrow \frac{dy}{dx} = 3\left(-\frac{3}{2}\right)(5x^2 + \sin 2x)^{-5/2}[10x + (\cos 2x)(2)] = \frac{-9(5x + \cos 2x)}{(5x^2 + \sin 2x)^{5/2}}$

41. $y = 10e^{-x/5} \Rightarrow \frac{dy}{dx} = (10)\left(-\frac{1}{5}\right)e^{-x/5} = -2e^{-x/5}$

43. $y = \frac{1}{4}xe^{4x} - \frac{1}{16}e^{4x} \Rightarrow \frac{dy}{dx} = \frac{1}{4}[x(4e^{4x}) + e^{4x}(1)] - \frac{1}{16}(4e^{4x}) = xe^{4x} + \frac{1}{4}e^{4x} - \frac{1}{4}e^{4x} = xe^{4x}$

45. $y = \ln(\sin^2\theta) \Rightarrow \frac{dy}{d\theta} = \frac{2(\sin\theta)(\cos\theta)}{\sin^2\theta} = \frac{2\cos\theta}{\sin\theta} = 2\cot\theta$

47. $y = \log_2\left(\frac{x^2}{2}\right) = \frac{\ln\left(\frac{x^2}{2}\right)}{\ln 2} \Rightarrow \frac{dy}{dx} = \frac{1}{\ln 2}\left(\frac{x}{\left(\frac{x^2}{2}\right)}\right) = \frac{2}{(\ln 2)x}$

49. $y = 8^{-t} \Rightarrow \frac{dy}{dt} = 8^{-t}(\ln 8)(-1) = -8^{-t}(\ln 8)$ 51. $y = 5x^{3.6} \Rightarrow \frac{dy}{dx} = 5(3.6)x^{2.6} = 18x^{2.6}$

53. $y = (x + 2)^{x+2} \Rightarrow \ln y = \ln(x + 2)^{x+2} = (x + 2)\ln(x + 2) \Rightarrow \frac{y'}{y} = (x + 2)\left(\frac{1}{x+2}\right) + (1)\ln(x + 2)$

$\Rightarrow \frac{dy}{dx} = (x + 2)^{x+2}[\ln(x + 2) + 1]$

55. $y = \sin^{-1}\sqrt{1 - u^2} = \sin^{-1}(1 - u^2)^{1/2} \Rightarrow \frac{dy}{du} = \frac{\frac{1}{2}(1 - u^2)^{-1/2}(-2u)}{\sqrt{1 - \left[(1 - u^2)^{1/2}\right]^2}} = \frac{-u}{\sqrt{1 - u^2}\sqrt{1 - (1 - u^2)}} = \frac{-u}{|u|\sqrt{1 - u^2}}$

$= \frac{-u}{u\sqrt{1 - u^2}} = \frac{-1}{\sqrt{1 - u^2}}, 0 < u < 1$

57. $y = \ln(\cos^{-1}x) \Rightarrow y' = \frac{\left(\frac{-1}{\sqrt{1 - x^2}}\right)}{\cos^{-1}x} = \frac{-1}{\sqrt{1 - x^2}\cos^{-1}x}$

59. $y = t\tan^{-1}t - \left(\frac{1}{2}\right)\ln t \Rightarrow \frac{dy}{dt} = \tan^{-1}t + t\left(\frac{1}{1+t^2}\right) - \left(\frac{1}{2}\right)\left(\frac{1}{t}\right) = \tan^{-1}t + \frac{t}{1+t^2} - \frac{1}{2t}$

61. $y = z\sec^{-1}z - \sqrt{z^2 - 1} = z\sec^{-1}z - (z^2 - 1)^{1/2} \Rightarrow \frac{dy}{dz} = z\left(\frac{1}{|z|\sqrt{z^2 - 1}}\right) + (\sec^{-1}z)(1) - \frac{1}{2}(z^2 - 1)^{-1/2}(2z)$

$= \frac{z}{|z|\sqrt{z^2 - 1}} - \frac{z}{\sqrt{z^2 - 1}} + \sec^{-1}z = \frac{1-z}{\sqrt{z^2 - 1}} + \sec^{-1}z, z > 1$

63. $y = \csc^{-1}(\sec\theta) \Rightarrow \frac{dy}{d\theta} = \frac{-\sec\theta\tan\theta}{|\sec\theta|\sqrt{\sec^2\theta - 1}} = -\frac{\tan\theta}{|\tan\theta|} = -1, 0 < \theta < \frac{\pi}{2}$

65. $xy + 2x + 3y = 1 \Rightarrow (xy' + y) + 2 + 3y' = 0 \Rightarrow xy' + 3y' = -2 - y \Rightarrow y'(x + 3) = -2 - y \Rightarrow y' = -\frac{y+2}{x+3}$

67. $x^3 + 4xy - 3y^{4/3} = 2x \Rightarrow 3x^2 + \left(4x\frac{dy}{dx} + 4y\right) - 4y^{1/3}\frac{dy}{dx} = 2 \Rightarrow 4x\frac{dy}{dx} - 4y^{1/3}\frac{dy}{dx} = 2 - 3x^2 - 4y$

$\Rightarrow \frac{dy}{dx}\left(4x - 4y^{1/3}\right) = 2 - 3x^2 - 4y \Rightarrow \frac{dy}{dx} = \frac{2 - 3x^2 - 4y}{4x - 4y^{1/3}}$

69. $(xy)^{1/2} = 1 \Rightarrow \frac{1}{2}(xy)^{-1/2}\left(x\frac{dy}{dx} + y\right) = 0 \Rightarrow x^{1/2}y^{-1/2}\frac{dy}{dx} = -x^{-1/2}y^{1/2} \Rightarrow \frac{dy}{dx} = -x^{-1}y \Rightarrow \frac{dy}{dx} = -\frac{y}{x}$

71. $y^2 = \frac{x}{x+1} \Rightarrow 2y\frac{dy}{dx} = \frac{(x+1)(1) - (x)(1)}{(x+1)^2} \Rightarrow \frac{dy}{dx} = \frac{1}{2y(x+1)^2}$

73. $e^{x+2y} = 1 \Rightarrow e^{x+2y}\left(1 + 2\frac{dy}{dx}\right) = 0 \Rightarrow \frac{dy}{dx} = -\frac{1}{2}$

75. $\ln\left(\frac{x}{y}\right) = 1 \Rightarrow \frac{1}{x/y}\frac{d}{dx}\left(\frac{x}{y}\right) = 0 \Rightarrow \frac{y(1) - x\frac{dy}{dx}}{y^2} = 0 \Rightarrow \frac{dy}{dx} = \frac{y}{x}$

77. $y e^{\tan^{-1}x} = 2 \Rightarrow y = 2e^{-\tan^{-1}x} \Rightarrow \frac{dy}{dx} = 2e^{-\tan^{-1}x}\frac{d}{dx}(-\tan^{-1}x) = -2e^{-\tan^{-1}x}\left(\frac{1}{1+x^2}\right) = -\frac{2e^{-\tan^{-1}x}}{1+x^2}$

79. $p^3 + 4pq - 3q^2 = 2 \Rightarrow 3p^2\frac{dp}{dq} + 4\left(p + q\frac{dp}{dq}\right) - 6q = 0 \Rightarrow 3p^2\frac{dp}{dq} + 4q\frac{dp}{dq} = 6q - 4p \Rightarrow \frac{dp}{dq}(3p^2 + 4q) = 6q - 4p$

$\Rightarrow \frac{dp}{dq} = \frac{6q - 4p}{3p^2 + 4q}$

81. $r\cos 2s + \sin^2 s = \pi \Rightarrow r(-\sin 2s)(2) + (\cos 2s)\left(\frac{dr}{ds}\right) + 2\sin s\cos s = 0 \Rightarrow \frac{dr}{ds}(\cos 2s) = 2r\sin 2s - 2\sin s\cos s$

$\Rightarrow \frac{dr}{ds} = \frac{2r\sin 2s - \sin 2s}{\cos 2s} = \frac{(2r-1)(\sin 2s)}{\cos 2s} = (2r-1)(\tan 2s)$

83. (a) $x^3 + y^3 = 1 \Rightarrow 3x^2 + 3y^2\frac{dy}{dx} = 0 \Rightarrow \frac{dy}{dx} = -\frac{x^2}{y^2} \Rightarrow \frac{d^2y}{dx^2} = \frac{y^2(-2x) - (-x^2)\left(2y\frac{dy}{dx}\right)}{y^4}$

$\Rightarrow \frac{d^2y}{dx^2} = \frac{-2xy^2 + (2yx^2)\left(-\frac{x^2}{y^2}\right)}{y^4} = \frac{-2xy^2 - \frac{2x^4}{y}}{y^4} = \frac{-2xy^3 - 2x^4}{y^5}$

(b) $y^2 = 1 - \frac{2}{x} \Rightarrow 2y\frac{dy}{dx} = \frac{2}{x^2} \Rightarrow \frac{dy}{dx} = \frac{1}{yx^2} \Rightarrow \frac{dy}{dx} = (yx^2)^{-1} \Rightarrow \frac{d^2y}{dx^2} = -(yx^2)^{-2}\left[y(2x) + x^2\frac{dy}{dx}\right]$

$\Rightarrow \frac{d^2y}{dx^2} = \frac{-2xy - x^2\left(\frac{1}{yx^2}\right)}{y^2x^4} = \frac{-2xy^2 - 1}{y^3x^4}$

85. (a) Let $h(x) = 6f(x) - g(x) \Rightarrow h'(x) = 6f'(x) - g'(x) \Rightarrow h'(1) = 6f'(1) - g'(1) = 6\left(\frac{1}{2}\right) - (-4) = 7$

(b) Let $h(x) = f(x)g^2(x) \Rightarrow h'(x) = f(x)(2g(x))g'(x) + g^2(x)f'(x) \Rightarrow h'(0) = 2f(0)g(0)g'(0) + g^2(0)f'(0)$

$= 2(1)(1)\left(\frac{1}{2}\right) + (1)^2(-3) = -2$

(c) Let $h(x) = \frac{f(x)}{g(x)+1} \Rightarrow h'(x) = \frac{(g(x)+1)f'(x) - f(x)g'(x)}{(g(x)+1)^2} \Rightarrow h'(1) = \frac{(g(1)+1)f'(1) - f(1)g'(1)}{(g(1)+1)^2} = \frac{(5+1)\left(\frac{1}{2}\right) - 3(-4)}{(5+1)^2} = \frac{5}{12}$

(d) Let $h(x) = f(g(x)) \Rightarrow h'(x) = f'(g(x))g'(x) \Rightarrow h'(0) = f'(g(0))g'(0) = f'(1)\left(\frac{1}{2}\right) = \left(\frac{1}{2}\right)\left(\frac{1}{2}\right) = \frac{1}{4}$

(e) Let $h(x) = g(f(x)) \Rightarrow h'(x) = g'(f(x))f'(x) \Rightarrow h'(0) = g'(f(0))f'(0) = g'(1)f'(0) = (-4)(-3) = 12$

(f) Let $h(x) = (x + f(x))^{3/2} \Rightarrow h'(x) = \frac{3}{2}(x + f(x))^{1/2}(1 + f'(x)) \Rightarrow h'(1) = \frac{3}{2}(1 + f(1))^{1/2}(1 + f'(1))$

$= \frac{3}{2}(1 + 3)^{1/2}\left(1 + \frac{1}{2}\right) = \frac{9}{2}$

(g) Let $h(x) = f(x + g(x)) \Rightarrow h'(x) = f'(x + g(x))(1 + g'(x)) \Rightarrow h'(0) = f'(g(0))(1 + g'(0)) = f'(1)\left(1 + \frac{1}{2}\right) = \left(\frac{1}{2}\right)\left(\frac{3}{2}\right)$

$= \frac{3}{4}$

87. $x = t^2 + \pi \Rightarrow \frac{dx}{dt} = 2t; \; y = 3\sin 2x \Rightarrow \frac{dy}{dx} = 3(\cos 2x)(2) = 6\cos 2x = 6\cos(2t^2 + 2\pi) = 6\cos(2t^2)$; thus,

$\frac{dy}{dt} = \frac{dy}{dx} \cdot \frac{dx}{dt} = 6\cos(2t^2) \cdot 2t \Rightarrow \frac{dy}{dt}\Big|_{t=0} = 6\cos(0) \cdot 0 = 0$

89. $\frac{dw}{ds} = \frac{dw}{dr} \cdot \frac{dr}{ds} = \left[\cos\left(e^{\sqrt{r}}\right)\left(e^{\sqrt{r}}\frac{1}{2\sqrt{r}}\right)\right]\left[3\cos\left(s + \frac{\pi}{6}\right)\right]$ at $x = 0$, $r = 3\sin\frac{\pi}{6} = \frac{3}{2}$

$\Rightarrow \frac{dw}{ds} = \cos\left(e^{\sqrt{3/2}}\right)\left(e^{\sqrt{3/2}}\frac{1}{2\sqrt{3/2}}\right)\left(3\cos\left(\frac{\pi}{6}\right)\right) = \frac{3\sqrt{3}\,e^{\sqrt{3/2}}}{4\sqrt{3/2}}\cos\left(e^{\sqrt{3/2}}\right) = \frac{3\sqrt{2}\,e^{\sqrt{3/2}}}{4}\cos\left(e^{\sqrt{3/2}}\right)$

91. $y^3 + y = 2\cos x \Rightarrow 3y^2\frac{dy}{dx} + \frac{dy}{dx} = -2\sin x \Rightarrow \frac{dy}{dx}(3y^2 + 1) = -2\sin x \Rightarrow \frac{dy}{dx} = \frac{-2\sin x}{3y^2 + 1} \Rightarrow \frac{dy}{dx}\Big|_{(0,1)}$

$= \frac{-2\sin(0)}{3+1} = 0; \frac{d^2y}{dx^2} = \frac{(3y^2+1)(-2\cos x) - (-2\sin x)\left(6y\frac{dy}{dx}\right)}{(3y^2+1)^2} \Rightarrow \frac{d^2y}{dx^2}\Big|_{(0,1)} = \frac{(3+1)(-2\cos 0) - (-2\sin 0)(6\cdot 0)}{(3+1)^2} = -\frac{1}{2}$

93. $f(t) = \frac{1}{2t+1}$ and $f(t+h) = \frac{1}{2(t+h)+1} \Rightarrow \frac{f(t+h) - f(t)}{h} = \frac{\frac{1}{2(t+h)+1} - \frac{1}{2t+1}}{h} = \frac{2t+1 - (2t+2h+1)}{(2t+2h+1)(2t+1)h}$

$= \frac{-2h}{(2t+2h+1)(2t+1)h} = \frac{-2}{(2t+2h+1)(2t+1)} \Rightarrow f'(t) = \lim_{h \to 0}\frac{f(t+h)-f(t)}{h} = \lim_{h \to 0}\frac{-2}{(2t+2h+1)(2t+1)} = \frac{-2}{(2t+1)^2}$

95. (a)

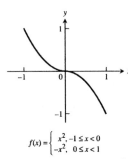

$f(x) = \begin{cases} x^2, & -1 \le x < 0 \\ -x^2, & 0 \le x < 1 \end{cases}$

(b) $\lim_{x \to 0^-} f(x) = \lim_{x \to 0^-} x^2 = 0$ and $\lim_{x \to 0^+} f(x) = \lim_{x \to 0^+} -x^2 = 0 \Rightarrow \lim_{x \to 0} f(x) = 0$. Since $\lim_{x \to 0} f(x) = 0 = f(0)$ it follows that f is continuous at $x = 0$.

(c) $\lim_{x \to 0^-} f'(x) = \lim_{x \to 0^-} (2x) = 0$ and $\lim_{x \to 0^+} f'(x) = \lim_{x \to 0^+} (-2x) = 0 \Rightarrow \lim_{x \to 0} f'(x) = 0$. Since this limit exists, it follows that f is differentiable at $x = 0$.

97. $y = \frac{x}{2} + \frac{1}{2x-4} = \frac{1}{2}x + (2x-4)^{-1} \Rightarrow \frac{dy}{dx} = \frac{1}{2} - 2(2x-4)^{-2}$; the slope of the tangent is $-\frac{3}{2} \Rightarrow -\frac{3}{2}$

$= \frac{1}{2} - 2(2x-4)^{-2} \Rightarrow -2 = -2(2x-4)^{-2} \Rightarrow 1 = \frac{1}{(2x-4)^2} \Rightarrow (2x-4)^2 = 1 \Rightarrow 4x^2 - 16x + 16 = 1$

$\Rightarrow 4x^2 - 16x + 15 = 0 \Rightarrow (2x-5)(2x-3) = 0 \Rightarrow x = \frac{5}{2}$ or $x = \frac{3}{2} \Rightarrow \left(\frac{5}{2}, \frac{9}{4}\right)$ and $\left(\frac{3}{2}, -\frac{1}{4}\right)$ are points on the curve where the slope is $-\frac{3}{2}$.

99. $y = 2x^3 - 3x^2 - 12x + 20 \Rightarrow \frac{dy}{dx} = 6x^2 - 6x - 12$

(a) The tangent is perpendicular to the line $y = 1 - \frac{x}{24}$ when $\frac{dy}{dx} = -\left(\frac{1}{-\left(\frac{1}{24}\right)}\right) = 24$; $6x^2 - 6x - 12 = 24$

$\Rightarrow x^2 - x - 2 = 4 \Rightarrow x^2 - x - 6 = 0 \Rightarrow (x-3)(x+2) = 0 \Rightarrow x = -2$ or $x = 3 \Rightarrow (-2, 16)$ and $(3, 11)$ are points where the tangent is perpendicular to $y = 1 - \frac{x}{24}$.

(b) The tangent is parallel to the line $y = \sqrt{2} - 12x$ when $\frac{dy}{dx} = -12 \Rightarrow 6x^2 - 6x - 12 = -12 \Rightarrow x^2 - x = 0$

$\Rightarrow x(x-1) = 0 \Rightarrow x = 0$ or $x = 1 \Rightarrow (0, 20)$ and $(1, 7)$ are points where the tangent is parallel to $y = \sqrt{2} - 12x$.

101. $y = x^2 + C \Rightarrow \frac{dy}{dx} = 2x$ and $y = x \Rightarrow \frac{dy}{dx} = 1$; the parabola is tangent to $y = x$ when $2x = 1 \Rightarrow x = \frac{1}{2} \Rightarrow y = \frac{1}{2}$;

thus, $\frac{1}{2} = \left(\frac{1}{2}\right)^2 + C \Rightarrow C = \frac{1}{4}$

103. $x^2 + 2y^2 = 9 \Rightarrow 2x + 4y\frac{dy}{dx} = 0 \Rightarrow \frac{dy}{dx} = -\frac{x}{2y} \Rightarrow \frac{dy}{dx}\Big|_{(1,2)} = -\frac{1}{4} \Rightarrow$ the tangent line is $y = 2 - \frac{1}{4}(x-1) = -\frac{1}{4}x + \frac{9}{4}$

and the normal line is $y = 2 + 4(x-1) = 4x - 2$.

105. $xy + 2x - 5y = 2 \Rightarrow \left(x \frac{dy}{dx} + y\right) + 2 - 5\frac{dy}{dx} = 0 \Rightarrow \frac{dy}{dx}(x - 5) = -y - 2 \Rightarrow \frac{dy}{dx} = \frac{-y-2}{x-5} \Rightarrow \frac{dy}{dx}\Big|_{(3,2)} = 2$

$\Rightarrow$ the tangent line is $y = 2 + 2(x - 3) = 2x - 4$ and the normal line is $y = 2 + \frac{-1}{2}(x - 3) = -\frac{1}{2}x + \frac{7}{2}$.

107. $\lim\limits_{x \to 0} \frac{\sin x}{2x^2 - x} = \lim\limits_{x \to 0} \left[\left(\frac{\sin x}{x}\right) \cdot \frac{1}{(2x - 1)}\right] = (1)\left(\frac{1}{-1}\right) = -1$

109. $\lim\limits_{r \to 0} \frac{\sin r}{\tan 2r} = \lim\limits_{r \to 0} \left(\frac{\sin r}{r} \cdot \frac{2r}{\tan 2r} \cdot \frac{1}{2}\right) = \left(\frac{1}{2}\right)(1) \lim\limits_{r \to 0} \frac{\cos 2r}{\left(\frac{\sin 2r}{2r}\right)} = \left(\frac{1}{2}\right)(1)\left(\frac{1}{1}\right) = \frac{1}{2}$

111. $\lim\limits_{\theta \to \left(\frac{\pi}{2}\right)^-} \frac{4\tan^2\theta + \tan\theta + 1}{\tan^2\theta + 5} = \lim\limits_{\theta \to \left(\frac{\pi}{2}\right)^-} \frac{\left(4 + \frac{1}{\tan\theta} + \frac{1}{\tan^2\theta}\right)}{\left(1 + \frac{5}{\tan^2\theta}\right)} = \frac{(4 + 0 + 0)}{(1 + 0)} = 4$

113. $y = \frac{2(x^2+1)}{\sqrt{\cos 2x}} \Rightarrow \ln y = \ln\left(\frac{2(x^2+1)}{\sqrt{\cos 2x}}\right) = \ln(2) + \ln(x^2 + 1) - \frac{1}{2}\ln(\cos 2x) \Rightarrow \frac{y'}{y} = 0 + \frac{2x}{x^2+1} - \left(\frac{1}{2}\right)\frac{(-2\sin 2x)}{\cos 2x}$

$\Rightarrow y' = \left(\frac{2x}{x^2+1} + \tan 2x\right)y = \frac{2(x^2+1)}{\sqrt{\cos 2x}}\left(\frac{2x}{x^2+1} + \tan 2x\right)$

115. $y = \left[\frac{(t+1)(t-1)}{(t-2)(t+3)}\right]^5 \Rightarrow \ln y = 5\left[\ln(t+1) + \ln(t-1) - \ln(t-2) - \ln(t+3)\right] \Rightarrow \left(\frac{1}{y}\right)\left(\frac{dy}{dt}\right)$

$= 5\left(\frac{1}{t+1} + \frac{1}{t-1} - \frac{1}{t-2} - \frac{1}{t+3}\right) \Rightarrow \frac{dy}{dt} = 5\left[\frac{(t+1)(t-1)}{(t-2)(t+3)}\right]^5 \left(\frac{1}{t+1} + \frac{1}{t-1} - \frac{1}{t-2} - \frac{1}{t+3}\right)$

117. $y = (\sin\theta)^{\sqrt{\theta}} \Rightarrow \ln y = \sqrt{\theta}\ln(\sin\theta) \Rightarrow \left(\frac{1}{y}\right)\left(\frac{dy}{d\theta}\right) = \sqrt{\theta}\left(\frac{\cos\theta}{\sin\theta}\right) + \frac{1}{2}\theta^{-1/2}\ln(\sin\theta)$

$\Rightarrow \frac{dy}{d\theta} = (\sin\theta)^{\sqrt{\theta}}\left(\sqrt{\theta}\cot\theta + \frac{\ln(\sin\theta)}{2\sqrt{\theta}}\right)$

119. (a) $S = 2\pi r^2 + 2\pi rh$ and h constant $\Rightarrow \frac{dS}{dt} = 4\pi r\frac{dr}{dt} + 2\pi h\frac{dr}{dt} = (4\pi r + 2\pi h)\frac{dr}{dt}$

(b) $S = 2\pi r^2 + 2\pi rh$ and r constant $\Rightarrow \frac{dS}{dt} = 2\pi r\frac{dh}{dt}$

(c) $S = 2\pi r^2 + 2\pi rh \Rightarrow \frac{dS}{dt} = 4\pi r\frac{dr}{dt} + 2\pi\left(r\frac{dh}{dt} + h\frac{dr}{dt}\right) = (4\pi r + 2\pi h)\frac{dr}{dt} + 2\pi r\frac{dh}{dt}$

(d) S constant $\Rightarrow \frac{dS}{dt} = 0 \Rightarrow 0 = (4\pi r + 2\pi h)\frac{dr}{dt} + 2\pi r\frac{dh}{dt} \Rightarrow (2r + h)\frac{dr}{dt} = -r\frac{dh}{dt} \Rightarrow \frac{dr}{dt} = \frac{-r}{2r+h}\frac{dh}{dt}$

121. $\frac{dR_1}{dt} = -1$ ohm/sec, $\frac{dR_2}{dt} = 0.5$ ohm/sec; and $\frac{1}{R} = \frac{1}{R_1} + \frac{1}{R_2} \Rightarrow \frac{-1}{R^2}\frac{dR}{dt} = \frac{-1}{R_1^2}\frac{dR_1}{dt} - \frac{1}{R_2^2}\frac{dR_2}{dt}$. Also,

$R_1 = 75$ ohms and $R_2 = 50$ ohms $\Rightarrow \frac{1}{R} = \frac{1}{75} + \frac{1}{50} \Rightarrow R = 30$ ohms. Therefore, from the derivative equation,

$\frac{-1}{(30)^2}\frac{dR}{dt} = \frac{-1}{(75)^2}(-1) - \frac{1}{(50)^2}(0.5) = \left(\frac{1}{5625} - \frac{1}{5000}\right) \Rightarrow \frac{dR}{dt} = (-900)\left(\frac{5000-5625}{5625\cdot5000}\right) = \frac{9(625)}{50(5625)} = \frac{1}{50} = 0.02$ ohm/sec.

123. Given $\frac{dx}{dt} = 10$ m/sec and $\frac{dy}{dt} = 5$ m/sec, let D be the distance from the origin $\Rightarrow D^2 = x^2 + y^2 \Rightarrow 2D\frac{dD}{dt}$

$= 2x\frac{dx}{dt} + 2y\frac{dy}{dt} \Rightarrow D\frac{dD}{dt} = x\frac{dx}{dt} + y\frac{dy}{dt}$. When $(x, y) = (3, -4)$, $D = \sqrt{3^2 + (-4)^2} = 5$ and

$5\frac{dD}{dt} = (5)(10) + (12)(5) \Rightarrow \frac{dD}{dt} = \frac{110}{5} = 22$. Therefore, the particle is moving <u>away</u> <u>from</u> the origin at 22 m/sec (because the distance D is increasing).

125. (a) From the diagram we have $\frac{10}{h} = \frac{4}{r} \Rightarrow r = \frac{2}{5}h$.

(b) $V = \frac{1}{3}\pi r^2 h = \frac{1}{3}\pi\left(\frac{2}{5}h\right)^2 h = \frac{4\pi h^3}{75} \Rightarrow \frac{dV}{dt} = \frac{4\pi h^2}{25}\frac{dh}{dt}$, so $\frac{dV}{dt} = -5$ and $h = 6 \Rightarrow \frac{dh}{dt} = -\frac{125}{144\pi}$ ft/min.

127. (a) From the sketch in the text, $\frac{d\theta}{dt} = -0.6$ rad/sec and $x = \tan\theta$. Also $x = \tan\theta \Rightarrow \frac{dx}{dt} = \sec^2\theta\frac{d\theta}{dt}$; at

point A, $x = 0 \Rightarrow \theta = 0 \Rightarrow \frac{dx}{dt} = (\sec^2 0)(-0.6) = -0.6$. Therefore the speed of the light is $0.6 = \frac{3}{5}$ km/sec

when it reaches point A.

(b) $\frac{(3/5)\text{ rad}}{\text{sec}} \cdot \frac{1\text{ rev}}{2\pi\text{ rad}} \cdot \frac{60\text{ sec}}{\text{min}} = \frac{18}{\pi}$ revs/min

129. (a) If $f(x) = \tan x$ and $x = -\frac{\pi}{4}$, then $f'(x) = \sec^2 x$,

   $f\left(-\frac{\pi}{4}\right) = -1$ and $f'\left(-\frac{\pi}{4}\right) = 2$. The linearization of

   $f(x)$ is $L(x) = 2\left(x + \frac{\pi}{4}\right) + (-1) = 2x + \frac{\pi-2}{2}$.

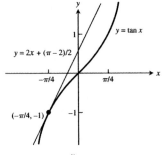

(b) If $f(x) = \sec x$ and $x = -\frac{\pi}{4}$, then $f'(x) = \sec x \tan x$,

   $f\left(-\frac{\pi}{4}\right) = \sqrt{2}$ and $f'\left(-\frac{\pi}{4}\right) = -\sqrt{2}$. The

   linearization of $f(x)$ is $L(x) = -\sqrt{2}\left(x + \frac{\pi}{4}\right) + \sqrt{2}$

   $= -\sqrt{2}x + \frac{\sqrt{2}(4-\pi)}{4}$.

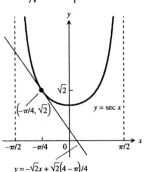

131. $C = 2\pi r \Rightarrow r = \frac{C}{2\pi}$, $S = 4\pi r^2 = \frac{C^2}{\pi}$, and $V = \frac{4}{3}\pi r^3 = \frac{C^3}{6\pi^2}$. It also follows that $dr = \frac{1}{2\pi}\,dC$, $dS = \frac{2C}{\pi}\,dC$ and

   $dV = \frac{C^2}{2\pi^2}\,dC$. Recall that $C = 10$ cm and $dC = 0.4$ cm.

   (a) $dr = \frac{0.4}{2\pi} = \frac{0.2}{\pi}$ cm $\Rightarrow \left(\frac{dr}{r}\right)(100\%) = \left(\frac{0.2}{\pi}\right)\left(\frac{2\pi}{10}\right)(100\%) = (.04)(100\%) = 4\%$

   (b) $dS = \frac{20}{\pi}(0.4) = \frac{8}{\pi}$ cm $\Rightarrow \left(\frac{dS}{S}\right)(100\%) = \left(\frac{8}{\pi}\right)\left(\frac{\pi}{100}\right)(100\%) = 8\%$

   (c) $dV = \frac{10^2}{2\pi^2}(0.4) = \frac{20}{\pi^2}$ cm $\Rightarrow \left(\frac{dV}{V}\right)(100\%) = \left(\frac{20}{\pi^2}\right)\left(\frac{6\pi^2}{1000}\right)(100\%) = 12\%$

CHAPTER 3 APPLICATIONS OF DERIVATIVES

3.1 EXTREME VALUES OF FUNCTIONS

1. An absolute minimum at $x = c_2$, an absolute maximum at $x = b$. Theorem 1 guarantees the existence of such extreme values because h is continuous on $[a, b]$.

3. No absolute minimum. An absolute maximum at $x = c$. Since the function's domain is an open interval, the function does not satisfy the hypotheses of Theorem 1 and need not have absolute extreme values.

5. An absolute minimum at $x = a$ and an absolute maximum at $x = c$. Note that $y = g(x)$ is not continuous but still has extrema. When the hypothesis of Theorem 1 is satisfied then extrema are guaranteed, but when the hypothesis is not satisfied, absolute extrema may or may not occur.

7. Local minimum at $(-1, 0)$, local maximum at $(1, 0)$

9. Maximum at $(0, 5)$. Note that there is no minimum since the endpoint $(2, 0)$ is excluded from the graph.

11. Graph (c), since this the only graph that has positive slope at c.

13. Graph (d), since this is the only graph representing a funtion that is differentiable at b but not at a.

15. $f(x) = \frac{2}{3} x - 5 \Rightarrow f'(x) = \frac{2}{3} \Rightarrow$ no critical points;
$f(-2) = -\frac{19}{3}, f(3) = -3 \Rightarrow$ the absolute maximum is -3 at $x = 3$ and the absolute minimum is $-\frac{19}{3}$ at $x = -2$

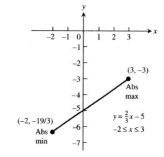

17. $F(x) = -\frac{1}{x^2} = -x^{-2} \Rightarrow F'(x) = 2x^{-3} = \frac{2}{x^3}$, however $x = 0$ is not a critical point since 0 is not in the domain; $F(0.5) = -4, F(2) = -0.25 \Rightarrow$ the absolute maximum is -0.25 at $x = 2$ and the absolute minimum is -4 at $x = 0.5$

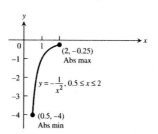

19. $h(x) = \sqrt[3]{x} = x^{1/3} \Rightarrow h'(x) = \frac{1}{3} x^{-2/3} \Rightarrow$ a critical point at $x = 0$; $h(-1) = -1, h(0) = 0, h(8) = 2 \Rightarrow$ the absolute maximum is 2 at $x = 8$ and the absolute minimum is -1 at $x = -1$

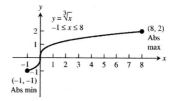

21. $f(\theta) = \sin \theta \Rightarrow f'(\theta) = \cos \theta \Rightarrow \theta = \frac{\pi}{2}$ is a critical point, but $\theta = \frac{-\pi}{2}$ is not a critical point because $\frac{-\pi}{2}$ is not interior to the domain; $f\left(\frac{-\pi}{2}\right) = -1$, $f\left(\frac{\pi}{2}\right) = 1$, $f\left(\frac{5\pi}{6}\right) = \frac{1}{2}$ $\Rightarrow$ the absolute maximum is 1 at $\theta = \frac{\pi}{2}$ and the absolute minimum is -1 at $\theta = \frac{-\pi}{2}$

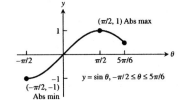

23. $g(x) = \csc x \Rightarrow g'(x) = -(\csc x)(\cot x) \Rightarrow$ a critical point at $x = \frac{\pi}{2}$; $g\left(\frac{\pi}{3}\right) = \frac{2}{\sqrt{3}}$, $g\left(\frac{\pi}{2}\right) = 1$, $g\left(\frac{2\pi}{3}\right) = \frac{2}{\sqrt{3}} \Rightarrow$ the absolute maximum is $\frac{2}{\sqrt{3}}$ at $x = \frac{\pi}{3}$ and $x = \frac{2\pi}{3}$, and the absolute minimum is 1 at $x = \frac{\pi}{2}$

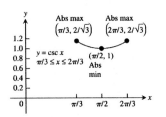

25. $g(x) = xe^{-x} \Rightarrow g'(x) = e^{-x} - xe^{-x}$ $\Rightarrow$ a critical point at $x = 1$; $g(-1) = -e$, and $g(1) = \frac{1}{e}$, $\Rightarrow$ the absolute maximum is $\frac{1}{e}$ at $x = 1$ and the absolute minimum is $-e$ at $x = -1$

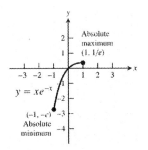

27. The first derivative $f'(x) = -\frac{1}{x^2} + \frac{1}{x}$ has a zero at $x = 1$. Critical point value: $f(1) = 1 + \ln 1 = 1$ Endpoint values: $f(0.5) = 2 + \ln 0.5 \approx 1.307$; $\qquad\qquad\quad f(4) = \frac{1}{4} + \ln 4 \approx 1.636$; Absolute maximum value is $\frac{1}{4} + \ln 4$ at $x = 4$; Absolute Minimum value is 1 at $x = 1$; Local maximum at $\left(\frac{1}{2}, 2 - \ln 2\right)$

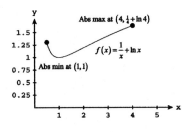

29. $f(x) = x^{4/3} \Rightarrow f'(x) = \frac{4}{3} x^{1/3} \Rightarrow$ a critical point at $x = 0$; $f(-1) = 1$, $f(0) = 0$, $f(8) = 16 \Rightarrow$ the absolute maximum is 16 at $x = 8$ and the absolute minimum is 0 at $x = 0$

31. Minimum value is 1 at $x = 2$.

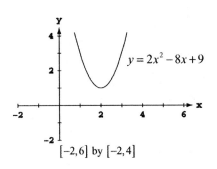

33. To find the exact values, note that that $y' = 3x^2 + 2x - 8$
    $= (3x - 4)(x + 2)$, which is zero when $x = -2$ or $x = \frac{4}{3}$.
    Local maximum at $(-2, 17)$; local minimum at $\left(\frac{4}{3}, -\frac{41}{27}\right)$

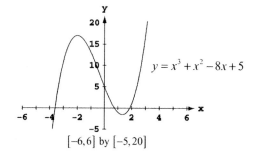

$y = x^3 + x^2 - 8x + 5$

$[-6, 6]$ by $[-5, 20]$

35. Minimum value is 0 when $x = -1$ or $x = 1$.

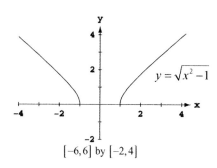

$y = \sqrt{x^2 - 1}$

$[-6, 6]$ by $[-2, 4]$

37. Maximum value is $\frac{1}{2}$ at $x = 1$;
    minimum value is $-\frac{1}{2}$ as $x = -1$.

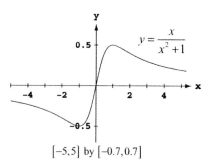

$y = \dfrac{x}{x^2 + 1}$

$[-5, 5]$ by $[-0.7, 0.7]$

39. $y = e^x + e^{-x} \Rightarrow y' = e^x - e^{-x} = e^x - \frac{1}{e^x} = \frac{e^{2x} - 1}{e^x}$,
    which is 0 at $x = 0$; an absolute minimum value is 2 at $x = 0$.

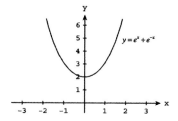

$y = e^x + e^{-x}$

41. $y = x \ln x \Rightarrow y' = x \cdot \frac{1}{x} + (1) \cdot \ln x = 1 + \ln x$, which is
    zero at $x = e^{-1}$; an absolute minimum value is $-\frac{1}{e}$ at $x = \frac{1}{e}$

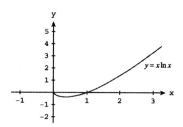

$y = x \ln x$

43. $y = \cos^{-1}(x^2) \Rightarrow y' = \dfrac{-1}{\sqrt{1-(x^2)^2}} \cdot (2x) = \dfrac{-2x}{\sqrt{1-x^4}}$,

which is zero at $x = 0$; an absolute maximum value is $\frac{\pi}{2}$ at
$x = 0$; an absolute minimum value is 0 at $x = 1$ and $x = -1$.

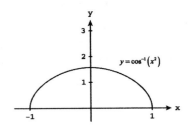

45. $y' = x^{2/3}(1) + \frac{2}{3}x^{-1/3}(x+2) = \dfrac{5x+4}{3\sqrt[3]{x}}$

crit. pt.	derivative	extremum	value
$x = -\frac{4}{5}$	0	local max	$\frac{12}{25}10^{1/3} = 1.034$
$x = 0$	undefined	local min	0

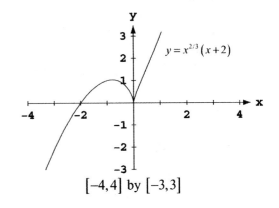

$[-4, 4]$ by $[-3, 3]$

47. $y' = x\dfrac{1}{2\sqrt{4-x^2}}(-2x) + (1)\sqrt{4-x^2}$

$\quad = \dfrac{-x^2 + (4-x^2)}{\sqrt{4-x^2}} = \dfrac{4-2x^2}{\sqrt{4-x^2}}$

crit. pt.	derivative	extremum	value
$x = -2$	undefined	local max	0
$x = -\sqrt{2}$	0	minimum	-2
$x = \sqrt{2}$	0	maximum	2
$x = 2$	undefined	local min	0

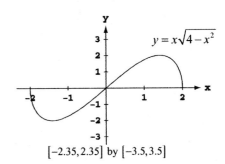

$[-2.35, 2.35]$ by $[-3.5, 3.5]$

49. $y' = \begin{cases} -2, & x < 1 \\ 1, & x > 1 \end{cases}$

crit. pt.	derivative	extremum	value
$x = 1$	undefined	minimum	2

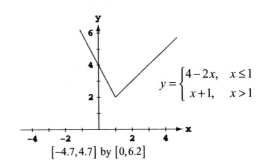

$[-4.7, 4.7]$ by $[0, 6.2]$

51. Yes, since $f(x) = |x| = \sqrt{x^2} = (x^2)^{1/2} \Rightarrow f'(x) = \frac{1}{2}(x^2)^{-1/2}(2x) = \dfrac{x}{(x^2)^{1/2}} = \dfrac{x}{|x|}$ is not defined at $x = 0$. Thus it

is not required that f' be zero at a local extreme point since f' may be undefined there.

53. (a) $V(x) = 160x - 52x^2 + 4x^3$

$\quad V'(x) = 160 - 104x + 12x^2 = 4(x-2)(3x-20)$

The only critical point in the interval $(0, 5)$ is at $x = 2$. The maximum value of $V(x)$ is 144 at $x = 2$.

(b) The largest possible volume of the box is 144 cubic units, and it occurs when $x = 2$ units.

55. $s = -\frac{1}{2}gt^2 + v_0t + s_0 \Rightarrow \frac{ds}{dt} = -gt + v_0 = 0 \Rightarrow t = \frac{v_0}{g}$. Now $s(t) = s_0 \Leftrightarrow t\left(-\frac{gt}{2} + v_0\right) = 0 \Leftrightarrow t = 0$ or $t = \frac{2v_0}{g}$.

Thus $s\left(\frac{v_0}{g}\right) = -\frac{1}{2}g\left(\frac{v_0}{g}\right)^2 + v_0\left(\frac{v_0}{g}\right) + s_0 = \frac{v_0^2}{2g} + s_0 > s_0$ is the <u>maximum</u> height over the interval $0 \le t \le \frac{2v_0}{g}$.

57. Maximum value is 11 at $x = 5$;
minimum value is 5 on the interval $[-3, 2]$;
local maximum at $(-5, 9)$

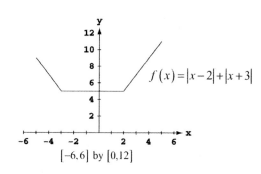

$$f(x) = |x - 2| + |x + 3|$$

$[-6, 6]$ by $[0, 12]$

3.2 THE MEAN VALUE THEOREM

1. When $f(x) = x^2 + 2x - 1$ for $0 \le x \le 1$, then $\frac{f(1) - f(0)}{1 - 0} = f'(c) \Rightarrow 3 = 2c + 2 \Rightarrow c = \frac{1}{2}$.

3. When $f(x) = \sin^{-1}(x)$ for $-1 \le x \le 1$, then $f'(c) = \frac{f(1) - f(-1)}{1 - (-1)} \Rightarrow \frac{1}{\sqrt{1 - c^2}} = \frac{\frac{\pi}{2} - \left(-\frac{\pi}{2}\right)}{2} \Rightarrow \sqrt{1 - c^2} = \frac{2}{\pi} \Rightarrow 1 - c^2 = \frac{4}{\pi^2}$

$\Rightarrow c^2 = 1 - \frac{4}{\pi^2} \Rightarrow c = \pm\sqrt{1 - \frac{4}{\pi^2}} \approx \pm 0.771$

5. Does not; $f(x)$ is not differentiable at $x = 0$ in $(-1, 8)$.

7. Does; $f(x)$ is continuous for every point of $[0, 1]$ and differentiable for every point in $(0, 1)$.

9. (a) i   ![number line with points at -2, 0, 2]
   ii   ![number line with points at -5, -4, -3]
   iii ![number line with points at -1, 0, 2]
   iv   ![number line with points at 0, 4, 9, 18, 24]

   (b) Let r_1 and r_2 be zeros of the polynomial $P(x) = x^n + a_{n-1}x^{n-1} + \ldots + a_1x + a_0$, then $P(r_1) = P(r_2) = 0$. Since polynomials are everywhere continuous and differentiable, by Rolle's Theorem $P'(r) = 0$ for some r between r_1 and r_2, where $P'(x) = nx^{n-1} + (n-1)a_{n-1}x^{n-2} + \ldots + a_1$.

11. Since f'' exists throughout $[a, b]$ the derivative function f' is continuous there. If f' has more than one zero in $[a, b]$, say $f'(r_1) = f'(r_2) = 0$ for $r_1 \ne r_2$, then by Rolle's Theorem there is a c between r_1 and r_2 such that $f''(c) = 0$, contrary to $f'' > 0$ throughout $[a, b]$. Therefore f' has at most one zero in $[a, b]$. The same argument holds if $f'' < 0$ throughout $[a, b]$.

13. By Corollary 1, $f'(x) = 0$ for all $x \Rightarrow f(x) = C$, where C is a constant. Since $f(-1) = 3$ we have $C = 3 \Rightarrow f(x) = 3$ for all x.

15. (a) $y = \frac{x^2}{2} + C$ (b) $y = \frac{x^3}{3} + C$ (c) $y = \frac{x^4}{4} + C$

17. (a) $y = \ln x + C$ if $x > 0$ and $y = \ln(-x) + C$ if $x < 0$, where C is a constant. (These functions can be combined as $y = \ln|x| + C$.)

    (b) $y = x - \ln x + C$ if $x > 0$ and $y = x - \ln(-x) + C$ if $x < 0$, where C is a constant. (These functions can be combined as $y = x - \ln|x| + C$.)

    (c) $y = 5x + \ln x + C$ if $x > 0$ and $y = 5x + \ln(-x) + C$ if $x < 0$, where C is a constant. (These functions can be combined as $y = 5x + \ln|x| + C$.)

19. $f(x) = x^2 - x + C$; $0 = f(0) = 0^2 - 0 + C \Rightarrow C = 0 \Rightarrow f(x) = x^2 - x$

21. $f(x) = \frac{e^{2x}}{2} + C$; $f(0) = \frac{3}{2} \Rightarrow \frac{e^{2(0)}}{2} + C = \frac{3}{2} \Rightarrow C = 1 \Rightarrow f(x) = 1 + \frac{e^{2x}}{2}$

23. $v = \frac{ds}{dt} = 9.8t + 5 \Rightarrow s = 4.9t^2 + 5t + C$; at $s = 10$ and $t = 0$ we have $C = 10 \Rightarrow s = 4.9t^2 + 5t + 10$

25. $a = \frac{dv}{dt} = e^t \Rightarrow v = e^t + C$; at $v = 20$ and $t = 0$ we have $C = 19 \Rightarrow v = e^t + 19$
    $v = \frac{ds}{dt} = e^t + 19 \Rightarrow s = e^t + 19t + C$; at $s = 5$ and $t = 0$ we have $C = 4 \Rightarrow s = e^t + 19t + 4$

27. If $T(t)$ is the temperature of the thermometer at time t, then $T(0) = -19°$ C and $T(14) = 100°$ C. From the Mean Value Theorem there exists a $0 < t_0 < 14$ such that $\frac{T(14) - T(0)}{14 - 0} = 8.5°$ C/sec $= T'(t_0)$, the rate at which the temperature was changing at $t = t_0$ as measured by the rising mercury on the thermometer.

29. The conclusion of the Mean Value Theorem yields $\frac{\frac{1}{b} - \frac{1}{a}}{b - a} = -\frac{1}{c^2} \Rightarrow c^2 \left(\frac{a - b}{ab}\right) = a - b \Rightarrow c = \sqrt{ab}$.

31. $f(x)$ must be zero at least once between a and b by the Intermediate Value Theorem. Now suppose that $f(x)$ is zero twice between a and b. Then by the Mean Value Theorem, $f'(x)$ would have to be zero at least once between the two zeros of $f(x)$, but this can't be true since we are given that $f'(x) \neq 0$ on this interval. Therefore, $f(x)$ is zero once and only once between a and b.

33. By the Mean Value Theorem we have $\frac{f(b) - f(a)}{b - a} = f'(c)$ for some point c between a and b. Since $b - a > 0$ and $f(b) < f(a)$, we have $f(b) - f(a) < 0 \Rightarrow f'(c) < 0$.

35. (a) Suppose $x < 1$, then by the Mean Value Theorem $\frac{f(x) - f(1)}{x - 1} < 0 \Rightarrow f(x) > f(1)$. Suppose $x > 1$, then by the Mean Value Theorem $\frac{f(x) - f(1)}{x - 1} > 0 \Rightarrow f(x) > f(1)$. Therefore $f(x) \geq 1$ for all x since $f(1) = 1$.
    (b) Yes. From part (a), $\lim_{x \to 1^-} \frac{f(x) - f(1)}{x - 1} \leq 0$ and $\lim_{x \to 1^+} \frac{f(x) - f(1)}{x - 1} \geq 0$. Since $f'(1)$ exists, these two one-sided limits are equal and have the value $f'(1) \Rightarrow f'(1) \leq 0$ and $f'(1) \geq 0 \Rightarrow f'(1) = 0$.

3.3 MONOTONIC FUNCTIONS AND THE FIRST DERIVATIVE TEST

1. (a) $f'(x) = x(x - 1) \Rightarrow$ critical points at 0 and 1
   (b) $f' = +++ \mid --- \mid +++ \Rightarrow$ increasing on $(-\infty, 0)$ and $(1, \infty)$, decreasing on $(0, 1)$
   $\quad\quad\quad\quad 0 \quad\quad 1$
   (c) Local maximum at $x = 0$ and a local minimum at $x = 1$

3. (a) $f'(x) = (x - 1)^2(x + 2) \Rightarrow$ critical points at -2 and 1
   (b) $f' = --- \mid +++ \mid +++ \Rightarrow$ increasing on $(-2, 1)$ and $(1, \infty)$, decreasing on $(-\infty, -2)$
   $\quad\quad\quad\quad -2 \quad\quad 1$
   (c) No local maximum and a local minimum at $x = -2$

5. (a) $f'(x) = (x - 1)e^{-x} \Rightarrow$ critical point at $x = 1$
   (b) $f' = ---- \mid ++++ \Rightarrow$ decreasing on $(-\infty, 1]$, increasing on $[1, \infty)$
   $\quad\quad\quad\quad\quad 1$
   (c) Local (and absolute) minimum at $x = 1$

7. (a) $f'(x) = x^{-1/3}(x + 2) \Rightarrow$ critical points at -2 and 0
   (b) $f' = +++ \mid ---)(+++ \Rightarrow$ increasing on $(-\infty, -2)$ and $(0, \infty)$, decreasing on $(-2, 0)$
   $\quad\quad\quad\quad -2 \quad\quad 0$
   (c) Local maximum at $x = -2$, local minimum at $x = 0$

9. (a) $g(t) = -t^2 - 3t + 3 \Rightarrow g'(t) = -2t - 3 \Rightarrow$ a critical point at $t = -\frac{3}{2}$; $g' = +++ \mid \underset{-3/2}{\quad} --- $, increasing on

$\left(-\infty, -\frac{3}{2}\right)$, decreasing on $\left(-\frac{3}{2}, \infty\right)$

(b) local maximum value of $g\left(-\frac{3}{2}\right) = \frac{21}{4}$ at $t = -\frac{3}{2}$

(c) absolute maximum is $\frac{21}{4}$ at $t = -\frac{3}{2}$

(d)

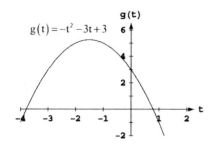

11. (a) $h(x) = -x^3 + 2x^2 \Rightarrow h'(x) = -3x^2 + 4x = x(4 - 3x) \Rightarrow$ critical points at $x = 0, \frac{4}{3}$

$\Rightarrow h' = --- \mid +++ \mid --- $, increasing on $\left(0, \frac{4}{3}\right)$, decreasing on $(-\infty, 0)$ and $\left(\frac{4}{3}, \infty\right)$
$\quad\quad\quad 0 \quad\quad 4/3$

(b) local maximum value of $h\left(\frac{4}{3}\right) = \frac{32}{27}$ at $x = \frac{4}{3}$; local minimum value of $h(0) = 0$ at $x = 0$

(c) no absolute extrema

(d)

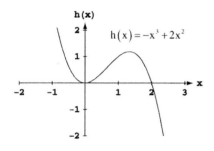

13. (a) $f(\theta) = 3\theta^2 - 4\theta^3 \Rightarrow f'(\theta) = 6\theta - 12\theta^2 = 6\theta(1 - 2\theta) \Rightarrow$ critical points at $\theta = 0, \frac{1}{2} \Rightarrow f' = --- \mid +++ \mid ---,$
$\quad\quad\quad\quad\quad\quad\quad\quad\quad\quad\quad\quad\quad\quad\quad\quad\quad\quad\quad\quad\quad\quad\quad\quad\quad\quad 0 \quad\quad 1/2$

increasing on $\left(0, \frac{1}{2}\right)$, decreasing on $(-\infty, 0)$ and $\left(\frac{1}{2}, \infty\right)$

(b) a local maximum is $f\left(\frac{1}{2}\right) = \frac{1}{4}$ at $\theta = \frac{1}{2}$, a local minimum is $f(0) = 0$ at $\theta = 0$

(c) no absolute extrema

(d)

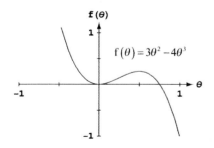

15. (a) $f(r) = 3r^3 + 16r \Rightarrow f'(r) = 9r^2 + 16 \Rightarrow$ no critical points $\Rightarrow f' = +++++$, increasing on $(-\infty, \infty)$, never decreasing

(b) no local extrema

(c) no absolute extrema

(d)

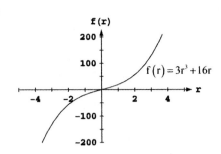

17. (a) $f(x) = x^4 - 8x^2 + 16 \Rightarrow f'(x) = 4x^3 - 16x = 4x(x+2)(x-2) \Rightarrow$ critical points at $x = 0$ and $x = \pm 2$

$\Rightarrow f' = --- \mid +++ \mid --- \mid +++$, increasing on $(-2, 0)$ and $(2, \infty)$, decreasing on $(-\infty, -2)$ and $(0, 2)$
$\quad\quad\quad\quad -2 \quad\; 0 \quad\; 2$

(b) a local maximum is $f(0) = 16$ at $x = 0$, local minima are $f(\pm 2) = 0$ at $x = \pm 2$

(c) no absolute maximum; absolute minimum is 0 at $x = \pm 2$

(d)

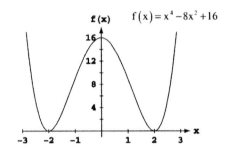

19. (a) $H(t) = \frac{3}{2} t^4 - t^6 \Rightarrow H'(t) = 6t^3 - 6t^5 = 6t^3(1+t)(1-t) \Rightarrow$ critical points at $t = 0, \pm 1$

$\Rightarrow H' = +++ \mid --- \mid +++ \mid ---$, increasing on $(-\infty, -1)$ and $(0, 1)$, decreasing on $(-1, 0)$ and $(1, \infty)$
$\quad\quad\quad\quad -1 \quad\; 0 \quad\; 1$

(b) the local maxima are $H(-1) = \frac{1}{2}$ at $t = -1$ and $H(1) = \frac{1}{2}$ at $t = 1$, the local minimum is $H(0) = 0$ at $t = 0$

(c) absolute maximum is $\frac{1}{2}$ at $t = \pm 1$; no absolute minimum

(d)

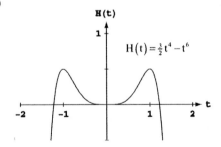

21. (a) $g(x) = x\sqrt{8 - x^2} = x(8 - x^2)^{1/2} \Rightarrow g'(x) = (8 - x^2)^{1/2} + x\left(\frac{1}{2}\right)(8 - x^2)^{-1/2}(-2x) = \dfrac{2(2-x)(2+x)}{\sqrt{\left(2\sqrt{2} - x\right)\left(2\sqrt{2} + x\right)}}$

$\Rightarrow$ critical points at $x = \pm 2, \pm 2\sqrt{2} \Rightarrow g' = (\quad --- \mid +++ \mid ---)$, increasing on $(-2, 2)$, decreasing on
$\quad\quad\quad\quad\quad\quad\quad\quad\quad\quad\quad\quad\quad -2\sqrt{2} \quad -2 \quad 2 \quad 2\sqrt{2}$
$\left(-2\sqrt{2}, -2\right)$ and $\left(2, 2\sqrt{2}\right)$

(b) local maxima are $g(2) = 4$ at $x = 2$ and $g\left(-2\sqrt{2}\right) = 0$ at $x = -2\sqrt{2}$, local minima are $g(-2) = -4$ at
$x = -2$ and $g\left(2\sqrt{2}\right) = 0$ at $x = 2\sqrt{2}$

(c) absolute maximum is 4 at $x = 2$; absolute minimum is -4 at $x = -2$

(d)

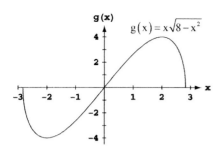

23. (a) $f(x) = \frac{x^2-3}{x-2} \Rightarrow f'(x) = \frac{2x(x-2)-(x^2-3)(1)}{(x-2)^2} = \frac{(x-3)(x-1)}{(x-2)^2} \Rightarrow$ critical points at $x = 1, 3$

$\Rightarrow f' = +++\ |\ ---\ )(---\ |\ +++$, increasing on $(-\infty, 1)$ and $(3, \infty)$, decreasing on $(1, 2)$ and $(2, 3)$,
123
discontinuous at $x = 2$

(b) a local maximum is $f(1) = 2$ at $x = 1$, a local minimum is $f(3) = 6$ at $x = 3$

(c) no absolute extrema

(d)

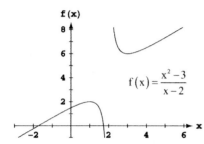

25. (a) $f(x) = x^{1/3}(x+8) = x^{4/3} + 8x^{1/3} \Rightarrow f'(x) = \frac{4}{3}x^{1/3} + \frac{8}{3}x^{-2/3} = \frac{4(x+2)}{3x^{2/3}} \Rightarrow$ critical points at $x = 0, -2$

$\Rightarrow f' = ---\ |\ +++\ )(+++$, increasing on $(-2, 0) \cup (0, \infty)$, decreasing on $(-\infty, -2)$
-20

(b) no local maximum, a local minimum is $f(-2) = -6\sqrt[3]{2} \approx -7.56$ at $x = -2$

(c) no absolute maximum; absolute minimum is $-6\sqrt[3]{2}$ at $x = -2$

(d)

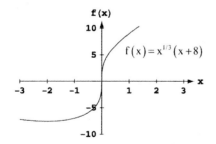

27. (a) $h(x) = x^{1/3}(x^2-4) = x^{7/3} - 4x^{1/3} \Rightarrow h'(x) = \frac{7}{3}x^{4/3} - \frac{4}{3}x^{-2/3} = \frac{\left(\sqrt{7}x+2\right)\left(\sqrt{7}x-2\right)}{3\sqrt[3]{x^2}} \Rightarrow$ critical points at

$x = 0, \frac{\pm 2}{\sqrt{7}} \Rightarrow h' = +++\ |\ ---)(---\ |\ +++$, increasing on $\left(-\infty, \frac{-2}{\sqrt{7}}\right)$ and $\left(\frac{2}{\sqrt{7}}, \infty\right)$, decreasing on
$\phantom{x = 0, \frac{\pm 2}{\sqrt{7}} \Rightarrow h' = +++\ |}-2/\sqrt{7}02/\sqrt{7}$
$\left(\frac{-2}{\sqrt{7}}, 0\right)$ and $\left(0, \frac{2}{\sqrt{7}}\right)$

(b) local maximum is $h\left(\frac{-2}{\sqrt{7}}\right) = \frac{24\sqrt[3]{2}}{7^{7/6}} \approx 3.12$ at $x = \frac{-2}{\sqrt{7}}$, the local minimum is $h\left(\frac{2}{\sqrt{7}}\right) = -\frac{24\sqrt[3]{2}}{7^{7/6}} \approx -3.12$

(c) no absolute extrema

(d)

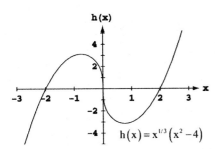

$$h(x) = x^{1/3}(x^2 - 4)$$

29. (a) $f(x) = e^{2x} + e^{-x} \Rightarrow f'(x) = 2e^{2x} - e^{-x} = 0 \Rightarrow e^{3x} = \frac{1}{2} \Rightarrow$ a critical point at $x = \frac{1}{3}\ln\left(\frac{1}{2}\right)$

$\Rightarrow f' = ---- \,|++++$, increasing on $\left(\frac{1}{3}\ln\left(\frac{1}{2}\right), \infty\right)$, decreasing on $\left(-\infty, \frac{1}{3}\ln\left(\frac{1}{2}\right)\right)$
$\quad\quad\quad \frac{1}{3}\ln\left(\frac{1}{2}\right)$

(b) a local minimum is $\frac{3}{2^{2/3}}$ at $x = \frac{1}{3}\ln\left(\frac{1}{2}\right)$; no local maximum

(c) an absolute minimum $\frac{3}{2^{2/3}}$ at $x = \frac{1}{3}\ln\left(\frac{1}{2}\right)$; no absolute maximum

(d)

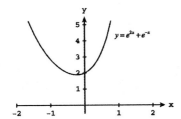

31. (a) $f(x) = x \ln x \Rightarrow f'(x) = 1 + \ln x \Rightarrow$ a critical point at $x = e^{-1} \Rightarrow f' = [\, --- \,|+++$, increasing on (e^{-1}, ∞),
$\quad\quad\quad\quad\quad\quad\quad\quad 0 \quad\quad e^{-1}$

decreasing on $(0, e^{-1})$

(b) A local minimum is $-e^{-1}$ at $x = e^{-1}$, no local maximum

(c) An absolute minimum is $-e^{-1}$ at $x = e^{-1}$, no absolute maximum

(d)

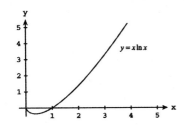

33. (a) $f(x) = 2x - x^2 \Rightarrow f'(x) = 2 - 2x \Rightarrow$ a critical point at $x = 1 \Rightarrow f' = +++\,|\,---\,]$ and $f(1) = 1$ and $f(2) = 0$
$\quad\quad\quad\quad\quad\quad\quad\quad\quad\quad\quad\quad\quad\quad\quad\quad\quad\quad\quad\quad 1 \quad\quad 2$

a local maximum is 1 at $x = 1$, a local minimum is 0 at $x = 2$.

(b) There is an absolute maximum of 1 at $x = 1$; no absolute minimum.

(c)

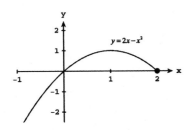

35. (a) $g(x) = x^2 - 4x + 4 \Rightarrow g'(x) = 2x - 4 = 2(x - 2) \Rightarrow$ a critical point at $x = 2 \Rightarrow g' = [\, --- \,|+++$ and
$\quad\quad\quad\quad\quad\quad\quad\quad\quad\quad\quad\quad\quad\quad\quad\quad\quad\quad\quad\quad\quad\quad\quad 1 \quad\quad 2$

$g(1) = 1$, $g(2) = 0 \Rightarrow$ a local maximum is 1 at $x = 1$, a local minimum is $g(2) = 0$ at $x = 2$

(b) no absolute maximum; absolute minimum is 0 at $x = 2$

(c)

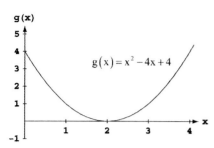

37. (a) $f(t) = 12t - t^3 \Rightarrow f'(t) = 12 - 3t^2 = 3(2 + t)(2 - t) \Rightarrow$ critical points at $t = \pm 2 \Rightarrow f' = [\; ---\; |\; +++\; |\; ---$
$ -3 \quad\;\; -2 \quad\;\; 2$

and $f(-3) = -9$, $f(-2) = -16$, $f(2) = 16 \Rightarrow$ local maxima are -9 at $t = -3$ and 16 at $t = -2$, a local minimum is -16 at $t = -2$

(b) absolute maximum is 16 at $t = 2$; no absolute minimum

(c)

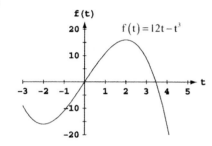

39. (a) $h(x) = \frac{x^3}{3} - 2x^2 + 4x \Rightarrow h'(x) = x^2 - 4x + 4 = (x - 2)^2 \Rightarrow$ a critical point at $x = 2 \Rightarrow h' = [\; +++\; |\; +++$ and
$ 0 \quad\quad 2$

$h(0) = 0 \Rightarrow$ no local maximum, a local minimum is 0 at $x = 0$

(b) no absolute maximum; absolute minimum is 0 at $x = 0$

(c)

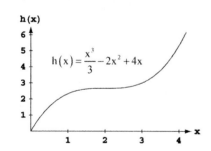

41. (a) $f(x) = \frac{x}{2} - 2 \sin\left(\frac{x}{2}\right) \Rightarrow f'(x) = \frac{1}{2} - \cos\left(\frac{x}{2}\right)$, $f'(x) = 0 \Rightarrow \cos\left(\frac{x}{2}\right) = \frac{1}{2} \Rightarrow$ a critical point at $x = \frac{2\pi}{3}$

$\Rightarrow f' = [\; ---\; |\; +++\;]$ and $f(0) = 0$, $f\left(\frac{2\pi}{3}\right) = \frac{\pi}{3} - \sqrt{3}$, $f(2\pi) = \pi \Rightarrow$ local maxima are 0 at $x = 0$ and π
$ 0 \quad 2\pi/3 \quad\; 2\pi$

at $x = 2\pi$, a local minimum is $\frac{\pi}{3} - \sqrt{3}$ at $x = \frac{2\pi}{3}$

(b) The graph of f rises when $f' > 0$, falls when $f' < 0$, and has a local minimum value at the point where f' changes from negative to positive.

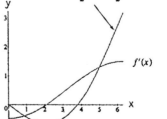

43. (a) $f(x) = \csc^2 x - 2\cot x \Rightarrow f'(x) = 2(\csc x)(-\csc x)(\cot x) - 2(-\csc^2 x) = -2(\csc^2 x)(\cot x - 1) \Rightarrow$ a critical

point at $x = \frac{\pi}{4} \Rightarrow f' = (\,---\,|\,+++\,)$ and $f\left(\frac{\pi}{4}\right) = 0 \Rightarrow$ no local maximum, a local minimum is 0 at $x = \frac{\pi}{4}$
$$0\pi/4\pi$$

(b) The graph of f rises when $f' > 0$, falls when $f' < 0$, and has a local minimum value at the point where $f' = 0$ and the values of f' change from negative to positive. The graph of f steepens as $f'(x) \to \pm\infty$.

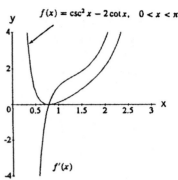

45. $h(\theta) = 3\cos\left(\frac{\theta}{2}\right) \Rightarrow h'(\theta) = -\frac{3}{2}\sin\left(\frac{\theta}{2}\right) \Rightarrow h' = [\,---\,]$, $(0,3)$ and $(2\pi, -3) \Rightarrow$ a local maximum is 3 at $\theta = 0$,
$$02\pi$$
a local minimum is -3 at $\theta = 2\pi$

47. (a)   (b)   (c)   (d)

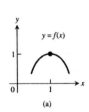   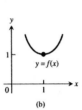   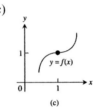   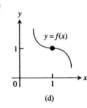

    (a)   (b)   (c)   (d)

49. (a)   (b)

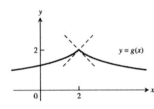   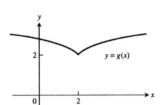

51. (a) $f(x) = \ln(\cos x) \Rightarrow f'(x) = -\frac{\sin x}{\cos x} = -\tan x = 0 \Rightarrow x = 0; f'(x) > 0$ for $-\frac{\pi}{4} \leq x < 0$ and $f'(x) < 0$ for

$0 < x \leq \frac{\pi}{3} \Rightarrow$ there is a relative maximum at $x = 0$ with $f(0) = \ln(\cos 0) = \ln 1 = 0; f\left(-\frac{\pi}{4}\right) = \ln\left(\cos\left(-\frac{\pi}{4}\right)\right)$

$= \ln\left(\frac{1}{\sqrt{2}}\right) = -\frac{1}{2}\ln 2$ and $f\left(\frac{\pi}{3}\right) = \ln\left(\cos\left(\frac{\pi}{3}\right)\right) = \ln\frac{1}{2} = -\ln 2$. Therefore, the absolute minimum occurs at

$x = \frac{\pi}{3}$ with $f\left(\frac{\pi}{3}\right) = -\ln 2$ and the absolute maximum occurs at $x = 0$ with $f(0) = 0$.

(b) $f(x) = \cos(\ln x) \Rightarrow f'(x) = \frac{-\sin(\ln x)}{x} = 0 \Rightarrow x = 1; f'(x) > 0$ for $\frac{1}{2} \leq x < 1$ and $f'(x) < 0$ for $1 < x \leq 2$

$\Rightarrow$ there is a relative maximum at $x = 1$ with $f(1) = \cos(\ln 1) = \cos 0 = 1; f\left(\frac{1}{2}\right) = \cos\left(\ln\left(\frac{1}{2}\right)\right) = \cos(-\ln 2)$

$= \cos(\ln 2)$ and $f(2) = \cos(\ln 2)$. Therefore, the absolute minimum occurs at $x = \frac{1}{2}$ and $x = 2$ with

$f\left(\frac{1}{2}\right) = f(2) = \cos(\ln 2)$, and the absolute maximum occurs at $x = 1$ with $f(1) = 1$.

53. $f(x) = e^x - 2x \Rightarrow f'(x) = e^x - 2; f'(x) = 0 \Rightarrow e^x = 2 \Rightarrow x = \ln 2; f(0) = 1$, the absolute maximum;
$f(\ln 2) = 2 - 2\ln 2 \approx 0.613706$, the absolute minimum; $f(1) = e - 2 \approx 0.71828$, a relative or local maximum since
$f''(x) = e^x$ is always positive.

55. $f(x) = x^2 \ln\frac{1}{x} \Rightarrow f'(x) = 2x \ln\frac{1}{x} + x^2\left(\frac{1}{\frac{1}{x}}\right)(-x^{-2}) = 2x \ln\frac{1}{x} - x = -x(2\ln x + 1); f'(x) = 0 \Rightarrow x = 0$ or $\ln x = -\frac{1}{2}$.

Since $x = 0$ is not in the domain of f, $x = e^{-1/2} = \frac{1}{\sqrt{e}}$. Also, $f'(x) > 0$ for $0 < x < \frac{1}{\sqrt{e}}$ and $f'(x) < 0$ for $x > \frac{1}{\sqrt{e}}$.

Therefore, $f\left(\frac{1}{\sqrt{e}}\right) = \frac{1}{e}\ln\sqrt{e} = \frac{1}{e}\ln e^{1/2} = \frac{1}{2e}\ln e = \frac{1}{2e}$ is the absolute maximum value of f assumed at $x = \frac{1}{\sqrt{e}}$.

57. $f(x) = x^3 - 3x + 2 \Rightarrow f'(x) = 3x^2 - 3 = 3(x - 1)(x + 1) \Rightarrow f' = {+++} \mid {---} \mid {+++} \Rightarrow$ rising for $x = c = 2$ since

$f'(x) > 0$ for $x = c = 2$.

59. Let $x_1 \neq x_2$ be two numbers in the domain of an increasing function f. Then, either $x_1 < x_2$ or $x_1 > x_2$ which implies $f(x_1) < f(x_2)$ or $f(x_1) > f(x_2)$, since $f(x)$ is increasing. In either case, $f(x_1) \neq f(x_2)$ and f is one-to-one. Similar arguments hold if f is decreasing.

61. $f(x)$ is increasing since $x_2 > x_1 \Rightarrow 27x_2^3 > 27x_1^3;\ y = 27x^3 \Rightarrow x = \frac{1}{3}y^{1/3} \Rightarrow f^{-1}(x) = \frac{1}{3}x^{1/3};\ \frac{df}{dx} = 81x^2$

$\Rightarrow \frac{df^{-1}}{dx} = \frac{1}{81x^2}\Big|_{\frac{1}{3}x^{1/3}} = \frac{1}{9x^{2/3}} = \frac{1}{9}x^{-2/3}$

63. $f(x)$ is decreasing since $x_2 > x_1 \Rightarrow (1 - x_2)^3 < (1 - x_1)^3;\ y = (1 - x)^3 \Rightarrow x = 1 - y^{1/3} \Rightarrow f^{-1}(x) = 1 - x^{1/3};$

$\frac{df}{dx} = -3(1 - x)^2 \Rightarrow \frac{df^{-1}}{dx} = \frac{1}{-3(1 - x)^2}\Big|_{1 - x^{1/3}} = \frac{-1}{3x^{2/3}} = -\frac{1}{3}x^{-2/3}$

3.4 CONCAVITY AND CURVE SKETCHING

1. $y = \frac{x^3}{3} - \frac{x^2}{2} - 2x + \frac{1}{3} \Rightarrow y' = x^2 - x - 2 = (x - 2)(x + 1) \Rightarrow y'' = 2x - 1 = 2\left(x - \frac{1}{2}\right)$. The graph is rising on $(-\infty, -1)$ and $(2, \infty)$, falling on $(-1, 2)$, concave up on $\left(\frac{1}{2}, \infty\right)$ and concave down on $\left(-\infty, \frac{1}{2}\right)$. Consequently, a local maximum is $\frac{3}{2}$ at $x = -1$, a local minimum is -3 at $x = 2$, and $\left(\frac{1}{2}, -\frac{3}{4}\right)$ is a point of inflection.

3. $y = \frac{3}{4}(x^2 - 1)^{2/3} \Rightarrow y' = \left(\frac{3}{4}\right)\left(\frac{2}{3}\right)(x^2 - 1)^{-1/3}(2x) = x(x^2 - 1)^{-1/3},\ y' = {---})(+++\mid{---})(+++$

$\Rightarrow$ the graph is rising on $(-1, 0)$ and $(1, \infty)$, falling on $(-\infty, -1)$ and $(0, 1) \Rightarrow$ a local maximum is $\frac{3}{4}$ at $x = 0$, local minima are 0 at $x = \pm 1;\ y'' = (x^2 - 1)^{-1/3} + (x)\left(-\frac{1}{3}\right)(x^2 - 1)^{-4/3}(2x) = \frac{x^2 - 3}{3\sqrt[3]{(x^2 - 1)^4}}$,

$y'' = {+++}\mid{---})({---})({---}\mid{+++} \Rightarrow$ the graph is concave up on $\left(-\infty, -\sqrt{3}\right)$ and $\left(\sqrt{3}, \infty\right)$, concave

down on $\left(-\sqrt{3}, \sqrt{3}\right) \Rightarrow$ points of inflection at $\left(\pm\sqrt{3}, \frac{3\sqrt[3]{4}}{4}\right)$

5. $y = x + \sin 2x \Rightarrow y' = 1 + 2\cos 2x,\ y' = [{---}\mid{+++}\mid{---}] \Rightarrow$ the graph is rising on $\left(-\frac{\pi}{3}, \frac{\pi}{3}\right)$, falling
   $ -2\pi/3\ -\pi/3 \quad \pi/3 \quad 2\pi/3$

on $\left(-\frac{2\pi}{3}, -\frac{\pi}{3}\right)$ and $\left(\frac{\pi}{3}, \frac{2\pi}{3}\right) \Rightarrow$ local maxima are $-\frac{2\pi}{3} + \frac{\sqrt{3}}{2}$ at $x = -\frac{2\pi}{3}$ and $\frac{\pi}{3} + \frac{\sqrt{3}}{2}$ at $x = \frac{\pi}{3}$, local minima are

$-\frac{\pi}{3} - \frac{\sqrt{3}}{2}$ at $x = -\frac{\pi}{3}$ and $\frac{2\pi}{3} - \frac{\sqrt{3}}{2}$ at $x = \frac{2\pi}{3};\ y'' = -4\sin 2x,\ y'' = [\ {---}\mid{+++}\mid{---}\mid{+++}\ ] \Rightarrow$ the
$ -2\pi/3 \quad -\pi/2 \quad 0 \quad \pi/2 \quad 2\pi/3$

graph is concave up on $\left(-\frac{\pi}{2}, 0\right)$ and $\left(\frac{\pi}{2}, \frac{2\pi}{3}\right)$, concave down on $\left(-\frac{2\pi}{3}, -\frac{\pi}{2}\right)$ and $\left(0, \frac{\pi}{2}\right) \Rightarrow$ points of inflection at $\left(-\frac{\pi}{2}, -\frac{\pi}{2}\right),\ (0, 0),$ and $\left(\frac{\pi}{2}, \frac{\pi}{2}\right)$

7. If $x \geq 0$, $\sin |x| = \sin x$ and if $x < 0$, $\sin |x| = \sin (-x)$
   $= -\sin x$. From the sketch the graph is rising on
   $\left(-\frac{3\pi}{2}, -\frac{\pi}{2}\right),\ \left(0, \frac{\pi}{2}\right)$ and $\left(\frac{3\pi}{2}, 2\pi\right)$, falling on $\left(-2\pi, -\frac{3\pi}{2}\right),$
   $\left(-\frac{\pi}{2}, 0\right)$ and $\left(\frac{\pi}{2}, \frac{3\pi}{2}\right)$; local minima are -1 at $x = \pm \frac{3\pi}{2}$
   and 0 at $x = 0$; local maxima are 1 at $x = \pm \frac{\pi}{2}$ and 0 at
   $x = \pm 2\pi$; concave up on $(-2\pi, -\pi)$ and $(\pi, 2\pi)$, and
   concave down on $(-\pi, 0)$ and $(0, \pi) \Rightarrow$ points of inflection
   are $(-\pi, 0)$ and $(\pi, 0)$

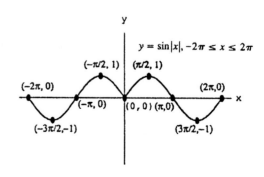

9. When $y = x^2 - 4x + 3$, then $y' = 2x - 4 = 2(x - 2)$ and $y'' = 2$. The curve rises on $(2, \infty)$ and falls on $(-\infty, 2)$. At $x = 2$ there is a minimum. Since $y'' > 0$, the curve is concave up for all x.

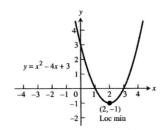

11. When $y = x^3 - 3x + 3$, then $y' = 3x^2 - 3 = 3(x - 1)(x + 1)$ and $y'' = 6x$. The curve rises on $(-\infty, -1) \cup (1, \infty)$ and falls on $(-1, 1)$. At $x = -1$ there is a local maximum and at $x = 1$ a local minimum. The curve is concave down on $(-\infty, 0)$ and concave up on $(0, \infty)$. There is a point of inflection at $x = 0$.

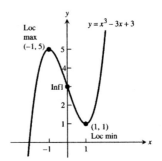

13. When $y = -2x^3 + 6x^2 - 3$, then $y' = -6x^2 + 12x = -6x(x - 2)$ and $y'' = -12x + 12 = -12(x - 1)$. The curve rises on $(0, 2)$ and falls on $(-\infty, 0)$ and $(2, \infty)$. At $x = 0$ there is a local minimum and at $x = 2$ a local maximum. The curve is concave up on $(-\infty, 1)$ and concave down on $(1, \infty)$. At $x = 1$ there is a point of inflection.

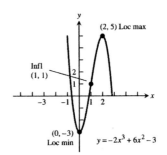

15. When $y = x^4 - 2x^2$, then $y' = 4x^3 - 4x = 4x(x + 1)(x - 1)$ and $y'' = 12x^2 - 4 = 12\left(x + \frac{1}{\sqrt{3}}\right)\left(x - \frac{1}{\sqrt{3}}\right)$. The curve rises on $(-1, 0)$ and $(1, \infty)$ and falls on $(-\infty, -1)$ and $(0, 1)$. At $x = \pm 1$ there are local minima and at $x = 0$ a local maximum. The curve is concave up on $\left(-\infty, -\frac{1}{\sqrt{3}}\right)$ and $\left(\frac{1}{\sqrt{3}}, \infty\right)$ and concave down on $\left(-\frac{1}{\sqrt{3}}, \frac{1}{\sqrt{3}}\right)$. At $x = \frac{\pm 1}{\sqrt{3}}$ there are points of inflection.

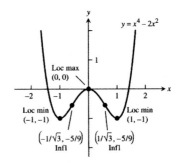

17. When $y = 4x^3 - x^4$, then $y' = 12x^2 - 4x^3 = 4x^2(3 - x)$ and $y'' = 24x - 12x^2 = 12x(2 - x)$. The curve rises on $(-\infty, 3)$ and falls on $(3, \infty)$. At $x = 3$ there is a local maximum, but there is no local minimum. The graph is concave up on $(0, 2)$ and concave down on $(-\infty, 0)$ and $(2, \infty)$. There are inflection points at $x = 0$ and $x = 2$.

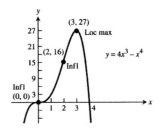

19. When $y = x^5 - 5x^4$, then $y' = 5x^4 - 20x^3 = 5x^3(x - 4)$ and $y'' = 20x^3 - 60x^2 = 20x^2(x - 3)$. The curve rises on $(-\infty, 0)$ and $(4, \infty)$, and falls on $(0, 4)$. There is a local maximum at $x = 0$, and a local minimum at $x = 4$. The curve is concave down on $(-\infty, 3)$ and concave up on $(3, \infty)$. At $x = 3$ there is a point of inflection.

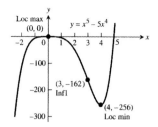

21. When $y = x + \sin x$, then $y' = 1 + \cos x$ and $y'' = -\sin x$. The curve rises on $(0, 2\pi)$. At $x = 0$ there is a local and absolute minimum and at $x = 2\pi$ there is a local and absolute maximum. The curve is concave down on $(0, \pi)$ and concave up on $(\pi, 2\pi)$. At $x = \pi$ there is a point of inflection.

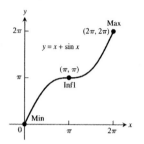

23. When $y = x^{1/5}$, then $y' = \frac{1}{5}x^{-4/5}$ and $y'' = -\frac{4}{25}x^{-9/5}$. The curve rises on $(-\infty, \infty)$ and there are no extrema. The curve is concave up on $(-\infty, 0)$ and concave down on $(0, \infty)$. At $x = 0$ there is a point of inflection.

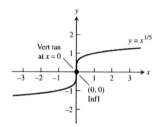

25. When $y = 2x - 3x^{2/3}$, then $y' = 2 - 2x^{-1/3}$ and $y'' = \frac{2}{3}x^{-4/3}$. The curve is rising on $(-\infty, 0)$ and $(1, \infty)$, and falling on $(0, 1)$. There is a local maximum at $x = 0$ and a local minimum at $x = 1$. The curve is concave up on $(-\infty, 0)$ and $(0, \infty)$. There are no points of inflection, but a cusp exists at $x = 0$.

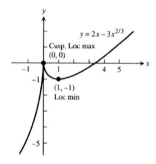

27. When $y = x\sqrt{8 - x^2} = x\left(8 - x^2\right)^{1/2}$, then

$y' = \left(8 - x^2\right)^{1/2} + (x)\left(\frac{1}{2}\right)\left(8 - x^2\right)^{-1/2}(-2x)$

$= \left(8 - x^2\right)^{-1/2}\left(8 - 2x^2\right) = \dfrac{2(2 - x)(2 + x)}{\sqrt{\left(2\sqrt{2} + x\right)\left(2\sqrt{2} - x\right)}}$ and

$y'' = \left(-\frac{1}{2}\right)\left(8 - x^2\right)^{-\frac{3}{2}}(-2x)(8 - 2x^2) + \left(8 - x^2\right)^{-\frac{1}{2}}(-4x)$

$= \dfrac{2x\left(x^2 - 12\right)}{\sqrt{\left(8 - x^2\right)^3}}$. The curve is rising on $(-2, 2)$, and falling

on $\left(-2\sqrt{2}, -2\right)$ and $\left(2, 2\sqrt{2}\right)$. There are local minima

$x = -2$ and $x = 2\sqrt{2}$, and local maxima at $x = -2\sqrt{2}$ and

$x = 2$. The curve is concave up on $\left(-2\sqrt{2}, 0\right)$ and

concave down on $\left(0, 2\sqrt{2}\right)$. There is a point of inflection

at $x = 0$.

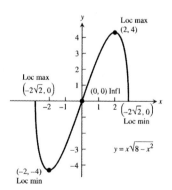

29. When $y = \frac{x^2 - 3}{x - 2}$, then $y' = \frac{2x(x - 2) - (x^2 - 3)(1)}{(x-2)^2}$

    $= \frac{(x - 3)(x - 1)}{(x - 2)^2}$ and

    $y'' = \frac{(2x - 4)(x - 2)^2 - (x^2 - 4x + 3)2(x - 2)}{(x - 2)^4} = \frac{2}{(x - 2)^3}$.

The curve is rising on $(-\infty, 1)$ and $(3, \infty)$, and falling on $(1, 2)$ and $(2, 3)$. There is a local maximum at $x = 1$ and a local minimum at $x = 3$. The curve is concave down on $(-\infty, 2)$ and concave up on $(2, \infty)$. There are no points of inflection because $x = 2$ is not in the domain.

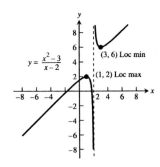

31. When $y = |x^2 - 1| = \begin{cases} x^2 - 1, & |x| \geq 1 \\ 1 - x^2, & |x| < 1 \end{cases}$, then

    $y' = \begin{cases} 2x, & |x| > 1 \\ -2x, & |x| < 1 \end{cases}$ and $y'' = \begin{cases} 2, & |x| > 1 \\ -2, & |x| < 1 \end{cases}$. The

curve rises on $(-1, 0)$ and $(1, \infty)$ and falls on $(-\infty, -1)$ and $(0, 1)$. There is a local maximum at $x = 0$ and local minima at $x = \pm 1$. The curve is concave up on $(-\infty, -1)$ and $(1, \infty)$, and concave down on $(-1, 1)$. There are no points of inflection because y is not differentiable at $x = \pm 1$ (so there is no tangent line at those points).

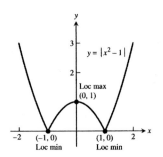

33. When $y = x\,e^{1/x}$, then $y' = -\frac{x\,e^{1/x}}{x^2} + e^{1/x} = e^{1/x}\left(1 - \frac{1}{x}\right)$

    and $y'' = e^{1/x}\left(\frac{1}{x^2}\right) + \left(1 - \frac{1}{x}\right)\left(-\frac{e^{1/x}}{x^2}\right) = \frac{e^{1/x}}{x^2}\left(\frac{1}{x}\right) = \frac{e^{1/x}}{x^3}$.

The curve is rising on $(1, \infty)$ and $(-\infty, 0)$ and falling on $(0, 1)$. The curve is concave down on $(-\infty, 0)$ and concave up on $(0, \infty)$. There is a local minimum of e at $x = 1$, but there are no inflection points.

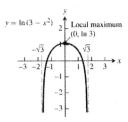

35. $y = \ln(3 - x^2), \Rightarrow y' = \frac{-2x}{3 - x^2} = \frac{2x}{x^2 - 3}$

    $\Rightarrow y' = (\underset{-\sqrt{3}}{\,+\!+\!+}\,\underset{0}{|}\,\underset{\sqrt{3}}{-\!-\!-}\,) \Rightarrow$ the graph is rising on

    $\left(-\sqrt{3}, 0\right)$, falling on $\left(0, \sqrt{3}\right)$; a local minimum is ln 3 at

    $x = 0$; $y'' = \frac{(x^2 - 3)(2) - (2x)(2x)}{(x^2 - 3)^2} = \frac{-2(x^2 + 3)}{(x^2 - 3)^2}$

    $\Rightarrow y'' = (\underset{-\sqrt{3}}{\quad}\,-\!-\!-\,\underset{\sqrt{3}}{\quad}) \Rightarrow$ the graph is concave down on

    $\left(-\sqrt{3}, \sqrt{3}\right)$.

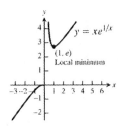

37. $y = e^x - 2e^{-x} - 3x \Rightarrow y' = e^x + 2e^{-x} - 3 = \frac{(e^x)^2 - 3e^x + 2}{e^x}$

$= \frac{(e^x - 2)(e^x - 1)}{e^x} \Rightarrow y' = \; +++ \; | \; --- \; | \; +++ \; \Rightarrow$
$ 0 \quad \ln 2$

the graph is increasing on $(-\infty, 0)$ and $(\ln 2, \infty)$,

decreasing on $(0, \ln 2)$; a local maximum is -1 at $x = 0$ and

a local minimum is $1 - 3\ln 2$ at $x = \ln 2$; $y'' = e^x - 2e^{-x}$

$= \frac{(e^x)^2 - 2}{e^x} \Rightarrow y'' = \quad --- \; | \; +++ \; \Rightarrow$ the graph is
$ \frac{1}{2}\ln 2$

concave up on $\left(\frac{1}{2}\ln 2, \infty\right)$, concave down on $\left(-\infty, \frac{1}{2}\ln 2\right)$

$\Rightarrow$ point of inflection at $\left(\frac{1}{2}\ln 2, -\frac{3}{2}\ln 2\right)$.

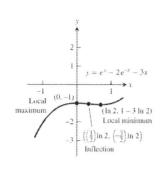

39. $y = \ln(\cos x) \Rightarrow y' = \frac{-\sin x}{\cos x} = -\tan x$

$\Rightarrow y' = \; \ldots \;) \; \text{none} \; (\; +++ \; | \; --- \;) \; \text{none} \; (\; +++ \; | ---) \; \text{none} \; (\; +++ \; | \; --- \;) \; \text{none} \; (\ldots$
$ -\frac{7\pi}{2} \quad -\frac{5\pi}{2} \quad -2\pi \quad -\frac{3\pi}{2} \quad -\frac{\pi}{2} \quad 0 \quad \frac{\pi}{2} \quad \frac{3\pi}{2} \quad 2\pi \quad \frac{5\pi}{2} \quad \frac{7\pi}{2}$

$\Rightarrow$ the graph is increasing on $\ldots, \left(-\frac{5\pi}{2}, -2\pi\right), \left(-\frac{\pi}{2}, 0\right),$

$\left(\frac{3\pi}{2}, 2\pi\right), \ldots$, decreasing on $\left(-2\pi, -\frac{3\pi}{2}\right), \left(0, \frac{\pi}{2}\right), \left(2\pi, \frac{5\pi}{2}\right);$

local maxima are 0 at $x = 0, \; \pm 2\pi, \pm 4\pi, \ldots$; $y'' = -\sec^2 x$

$\frac{-1}{\cos^2 x} \Rightarrow$ the graph is concave down on $\left(-\frac{5\pi}{2}, -\frac{3\pi}{2}\right), \left(-\frac{\pi}{2}, \frac{\pi}{2}\right),$

$\left(\frac{3\pi}{2}, \frac{5\pi}{2}\right), \ldots$

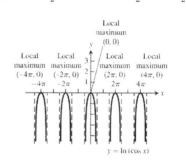

41. $y = \frac{1}{1 + e^{-x}} = \frac{e^x}{e^x + 1} \Rightarrow y' = \frac{(e^x + 1)e^x - e^x \cdot e^x}{(e^x + 1)^2} = \frac{e^x}{(e^x + 1)^2}$

$\Rightarrow y' = \; +++ \; \Rightarrow$ the graph is increasing on $(-\infty, \infty)$;

$y'' = \frac{(e^x + 1)^2 \cdot e^x - e^x \cdot 2(e^x + 1)e^x}{(e^x + 1)^4} = \frac{e^x(1 - e^x)}{(e^x + 1)^3}$

$\Rightarrow y'' = \; +++ \; | \; --- \; \Rightarrow$ the graph is concave up on
$ 0$

$(-\infty, 0)$, concave down on $(0, \infty) \Rightarrow$ point of inflection

is $\left(0, \frac{1}{2}\right)$.

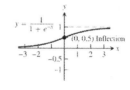

43. $y' = 2 + x - x^2 = (1 + x)(2 - x)$, $y' = \; --- \; | \; +++ \; | \; ---$
$ -1 2$

$\Rightarrow$ rising on $(-1, 2)$, falling on $(-\infty, -1)$ and $(2, \infty)$

$\Rightarrow$ there is a local maximum at $x = 2$ and a local minimum

at $x = -1$; $y'' = 1 - 2x$, $y'' = \; +++ \; | \; ---$
$ 1/2$

$\Rightarrow$ concave up on $\left(-\infty, \frac{1}{2}\right)$, concave down on $\left(\frac{1}{2}, \infty\right)$

$\Rightarrow$ a point of inflection at $x = \frac{1}{2}$

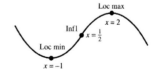

45. $y' = x(x - 3)^2$, $y' = \; --- \; | \; +++ \; | \; +++ \; \Rightarrow$ rising on
$ 0 3$

$(0, \infty)$, falling on $(-\infty, 0) \Rightarrow$ no local maximum, but there

is a local minimum at $x = 0$; $y'' = (x - 3)^2 + x(2)(x - 3)$

$= 3(x - 3)(x - 1)$, $y'' = \; +++ \; | \; --- \; | \; +++ \; \Rightarrow$ concave
$ 1 3$

up on $(-\infty, 1)$ and $(3, \infty)$, concave down on $(1, 3) \Rightarrow$

points of inflection at $x = 1$ and $x = 3$

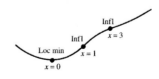

47. $y' = x(x^2 - 12) = x(x - 2\sqrt{3})(x + 2\sqrt{3})$,

$y' = \underset{-2\sqrt{3}}{---} \mid \underset{0}{+++} \mid \underset{2\sqrt{3}}{---} \mid +++ \Rightarrow$ rising on

$(-2\sqrt{3}, 0)$ and $(2\sqrt{3}, \infty)$, falling on $(-\infty, -2\sqrt{3})$

and $(0, 2\sqrt{3}) \Rightarrow$ a local maximum at $x = 0$, local minima

at $x = \pm 2\sqrt{3}$; $y'' = (1)(x^2 - 12) + (x)(2x)$

$= 3(x - 2)(x + 2)$, $y'' = +++ \mid \underset{-2}{---} \mid \underset{2}{+++}$

$\Rightarrow$ concave up on $(-\infty, -2)$ and $(2, \infty)$, concave down on

$(-2, 2) \Rightarrow$ points of inflection at $x = \pm 2$

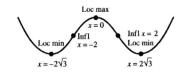

49. $y' = \sec^2 x$, $y' = (\underset{-\pi/2}{\quad} +++ \underset{\pi/2}{\;}) \Rightarrow$ rising on $(-\frac{\pi}{2}, \frac{\pi}{2})$,

never falling $\Rightarrow$ no local extrema;

$y'' = 2(\sec x)(\sec x)(\tan x) = 2(\sec^2 x)(\tan x)$,

$y'' = (\underset{-\pi/2}{\quad} --- \mid \underset{0}{\;} +++ \underset{\pi/2}{\;}) \Rightarrow$ concave up on $(0, \frac{\pi}{2})$,

concave down on $(-\frac{\pi}{2}, 0)$, 0 is a opoint of inflection.

51. $y' = \cot \frac{\theta}{2}$, $y' = (\underset{0}{+++} \mid \underset{\pi}{---} \underset{2\pi}{\;}) \Rightarrow$ rising on $(0, \pi)$,

falling on $(\pi, 2\pi) \Rightarrow$ a local maximum at $\theta = \pi$, no local

minimum; $y'' = -\frac{1}{2} \csc^2 \frac{\theta}{2}$, $y'' = (\underset{0}{---} \underset{2\pi}{\;}) \Rightarrow$ never

concave up, concave down on $(0, 2\pi) \Rightarrow$ no points of

inflection

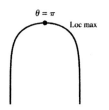

53. $y' = \tan^2 \theta - 1 = (\tan \theta - 1)(\tan \theta + 1)$,

$y' = (\underset{-\pi/2}{\quad} +++ \mid \underset{-\pi/4}{\;} --- \mid \underset{\pi/4}{\;} +++ \underset{\pi/2}{\;}) \Rightarrow$ rising on

$(-\frac{\pi}{2}, -\frac{\pi}{4})$ and $(\frac{\pi}{4}, \frac{\pi}{2})$, falling on $(-\frac{\pi}{4}, \frac{\pi}{4})$

$\Rightarrow$ a local maximum at $\theta = -\frac{\pi}{4}$, a local minimum at $\theta = \frac{\pi}{4}$;

$y'' = 2\tan\theta \sec^2\theta$, $y'' = (\underset{-\pi/2}{\quad} --- \mid \underset{0}{\;} +++ \underset{\pi/2}{\;})$

$\Rightarrow$ concave up on $(0, \frac{\pi}{2})$, concave down on $(-\frac{\pi}{2}, 0)$

$\Rightarrow$ a point of inflection at $\theta = 0$

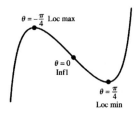

55. $y' = \cos t$, $y' = [\underset{0}{+++} \mid \underset{\pi/2}{\;} --- \mid \underset{3\pi/2}{\;} +++ \underset{2\pi}{\;}] \Rightarrow$ rising on

$(0, \frac{\pi}{2})$ and $(\frac{3\pi}{2}, 2\pi)$, falling on $(\frac{\pi}{2}, \frac{3\pi}{2}) \Rightarrow$ local maxima at

$t = \frac{\pi}{2}$ and $t = 2\pi$, local minima at $t = 0$ and $t = \frac{3\pi}{2}$;

$y'' = -\sin t$, $y'' = [\underset{0}{---} \mid \underset{\pi}{\;} +++ \underset{2\pi}{\;}]$

$\Rightarrow$ concave up on $(\pi, 2\pi)$, concave down

on $(0, \pi) \Rightarrow$ a point of inflection at $t = \pi$

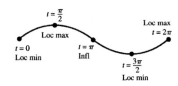

57. $y' = (x + 1)^{-2/3}$, $y' = +++ \underset{-1}{)} \ (+++ \ \Rightarrow \ $ rising on

$(-\infty, \infty)$, never falling $\Rightarrow$ no local extrema;

$y'' = -\frac{2}{3}(x + 1)^{-5/3}$, $y'' = +++ \underset{-1}{)} \ (---$

$\Rightarrow$ concave up on $(-\infty, -1)$, concave down on $(-1, \infty)$

$\Rightarrow$ a point of inflection and vertical tangent at $x = -1$

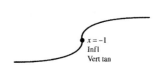

59. $y' = x^{-2/3}(x - 1)$, $y' = --- \underset{0}{)}(--- \underset{1}{|} +++ \ \Rightarrow \ $ rising on

$(1, \infty)$, falling on $(-\infty, 1)$ $\Rightarrow$ no local maximum, but a

local minimum at $x = 1$; $y'' = \frac{1}{3}x^{-2/3} + \frac{2}{3}x^{-5/3}$

$= \frac{1}{3}x^{-5/3}(x + 2)$, $y'' = +++ \underset{-2}{|} --- \underset{0}{)}(+++$

$\Rightarrow$ concave up on $(-\infty, -2)$ and $(0, \infty)$, concave down on

$(-2, 0)$ $\Rightarrow$ points of inflection at $x = -2$ and $x = 0$, and a

vertical tangent at $x = 0$

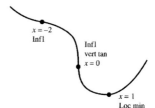

61. $y' = \begin{cases} -2x, \ x \le 0 \\ 2x, \ x > 0 \end{cases}$, $y' = +++ \underset{0}{|} +++ \ \Rightarrow \ $ rising on

$(-\infty, \infty)$ $\Rightarrow$ no local extrema; $y'' = \begin{cases} -2, \ x < 0 \\ 2, \ x > 0 \end{cases}$,

$y'' = --- \underset{0}{)}(+++ \ \Rightarrow \ $ concave up on $(0, \infty)$, concave

down on $(-\infty, 0)$ $\Rightarrow$ a point of inflection at $x = 0$

63.

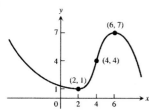

65. Graphs printed in color can shift during a press run, so your values may differ somewhat from those given here.

   (a) The body is moving away from the origin when |displacement| is increasing as t increases, $0 < t < 2$ and $6 < t < 9.5$; the body is moving toward the origin when |displacement| is decreasing as t increases, $2 < t < 6$ and $9.5 < t < 15$

   (b) The velocity will be zero when the slope of the tangent line for $y = s(t)$ is horizontal. The velocity is zero when t is approximately 2, 6, or 9.5 sec.

   (c) The acceleration will be zero at those values of t where the curve $y = s(t)$ has points of inflection. The acceleration is zero when t is approximately 4, 7.5, or 12.5 sec.

   (d) The acceleration is positive when the concavity is up, $4 < t < 7.5$ and $12.5 < t < 15$; the acceleration is negative when the concavity is down, $0 < t < 4$ and $7.5 < t < 12.5$

67. When $y' = (x - 1)^2(x - 2)$, then $y'' = 2(x - 1)(x - 2) + (x - 1)^2$. The curve falls on $(-\infty, 2)$ and rises on $(2, \infty)$. At $x = 2$ there is a local minimum. There is no local maximum. The curve is concave upward on $(-\infty, 1)$ and $\left(\frac{5}{3}, \infty\right)$, and concave downward on $\left(1, \frac{5}{3}\right)$. At $x = 1$ or $x = \frac{5}{3}$ there are inflection points.

69. (a) $f(x) = ax^2 + bx + c = a\left(x^2 + \frac{b}{a}x\right) + c = a\left(x^2 + \frac{b}{a}x + \frac{b^2}{4a^2}\right) - \frac{b^2}{4a} + c = a\left(x + \frac{b}{2a}\right)^2 - \frac{b^2 - 4ac}{4a}$ a parabola

   whose vertex is at $x = -\frac{b}{2a}$ $\Rightarrow$ the coordinates of the vertex are $\left(-\frac{b}{2a}, -\frac{b^2 - 4ac}{4a}\right)$

(b) The second derivative, $f''(x) = 2a$, describes concavity $\Rightarrow$ when $a > 0$ the parabola is concave up and when $a < 0$ the parabola is concave down.

71. A quadratic curve never has an inflection point. If $y = ax^2 + bx + c$ where $a \neq 0$, then $y' = 2ax + b$ and $y'' = 2a$. Since $2a$ is a constant, it is not possible for y'' to change signs.

73. If $y = x^5 - 5x^4 - 240$, then $y' = 5x^3(x - 4)$ and $y'' = 20x^2(x - 3)$. The zeros of y' are extrema, and there is a point of inflection at $x = 3$.

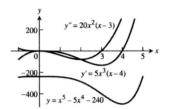

75. If $y = \frac{4}{5}x^5 + 16x^2 - 25$, then $y' = 4x(x^3 + 8)$ and $y'' = 16(x^3 + 2)$. The zeros of y' and y'' are extrema and points of inflection, respectively.

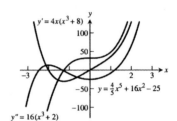

77. (a) If $y = x^{2/3}(x^2 - 2)$, then $y' = \frac{4}{3}x^{-1/3}(2x^2 - 1)$ and $y'' = \frac{4}{9}x^{-4/3}(10x^2 + 1)$. The curve rises on $\left(-\frac{1}{\sqrt{2}}, 0\right)$ and $\left(\frac{1}{\sqrt{2}}, \infty\right)$ and falls on $\left(-\infty, -\frac{1}{\sqrt{2}}\right)$ and $\left(0, \frac{1}{\sqrt{2}}\right)$. The curve is concave up on $(-\infty, 0)$ and $(0, \infty)$.

(b) A cusp since $\lim\limits_{x \to 0^-} y' = \infty$ and $\lim\limits_{x \to 0^+} y' = -\infty$.

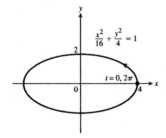

3.5 PARAMETRIZATIONS OF PLANE CURVES

1. $x = \cos 2t, \ y = \sin 2t, \ 0 \leq t \leq \pi$
   $\Rightarrow \cos^2 2t + \sin^2 2t = 1 \Rightarrow x^2 + y^2 = 1$

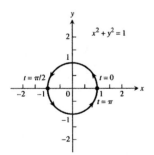

3. $x = 4 \cos t, \ y = 2 \sin t, \ 0 \leq t \leq 2\pi$
   $\Rightarrow \frac{16 \cos^2 t}{16} + \frac{4 \sin^2 t}{4} = 1 \Rightarrow \frac{x^2}{16} + \frac{y^2}{4} = 1$

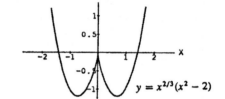

5. $x = 3t, y = 9t^2, -\infty < t < \infty \Rightarrow y = x^2$

7. $x = 2t - 5, y = 4t - 7, -\infty < t < \infty$

   $\Rightarrow x + 5 = 2t \Rightarrow 2(x + 5) = 4t$

   $\Rightarrow y = 2(x + 5) - 7 \Rightarrow y = 2x + 3$

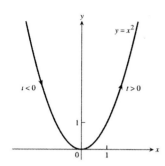

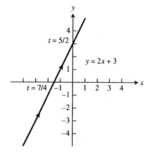

9. $x = t, y = \sqrt{1 - t^2}, -1 \le t \le 0$

   $\Rightarrow y = \sqrt{1 - x^2}$

11. $x = \sec^2 t - 1, y = \tan t, -\frac{\pi}{2} < t < \frac{\pi}{2}$

   $\Rightarrow \sec^2 t - 1 = \tan^2 t \Rightarrow x = y^2$

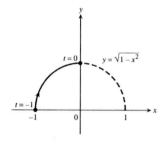

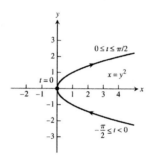

13. (a) $x = a \cos t, y = -a \sin t, 0 \le t \le 2\pi$

   (c) $x = a \cos t, y = -a \sin t, 0 \le t \le 4\pi$

   (b) $x = a \cos t, y = a \sin t, 0 \le t \le 2\pi$

   (d) $x = a \cos t, y = a \sin t, 0 \le t \le 4\pi$

15. Using $(-1, -3)$ we create the parametric equations $x = -1 + at$ and $y = -3 + bt$, representing a line which goes
    through $(-1, -3)$ at $t = 0$. We determine a and b so that the line goes through $(4, 1)$ when $t = 1$.
    Since $4 = -1 + a \Rightarrow a = 5$. Since $1 = -3 + b \Rightarrow b = 4$. Therefore, one possible parameterization is $x = -1 + 5t$,
    $y = -3 + 4t, 0 \le t \le 1$.

17. The lower half of the parabola is given by $x = y^2 + 1$ for $y \le 0$. Substituting t for y, we obtain one possible
    parameterization $x = t^2 + 1, y = t, t \le 0$.

19. For simplicity, we assume that x and y are linear functions of t and that the point(x, y) starts at $(2, 3)$ for $t = 0$ and passes
    through $(-1, -1)$ at $t = 1$. Then $x = f(t)$, where $f(0) = 2$ and $f(1) = -1$.
    Since slope $= \frac{\Delta x}{\Delta t} = \frac{-1-2}{1-0} = -3, x = f(t) = -3t + 2 = 2 - 3t$. Also, $y = g(t)$, where $g(0) = 3$ and $g(1) = -1$.
    Since slope $= \frac{\Delta y}{\Delta t} = \frac{-1-3}{1-0} = -4. y = g(t) = -4t + 3 = 3 - 4t$.
    One possible parameterization is: $x = 2 - 3t, y = 3 - 4t, t \ge 0$.

21. $t = \frac{\pi}{4} \Rightarrow x = 2 \cos \frac{\pi}{4} = \sqrt{2}, y = 2 \sin \frac{\pi}{4} = \sqrt{2}; \frac{dx}{dt} = -2 \sin t, \frac{dy}{dt} = 2 \cos t \Rightarrow \frac{dy}{dx} = \frac{dy/dt}{dx/dt} = \frac{2 \cos t}{-2 \sin t} = -\cot t$

   $\Rightarrow \frac{dy}{dx}\Big|_{t=\frac{\pi}{4}} = -\cot \frac{\pi}{4} = -1$; tangent line is $y - \sqrt{2} = -1\left(x - \sqrt{2}\right)$ or $y = -x + 2\sqrt{2}; \frac{dy'}{dt} = \csc^2 t$

   $\Rightarrow \frac{d^2y}{dx^2} = \frac{dy'/dt}{dx/dt} = \frac{\csc^2 t}{-2 \sin t} = -\frac{1}{2 \sin^3 t} \Rightarrow \frac{d^2y}{dx^2}\Big|_{t=\frac{\pi}{4}} = -\sqrt{2}$

23. $t = \frac{1}{4} \Rightarrow x = \frac{1}{4}, y = \frac{1}{2}; \frac{dx}{dt} = 1, \frac{dy}{dt} = \frac{1}{2\sqrt{t}} \Rightarrow \frac{dy}{dx} = \frac{dy/dt}{dx/dt} = \frac{1}{2\sqrt{t}} \Rightarrow \frac{dy}{dx}\Big|_{t=\frac{1}{4}} = \frac{1}{2\sqrt{\frac{1}{4}}} = 1$; tangent line is

$y - \frac{1}{2} = 1 \cdot \left(x - \frac{1}{4}\right)$ or $y = x + \frac{1}{4}; \frac{dy'}{dt} = -\frac{1}{4}t^{-3/2} \Rightarrow \frac{d^2y}{dx^2} = \frac{dy'/dt}{dx/dt} = -\frac{1}{4}t^{-3/2} \Rightarrow \frac{d^2y}{dx^2}\Big|_{t=\frac{1}{4}} = -2$

25. $t = -1 \Rightarrow x = 5, y = 1; \frac{dx}{dt} = 4t, \frac{dy}{dt} = 4t^3 \Rightarrow \frac{dy}{dx} = \frac{dy/dt}{dx/dt} = \frac{4t^3}{4t} = t^2 \Rightarrow \frac{dy}{dx}\Big|_{t=-1} = (-1)^2 = 1$; tangent line is

$y - 1 = 1 \cdot (x - 5)$ or $y = x - 4; \frac{dy'}{dt} = 2t \Rightarrow \frac{d^2y}{dx^2} = \frac{dy'/dt}{dx/dt} = \frac{2t}{4t} = \frac{1}{2} \Rightarrow \frac{d^2y}{dx^2}\Big|_{t=-1} = \frac{1}{2}$

27. $t = \frac{\pi}{2} \Rightarrow x = \cos\frac{\pi}{2} = 0, y = 1 + \sin\frac{\pi}{2} = 2; \frac{dx}{dt} = -\sin t, \frac{dy}{dt} = \cos t \Rightarrow \frac{dy}{dx} = \frac{\cos t}{-\sin t} = -\cot t$

$\Rightarrow \frac{dy}{dx}\Big|_{t=\frac{\pi}{2}} = -\cot\frac{\pi}{2} = 0$; tangent line is $y = 2; \frac{dy'}{dt} = \csc^2 t \Rightarrow \frac{d^2y}{dx^2} = \frac{\csc^2 t}{-\sin t} = -\csc^3 t \Rightarrow \frac{d^2y}{dx^2}\Big|_{t=\frac{\pi}{2}} = -1$

29. $x^3 + 2t^2 = 9 \Rightarrow 3x^2\frac{dx}{dt} + 4t = 0 \Rightarrow 3x^2\frac{dx}{dt} = -4t \Rightarrow \frac{dx}{dt} = \frac{-4t}{3x^2};$

$2y^3 - 3t^2 = 4 \Rightarrow 6y^2\frac{dy}{dt} - 6t = 0 \Rightarrow \frac{dy}{dt} = \frac{6t}{6y^2} = \frac{t}{y^2};$ thus $\frac{dy}{dx} = \frac{dy/dt}{dx/dt} = \frac{\left(\frac{t}{y^2}\right)}{\left(\frac{-4t}{3x^2}\right)} = \frac{t(3x^2)}{y^2(-4t)} = \frac{3x^2}{-4y^2}; t = 2$

$\Rightarrow x^3 + 2(2)^2 = 9 \Rightarrow x^3 + 8 = 9 \Rightarrow x^3 = 1 \Rightarrow x = 1; t = 2 \Rightarrow 2y^3 - 3(2)^2 = 4$

$\Rightarrow 2y^3 = 16 \Rightarrow y^3 = 8 \Rightarrow y = 2$; therefore $\frac{dy}{dx}\Big|_{t=2} = \frac{3(1)^2}{-4(2)^2} = -\frac{3}{16}$

31. $x + 2x^{3/2} = t^2 + t \Rightarrow \frac{dx}{dt} + 3x^{1/2}\frac{dx}{dt} = 2t + 1 \Rightarrow \left(1 + 3x^{1/2}\right)\frac{dx}{dt} = 2t + 1 \Rightarrow \frac{dx}{dt} = \frac{2t+1}{1+3x^{1/2}}; y\sqrt{t+1} + 2t\sqrt{y} = 4$

$\Rightarrow \frac{dy}{dt}\sqrt{t+1} + y\left(\frac{1}{2}\right)(t+1)^{-1/2} + 2\sqrt{y} + 2t\left(\frac{1}{2}y^{-1/2}\right)\frac{dy}{dt} = 0 \Rightarrow \frac{dy}{dt}\sqrt{t+1} + \frac{y}{2\sqrt{t+1}} + 2\sqrt{y} + \left(\frac{t}{\sqrt{y}}\right)\frac{dy}{dt} = 0$

$\Rightarrow \left(\sqrt{t+1} + \frac{t}{\sqrt{y}}\right)\frac{dy}{dt} = \frac{-y}{2\sqrt{t+1}} - 2\sqrt{y} \Rightarrow \frac{dy}{dt} = \frac{\left(\frac{-y}{2\sqrt{t+1}} - 2\sqrt{y}\right)}{\left(\sqrt{t+1} + \frac{t}{\sqrt{y}}\right)} = \frac{-y\sqrt{y} - 4y\sqrt{t+1}}{2\sqrt{y}(t+1) + 2t\sqrt{t+1}};$ thus

$\frac{dy}{dx} = \frac{dy/dt}{dx/dt} = \frac{\left(\frac{-y\sqrt{y}-4y\sqrt{t+1}}{2\sqrt{y}(t+1)+2t\sqrt{t+1}}\right)}{\left(\frac{2t+1}{1+3x^{1/2}}\right)}; t = 0 \Rightarrow x + 2x^{3/2} = 0 \Rightarrow x\left(1 + 2x^{1/2}\right) = 0 \Rightarrow x = 0; t = 0$

$\Rightarrow y\sqrt{0+1} + 2(0)\sqrt{y} = 4 \Rightarrow y = 4$; therefore $\frac{dy}{dx}\Big|_{t=0} = \frac{\left(\frac{-4\sqrt{4}-4(4)\sqrt{0+1}}{2\sqrt{4(0+1)}+2(0)\sqrt{0+1}}\right)}{\left(\frac{2(0)+1}{1+3(0)^{1/2}}\right)} = -6$

33. $\frac{dx}{dt} = \cos t$ and $\frac{dy}{dt} = 2\cos 2t \Rightarrow \frac{dy}{dx} = \frac{dy/dt}{dx/dt} = \frac{2\cos 2t}{\cos t} = \frac{2(2\cos^2 t - 1)}{\cos t};$ then $\frac{dy}{dx} = 0 \Rightarrow \frac{2(2\cos^2 t - 1)}{\cos t} = 0$

$\Rightarrow 2\cos^2 t - 1 = 0 \Rightarrow \cos t = \pm\frac{1}{\sqrt{2}} \Rightarrow t = \frac{\pi}{4}, \frac{3\pi}{4}, \frac{5\pi}{4}, \frac{7\pi}{4}.$ In the 1st quadrant: $t = \frac{\pi}{4} \Rightarrow x = \sin\frac{\pi}{4} = \frac{\sqrt{2}}{2}$ and

$y = \sin 2\left(\frac{\pi}{4}\right) = 1 \Rightarrow \left(\frac{\sqrt{2}}{2}, 1\right)$ is the point where the tangent line is horizontal. At the origin: $x = 0$ and $y = 0$

$\Rightarrow \sin t = 0 \Rightarrow t = 0$ or $t = \pi$ and $\sin 2t = 0 \Rightarrow t = 0, \frac{\pi}{2}, \pi, \frac{3\pi}{2}$; thus $t = 0$ and $t = \pi$ give the tangent lines at

the origin. Tangents at origin: $\frac{dy}{dx}\Big|_{t=0} = 2 \Rightarrow y = 2x$ and $\frac{dy}{dx}\Big|_{t=\pi} = -2 \Rightarrow y = -2x$

3.6 APPLIED OPTIMIZATION

1. Let ℓ and w represent the length and width of the rectangle, respectively. With an area of 16 in.^{2}, we have
   that $(\ell)(w) = 16 \Rightarrow w = 16\ell^{-1} \Rightarrow$ the perimeter is $P = 2\ell + 2w = 2\ell + 32\ell^{-1}$ and $P'(\ell) = 2 - \frac{32}{\ell^2} = \frac{2(\ell^2 - 16)}{\ell^2}$.
   Solving $P'(\ell) = 0 \Rightarrow \frac{2(\ell+4)(\ell-4)}{\ell^2} = 0 \Rightarrow \ell = -4, 4$. Since $\ell > 0$ for the length of a rectangle, ℓ must be 4 and
   $w = 4 \Rightarrow$ the perimeter is 16 in., a minimum since $P''(\ell) = \frac{16}{\ell^3} > 0$.

3. (a) The line containing point P also contains the points $(0, 1)$ and $(1, 0) \Rightarrow$ the line containing P is $y = 1 - x$
       $\Rightarrow$ a general point on that line is $(x, 1 - x)$.
   (b) The area $A(x) = 2x(1 - x)$, where $0 \le x \le 1$.

(c) When $A(x) = 2x - 2x^2$, then $A'(x) = 0 \Rightarrow 2 - 4x = 0 \Rightarrow x = \frac{1}{2}$. Since $A(0) = 0$ and $A(1) = 0$, we conclude that $A\left(\frac{1}{2}\right) = \frac{1}{2}$ sq units is the largest area. The dimensions are 1 unit by $\frac{1}{2}$ unit.

5. The volume of the box is $V(x) = x(15 - 2x)(8 - 2x)$
   $= 120x - 46x^2 + 4x^3$, where $0 \le x \le 4$. Solving $V'(x) = 0$
   $\Rightarrow 120 - 92x + 12x^2 = 4(6 - x)(5 - 3x) = 0 \Rightarrow x = \frac{5}{3}$
   or 6, but 6 is not in the domain. Since $V(0) = V(4) = 0$,
   $V\left(\frac{5}{3}\right) = \frac{2450}{27} \approx 91$ in^3 must be the maximum volume of
   the box with dimensions $\frac{14}{3} \times \frac{35}{3} \times \frac{5}{3}$ inches.

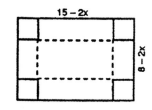

7. The area is $A(x) = x(800 - 2x)$, where $0 \le x \le 400$.
   Solving $A'(x) = 800 - 4x = 0 \Rightarrow x = 200$. With
   $A(0) = A(400) = 0$, the maximum area is
   $A(200) = 80,000$ m^2. The dimensions are 200 m by 400 m.

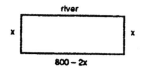

9. (a) We minimize the weight $= tS$ where S is the surface area, and t is the thickness of the steel walls of the tank. The surace area is $S = x^2 + 4xy$ where x is the length of a side of the square base of the tank, and y is its depth. The volume of the tank must be 500ft$^3 \Rightarrow y = \frac{500}{x^2}$. Therefore, the weight of the tank is $w(x) = t\left(x^2 + \frac{2000}{x}\right)$. Treating the thickness as a constant gives $w'(x) = t\left(2x - \frac{2000}{x^2}\right)$ for x.0. The critical value is at $x = 10$. Since $w''(10) = t\left(2 + \frac{4000}{10^3}\right) > 0$, there is a minimum at $x = 10$. Therefore, the optimum dimensions of the tank are 10 ft on the base edges and 5 ft deep.

   (b) Minimizing the surface area of the tank minimizes its weight for a given wall thickness. The thickness of the steel walls would likely be determined by other considerations such as structural requirements.

11. The area of the printing is $(y - 4)(x - 8) = 50$.
    Consequently, $y = \left(\frac{50}{x-8}\right) + 4$. The area of the paper is
    $A(x) = x\left(\frac{50}{x-8} + 4\right)$, where $8 < x$. Then
    $A'(x) = \left(\frac{50}{x-8} + 4\right) - x\left(\frac{50}{(x-8)^2}\right) = \frac{4(x-8)^2 - 400}{(x-8)^2} = 0$
    $\Rightarrow$ the critical points are -2 and 18, but -2 is not in the
    domain. Thus $A''(18) > 0 \Rightarrow$ at $x = 18$ we have
    a minimum. Therefore the dimensions 18 by 9 inches
    minimize the amount of paper.

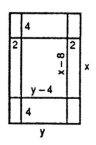

13. The area of the triangle is $A(\theta) = \frac{ab \sin \theta}{2}$, where $0 < \theta < \pi$.
    Solving $A'(\theta) = 0 \Rightarrow \frac{ab \cos \theta}{2} = 0 \Rightarrow \theta = \frac{\pi}{2}$. Since $A''(\theta)$
    $= -\frac{ab \sin \theta}{2} \Rightarrow A''\left(\frac{\pi}{2}\right) < 0$, there is a maximum at $\theta = \frac{\pi}{2}$.

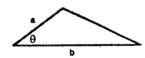

15. With a volume of 1000 cm and $V = \pi r^2 h$, then $h = \frac{1000}{\pi r^2}$. The amount of aluminum used per can is
    $A = 8r^2 + 2\pi rh = 8r^2 + \frac{2000}{r}$. Then $A'(r) = 16r - \frac{2000}{r^2} = 0 \Rightarrow \frac{8r^3 - 1000}{r^2} = 0 \Rightarrow$ the critical points are 0 and 5,
    but $r = 0$ results in no can. Since $A''(r) = 16 + \frac{1000}{r^3} > 0$ we have a minimum at $r = 5 \Rightarrow h = \frac{40}{\pi}$ and h:r = 8:π.

17. Let the radius of the cylinder be r cm, $0 < r < 10$. Then the height is $2\sqrt{100 - r^2}$ and the volume is
    $V(r) = 2\pi r^2 \sqrt{100 - r^2}$ cm^3. Then, $V'(r) = 2\pi r^2 \left(\frac{1}{\sqrt{100-r^2}}\right)(-2r) + \left(2\pi \sqrt{100 - r^2}\right)(2r)$
    $= \frac{-2\pi r^3 + 4\pi r(100 - r^2)}{\sqrt{100 - r^2}} = \frac{2\pi r(200 - 3r^2)}{\sqrt{100 - r^2}}$. The critical point for $0 < r < 10$ occurs at $r = \sqrt{\frac{200}{3}} = 10\sqrt{\frac{2}{3}}$. Since $V'(r) > 0$ for

$0 < r < 10\sqrt{\frac{2}{3}}$ and $V'(r) < 0$ for $10\sqrt{\frac{2}{3}} < r < 10$, the critical point corresponds to the maximum volume. The

dimensions are $r = 10\sqrt{\frac{2}{3}} \approx 8.16$ cm and $h = \frac{20}{\sqrt{3}} \approx 11.55$ cm, and the volume is $\frac{4000\pi}{3\sqrt{3}} \approx 2418.40$ cm^3.

19. The fixed volume is $V = \pi r^2 h + \frac{2}{3}\pi r^3 \Rightarrow h = \frac{V}{\pi r^2} - \frac{2r}{3}$, where h is the height of the cylinder and r is the radius

of the hemisphere. To minimize the cost we must minimize surface area of the cylinder added to twice the

surface area of the hemisphere. Thus, we minimize $C = 2\pi rh + 4\pi r^2 = 2\pi r\left(\frac{V}{\pi r^2} - \frac{2r}{3}\right) + 4\pi r^2 = \frac{2V}{r} + \frac{8}{3}\pi r^2$.

Then $\frac{dC}{dr} = -\frac{2V}{r^2} + \frac{16}{3}\pi r = 0 \Rightarrow V = \frac{8}{3}\pi r^3 \Rightarrow r = \left(\frac{3V}{8\pi}\right)^{1/3}$. From the volume equation, $h = \frac{V}{\pi r^2} - \frac{2r}{3}$

$= \frac{4V^{1/3}}{\pi^{1/3}\cdot 3^{2/3}} - \frac{2\cdot 3^{1/3}\cdot V^{1/3}}{3\cdot 2\cdot \pi^{1/3}} = \frac{3^{1/3}\cdot 2\cdot 4\cdot V^{1/3} - 2\cdot 3^{1/3}\cdot V^{1/3}}{3\cdot 2\cdot \pi^{1/3}} = \left(\frac{3V}{\pi}\right)^{1/3}$. Since $\frac{d^2C}{dr^2} = \frac{4V}{r^3} + \frac{16}{3}\pi > 0$, these

dimensions do minimize the cost.

21. Note that $h^2 + r^2 = 3$ and so $r = \sqrt{3 - h^2}$. Then the volume is given by $V = \frac{\pi}{3}r^2 h = \frac{\pi}{3}(3 - h^2)h = \pi h - \frac{\pi}{3}h^3$ for

$0 < h < \sqrt{3}$, and so $\frac{dV}{dh} = \pi - \pi r^2 = \pi(1 - r^2)$. The critical point (for $h > 0$) occurs at $h = 1$. Since $\frac{dV}{dh} > 0$ for

$0 < h < 1$, and $\frac{dV}{dh} < 0$ for $1 < h < \sqrt{3}$, the critical point corresponds to the maximum volume. The cone of greatest

volume has radius $\sqrt{2}$ m, height 1m, and volume $\frac{2\pi}{3}$ m^3.

23. (a) $s(t) = -16t^2 + 96t + 112 \Rightarrow v(t) = s'(t) = -32t + 96$. At $t = 0$, the velocity is $v(0) = 96$ ft/sec.

(b) The maximum height ocurs when $v(t) = 0$, when $t = 3$. The maximum height is $s(3) = 256$ ft and it occurs at $t = 3$

sec.

(c) Note that $s(t) = -16t^2 + 96t + 112 = -16(t + 1)(t - 7)$, so $s = 0$ at $t = -1$ or $t = 7$. Choosing the positive value

of t, the velocity when $s = 0$ is $v(7) = -128$ ft/sec.

25. $\frac{8}{x} = \frac{h}{x + 27} \Rightarrow h = 8 + \frac{216}{x}$ and $L(x) = \sqrt{h^2 + (x + 27)^2}$

$= \sqrt{\left(8 + \frac{216}{x}\right)^2 + (x + 27)^2}$ when $x \geq 0$. Note that $L(x)$ is

minimized when $f(x) = \left(8 + \frac{216}{x}\right)^2 + (x + 27)^2$ is

minimized. If $f'(x) = 0$, then

$2\left(8 + \frac{216}{x}\right)\left(-\frac{216}{x^2}\right) + 2(x + 27) = 0$

$\Rightarrow (x + 27)\left(1 - \frac{1728}{x^3}\right) = 0 \Rightarrow x = -27$ (not acceptable

since distance is never negative or $x = 12$. Then $L(12) = \sqrt{2197} \approx 46.87$ ft.

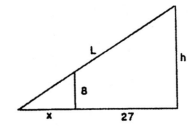

27. (a) $s = 10\cos(\pi t) \Rightarrow v = -10\pi\sin(\pi t) \Rightarrow$ speed $= |10\pi\sin(\pi t)| = 10\pi|\sin(\pi t)| \Rightarrow$ the maximum speed is

$10\pi \approx 31.42$ cm/sec since the maximum value of $|\sin(\pi t)|$ is 1; the cart is moving the fastest at $t = 0.5$ sec,

1.5 sec, 2.5 sec and 3.5 sec when $|\sin(\pi t)|$ is 1. At these times the distance is $s = 10\cos\left(\frac{\pi}{2}\right) = 0$ cm and

$a = -10\pi^2\cos(\pi t) \Rightarrow |a| = 10\pi^2|\cos(\pi t)| \Rightarrow |a| = 0$ cm/sec^2

(b) $|a| = 10\pi^2|\cos(\pi t)|$ is greatest at $t = 0.0$ sec, 1.0 sec, 2.0 sec, 3.0 sec and 4.0 sec, and at these times the

magnitude of the cart's position is $|s| = 10$ cm from the rest position and the speed is 0 cm/sec.

29.

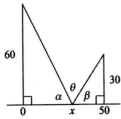

From the diagram above we have $\alpha + \beta + \theta = \pi \Rightarrow \theta = \pi - \alpha - \beta$. From the left triangle we have

$\cot\alpha = \frac{x}{60} \Rightarrow \cot^{-1}\left(\frac{x}{60}\right) = \alpha$, and from the triangle on the right side we have $\cot\beta = \frac{50-x}{60} \Rightarrow \cot^{-1}\left(\frac{50-x}{30}\right) = \beta$

$\Rightarrow \theta = \pi - \cot^{-1}\left(\frac{x}{60}\right) - \cot^{-1}\left(\frac{50-x}{30}\right) \Rightarrow \frac{d\theta}{dx} = 0 - \left(-\frac{1}{1+\left(\frac{x}{60}\right)^2} \cdot \frac{1}{60}\right) - \left(-\frac{1}{1+\left(\frac{50-x}{30}\right)^2} \cdot \frac{-1}{30}\right)$

$\frac{d\theta}{dx} = \frac{1}{60\left(1+\left(\frac{x}{60}\right)^2\right)} - \frac{1}{30\left(1+\left(\frac{50-x}{30}\right)^2\right)} = \frac{1}{60+\frac{x^2}{60}} - \frac{1}{30+\frac{(50-x)^2}{30}} = \frac{60}{60^2+x^2} - \frac{30}{30^2+(50-x)^2}$

$\frac{60}{60^2+x^2} - \frac{30}{30^2+(50-x)^2} = 0 \Rightarrow 60\left(30^2+(50-x)^2\right) - 30(60^2+x^2) \Rightarrow 30x^2 - 6000x + 96000 = 0$

$\Rightarrow x = 20\left(5 \pm \sqrt{17}\right)$. Since $0 \le x \le 50 \Rightarrow x = 20\left(5 - \sqrt{17}\right) \approx 17.54$. $\theta'(10) = \frac{60}{60^2+10^2} - \frac{30}{30^2+(50-10)^2}$

$= \frac{60}{3700} - \frac{30}{2500} = \frac{39}{9250} > 0$ and $\theta'(20) = \frac{60}{60^2+20^2} - \frac{30}{30^2+(50-20)^2} = \frac{60}{4000} - \frac{30}{1800} = -\frac{1}{600}$.

$\Rightarrow$ the maximum angle is $\theta = \pi - \cot^{-1}\left(\frac{20\left(5-\sqrt{17}\right)}{60}\right) - \cot^{-1}\left(\frac{50-20\left(5-\sqrt{17}\right)}{30}\right) \approx 1.10917$ rad or

63.55° when the solar station is ≈ 17.54 m west of the left (60 m) building.

31. The profit is $p = nx - nc = n(x-c) = \left[a(x-c)^{-1} + b(100-x)\right](x-c) = a + b(100-x)(x-c)$
    $= a + (bc + 100b)x - 100bc - bx^2$. Then $p'(x) = bc + 100b - 2bx$ and $p''(x) = -2b$. Solving $p'(x) = 0 \Rightarrow x = \frac{c}{2} + 50$.
    At $x = \frac{c}{2} + 50$ there is a maximum profit since $p''(x) = -2b < 0$ for all x.

33. (a) $A(q) = kmq^{-1} + cm + \frac{h}{2}q$, where $q > 0 \Rightarrow A'(q) = -kmq^{-2} + \frac{h}{2} = \frac{hq^2 - 2km}{2q^2}$ and $A''(q) = 2kmq^{-3}$. The
    critical points are $-\sqrt{\frac{2km}{h}}$, 0, and $\sqrt{\frac{2km}{h}}$, but only $\sqrt{\frac{2km}{h}}$ is in the domain. Then $A''\left(\sqrt{\frac{2km}{h}}\right) > 0 \Rightarrow$ at
    $q = \sqrt{\frac{2km}{h}}$ there is a minimum average weekly cost.

    (b) $A(q) = \frac{(k+bq)m}{q} + cm + \frac{h}{2}q = kmq^{-1} + bm + cm + \frac{h}{2}q$, where $q > 0 \Rightarrow A'(q) = 0$ at $q = \sqrt{\frac{2km}{h}}$ as in (a).
    Also $A''(q) = 2kmq^{-3} > 0$ so the most economical quantity to order is still $q = \sqrt{\frac{2km}{h}}$ which minimizes the
    average weekly cost.

35. We have $\frac{dR}{dM} = CM - M^2$. Solving $\frac{d^2R}{dM^2} = C - 2M = 0 \Rightarrow M = \frac{C}{2}$. Also, $\frac{d^3R}{dM^3} = -2 < 0 \Rightarrow$ at $M = \frac{C}{2}$ there is a
    maximum.

37. If $x > 0$, then $(x-1)^2 \ge 0 \Rightarrow x^2 + 1 \ge 2x \Rightarrow \frac{x^2+1}{x} \ge 2$. In particular if a, b, c and d are positive integers,
    then $\left(\frac{a^2+1}{a}\right)\left(\frac{b^2+1}{b}\right)\left(\frac{c^2+1}{c}\right)\left(\frac{d^2+1}{d}\right) \ge 16$.

39. At $x = c$, the tangents to the curves are parallel. Justification: The vertical distance between the curves is
    $D(x) = f(x) - g(x)$, so $D'(x) = f'(x) - g'(x)$. The maximum value of D will occur at a point c where $D' = 0$. At
    such a point, $f'(c) - g'(c) = 0$, or $f'(c) = g'(c)$.

41. (a) If $y = \cot x - \sqrt{2}\csc x$ where $0 < x < \pi$, then $y' = (\csc x)\left(\sqrt{2}\cot x - \csc x\right)$. Solving $y' = 0 \Rightarrow \cos x = \frac{1}{\sqrt{2}}$
    $\Rightarrow x = \frac{\pi}{4}$. For $0 < x < \frac{\pi}{4}$ we have $y' > 0$, and $y' < 0$ when $\frac{\pi}{4} < x < \pi$. Therefore, at $x = \frac{\pi}{4}$ there is a maximum
    value of $y = -1$.

(b)

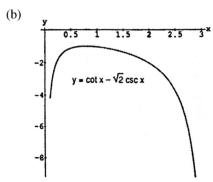

The graph confirms the findings in (a).

43. (a) The square of the distance is $D(x) = \left(x - \frac{3}{2}\right)^2 + \left(\sqrt{x} + 0\right)^2 = x^2 - 2x + \frac{9}{4}$, so $D'(x) = 2x - 2$ and the critical

point occurs at $x = 1$. Since $D'(x) < 0$ for $x < 1$ and $D'(x) > 0$ for $x > 1$, the critical point corresponds to the

minimum distance. The minimum distance is $\sqrt{D(1)} = \frac{\sqrt{5}}{2}$.

(b)

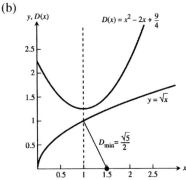

The minimum distance is from the point $\left(\frac{3}{2}, 0\right)$ to the point $(1, 1)$ on the graph of $y = \sqrt{x}$, and this occurs at the

value $x = 1$ where $D(x)$, the distance squared, has its minimum value.

3.7 INDETERMINATE FORMS AND L'HÔPITAL'S RULE

1. l'Hôpital: $\lim\limits_{x \to 2} \frac{x-2}{x^2-4} = \frac{1}{2x}\Big|_{x=2} = \frac{1}{4}$ or $\lim\limits_{x \to 2} \frac{x-2}{x^2-4} = \lim\limits_{x \to 2} \frac{x-2}{(x-2)(x+2)} = \lim\limits_{x \to 2} \frac{1}{x+2} = \frac{1}{4}$

3. l'Hôpital: $\lim\limits_{x \to \infty} \frac{5x^2-3x}{7x^2+1} = \lim\limits_{x \to \infty} \frac{10x-3}{14x} = \lim\limits_{x \to \infty} \frac{10}{14} = \frac{5}{7}$ or $\lim\limits_{x \to \infty} \frac{5x^2-3x}{7x^2+1} = \lim\limits_{x \to \infty} \frac{5-\frac{3}{x}}{7+\frac{1}{x}} = \frac{5}{7}$

5. l'Hôpital: $\lim\limits_{x \to 0} \frac{1-\cos x}{x^2} = \lim\limits_{x \to 0} \frac{\sin x}{2x} = \lim\limits_{x \to 0} \frac{\cos x}{2} = \frac{1}{2}$ or $\lim\limits_{x \to 0} \frac{1-\cos x}{x^2} = \lim\limits_{x \to 0} \left[\left(\frac{1-\cos x}{x^2}\right)\left(\frac{1+\cos x}{1+\cos x}\right)\right]$

$= \lim\limits_{x \to 0} \frac{\sin^2 x}{x^2(1+\cos x)} = \lim\limits_{x \to 0} \left[\left(\frac{\sin x}{x}\right)\left(\frac{\sin x}{x}\right)\left(\frac{1}{1+\cos x}\right)\right] = \frac{1}{2}$

7. $\lim\limits_{x \to 2} \frac{x-2}{x^2-4} = \lim\limits_{x \to 2} \frac{1}{2x} = \frac{1}{4}$

9. $\lim\limits_{t \to -3} \frac{t^3-4t+15}{t^2-t-12} = \lim\limits_{t \to -3} \frac{3t^2-4}{2t-1} = \frac{3(-3)^2-4}{2(-3)-1} = -\frac{23}{7}$

11. $\lim\limits_{x \to \infty} \frac{5x^3-2x}{7x^3+3} = \lim\limits_{x \to \infty} \frac{15x^2-2}{21x^2} = \lim\limits_{x \to \infty} \frac{30x}{42x} = \lim\limits_{x \to \infty} \frac{30}{42} = \frac{5}{7}$

13. $\lim\limits_{t \to 0} \frac{\sin t^2}{t} = \lim\limits_{t \to 0} \frac{(\cos t^2)(2t)}{1} = 0$

15. $\lim\limits_{x \to 0} \frac{8x^2}{\cos x - 1} = \lim\limits_{x \to 0} \frac{16x}{-\sin x} = \lim\limits_{x \to 0} \frac{16}{-\cos x} = \frac{16}{-1} = -16$

17. $\lim\limits_{\theta \to \pi/2} \frac{2\theta - \pi}{\cos(2\pi - \theta)} = \lim\limits_{\theta \to \pi/2} \frac{2}{\sin(2\pi - \theta)} = \frac{2}{\sin\left(\frac{3\pi}{2}\right)} = -2$

19. $\lim\limits_{\theta \to \pi/2} \frac{1 - \sin\theta}{1 + \cos 2\theta} = \lim\limits_{\theta \to \pi/2} \frac{-\cos\theta}{-2\sin 2\theta} = \lim\limits_{\theta \to \pi/2} \frac{\sin\theta}{-4\cos 2\theta} = \frac{1}{(-4)(-1)} = \frac{1}{4}$

21. $\lim\limits_{x \to 0} \frac{x^2}{\ln(\sec x)} = \lim\limits_{x \to 0} \frac{2x}{\left(\frac{\sec x \tan x}{\sec x}\right)} = \lim\limits_{x \to 0} \frac{2x}{\tan x} = \lim\limits_{x \to 0} \frac{2}{\sec^2 x} = \frac{2}{1^2} = 2$

23. $\lim\limits_{t \to 0} \frac{t(1 - \cos t)}{t - \sin t} = \lim\limits_{t \to 0} \frac{(1 - \cos t) + t(\sin t)}{1 - \cos t} = \lim\limits_{t \to 0} \frac{\sin t + (\sin t + t\cos t)}{\sin t}$

$= \lim\limits_{t \to 0} \frac{\cos t + \cos t + \cos t - t\sin t}{\cos t} = \frac{1 + 1 + 1 - 0}{1} = 3$

25. $\lim\limits_{x \to (\pi/2)^-} \left(x - \frac{\pi}{2}\right)\sec x = \lim\limits_{x \to (\pi/2)^-} \frac{\left(x - \frac{\pi}{2}\right)}{\cos x} = \lim\limits_{x \to (\pi/2)^-} \left(\frac{1}{-\sin x}\right) = \frac{1}{-1} = -1$

27. $\lim\limits_{\theta \to 0} \frac{3^{\sin\theta} - 1}{\theta} = \lim\limits_{\theta \to 0} \frac{3^{\sin\theta}(\ln 3)(\cos\theta)}{1} = \frac{(3^0)(\ln 3)(1)}{1} = \ln 3$

29. $\lim\limits_{x \to 0} \frac{x\,2^x}{2^x - 1} = \lim\limits_{x \to 0} \frac{(1)(2^x) + (x)(\ln 2)(2^x)}{(\ln 2)(2^x)} = \frac{1 \cdot 2^0 + 0}{(\ln 2) \cdot 2^0} = \frac{1}{\ln 2}$

31. $\lim\limits_{x \to \infty} \frac{\ln(x+1)}{\log_2 x} = \lim\limits_{x \to \infty} \frac{\ln(x+1)}{\left(\frac{\ln x}{\ln 2}\right)} = (\ln 2)\lim\limits_{x \to \infty} \frac{\left(\frac{1}{x+1}\right)}{\left(\frac{1}{x}\right)} = (\ln 2)\lim\limits_{x \to \infty} \frac{x}{x+1} = (\ln 2)\lim\limits_{x \to \infty} \frac{1}{1} = \ln 2$

33. $\lim\limits_{x \to 0^+} \frac{\ln(x^2 + 2x)}{\ln x} = \lim\limits_{x \to 0^+} \frac{\left(\frac{2x+2}{x^2+2x}\right)}{\left(\frac{1}{x}\right)} = \lim\limits_{x \to 0^+} \frac{2x^2 + 2x}{x^2 + 2x} = \lim\limits_{x \to 0^+} \frac{4x+2}{2x+2} = \lim\limits_{x \to 0^+} \frac{2}{2} = 1$

35. $\lim\limits_{y \to 0} \frac{\sqrt{5y + 25} - 5}{y} = \lim\limits_{y \to 0} \frac{(5y + 25)^{1/2} - 5}{y} = \lim\limits_{y \to 0} \frac{\left(\frac{1}{2}\right)(5y + 25)^{-1/2}(5)}{1} = \lim\limits_{y \to 0} \frac{5}{2\sqrt{5y + 25}} = \frac{1}{2}$

37. $\lim\limits_{x \to \infty} [\ln 2x - \ln(x + 1)] = \lim\limits_{x \to \infty} \ln\left(\frac{2x}{x+1}\right) = \ln\left(\lim\limits_{x \to \infty} \frac{2x}{x+1}\right) = \ln\left(\lim\limits_{x \to \infty} \frac{2}{1}\right) = \ln 2$

39. $\lim\limits_{h \to 0} \frac{\sin(a + h) - \sin a}{h} = \lim\limits_{h \to 0} \frac{\cos(a + h) - 0}{1} = \cos a$

41. $\lim\limits_{x \to 1^+} \left(\frac{1}{x-1} - \frac{1}{\ln x}\right) = \lim\limits_{x \to 1^+} \left(\frac{\ln x - (x-1)}{(x-1)(\ln x)}\right) = \lim\limits_{x \to 1^+} \left(\frac{\frac{1}{x} - 1}{(\ln x) + (x-1)\left(\frac{1}{x}\right)}\right) = \lim\limits_{x \to 1^+} \left(\frac{1 - x}{(x\ln x) + x - 1}\right)$

$= \lim\limits_{x \to 1^+} \left(\frac{-1}{(\ln x + 1) + 1}\right) = \frac{-1}{(0 + 1) + 1} = -\frac{1}{2}$

43. $\lim\limits_{\theta \to 0} \frac{\cos\theta - 1}{e^\theta - \theta - 1} = \lim\limits_{\theta \to 0} \frac{-\sin\theta}{e^\theta - 1} = \lim\limits_{\theta \to 0} \frac{-\cos\theta}{e^\theta} = -1$

45. $\lim\limits_{t \to \infty} \frac{e^t + t^2}{e^t - 1} = \lim\limits_{t \to \infty} \frac{e^t + 2t}{e^t} = \lim\limits_{t \to \infty} \frac{e^t + 2}{e^t} = \lim\limits_{t \to \infty} \frac{e^t}{e^t} = 1$

47. The limit leads to the indeterminate form 1^∞. Let $f(x) = x^{1/(1-x)} \Rightarrow \ln f(x) = \ln\left(x^{1/(1-x)}\right) = \frac{\ln x}{1 - x}$. Now

$\lim\limits_{x \to 1^+} \ln f(x) = \lim\limits_{x \to 1^+} \frac{\ln x}{1 - x} = \lim\limits_{x \to 1^+} \frac{\left(\frac{1}{x}\right)}{-1} = -1$. Therefore $\lim\limits_{x \to 1^+} x^{1/(1-x)} = \lim\limits_{x \to 1^+} f(x) = \lim\limits_{x \to 1^+} e^{\ln f(x)} = e^{-1} = \frac{1}{e}$

49. The limit leads to the indeterminate form ∞^0. Let $f(x) = (\ln x)^{1/x} \Rightarrow \ln f(x) = \ln (\ln x)^{1/x} = \frac{\ln (\ln x)}{x}$. Now

$\lim\limits_{x \to \infty} \ln f(x) = \lim\limits_{x \to \infty} \frac{\ln (\ln x)}{x} = \lim\limits_{x \to \infty} \frac{\left(\frac{1}{x \ln x}\right)}{1} = 0$. Therefore $\lim\limits_{x \to \infty} (\ln x)^{1/x} = \lim\limits_{x \to \infty} f(x)$

$= \lim\limits_{x \to \infty} e^{\ln f(x)} = e^0 = 1$

51. The limit leads to the indeterminate form 0^0. Let $f(x) = x^{-1/\ln x} \Rightarrow \ln f(x) = -\frac{\ln x}{\ln x} = -1$. Therefore

$\lim\limits_{x \to 0^+} x^{-1/\ln x} = \lim\limits_{x \to 0^+} f(x) = \lim\limits_{x \to 0^+} e^{\ln f(x)} = e^{-1} = \frac{1}{e}$

53. The limit leads to the indeterminate form ∞^0. Let $f(x) = (1 + 2x)^{1/(2 \ln x)} \Rightarrow \ln f(x) = \frac{\ln (1 + 2x)}{2 \ln x}$

$\Rightarrow \lim\limits_{x \to \infty} \ln f(x) = \lim\limits_{x \to \infty} \frac{\ln (1 + 2x)}{2 \ln x} = \lim\limits_{x \to \infty} \frac{x}{1 + 2x} = \lim\limits_{x \to \infty} \frac{1}{2} = \frac{1}{2}$. Therefore $\lim\limits_{x \to \infty} (1 + 2x)^{1/(2 \ln x)}$

$= \lim\limits_{x \to \infty} f(x) = \lim\limits_{x \to \infty} e^{\ln f(x)} = e^{1/2}$

55. The limit leads to the indeterminate form 0^0. Let $f(x) = x^x \Rightarrow \ln f(x) = x \ln x \Rightarrow \ln f(x) = \frac{\ln x}{\left(\frac{1}{x}\right)}$

$= \lim\limits_{x \to 0^+} \ln f(x) = \lim\limits_{x \to 0^+} \frac{\ln x}{\left(\frac{1}{x}\right)} = \lim\limits_{x \to 0^+} \frac{\left(\frac{1}{x}\right)}{\left(-\frac{1}{x^2}\right)} = \lim\limits_{x \to 0^+} (-x) = 0$. Therefore $\lim\limits_{x \to 0^+} x^x = \lim\limits_{x \to 0^+} f(x)$

$= \lim\limits_{x \to 0^+} e^{\ln f(x)} = e^0 = 1$

57. $\lim\limits_{x \to \infty} \frac{\sqrt{9x + 1}}{\sqrt{x + 1}} = \sqrt{\lim\limits_{x \to \infty} \frac{9x + 1}{x + 1}} = \sqrt{\lim\limits_{x \to \infty} \frac{9}{1}} = \sqrt{9} = 3$

59. $\lim\limits_{x \to \pi/2^-} \frac{\sec x}{\tan x} = \lim\limits_{x \to \pi/2^-} \left(\frac{1}{\cos x}\right)\left(\frac{\cos x}{\sin x}\right) = \lim\limits_{x \to \pi/2^-} \frac{1}{\sin x} = 1$

61. Part (b) is correct because part (a) is neither in the $\frac{0}{0}$ nor $\frac{\infty}{\infty}$ form and so l'Hôpital's rule may not be used.

63. Part (d) is correct, the other parts are indeterminate forms and cannot be calculated by the incorrect arithmetic

65. The graph indicates a limit near -1. The limit leads to the

indeterminate form $\frac{0}{0}$: $\lim\limits_{x \to 1} \frac{2x^2 - (3x + 1)\sqrt{x} + 2}{x - 1}$

$= \lim\limits_{x \to 1} \frac{2x^2 - 3x^{3/2} - x^{1/2} + 2}{x - 1} = \lim\limits_{x \to 1} \frac{4x - \frac{9}{2}x^{1/2} - \frac{1}{2}x^{-1/2}}{1}$

$= \frac{4 - \frac{9}{2} - \frac{1}{2}}{1} = \frac{4 - 5}{1} = -1$

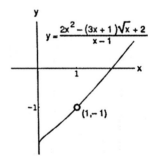

67. Let $f(k) = \left(1 + \frac{r}{k}\right)^k \Rightarrow \ln f(k) = \frac{\ln (1 + rk^{-1})}{k^{-1}} \Rightarrow \lim\limits_{k \to \infty} \frac{\ln (1 + rk^{-1})}{k^{-1}} = \lim\limits_{k \to \infty} \frac{\left(\frac{-rk^{-2}}{1 + rk^{-1}}\right)}{-k^{-2}} = \lim\limits_{k \to \infty} \frac{r}{1 + rk^{-1}}$

$= \lim\limits_{k \to \infty} \frac{rk}{k + r} = \lim\limits_{k \to \infty} \frac{r}{1} = r$. Therefore $\lim\limits_{k \to \infty} \left(1 + \frac{r}{k}\right)^k = \lim\limits_{k \to \infty} f(k) = \lim\limits_{k \to \infty} e^{\ln f(k)} = e^r$.

3.8 NEWTON'S METHOD

1. $y = x^2 + x - 1 \Rightarrow y' = 2x + 1 \Rightarrow x_{n+1} = x_n - \frac{x_n^2 + x_n - 1}{2x_n + 1}$; $x_0 = 1 \Rightarrow x_1 = 1 - \frac{1 + 1 - 1}{2 + 1} = \frac{2}{3}$

$\Rightarrow x_2 = \frac{2}{3} - \frac{\frac{4}{9} + \frac{2}{3} - 1}{\frac{4}{3} + 1} \Rightarrow x_2 = \frac{2}{3} - \frac{4 + 6 - 9}{12 + 9} = \frac{2}{3} - \frac{1}{21} = \frac{13}{21} \approx .61905$; $x_0 = -1 \Rightarrow x_1 = 1 - \frac{1 - 1 - 1}{-2 + 1} = -2$

$\Rightarrow x_2 = -2 - \frac{4 - 2 - 1}{-4 + 1} = -\frac{5}{3} \approx -1.66667$

3. $y = x^4 + x - 3 \Rightarrow y' = 4x^3 + 1 \Rightarrow x_{n+1} = x_n - \frac{x_n^4 + x_n - 3}{4x_n^3 + 1}$; $x_0 = 1 \Rightarrow x_1 = 1 - \frac{1 + 1 - 3}{4 + 1} = \frac{6}{5}$

$\Rightarrow x_2 = \frac{6}{5} - \frac{\frac{1296}{625} + \frac{6}{5} - 3}{\frac{864}{125} + 1} = \frac{6}{5} - \frac{1296 + 750 - 1875}{4320 + 625} = \frac{6}{5} - \frac{171}{4945} = \frac{5763}{4945} \approx 1.16542$; $x_0 = -1 \Rightarrow x_1 = -1 - \frac{1 - 1 - 3}{-4 + 1}$

$= -2 \Rightarrow x_2 = -2 - \frac{16 - 2 - 3}{-32 + 1} = -2 + \frac{11}{31} = -\frac{51}{31} \approx -1.64516$

5. One obvious root is $x = 0$. Graphing e^{-x} and $2x + 1$ shows that $x = 0$ is the only root. Taking a naive approach we can use Newton's Method to estimate the root as follows: Let $f(x) = e^{-x} - 2x - 1$, $x_0 = 1$, and $x_{n+1} = x_n - \frac{f(x_n)}{f'(x_n)}$ $= x_n + \frac{e^{-x_n} - 2x_n - 1}{e^{-x_n} + 2}$. Performing iterations on a calculator, spreadsheet, or CAS gives $x_1 = -0.111594$, $x_2 = -0.00215192$, $x_3 = -0.000000773248$. You may get different results depending upon what you select for $f(x)$ and x_0, and what calculator or computer you may use.

7. $f(x_0) = 0$ and $f'(x_0) \neq 0 \Rightarrow x_{n+1} = x_n - \frac{f(x_n)}{f'(x_n)}$ gives $x_1 = x_0 \Rightarrow x_2 = x_0 \Rightarrow x_n = x_0$ for all $n \geq 0$. That is, all of the approximations in Newton's method will be the root of $f(x) = 0$.

9. If $x_0 = h > 0 \Rightarrow x_1 = x_0 - \frac{f(x_0)}{f'(x_0)} = h - \frac{f(h)}{f'(h)}$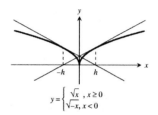

$= h - \frac{\sqrt{h}}{\left(\frac{1}{2\sqrt{h}}\right)} = h - \left(\sqrt{h}\right)\left(2\sqrt{h}\right) = -h$;

if $x_0 = -h < 0 \Rightarrow x_1 = x_0 - \frac{f(x_0)}{f'(x_0)} = -h - \frac{f(-h)}{f'(-h)}$

$= -h - \frac{\sqrt{h}}{\left(\frac{-1}{2\sqrt{h}}\right)} = -h + \left(\sqrt{h}\right)\left(2\sqrt{h}\right) = h$.

$y = \begin{cases} \sqrt{x}, & x \geq 0 \\ \sqrt{-x}, & x < 0 \end{cases}$

11. i) is equivalent to solving $x^3 - 3x - 1 = 0$.
    ii) is equivalent to solving $x^3 - 3x - 1 = 0$.
    iii) is equivalent to solving $x^3 - 3x - 1 = 0$.
    iv) is equivalent to solving $x^3 - 3x - 1 = 0$.
    All four equations are equivalent.

13. $f(x) = \tan x - 2x \Rightarrow f'(x) = \sec^2 x - 2 \Rightarrow x_{n+1} = x_n - \frac{\tan(x_n) - 2x_n}{\sec^2(x_n)}$; $x_0 = 1 \Rightarrow x_1 = 12920445$

$\Rightarrow x_2 = 1.155327774 \Rightarrow x_{16} = x_{17} = 1.165561185$

15. (a) The graph of $f(x) = \sin 3x - 0.99 + x^2$ in the window $-2 \leq x \leq 2, -2 \leq y \leq 3$ suggests three roots. However, when you zoom in on the x-axis near $x = 1.2$, you can see that the graph lies above the axis there. There are only two roots, one near $x = -1$, the other near $x = 0.4$.

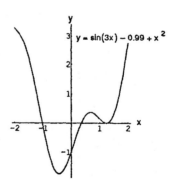

(b) $f(x) = \sin 3x - 0.99 + x^2 \Rightarrow f'(x) = 3 \cos 3x + 2x$
$\Rightarrow x_{n+1} = x_n - \frac{\sin(3x_n) - 0.99 + x_n^2}{3\cos(3x_n) + 2x_n}$ and the solutions are approximately 0.35003501505249 and -1.0261731615301

17. Graphing e^{-x^2} and $x^2 - x + 1$ shows that there are two places where the curves intersect, one at $x = 0$ and the other between $x = 0.5$ and $x = 0.6$. Let $f(x) = e^{-x^2} - x^2 + x - 1$, $x_0 = 0.5$, and $x_{n+1} = x_n - \frac{f(x_n)}{f'(x_n)} = x_n - \frac{e^{-x_n^2} - x_n^2 + x_n - 1}{1 - 2x_n - 2x_n e^{-x_n^2}}$.
Performing iterations on a calculator, spreadsheet, or CAS gives $x_1 = 0.536981$, $x_2 = 0.534856$, $x_3 = 0.53485$, $x_4 = 0.53485$. (You may get different results depending upon what you select for $f(x)$ and x_0, and what calculator or computer you may use.) Therefore, the two curves intersect at $x = 0$ and $x = 0.53485$.

19. $f(x) = 4x^4 - 4x^2 \Rightarrow f'(x) = 16x^3 - 8x \Rightarrow x_{i+1} = x_i - \frac{f(x_i)}{f'(x_i)} = x_i - \frac{x_i^3 - x_i}{4x_i^2 - 2}$. Iterations are performed using the

procedure in problem 13 in this section.

(a) For $x_0 = -2$ or $x_0 = -0.8$, $x_i \to -1$ as i gets large.

(b) For $x_0 = -0.5$ or $x_0 = 0.25$, $x_i \to 0$ as i gets large.

(c) For $x_0 = 0.8$ or $x_0 = 2$, $x_i \to 1$ as i gets large.

(d) (If your calculator has a CAS, put it in exact mode, otherwise approximate the radicals with a decimal value.)

For $x_0 = -\frac{\sqrt{21}}{7}$ or $x_0 = -\frac{\sqrt{21}}{7}$, Newton's method does not converge. The values of x_i alternate between

$x_0 = -\frac{\sqrt{21}}{7}$ or $x_0 = -\frac{\sqrt{21}}{7}$ as i increases.

3.9 HYPERBOLIC FUNCTIONS

1. $\sinh x = -\frac{3}{4} \Rightarrow \cosh x = \sqrt{1 + \sinh^2 x} = \sqrt{1 + \left(-\frac{3}{4}\right)^2} = \sqrt{1 + \frac{9}{16}} = \sqrt{\frac{25}{16}} = \frac{5}{4}$, $\tanh x = \frac{\sinh x}{\cosh x} = \frac{\left(-\frac{3}{4}\right)}{\left(\frac{5}{4}\right)} = -\frac{3}{5}$,

$\coth x = \frac{1}{\tanh x} = -\frac{5}{3}$, $\operatorname{sech} x = \frac{1}{\cosh x} = \frac{4}{5}$, and $\operatorname{csch} x = \frac{1}{\sin x} = -\frac{4}{3}$

3. $\cosh x = \frac{17}{15}$, $x > 0 \Rightarrow \sinh x = \sqrt{\cosh^2 x - 1} = \sqrt{\left(\frac{17}{15}\right)^2 - 1} = \sqrt{\frac{289}{225} - 1} = \sqrt{\frac{64}{225}} = \frac{8}{15}$, $\tanh x = \frac{\sinh x}{\cosh x} = \frac{\left(\frac{8}{15}\right)}{\left(\frac{17}{15}\right)}$

$= \frac{8}{17}$, $\coth x = \frac{1}{\tanh x} = \frac{17}{8}$, $\operatorname{sech} x = \frac{1}{\cosh x} = \frac{15}{17}$, and $\operatorname{csch} x = \frac{1}{\sinh x} = \frac{15}{8}$

5. $2 \cosh(\ln x) = 2\left(\frac{e^{\ln x} + e^{-\ln x}}{2}\right) = e^{\ln x} + \frac{1}{e^{\ln x}} = x + \frac{1}{x}$

7. $\cosh 5x + \sinh 5x = \frac{e^{5x} + e^{-5x}}{2} + \frac{e^{5x} - e^{-5x}}{2} = e^{5x}$

9. $(\sinh x + \cosh x)^4 = \left(\frac{e^x - e^{-x}}{2} + \frac{e^x + e^{-x}}{2}\right)^4 = (e^x)^4 = e^{4x}$

11. (a) $\sinh 2x = \sinh(x + x) = \sinh x \cosh x + \cosh x \sinh x = 2 \sinh x \cosh x$

(b) $\cosh 2x = \cosh(x + x) = \cosh x \cosh x + \sinh x \sin x = \cosh^2 x + \sinh^2 x$

13. $y = 6 \sinh \frac{x}{3} \Rightarrow \frac{dy}{dx} = 6\left(\cosh \frac{x}{3}\right)\left(\frac{1}{3}\right) = 2 \cosh \frac{x}{3}$

15. $y = 2\sqrt{t} \tanh \sqrt{t} = 2t^{1/2} \tanh t^{1/2} \Rightarrow \frac{dy}{dt} = \left[\operatorname{sech}^2\left(t^{1/2}\right)\right]\left(\frac{1}{2}t^{-1/2}\right)\left(2t^{1/2}\right) + \left(\tanh t^{1/2}\right)\left(t^{-1/2}\right)$

$= \operatorname{sech}^2 \sqrt{t} + \frac{\tanh \sqrt{t}}{\sqrt{t}}$

17. $y = \ln(\sinh z) \Rightarrow \frac{dy}{dz} = \frac{\cosh z}{\sinh z} = \coth z$

19. $y = (\operatorname{sech} \theta)(1 - \ln \operatorname{sech} \theta) \Rightarrow \frac{dy}{d\theta} = \left(-\frac{-\operatorname{sech} \theta \tanh \theta}{\operatorname{sech} \theta}\right)(\operatorname{sech} \theta) + (-\operatorname{sech} \theta \tanh \theta)(1 - \ln \operatorname{sech} \theta)$

$= \operatorname{sech} \theta \tanh \theta - (\operatorname{sech} \theta \tanh \theta)(1 - \ln \operatorname{sech} \theta) = (\operatorname{sech} \theta \tanh \theta)[1 - (1 - \ln \operatorname{sech} \theta)]$

$= (\operatorname{sech} \theta \tanh \theta)(\ln \operatorname{sech} \theta)$

21. $y = \ln \cosh v - \frac{1}{2} \tanh^2 v \Rightarrow \frac{dy}{dv} = \frac{\sinh v}{\cosh v} - \left(\frac{1}{2}\right)(2 \tanh v)(\operatorname{sech}^2 v) = \tanh v - (\tanh v)(\operatorname{sech}^2 v)$

$= (\tanh v)(1 - \operatorname{sech}^2 v) = (\tanh v)(\tanh^2 v) = \tanh^3 v$

23. $y = (x^2 + 1) \operatorname{sech}(\ln x) = (x^2 + 1)\left(\frac{2}{e^{\ln x} + e^{-\ln x}}\right) = (x^2 + 1)\left(\frac{2}{x + x^{-1}}\right) = (x^2 + 1)\left(\frac{2x}{x^2 + 1}\right) = 2x \Rightarrow \frac{dy}{dx} = 2$

25. $y = \sinh^{-1} \sqrt{x} = \sinh^{-1}\left(x^{1/2}\right) \Rightarrow \frac{dy}{dx} = \frac{\left(\frac{1}{2}\right)x^{-1/2}}{\sqrt{1 + (x^{1/2})^2}} = \frac{1}{2\sqrt{x}\sqrt{1 + x}} = \frac{1}{2\sqrt{x(1 + x)}}$

27. $y = (1 - \theta) \tanh^{-1} \theta \Rightarrow \frac{dy}{d\theta} = (1 - \theta) \left(\frac{1}{1 - \theta^2}\right) + (-1) \tanh^{-1} \theta = \frac{1}{1 + \theta} - \tanh^{-1} \theta$

29. $y = (1 - t) \coth^{-1} \sqrt{t} = (1 - t) \coth^{-1} (t^{1/2}) \Rightarrow \frac{dy}{dt} = (1 - t) \left[\frac{\left(\frac{1}{2}\right) t^{-1/2}}{1 - (t^{1/2})^2}\right] + (-1) \coth^{-1} (t^{1/2}) = \frac{1}{2\sqrt{t}} - \coth^{-1} \sqrt{t}$

31. $y = \cos^{-1} x - x \operatorname{sech}^{-1} x \Rightarrow \frac{dy}{dx} = \frac{-1}{\sqrt{1 - x^2}} - \left[x \left(\frac{-1}{x\sqrt{1 - x^2}}\right) + (1) \operatorname{sech}^{-1} x\right] = \frac{-1}{\sqrt{1 - x^2}} + \frac{1}{\sqrt{1 - x^2}} - \operatorname{sech}^{-1} x$
    $= -\operatorname{sech}^{-1} x$

33. $y = \operatorname{csch}^{-1} \left(\frac{1}{2}\right)^\theta \Rightarrow \frac{dy}{d\theta} = -\frac{\left[\ln \left(\frac{1}{2}\right)\right] \left(\frac{1}{2}\right)^\theta}{\left(\frac{1}{2}\right)^\theta \sqrt{1 + \left[\left(\frac{1}{2}\right)^\theta\right]^2}} = -\frac{\ln (1) - \ln (2)}{\sqrt{1 + \left(\frac{1}{2}\right)^{2\theta}}} = \frac{\ln 2}{\sqrt{1 + \left(\frac{1}{2}\right)^{2\theta}}}$

35. $y = \sinh^{-1} (\tan x) \Rightarrow \frac{dy}{dx} = \frac{\sec^2 x}{\sqrt{1 + (\tan x)^2}} = \frac{\sec^2 x}{\sqrt{\sec^2 x}} = \frac{\sec^2 x}{|\sec x|} = \frac{|\sec x| \, |\sec x|}{|\sec x|} = |\sec x|$

37. $\sinh^{-1} \left(\frac{-5}{12}\right) = \ln \left(-\frac{5}{12} + \sqrt{\frac{25}{144} + 1}\right) = \ln \left(\frac{2}{3}\right)$ 39. $\tanh^{-1} \left(-\frac{1}{2}\right) = \frac{1}{2} \ln \left(\frac{1 - (1/2)}{1 + (1/2)}\right) = -\frac{\ln 3}{2}$

41. $\operatorname{sech}^{-1} \left(\frac{3}{5}\right) = \ln \left(\frac{1 + \sqrt{1 - (9/25)}}{(3/5)}\right) = \ln 3$

43. (a) $\lim\limits_{x \to \infty} \tanh x = \lim\limits_{x \to \infty} \frac{e^x - e^{-x}}{e^x + e^{-x}} = \lim\limits_{x \to \infty} \frac{e^x - \frac{1}{e^x}}{e^x + \frac{1}{e^x}} = \lim\limits_{x \to \infty} \frac{\left(e^x - \frac{1}{e^x}\right)}{\left(e^x + \frac{1}{e^x}\right)} \cdot \frac{\frac{1}{e^x}}{\frac{1}{e^x}} = \lim\limits_{x \to \infty} \frac{1 - \frac{1}{e^{2x}}}{1 + \frac{1}{e^{2x}}} = \frac{1 - 0}{1 + 0} = 1$

   (b) $\lim\limits_{x \to -\infty} \tanh x = \lim\limits_{x \to -\infty} \frac{e^x - e^{-x}}{e^x + e^{-x}} = \lim\limits_{x \to -\infty} \frac{e^x - \frac{1}{e^x}}{e^x + \frac{1}{e^x}} = \lim\limits_{x \to -\infty} \frac{\left(e^x - \frac{1}{e^x}\right)}{\left(e^x + \frac{1}{e^x}\right)} \cdot \frac{e^x}{e^x} = \lim\limits_{x \to -\infty} \frac{e^{2x} - 1}{e^{2x} + 1} = \frac{0 - 1}{0 + 1} = -1$

   (c) $\lim\limits_{x \to \infty} \sinh x = \lim\limits_{x \to \infty} \frac{e^x - e^{-x}}{2} = \lim\limits_{x \to \infty} \frac{e^x - \frac{1}{e^x}}{2} = \lim\limits_{x \to \infty} \left(\frac{e^x}{2} - \frac{1}{2e^x}\right) = \infty - 0 = \infty$

   (d) $\lim\limits_{x \to -\infty} \sinh x = \lim\limits_{x \to -\infty} \frac{e^x - e^{-x}}{2} = \lim\limits_{x \to -\infty} \left(\frac{e^x}{2} - \frac{e^{-x}}{2}\right) = 0 - \infty = -\infty$

   (e) $\lim\limits_{x \to \infty} \operatorname{sech} x = \lim\limits_{x \to \infty} \frac{2}{e^x + e^{-x}} = \lim\limits_{x \to \infty} \frac{2}{e^x + \frac{1}{e^x}} \cdot \frac{\frac{1}{e^x}}{\frac{1}{e^x}} = \lim\limits_{x \to \infty} \frac{\frac{2}{e^x}}{1 + \frac{1}{e^{2x}}} = \frac{0}{1 + 0} = 0$

   (f) $\lim\limits_{x \to \infty} \coth x = \lim\limits_{x \to \infty} \frac{e^x + e^{-x}}{e^x - e^{-x}} = \lim\limits_{x \to \infty} \frac{e^x + \frac{1}{e^x}}{e^x - \frac{1}{e^x}} = \lim\limits_{x \to \infty} \frac{\left(e^x + \frac{1}{e^x}\right)}{\left(e^x - \frac{1}{e^x}\right)} \cdot \frac{\frac{1}{e^x}}{\frac{1}{e^x}} = \lim\limits_{x \to \infty} \frac{1 + \frac{1}{e^{2x}}}{1 - \frac{1}{e^{2x}}} = \frac{1 + 0}{1 - 0} = 1$

   (g) $\lim\limits_{x \to 0^+} \coth x = \lim\limits_{x \to 0^+} \frac{e^x + e^{-x}}{e^x - e^{-x}} = \lim\limits_{x \to 0^+} \frac{e^x + \frac{1}{e^x}}{e^x - \frac{1}{e^x}} \cdot \frac{e^x}{e^x} = \lim\limits_{x \to 0^+} \frac{e^{2x} + 1}{e^{2x} - 1} = +\infty$

   (h) $\lim\limits_{x \to 0^-} \coth x = \lim\limits_{x \to 0^-} \frac{e^x + e^{-x}}{e^x - e^{-x}} = \lim\limits_{x \to 0^-} \frac{e^x + \frac{1}{e^x}}{e^x - \frac{1}{e^x}} \cdot \frac{e^x}{e^x} = \lim\limits_{x \to 0^-} \frac{e^{2x} + 1}{e^{2x} - 1} = -\infty$

   (i) $\lim\limits_{x \to -\infty} \operatorname{csch} x = \lim\limits_{x \to -\infty} \frac{2}{e^x - e^{-x}} = \lim\limits_{x \to -\infty} \frac{2}{e^x - \frac{1}{e^x}} \cdot \frac{e^x}{e^x} = \lim\limits_{x \to -\infty} \frac{2e^x}{e^{2x} - 1} = \frac{0}{0 - 1} = 0$

45. (a) $v = \sqrt{\frac{mg}{k}} \tanh \left(\sqrt{\frac{gk}{m}} \, t\right) \Rightarrow \frac{dv}{dt} = \sqrt{\frac{mg}{k}} \left[\operatorname{sech}^2 \left(\sqrt{\frac{gk}{m}} \, t\right)\right] \left(\sqrt{\frac{gk}{m}}\right) = g \operatorname{sech}^2 \left(\sqrt{\frac{gk}{m}} \, t\right)$.

   Thus $m\frac{dv}{dt} = mg \operatorname{sech}^2 \left(\sqrt{\frac{gk}{m}} \, t\right) = mg \left(1 - \tanh^2 \left(\sqrt{\frac{gk}{m}} \, t\right)\right) = mg - kv^2$. Also, since $\tanh x = 0$ when $x = 0$, $v = 0$

   when $t = 0$.

   (b) $\lim\limits_{t \to \infty} v = \lim\limits_{t \to \infty} \sqrt{\frac{mg}{k}} \tanh \left(\sqrt{\frac{kg}{m}} \, t\right) = \sqrt{\frac{mg}{k}} \lim\limits_{t \to \infty} \tanh \left(\sqrt{\frac{kg}{m}} \, t\right) = \sqrt{\frac{mg}{k}} (1) = \sqrt{\frac{mg}{k}}$

   (c) $\sqrt{\frac{160}{0.005}} = \sqrt{\frac{160,000}{5}} = \frac{400}{\sqrt{5}} = 80\sqrt{5} \approx 178.89$ ft/sec

47. (a) $y = \frac{H}{w} \cosh \left(\frac{w}{H} x\right) \Rightarrow \tan \phi = \frac{dy}{dx} = \left(\frac{H}{w}\right) \left[\frac{w}{H} \sinh \left(\frac{w}{H} x\right)\right] = \sinh \left(\frac{w}{H} x\right)$

   (b) The tension at P is given by $T \cos \phi = H \Rightarrow T = H \sec \phi = H\sqrt{1 + \tan^2 \phi} = H\sqrt{1 + \left(\sinh \frac{w}{H} x\right)^2}$
       $= H \cosh \left(\frac{w}{H} x\right) = w \left(\frac{H}{w}\right) \cosh \left(\frac{w}{H} x\right) = wy$

CHAPTER 3 PRACTICE AND ADDITIONAL EXERCISES

1. No, since $f(x) = x^3 + 2x + \tan x \Rightarrow f'(x) = 3x^2 + 2 + \sec^2 x > 0 \Rightarrow f(x)$ is always increasing on its domain

3. No absolute minimum because $\lim\limits_{x \to \infty} (7 + x)(11 - 3x)^{1/3} = -\infty$. Next $f'(x) =$
    $(11 - 3x)^{1/3} - (7 + x)(11 - 3x)^{-2/3} = \frac{(11-3x)-(7+x)}{(11-3x)^{2/3}} = \frac{4(1-x)}{(11-3x)^{2/3}} \Rightarrow x = 1$ and $x = \frac{11}{3}$ are critical points.
    Since $f' > 0$ if $x < 1$ and $f' < 0$ if $x > 1$, $f(1) = 16$ is the absolute maximum.

5. $g(x) = e^x - x \Rightarrow g'(x) = e^x - 1 \Rightarrow g' = \underset{0}{\underline{\quad\quad} \mid +++} \Rightarrow$ the graph is decreasing on $(-\infty, 0)$, increasing on $(0, \infty)$;
    an absolute minimum value is 1 at $x = 0$; $x = 0$ is the only critical point of g; there is no absolute maximum value

7. $f(x) = x - 2 \ln x$ on $1 \le x \le 3 \Rightarrow f'(x) = 1 - \frac{2}{x} \Rightarrow f' = \underset{1}{\mid} \underset{2}{\underline{---} \mid +++} \underset{3}{\mid} \Rightarrow$ the graph is decreasing on $(1, 2)$,
    increasing on $(2, 3)$; an absolute minimum value is $2 - 2 \ln 2$ at $x = 2$; an absolute maximum value is 1 at $x = 1$.

9. Yes, because at each point of $[0, 1)$ except $x = 0$, the function's value is a local minimum value as well as a local
    maximum value. At $x = 0$ the function's value, 0, is not a local minimum value because each open interval around
    $x = 0$ on the x-axis contains points to the left of 0 where f equals -1.

11. No, because the interval $0 < x < 1$ fails to be closed. The Extreme Value Theorem says that if the function is continuous
    throughout a finite closed interval $a \le x \le b$ then the existence of absolute extrema is guaranteed on that interval.

13. (a) $g(t) = \sin^2 t - 3t \Rightarrow g'(t) = 2 \sin t \cos t - 3 = \sin(2t) - 3 \Rightarrow g' < 0 \Rightarrow g(t)$ is always falling and hence must
         decrease on every interval in its domain.
    (b) One, since $\sin^2 t - 3t - 5 = 0$ and $\sin^2 t - 3t = 5$ have the same solutions: $f(t) = \sin^2 t - 3t - 5$ has the same
         derivative as g(t) in part (a) and is always decreasing with $f(-3) > 0$ and $f(0) < 0$. The Intermediate Value
         Theorem guarantees the continuous function f has a root in $[-3, 0]$.

15. (a) $f(x) = x^4 + 2x^2 - 2 \Rightarrow f'(x) = 4x^3 + 4x$. Since $f(0) = -2 < 0$, $f(1) = 1 > 0$ and $f'(x) \ge 0$ for $0 \le x \le 1$, we
         may conclude from the Intermediate Value Theorem that f(x) has exactly one solution when $0 \le x \le 1$.
    (b) $x^2 = \frac{-2 \pm \sqrt{4+8}}{2} > 0 \Rightarrow x^2 = \sqrt{3} - 1$ and $x \ge 0 \Rightarrow x \approx \sqrt{.7320508076} \approx .8555996772$

17. Let V(t) represent the volume of the water in the reservoir at time t, in minutes, let $V(0) = a_0$ be the initial
    amount and $V(1440) = a_0 + (1400)(43,560)(7.48)$ gallons be the amount of water contained in the reservoir
    after the rain, where $24 \, hr = 1440 \, min$. Assume that V(t) is continuous on $[0, 1440]$ and differentiable on
    $(0, 1440)$. The Mean Value Theorem says that for some t_0 in $(0, 1440)$ we have $V'(t_0) = \frac{V(1440) - V(0)}{1440 - 0}$
    $= \frac{a_0 + (1400)(43,560)(7.48) - a_0}{1440} = \frac{456,160,320 \text{ gal}}{1440 \text{ min}} = 316,778$ gal/min. Therefore at t_0 the reservoir's volume
    was increasing at a rate in excess of 225,000 gal/min.

19. No, $\frac{x}{x+1} = 1 + \frac{-1}{x+1} \Rightarrow \frac{x}{x+1}$ differs from $\frac{-1}{x+1}$ by the constant 1. Both functions have the same derivative
    $\frac{d}{dx}\left(\frac{x}{x+1}\right) = \frac{(x+1)-x(1)}{(x+1)^2} = \frac{1}{(x+1)^2} = \frac{d}{dx}\left(\frac{-1}{x+1}\right).$

21. The global minimum value of $\frac{1}{2}$ occurs at $x = 2$.

23.

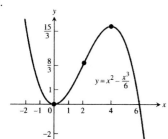

25.

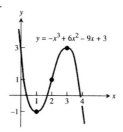

27.

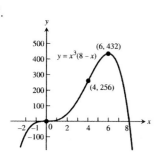

29.

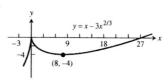

31.

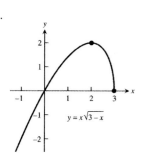

33.

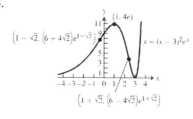

35.

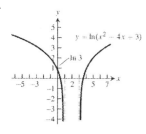

37.

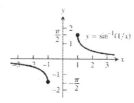

39. (a) $y' = 16 - x^2 \Rightarrow y' = --- \mid +++ \mid --- \Rightarrow$ the curve is rising on $(-4, 4)$, falling on $(-\infty, -4)$ and $(4, \infty)$
$\qquad\qquad\qquad\qquad\qquad\quad -4 \qquad 4$
$\Rightarrow$ a local maximum at $x = 4$ and a local minimum at $x = -4$; $y'' = -2x \Rightarrow y'' = +++ \mid --- \Rightarrow$ the curve
$\qquad\qquad\qquad\qquad\qquad\qquad\qquad\qquad\qquad\qquad\qquad\qquad\qquad\qquad\qquad\quad 0$
is concave up on $(-\infty, 0)$, concave down on $(0, \infty) \Rightarrow$ a point of inflection at $x = 0$

(b)

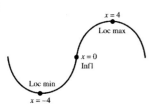

41. (a) $y' = 6x(x + 1)(x - 2) = 6x^3 - 6x^2 - 12x \Rightarrow y' = ---\underset{-1}{|} +++\underset{0}{|} ---\underset{2}{|} +++ \Rightarrow$ the graph is rising on $(-1, 0)$

and $(2, \infty)$, falling on $(-\infty, -1)$ and $(0, 2) \Rightarrow$ a local maximum at $x = 0$, local minima at $x = -1$ and

$x = 2$; $y'' = 18x^2 - 12x - 12 = 6(3x^2 - 2x - 2) = 6\left(x - \frac{1-\sqrt{7}}{3}\right)\left(x - \frac{1+\sqrt{7}}{3}\right) \Rightarrow$

$y'' = +++ \underset{\frac{1-\sqrt{7}}{3}}{|} --- \underset{\frac{1+\sqrt{7}}{3}}{|} +++ \Rightarrow$ the curve is concave up on $\left(-\infty, \frac{1-\sqrt{7}}{3}\right)$ and $\left(\frac{1+\sqrt{7}}{3}, \infty\right)$, concave down

on $\left(\frac{1-\sqrt{7}}{3}, \frac{1+\sqrt{7}}{3}\right) \Rightarrow$ points of inflection at $x = \frac{1 \pm \sqrt{7}}{3}$

(b)

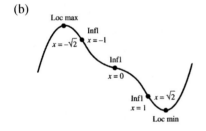

43. (a) $y' = x^4 - 2x^2 = x^2(x^2 - 2) \Rightarrow y' = +++ \underset{-\sqrt{2}}{|} --- \underset{0}{|} --- \underset{\sqrt{2}}{|} +++ \Rightarrow$ the curve is rising on $\left(-\infty, -\sqrt{2}\right)$ and

$\left(\sqrt{2}, \infty\right)$, falling on $\left(-\sqrt{2}, \sqrt{2}\right) \Rightarrow$ a local maximum at $x = -\sqrt{2}$ and a local minimum at $x = \sqrt{2}$;

$y'' = 4x^3 - 4x = 4x(x - 1)(x + 1) \Rightarrow y'' = --- \underset{-1}{|} +++ \underset{0}{|} --- \underset{1}{|} +++ \Rightarrow$ concave up on $(-1, 0)$ and $(1, \infty)$,

concave down on $(-\infty, -1)$ and $(0, 1) \Rightarrow$ points of inflection at $x = 0$ and $x = \pm 1$

(b)

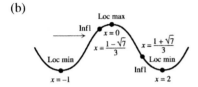

45. $y = \frac{x + 1}{x - 3} = 1 + \frac{4}{x - 3}$

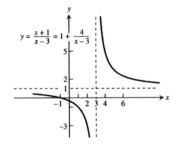

47. $y = \frac{x^2 + 1}{x} = x + \frac{1}{x}$

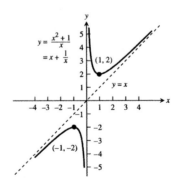

49. $y = \frac{x^3 + 2}{2x} = \frac{x^2}{2} + \frac{1}{x}$

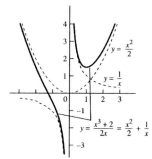

51. $\lim\limits_{x \to 1} \frac{x^2 + 3x - 4}{x - 1} = \lim\limits_{x \to 1} \frac{2x + 3}{1} = 5$

53. $\lim\limits_{x \to \pi} \frac{\tan x}{x} = \frac{\tan \pi}{\pi} = 0$

55. $\lim\limits_{x \to 0} \frac{\sin^2 x}{\tan(x^2)} = \lim\limits_{x \to 0} \frac{2\sin x \cdot \cos x}{2x \sec^2(x^2)} = \lim\limits_{x \to 0} \frac{\sin(2x)}{2x \sec^2(x^2)} = \lim\limits_{x \to 0} \frac{2\cos(2x)}{2x\,(2\sec^2(x^2)\tan(x^2)\cdot 2x) + 2\sec^2(x^2)} = \frac{2}{0 + 2\cdot 1} = 1$

57. $\lim\limits_{x \to \pi/2^-} \sec(7x)\cos(3x) = \lim\limits_{x \to \pi/2^-} \frac{\cos(3x)}{\cos(7x)} = \lim\limits_{x \to \pi/2^-} \frac{-3\sin(3x)}{-7\sin(7x)} = \frac{3}{7}$

59. $\lim\limits_{x \to 0} (\csc x - \cot x) = \lim\limits_{x \to 0} \frac{1 - \cos x}{\sin x} = \lim\limits_{x \to 0} \frac{\sin x}{\cos x} = \frac{0}{1} = 0$

61. The limit leads to the indeterminate form $\frac{0}{0}$: $\lim\limits_{x \to 0} \frac{10^x - 1}{x} = \lim\limits_{x \to 0} \frac{(\ln 10)10^x}{1} = \ln 10$

63. The limit leads to the indeterminate form $\frac{0}{0}$: $\lim\limits_{x \to 0} \frac{5 - 5\cos x}{e^x - x - 1} = \lim\limits_{x \to 0} \frac{5\sin x}{e^x - 1} = \lim\limits_{x \to 0} \frac{5\cos x}{e^x} = 5$

65. The limit leads to the indeterminate form $\frac{0}{0}$: $\lim\limits_{t \to 0^+} \frac{t - \ln(1 + 2t)}{t^2} = \lim\limits_{t \to 0^+} \frac{\left(1 - \frac{2}{1 + 2t}\right)}{2t} = -\infty$

67. $\lim\limits_{x \to \infty} \left(1 + \frac{b}{x}\right)^{kx} = \lim\limits_{x \to \infty} \left[\left(1 + \frac{1}{x/b}\right)^{x/b}\right]^{bk} = e^{bk}$

69. (a) Maximize $f(x) = \sqrt{x} - \sqrt{36 - x} = x^{1/2} - (36 - x)^{1/2}$ where $0 \le x \le 36 \Rightarrow f'(x) = \frac{1}{2}x^{-1/2} - \frac{1}{2}(36 - x)^{-1/2}(-1)$

   $= \frac{\sqrt{36 - x} + \sqrt{x}}{2\sqrt{x}\sqrt{36 - x}} \Rightarrow$ derivative fails to exist at 0 and 36; $f(0) = -6$, and $f(36) = 6 \Rightarrow$ the numbers are 0 and 36

   (b) Maximize $g(x) = \sqrt{x} + \sqrt{36 - x} = x^{1/2} + (36 - x)^{1/2}$ where $0 \le x \le 36$

   $\Rightarrow g'(x) = \frac{1}{2}x^{-1/2} + \frac{1}{2}(36 - x)^{-1/2}(-1) = \frac{\sqrt{36 - x} - \sqrt{x}}{2\sqrt{x}\sqrt{36 - x}} \Rightarrow$ critical points at 0, 18 and 36; $g(0) = 6$,

   $g(18) = 2\sqrt{18} = 6\sqrt{2}$ and $g(36) = 6 \Rightarrow$ the numbers are 18 and 18

71. $A(x) = \frac{1}{2}(2x)(27 - x^2)$ for $0 \le x \le \sqrt{27}$

   $\Rightarrow A'(x) = 3(3 + x)(3 - x)$ and $A''(x) = -6x$.
   The critical points are -3 and 3, but -3 is not in the
   domain. Since $A''(3) = -18 < 0$ and $A\left(\sqrt{27}\right) = 0$,
   the maximum occurs at $x = 3 \Rightarrow$ the largest area is
   $A(3) = 54$ sq units.

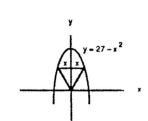

73. The dimensions will be x in. by $10 - 2x$ in. by $16 - 2x$ in., so $V(x) = x(10 - 2x)(16 - 2x) = 4x^3 - 52x^2 + 160x$ for $0 < x < 5$. Then $V'(x) = 12x^2 - 104x + 160 = 4(x - 2)(3x - 20)$, so the critical point in the correct domain is $x = 2$. This critical point corresponds to the maximum possible volume because $V'(x) > 0$ for $0 < x < 2$ and $V'(x) < 0$ for $2 < x < 5$. The box of largest volume has a height of 2 in. and a base measuring 6 in. by 12 in., and its volume is 144 in.[3] Graphical support:

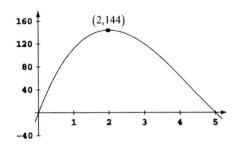

75. $g(x) = 3x - x^3 + 4 \Rightarrow g(2) = 2 > 0$ and $g(3) = -14 < 0 \Rightarrow g(x) = 0$ in the interval $[2, 3]$ by the Intermediate Value Theorem. Then $g'(x) = 3 - 3x^2 \Rightarrow x_{n+1} = x_n - \frac{3x_n - x_n^3 + 4}{3 - 3x_n^2}$; $x_0 = 2 \Rightarrow x_1 = 2.\overline{22} \Rightarrow x_2 = 2.196215$, and so forth to $x_5 = 2.195823345$.

77. $A = xy = xe^{-x^2} \Rightarrow \frac{dA}{dx} = e^{-x^2} + (x)(-2x)e^{-x^2} = e^{-x^2}(1 - 2x^2)$. Solving $\frac{dA}{dx} = 0 \Rightarrow 1 - 2x^2 = 0$ $\Rightarrow x = \frac{1}{\sqrt{2}}$; $\frac{dA}{dx} < 0$ for $x > \frac{1}{\sqrt{2}}$ and $\frac{dA}{dx} > 0$ for $0 < x < \frac{1}{\sqrt{2}} \Rightarrow$ absolute maximum of $\frac{1}{\sqrt{2}}e^{-1/2} = \frac{1}{\sqrt{2e}}$ at $x = \frac{1}{\sqrt{2}}$ units long by $y = e^{-1/2} = \frac{1}{\sqrt{e}}$ units high.

79. $y = x \ln 2x - x \Rightarrow y' = x\left(\frac{2}{2x}\right) + \ln(2x) - 1 = \ln 2x$; solving $y' = 0 \Rightarrow x = \frac{1}{2}$; $y' > 0$ for $x > \frac{1}{2}$ and $y' < 0$ for $x < \frac{1}{2} \Rightarrow$ relative minimum of $-\frac{1}{2}$ at $x = \frac{1}{2}$; $f\left(\frac{1}{2e}\right) = -\frac{1}{e}$ and $f\left(\frac{e}{2}\right) = 0 \Rightarrow$ absolute minimum is $-\frac{1}{2}$ at $x = \frac{1}{2}$ and the absolute maximum is 0 at $x = \frac{e}{2}$

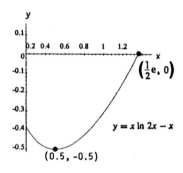

81. $f(x) = e^{x/\sqrt{x^4+1}}$ for all x in $(-\infty, \infty)$; $f'(x) = \left[\frac{\left(\sqrt{x^4+1}\right) \cdot 1 - x\left(\frac{2x^3}{\sqrt{x^4+1}}\right)}{\left(\sqrt{x^4+1}\right)^2}\right] e^{x/\sqrt{x^4+1}} = \frac{1 - x^4}{\left(\sqrt{x^4+1}\right)^3} e^{x/\sqrt{x^4+1}}$

$= \frac{(1-x^2)(1+x^2)}{(x^4+1)^{3/2}} e^{x/\sqrt{x^4+1}} = 0 \Rightarrow 1 - x^2 = 0 \Rightarrow x = \pm 1$ are critical points. Consider the behavior of f as $x \rightarrow \pm\infty$; $\lim\limits_{x \rightarrow \infty} e^{x/\sqrt{x^4+1}} = \lim\limits_{x \rightarrow -\infty} e^{x/\sqrt{x^4+1}} = 1$ as suggested by the following table (14 digit precision, 12 digits displayed):

x	$x/\sqrt{x^4+1}$	$e^{x/\sqrt{x^4+1}}$
$-\infty$	0	1
$\vdots$	$\vdots$	$\vdots$
-100000	$-0.0000\ 10000\ 0000\ 00000$	$0.9999\ 9000\ 0050$
-10000	$-0.0001\ 0000\ 0000\ 000$	$0.9999\ 0000\ 5000$
-1000	$-0.0010\ 0000\ 0000\ 00$	$0.9990\ 0049\ 9833$
-100	$-0.0099\ 9999\ 9950\ 00$	$0.9900\ 4983\ 3799$
-10	$-0.0999\ 9500\ 0375\ 0$	$0.9048\ 4194\ 1895$
0	0	1
10	$0.0999\ 9500\ 0375\ 0$	$1.1051\ 6539\ 265$
100	$0.0099\ 9999\ 9950\ 00$	$1.0100\ 5016\ 703$
1000	$0.0010\ 0000\ 0000\ 00$	$1.0010\ 0050\ 017$
10000	$0.0001\ 0000\ 0000\ 000$	$1.0001\ 0000\ 500$
100000	$0.0000\ 10000\ 0000\ 00000$	$1.0000\ 1000\ 005$
$\vdots$	$\vdots$	$\vdots$
∞	0	1

Therefore, $y = 1$ is a horizontal asymptote in both directions. Check the critical points for absolute extreme values: $f(-1) = e^{-\sqrt{2}/2} \approx 0.4931$, $f(1) = e^{\sqrt{2}/2} \approx 2.0281 \Rightarrow$ the absolute minimum value of the function is $e^{-\sqrt{2}/2}$ at $x = -1$, and the absolute maximum value is $e^{\sqrt{2}/2}$ at $x = 1$.

83. (a) $y = \frac{\ln x}{\sqrt{x}} \Rightarrow y' = \frac{1}{x\sqrt{x}} - \frac{\ln x}{2x^{3/2}} = \frac{2 - \ln x}{2x\sqrt{x}}$

$\Rightarrow y'' = -\frac{3}{4}x^{-5/2}(2 - \ln x) - \frac{1}{2}x^{-5/2} = x^{-5/2}\left(\frac{3}{4}\ln x - 2\right)$;

solving $y' = 0 \Rightarrow \ln x = 2 \Rightarrow x = e^2$; $y' < 0$ for $x > e^2$ and and $y' > 0$ for $x < e^2 \Rightarrow$ a maximum of $\frac{2}{e}$; $y'' = 0$

$\Rightarrow \ln x = \frac{8}{3} \Rightarrow x = e^{8/3}$; the curve is concave down on $\left(0, e^{8/3}\right)$ and concave up on $\left(e^{8/3}, \infty\right)$; so there is an inflection point at $\left(e^{8/3}, \frac{8}{3e^{4/3}}\right)$.

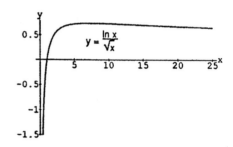

(b) $y = e^{-x^2} \Rightarrow y' = -2xe^{-x^2} \Rightarrow y'' = -2e^{-x^2} + 4x^2e^{-x^2}$

$= (4x^2 - 2)e^{-x^2}$; solving $y' = 0 \Rightarrow x = 0$; $y' < 0$ for $x > 0$ and $y' > 0$ for $x < 0 \Rightarrow$ a maximum at $x = 0$ of $e^0 = 1$; there are points of inflection at $x = \pm\frac{1}{\sqrt{2}}$; the curve is concave down for $-\frac{1}{\sqrt{2}} < x < \frac{1}{\sqrt{2}}$ and concave up otherwise.

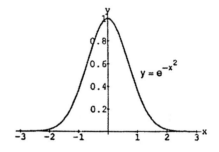

(c) $y = (1 + x)e^{-x} \Rightarrow y' = e^{-x} - (1 + x)e^{-x} = -xe^{-x}$

$\Rightarrow y'' = -e^{-x} + xe^{-x} = (x - 1)e^{-x}$; solving $y' = 0$

$\Rightarrow -xe^{-x} = 0 \Rightarrow x = 0$; $y' < 0$ for $x > 0$ and $y' > 0$ for $x < 0 \Rightarrow$ a maximum at $x = 0$ of $(1 + 0)e^0 = 1$; there is a point of inflection at $x = 1$ and the curve is concave up for $x > 1$ and concave down for $x < 1$.

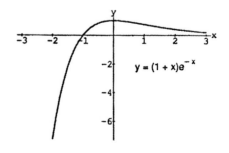

85. $x = \frac{1}{2}\tan t$, $y = \frac{1}{2}\sec t \Rightarrow \frac{dy}{dx} = \frac{dy/dt}{dx/dt} = \frac{\frac{1}{2}\sec t \tan t}{\frac{1}{2}\sec^2 t} = \frac{\tan t}{\sec t} = \sin t \Rightarrow \left.\frac{dy}{dx}\right|_{t=\pi/3} = \sin\frac{\pi}{3} = \frac{\sqrt{3}}{2}$; $t = \frac{\pi}{3}$

$\Rightarrow x = \frac{1}{2}\tan\frac{\pi}{3} = \frac{\sqrt{3}}{2}$ and $y = \frac{1}{2}\sec\frac{\pi}{3} = 1 \Rightarrow y = \frac{\sqrt{3}}{2}x + \frac{1}{4}$; $\frac{d^2y}{dx^2} = \frac{dy'/dt}{dx/dt} = \frac{\cos t}{\frac{1}{2}\sec^2 t} = 2\cos^3 t \Rightarrow \left.\frac{d^2y}{dx^2}\right|_{t=\pi/3}$

$= 2\cos^3\left(\frac{\pi}{3}\right) = \frac{1}{4}$

87. $y = \sinh^2 3x \Rightarrow y' = 2\sinh 3x \cosh 3x \cdot 3 = 6\sinh 3x \cosh 3x \Rightarrow y' = 3\sinh 6x$

89. $\sin^{-1} x = \operatorname{sech} y \Rightarrow \dfrac{1}{\sqrt{1-x^2}} = -\operatorname{sech} y \tanh y \dfrac{dy}{dx} \Rightarrow \dfrac{-1}{\operatorname{sech} y \tanh y \sqrt{1-x^2}} = \dfrac{dy}{dx} \Rightarrow \dfrac{dy}{dx} = -\dfrac{\cosh y \coth y}{\sqrt{1-x^2}}$

91. $\tan^{-1} y = \tanh^{-1} x \Rightarrow \dfrac{1}{1+y^2} \cdot \dfrac{dy}{dx} = \dfrac{1}{1-x^2} \Rightarrow \dfrac{dy}{dx} = \dfrac{1+y^2}{1-x^2}$

93. $x = \cosh(\ln y) \Rightarrow 1 = \sinh(\ln y) \cdot \dfrac{1}{y} \cdot \dfrac{dy}{dx} \Rightarrow 1 = \dfrac{\sinh(\ln y)}{y} \cdot \dfrac{dy}{dx} \Rightarrow \dfrac{y}{\sinh(\ln y)} = \dfrac{dy}{dx} \Rightarrow \dfrac{dy}{dx} = y \operatorname{csch}(\ln y)$

95. $y = \sinh^{-1}(\tan x) \Rightarrow \dfrac{dy}{dx} = \dfrac{1}{\sqrt{1 + (\tan x)^2}} \cdot \sec^2 x = \dfrac{\sec^2 x}{\sqrt{1 + \tan^2 x}} = \dfrac{\sec^2 x}{\sqrt{\sec^2 x}} = \sqrt{\sec^2 x} \Rightarrow \dfrac{dy}{dx} = |\sec x|$

CHAPTER 4 INTEGRATION

4.1 ANTIDERIVATIVES

1. (a) x^2 (b) $\frac{x^3}{3}$ (c) $\frac{x^3}{3} - x^2 + x$

3. (a) x^{-3} (b) $-\frac{x^{-3}}{3}$ (c) $-\frac{x^{-3}}{3} + x^2 + 3x$

5. (a) $\frac{-1}{x}$ (b) $\frac{-5}{x}$ (c) $2x + \frac{5}{x}$

7. (a) $\sqrt{x^3}$ (b) $\sqrt{x}$ (c) $\frac{2}{3}\sqrt{x^3} + 2\sqrt{x}$

9. (a) $x^{2/3}$ (b) $x^{1/3}$ (c) $x^{-1/3}$

11. (a) $\ln|x|$ (b) $7\ln|x|$ (c) $x - 5\ln|x|$

13. (a) $\cos(\pi x)$ (b) $-3\cos x$ (c) $\frac{-\cos(\pi x)}{\pi} + \cos(3x)$

15. (a) $\tan x$ (b) $2\tan\left(\frac{x}{3}\right)$ (c) $-\frac{2}{3}\tan\left(\frac{3x}{2}\right)$

17. (a) $-\csc x$ (b) $\frac{1}{5}\csc(5x)$ (c) $2\csc\left(\frac{\pi x}{2}\right)$

19. (a) $\frac{1}{3}e^{3x}$ (b) $-e^{-x}$ (c) $2e^{x/2}$

21. (a) $\frac{1}{\ln 3}\cdot 3^x$ (b) $\frac{-1}{\ln 2}\cdot 2^{-x}$ (c) $\frac{1}{\ln(5/3)}\cdot\left(\frac{5}{3}\right)^x$

23. (a) $2\sin^{-1}x$ (b) $\frac{1}{2}\tan^{-1}x$ (c) $\frac{1}{2}\tan^{-1}(2x)$

25. $\int (x+1)\,dx = \frac{x^2}{2} + x + C$ 27. $\int \left(3t^2 + \frac{1}{2}\right) dt = t^3 + \frac{t^2}{4} + C$

29. $\int (2x^3 - 5x + 7)\,dx = \frac{1}{2}x^4 - \frac{5}{2}x^2 + 7x + C$

31. $\int \left(\frac{1}{x^2} - x^2 - \frac{1}{3}\right) dx = \int \left(x^{-2} - x^2 - \frac{1}{3}\right) dx = \frac{x^{-1}}{-1} - \frac{x^3}{3} - \frac{1}{3}x + C = -\frac{1}{x} - \frac{x^3}{3} - \frac{x}{3} + C$

33. $\int x^{-1/3}\,dx = \frac{x^{2/3}}{\frac{2}{3}} + C = \frac{3}{2}x^{2/3} + C$

35. $\int \left(\sqrt{x} + \sqrt[3]{x}\right) dx = \int \left(x^{1/2} + x^{1/3}\right) dx = \frac{x^{3/2}}{\frac{3}{2}} + \frac{x^{4/3}}{\frac{4}{3}} + C = \frac{2}{3}x^{3/2} + \frac{3}{4}x^{4/3} + C$

37. $\int \left(8y - \frac{2}{y^{1/4}}\right) dy = \int \left(8y - 2y^{-1/4}\right) dy = \frac{8y^2}{2} - 2\left(\frac{y^{3/4}}{\frac{3}{4}}\right) + C = 4y^2 - \frac{8}{3}y^{3/4} + C$

39. $\int 2x\left(1 - x^{-3}\right) dx = \int \left(2x - 2x^{-2}\right) dx = \frac{2x^2}{2} - 2\left(\frac{x^{-1}}{-1}\right) + C = x^2 + \frac{2}{x} + C$

41. $\int \frac{t\sqrt{t}+\sqrt{t}}{t^2}\,dt = \int \left(\frac{t^{3/2}}{t^2}+\frac{t^{1/2}}{t^2}\right)\,dt = \int \left(t^{-1/2}+t^{-3/2}\right)\,dt = \frac{t^{1/2}}{\frac{1}{2}}+\left(\frac{t^{-1/2}}{-\frac{1}{2}}\right)+C = 2\sqrt{t}-\frac{2}{\sqrt{t}}+C$

43. $\int -2\cos t\,dt = -2\sin t + C$

45. $\int 7\sin\frac{\theta}{3}\,d\theta = -21\cos\frac{\theta}{3}+C$

47. $\int -3\csc^2 x\,dx = 3\cot x + C$

49. $\int \frac{\csc\theta\cot\theta}{2}\,d\theta = -\frac{1}{2}\csc\theta + C$

51. $\int \left(e^{3x}+5e^{-x}\right)dx = \frac{e^{3x}}{3}-5e^{-x}+C$

53. $\int \left(e^{-x}+4^x\right)dx = -e^{-x}+\frac{4^x}{\ln 4}+C$

55. $\int \left(4\sec x\tan x - 2\sec^2 x\right)dx = 4\sec x - 2\tan x + C$

57. $\int \left(\sin 2x - \csc^2 x\right)dx = -\frac{1}{2}\cos 2x + \cot x + C$

59. $\int \frac{1+\cos 4t}{2}\,dt = \int \left(\frac{1}{2}+\frac{1}{2}\cos 4t\right)dt = \frac{1}{2}t+\frac{1}{2}\left(\frac{\sin 4t}{4}\right)+C = \frac{t}{2}+\frac{\sin 4t}{8}+C$

61. $\int \left(\frac{1}{x}-\frac{5}{x^2+1}\right)dx = \ln|x| - 5\tan^{-1}x + C$

63. $\int 3x^{\sqrt 3}\,dx = \frac{3x^{(\sqrt 3 + 1)}}{\sqrt 3 + 1}+C$

65. $\int \left(1+\tan^2\theta\right)d\theta = \int \sec^2\theta\,d\theta = \tan\theta + C$

67. $\int \cot^2 x\,dx = \int \left(\csc^2 x - 1\right)dx = -\cot x - x + C$

69. $\int \cos\theta\,(\tan\theta + \sec\theta)\,d\theta = \int (\sin\theta + 1)\,d\theta = -\cos\theta + \theta + C$

71. $\frac{d}{dx}\left(\frac{(7x-2)^4}{28}+C\right) = \frac{4(7x-2)^3(7)}{28} = (7x-2)^3$

73. $\frac{d}{dx}\left(\frac{1}{5}\tan(5x-1)+C\right) = \frac{1}{5}\left(\sec^2(5x-1)\right)(5) = \sec^2(5x-1)$

75. $\frac{d}{dx}\left(\frac{-1}{x+1}+C\right) = (-1)(-1)(x+1)^{-2} = \frac{1}{(x+1)^2}$

77. $\frac{d}{dx}\left(\ln(x+1)+C\right) = \frac{1}{x+1}$

79. $\frac{d}{dx}\left(\frac{1}{a}\tan^{-1}\left(\frac{x}{a}\right)+C\right) = \frac{1}{a}\cdot\frac{1}{1+\left(\frac{x}{a}\right)^2}\cdot\frac{d}{dx}\left(\frac{x}{a}\right) = \frac{1}{a^2\left(1+\frac{x^2}{a^2}\right)} = \frac{1}{a^2+x^2}$

81. If $y = \ln x - \frac{1}{2}\ln\left(1+x^2\right) - \frac{\tan^{-1}x}{x}+C$, then $dy = \left[\frac{1}{x}-\frac{x}{1+x^2}-\frac{\left(\frac{x}{1+x^2}\right)-\tan^{-1}x}{x^2}\right]dx$

$= \left(\frac{1}{x}-\frac{x}{1+x^2}-\frac{1}{x(1+x^2)}+\frac{\tan^{-1}x}{x^2}\right)dx = \frac{x(1+x^2)-x^3-x+(\tan^{-1}x)(1+x^2)}{x^2(1+x^2)}\,dx = \frac{\tan^{-1}x}{x^2}\,dx,$

which verifies the formula

83. (a) Wrong: $\frac{d}{dx}\left(\frac{x^2}{2}\sin x + C\right) = \frac{2x}{2}\sin x + \frac{x^2}{2}\cos x = x\sin x + \frac{x^2}{2}\cos x \neq x\sin x$

  (b) Wrong: $\frac{d}{dx}(-x\cos x + C) = -\cos x + x\sin x \neq x\sin x$

  (c) Right: $\frac{d}{dx}(-x\cos x + \sin x + C) = -\cos x + x\sin x + \cos x = x\sin x$

85. $\frac{dy}{dx} = 2x - 7 \Rightarrow y = x^2 - 7x + C$; at $x = 2$ and $y = 0$ we have $0 = 2^2 - 7(2)+C \Rightarrow C = 10 \Rightarrow y = x^2 - 7x + 10$

87. $\frac{dy}{dx} = \frac{1}{x^2} + x = x^{-2} + x \Rightarrow y = -x^{-1} + \frac{x^2}{2} + C$; at $x = 2$ and $y = 1$ we have $1 = -2^{-1} + \frac{2^2}{2} + C \Rightarrow C = -\frac{1}{2}$
$\Rightarrow y = -x^{-1} + \frac{x^2}{2} - \frac{1}{2}$ or $y = -\frac{1}{x} + \frac{x^2}{2} - \frac{1}{2}$

89. $\frac{dy}{dx} = 3x^{-2/3} \Rightarrow y = \frac{3x^{1/3}}{\frac{1}{3}} + C = 9$; at $x = 9x^{1/3} + C$; at $x = -1$ and $y = -5$ we have $-5 = 9(-1)^{1/3} + C \Rightarrow C = 4$
$\Rightarrow y = 9x^{1/3} + 4$

91. $\frac{dr}{d\theta} = -\pi \sin \pi\theta \Rightarrow r = \cos(\pi\theta) + C$; at $r = 0$ and $\theta = 0$ we have $0 = \cos(\pi 0) + C \Rightarrow C = -1 \Rightarrow r = \cos(\pi\theta) - 1$

93. $\frac{dv}{dt} = \frac{1}{2} \sec t \tan t \Rightarrow v = \frac{1}{2} \sec t + C$; at $v = 1$ and $t = 0$ we have $1 = \frac{1}{2} \sec(0) + C \Rightarrow C = \frac{1}{2} \Rightarrow v = \frac{1}{2} \sec t + \frac{1}{2}$

95. $\frac{dv}{dt} = \frac{3}{t\sqrt{t^2-1}}, t > 1 \Rightarrow v = 3 \sec^{-1} t + C$; at $t = 2$ and $v = 0$ we have $0 = 3 \sec^{-1} 2 + C \Rightarrow C = -\pi \Rightarrow v = 3 \sec^{-1} t - \pi$

97. $\frac{d^2y}{dx^2} = 2 - 6x \Rightarrow \frac{dy}{dx} = 2x - 3x^2 + C_1$; at $\frac{dy}{dx} = 4$ and $x = 0$ we have $4 = 2(0) - 3(0)^2 + C_1 \Rightarrow C_1 = 4$
$\Rightarrow \frac{dy}{dx} = 2x - 3x^2 + 4 \Rightarrow y = x^2 - x^3 + 4x + C_2$; at $y = 1$ and $x = 0$ we have $1 = 0^2 - 0^3 + 4(0) + C_2 \Rightarrow C_2 = 1$
$\Rightarrow y = x^2 - x^3 + 4x + 1$

99. $\frac{d^3y}{dx^3} = 6 \Rightarrow \frac{d^2y}{dx^2} = 6x + C_1$; at $\frac{d^2y}{dx^2} = -8$ and $x = 0$ we have $-8 = 6(0) + C_1 \Rightarrow C_1 = -8 \Rightarrow \frac{d^2y}{dx^2} = 6x - 8$
$\Rightarrow \frac{dy}{dx} = 3x^2 - 8x + C_2$; at $\frac{dy}{dx} = 0$ and $x = 0$ we have $0 = 3(0)^2 - 8(0) + C_2 \Rightarrow C_2 = 0 \Rightarrow \frac{dy}{dx} = 3x^2 - 8x$
$\Rightarrow y = x^3 - 4x^2 + C_3$; at $y = 5$ and $x = 0$ we have $5 = 0^3 - 4(0)^2 + C_3 \Rightarrow C_3 = 5 \Rightarrow y = x^3 - 4x^2 + 5$

101. $m = y' = 3\sqrt{x} = 3x^{1/2} \Rightarrow y = 2x^{3/2} + C$; at $(9, 4)$ we have $4 = 2(9)^{3/2} + C \Rightarrow C = -50 \Rightarrow y = 2x^{3/2} - 50$

103. (a) $\frac{ds}{dt} = 9.8t - 3 \Rightarrow s = 4.9t^2 - 3t + C$; (i) at $s = 5$ and $t = 0$ we have $C = 5 \Rightarrow s = 4.9t^2 - 3t + 5$;
displacement $= s(3) - s(1) = ((4.9)(9) - 9 + 5) - (4.9 - 3 + 5) = 33.2$ units; (ii) at $s = -2$ and $t = 0$ we have
$C = -2 \Rightarrow s = 4.9t^2 - 3t - 2$; displacement $= s(3) - s(1) = ((4.9)(9) - 9 - 2) - (4.9 - 3 - 2) = 33.2$ units;
(iii) at $s = s_0$ and $t = 0$ we have $C = s_0 \Rightarrow s = 4.9t^2 - 3t + s_0$; displacement $= s(3) - s(1)$
$= ((4.9)(9) - 9 + s_0) - (4.9 - 3 + s_0) = 33.2$ units
(b) True. Given an antiderivative f(t) of the velocity function, we know that the body's position function is
$s = f(t) + C$ for some constant C. Therefore, the displacement from $t = a$ to $t = b$ is $(f(b) + C) - (f(a) + C)$
$= f(b) - f(a)$. Thus we can find the displacement from any antiderivative f as the numerical difference
$f(b) - f(a)$ without knowing the exact values of C and s.

105. (a) $v = \int a\, dt = \int \left(15t^{1/2} - 3t^{-1/2}\right) dt = 10t^{3/2} - 6t^{1/2} + C$; $\frac{ds}{dt}(1) = 4 \Rightarrow 4 = 10(1)^{3/2} - 6(1)^{1/2} + C \Rightarrow C = 0$
$\Rightarrow v = 10t^{3/2} - 6t^{1/2}$
(b) $s = \int v\, dt = \int \left(10t^{3/2} - 6t^{1/2}\right) dt = 4t^{5/2} - 4t^{3/2} + C$; $s(1) = 0 \Rightarrow 0 = 4(1)^{5/2} - 4(1)^{3/2} + C \Rightarrow C = 0$
$\Rightarrow s = 4t^{5/2} - 4t^{3/2}$

107. (a) $\int f(x)\, dx = 1 - \sqrt{x} + C_1 = -\sqrt{x} + C$ (b) $\int g(x)\, dx = x + 2 + C_1 = x + C$
(c) $\int -f(x)\, dx = -\left(1 - \sqrt{x}\right) + C_1 = \sqrt{x} + C$ (d) $\int -g(x)\, dx = -(x + 2) + C_1 = -x + C$
(e) $\int [f(x) + g(x)]\, dx = \left(1 - \sqrt{x}\right) + (x + 2) + C_1 = x - \sqrt{x} + C$
(f) $\int [f(x) - g(x)]\, dx = \left(1 - \sqrt{x}\right) - (x + 2) + C_1 = -x - \sqrt{x} + C$

4.2 ESTIMATING WITH FINITE SUMS

1. $f(x) = x^2$

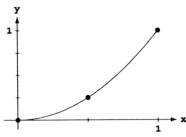

Since f is increasing on [0, 1], we use left endpoints to obtain lower sums and right endpoints to obtain upper sums.

(a) $\triangle x = \frac{1-0}{2} = \frac{1}{2}$ and $x_i = i\triangle x = \frac{i}{2} \Rightarrow$ a lower sum is $\sum_{i=0}^{1} \left(\frac{i}{2}\right)^2 \cdot \frac{1}{2} = \frac{1}{2}\left(0^2 + \left(\frac{1}{2}\right)^2\right) = \frac{1}{8}$

(b) $\triangle x = \frac{1-0}{4} = \frac{1}{4}$ and $x_i = i\triangle x = \frac{i}{4} \Rightarrow$ a lower sum is $\sum_{i=0}^{3} \left(\frac{i}{4}\right)^2 \cdot \frac{1}{4} = \frac{1}{4}\left(0^2 + \left(\frac{1}{4}\right)^2 + \left(\frac{1}{2}\right)^2 + \left(\frac{3}{4}\right)^2\right) = \frac{1}{4} \cdot \frac{7}{8} = \frac{7}{32}$

(c) $\triangle x = \frac{1-0}{2} = \frac{1}{2}$ and $x_i = i\triangle x = \frac{i}{2} \Rightarrow$ an upper sum is $\sum_{i=1}^{2} \left(\frac{i}{2}\right)^2 \cdot \frac{1}{2} = \frac{1}{2}\left(\left(\frac{1}{2}\right)^2 + 1^2\right) = \frac{5}{8}$

(d) $\triangle x = \frac{1-0}{4} = \frac{1}{4}$ and $x_i = i\triangle x = \frac{i}{4} \Rightarrow$ an upper sum is $\sum_{i=1}^{4} \left(\frac{i}{4}\right)^2 \cdot \frac{1}{4} = \frac{1}{4}\left(\left(\frac{1}{4}\right)^2 + \left(\frac{1}{2}\right)^2 + \left(\frac{3}{4}\right)^2 + 1^2\right) = \frac{1}{4} \cdot \left(\frac{30}{16}\right) = \frac{15}{32}$

3. $f(x) = \frac{1}{x}$

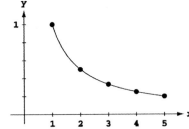

Since f is decreasing on [1, 5], we use left endpoints to obtain upper sums and right endpoints to obtain lower sums.

(a) $\triangle x = \frac{5-1}{2} = 2$ and $x_i = 1 + i\triangle x = 1 + 2i \Rightarrow$ a lower sum is $\sum_{i=1}^{2} \frac{1}{x_i} \cdot 2 = 2\left(\frac{1}{3} + \frac{1}{5}\right) = \frac{16}{15}$

(b) $\triangle x = \frac{5-1}{4} = 1$ and $x_i = 1 + i\triangle x = 1 + i \Rightarrow$ a lower sum is $\sum_{i=1}^{4} \frac{1}{x_i} \cdot 1 = 1\left(\frac{1}{2} + \frac{1}{3} + \frac{1}{4} + \frac{1}{5}\right) = \frac{77}{60}$

(c) $\triangle x = \frac{5-1}{2} = 2$ and $x_i = 1 + i\triangle x = 1 + 2i \Rightarrow$ an upper sum is $\sum_{i=0}^{1} \frac{1}{x_i} \cdot 2 = 2\left(1 + \frac{1}{3}\right) = \frac{8}{3}$

(d) $\triangle x = \frac{5-1}{4} = 1$ and $x_i = 1 + i\triangle x = 1 + i \Rightarrow$ an upper sum is $\sum_{i=0}^{3} \frac{1}{x_i} \cdot 1 = 1\left(1 + \frac{1}{2} + \frac{1}{3} + \frac{1}{4}\right) = \frac{25}{12}$

5. $f(x) = x^2$

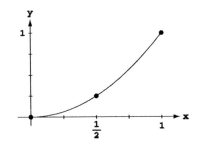

Using 2 rectangles $\Rightarrow \triangle x = \frac{1-0}{2} = \frac{1}{2} \Rightarrow \frac{1}{2}\left(f\left(\frac{1}{4}\right) + f\left(\frac{3}{4}\right)\right)$
$= \frac{1}{2}\left(\left(\frac{1}{4}\right)^2 + \left(\frac{3}{4}\right)^2\right) = \frac{10}{32} = \frac{5}{16}$

Using 4 rectangles $\Rightarrow \triangle x = \frac{1-0}{4} = \frac{1}{4}$
$\Rightarrow \frac{1}{4}\left(f\left(\frac{1}{8}\right) + f\left(\frac{3}{8}\right) + f\left(\frac{5}{8}\right) + f\left(\frac{7}{8}\right)\right)$
$= \frac{1}{4}\left(\left(\frac{1}{8}\right)^2 + \left(\frac{3}{8}\right)^2 + \left(\frac{5}{8}\right)^2 + \left(\frac{7}{8}\right)^2\right) = \frac{21}{64}$

7. $f(x) = \frac{1}{x}$

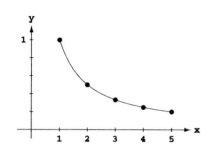

Using 2 rectangles $\Rightarrow \triangle x = \frac{5-1}{2} = 2 \Rightarrow 2(f(2) + f(4))$
$$= 2\left(\tfrac{1}{2} + \tfrac{1}{4}\right) = \tfrac{3}{2}$$

Using 4 rectangles $\Rightarrow \triangle x = \frac{5-1}{4} = 1$
$$\Rightarrow 1\left(f\left(\tfrac{3}{2}\right) + f\left(\tfrac{5}{2}\right) + f\left(\tfrac{7}{2}\right) + f\left(\tfrac{9}{2}\right)\right)$$
$$= 1\left(\tfrac{2}{3} + \tfrac{2}{5} + \tfrac{2}{7} + \tfrac{2}{9}\right) = \frac{1488}{3 \cdot 5 \cdot 7 \cdot 9} = \frac{496}{5 \cdot 7 \cdot 9} = \frac{496}{315}$$

9. (a) $D \approx (0)(1) + (12)(1) + (22)(1) + (10)(1) + (5)(1) + (13)(1) + (11)(1) + (6)(1) + (2)(1) + (6)(1) = 87$ inches
    (b) $D \approx (12)(1) + (22)(1) + (10)(1) + (5)(1) + (13)(1) + (11)(1) + (6)(1) + (2)(1) + (6)(1) + (0)(1) = 87$ inches

11. (a) Because the acceleration is decreasing, an upper estimate is obtained using left end-points in summing
         acceleration $\cdot \Delta t$. Thus, $\Delta t = 1$ and speed $\approx [32.00 + 19.41 + 11.77 + 7.14 + 4.33](1) = 74.65$ ft/sec
     (b) Using right end-points we obtain a lower estimate: speed $\approx [19.41 + 11.77 + 7.14 + 4.33 + 2.63](1)$
         $= 45.28$ ft/sec
     (c) Upper estimates for the speed at each second are:

t	0	1	2	3	4	5
v	0	32.00	51.41	63.18	70.32	74.65

   Thus, the distance fallen when $t = 3$ seconds is $s \approx [32.00 + 51.41 + 63.18](1) = 146.59$ ft.

13. Partition $[0, 2]$ into the four subintervals $[0, 0.5]$, $[0.5, 1]$, $[1, 1.5]$, and $[1.5, 2]$. The midpoints of these
    subintervals are $m_1 = 0.25$, $m_2 = 0.75$, $m_3 = 1.25$, and $m_4 = 1.75$. The heights of the four approximating
    rectangles are $f(m_1) = (0.25)^3 = \frac{1}{64}$, $f(m_2) = (0.75)^3 = \frac{27}{64}$, $f(m_3) = (1.25)^3 = \frac{125}{64}$, and $f(m_4) = (1.75)^3 = \frac{343}{64}$
    Notice that the average value is approximated by $\frac{1}{2}\left[\left(\tfrac{1}{4}\right)^3\left(\tfrac{1}{2}\right) + \left(\tfrac{3}{4}\right)^3\left(\tfrac{1}{2}\right) + \left(\tfrac{5}{4}\right)^3\left(\tfrac{1}{2}\right) + \left(\tfrac{7}{4}\right)^3\left(\tfrac{1}{2}\right)\right] = \frac{31}{16}$
    $= \frac{1}{\text{length of } [0,2]} \cdot \begin{bmatrix} \text{approximate area under} \\ \text{curve } f(x) = x^3 \end{bmatrix}$. We use this observation in solving the next several exercises.

15. Since the leakage is increasing, an upper estimate uses right endpoints and a lower estimate uses left
    endpoints:
     (a) upper estimate $= (70)(1) + (97)(1) + (136)(1) + (190)(1) + (265)(1) = 758$ gal,
         lower estimate $= (50)(1) + (70)(1) + (97)(1) + (136)(1) + (190)(1) = 543$ gal.
     (b) upper estimate $= (70 + 97 + 136 + 190 + 265 + 369 + 516 + 720) = 2363$ gal,
         lower estimate $= (50 + 70 + 97 + 136 + 190 + 265 + 369 + 516) = 1693$ gal.
     (c) worst case: $2363 + 720t = 25{,}000 \Rightarrow t \approx 31.4$ hrs;
         best case: $1693 + 720t = 25{,}000 \Rightarrow t \approx 32.4$ hrs

4.3 SIGMA NOTATION AND LIMITS OF FINITE SUMS

1. $\displaystyle\sum_{k=1}^{2} \frac{6k}{k+1} = \frac{6(1)}{1+1} + \frac{6(2)}{2+1} = \frac{6}{2} + \frac{12}{3} = 7$

3. $\displaystyle\sum_{k=1}^{4} \cos k\pi = \cos(1\pi) + \cos(2\pi) + \cos(3\pi) + \cos(4\pi) = -1 + 1 - 1 + 1 = 0$

5. $\displaystyle\sum_{k=1}^{3} (-1)^{k+1} \sin \frac{\pi}{k} = (-1)^{1+1} \sin \frac{\pi}{1} + (-1)^{2+1} \sin \frac{\pi}{2} + (-1)^{3+1} \sin \frac{\pi}{3} = 0 - 1 + \frac{\sqrt{3}}{2} = \frac{\sqrt{3}-2}{2}$

7. (a) $\sum\limits_{k=1}^{6} 2^{k-1} = 2^{1-1} + 2^{2-1} + 2^{3-1} + 2^{4-1} + 2^{5-1} + 2^{6-1} = 1 + 2 + 4 + 8 + 16 + 32$

(b) $\sum\limits_{k=0}^{5} 2^{k} = 2^{0} + 2^{1} + 2^{2} + 2^{3} + 2^{4} + 2^{5} = 1 + 2 + 4 + 8 + 16 + 32$

(c) $\sum\limits_{k=-1}^{4} 2^{k+1} = 2^{-1+1} + 2^{0+1} + 2^{1+1} + 2^{2+1} + 2^{3+1} + 2^{4+1} = 1 + 2 + 4 + 8 + 16 + 32$

All of them represent $1 + 2 + 4 + 8 + 16 + 32$

9. (a) $\sum\limits_{k=2}^{4} \frac{(-1)^{k-1}}{k-1} = \frac{(-1)^{2-1}}{2-1} + \frac{(-1)^{3-1}}{3-1} + \frac{(-1)^{4-1}}{4-1} = -1 + \frac{1}{2} - \frac{1}{3}$

(b) $\sum\limits_{k=0}^{2} \frac{(-1)^{k}}{k+1} = \frac{(-1)^{0}}{0+1} + \frac{(-1)^{1}}{1+1} + \frac{(-1)^{2}}{2+1} = 1 - \frac{1}{2} + \frac{1}{3}$

(c) $\sum\limits_{k=-1}^{1} \frac{(-1)^{k}}{k+2} = \frac{(-1)^{-1}}{-1+2} + \frac{(-1)^{0}}{0+2} + \frac{(-1)^{1}}{1+2} = -1 + \frac{1}{2} - \frac{1}{3}$

(a) and (c) are equivalent; (b) is not equivalent to the other two.

11. $\sum\limits_{k=1}^{6} k$

13. $\sum\limits_{k=1}^{4} \frac{1}{2^{k}}$

15. $\sum\limits_{k=1}^{5} (-1)^{k+1} \frac{1}{k}$

17. (a) $\sum\limits_{k=1}^{n} 3a_{k} = 3 \sum\limits_{k=1}^{n} a_{k} = 3(-5) = -15$

(b) $\sum\limits_{k=1}^{n} \frac{b_{k}}{6} = \frac{1}{6} \sum\limits_{k=1}^{n} b_{k} = \frac{1}{6}(6) = 1$

(c) $\sum\limits_{k=1}^{n} (a_{k} + b_{k}) = \sum\limits_{k=1}^{n} a_{k} + \sum\limits_{k=1}^{n} b_{k} = -5 + 6 = 1$

(d) $\sum\limits_{k=1}^{n} (a_{k} - b_{k}) = \sum\limits_{k=1}^{n} a_{k} - \sum\limits_{k=1}^{n} b_{k} = -5 - 6 = -11$

(e) $\sum\limits_{k=1}^{n} (b_{k} - 2a_{k}) = \sum\limits_{k=1}^{n} b_{k} - 2 \sum\limits_{k=1}^{n} a_{k} = 6 - 2(-5) = 16$

19. (a) $\sum\limits_{k=1}^{10} k = \frac{10(10+1)}{2} = 55$

(b) $\sum\limits_{k=1}^{10} k^{2} = \frac{10(10+1)(2(10)+1)}{6} = 385$

(c) $\sum\limits_{k=1}^{10} k^{3} = \left[\frac{10(10+1)}{2} \right]^{2} = 55^{2} = 3025$

21. $\sum\limits_{k=1}^{7} -2k = -2 \sum\limits_{k=1}^{7} k = -2 \left(\frac{7(7+1)}{2} \right) = -56$

23. $\sum\limits_{k=1}^{6} (3 - k^{2}) = \sum\limits_{k=1}^{6} 3 - \sum\limits_{k=1}^{6} k^{2} = 3(6) - \frac{6(6+1)(2(6)+1)}{6} = -73$

25. $\sum\limits_{k=1}^{5} k(3k + 5) = \sum\limits_{k=1}^{5} (3k^{2} + 5k) = 3 \sum\limits_{k=1}^{5} k^{2} + 5 \sum\limits_{k=1}^{5} k = 3 \left(\frac{5(5+1)(2(5)+1)}{6} \right) + 5 \left(\frac{5(5+1)}{2} \right) = 240$

27. $\sum\limits_{k=1}^{5} \frac{k^{3}}{225} + \left(\sum\limits_{k=1}^{5} k \right)^{3} = \frac{1}{225} \sum\limits_{k=1}^{5} k^{3} + \left(\sum\limits_{k=1}^{5} k \right)^{3} = \frac{1}{225} \left(\frac{5(5+1)}{2} \right)^{2} + \left(\frac{5(5+1)}{2} \right)^{3} = 3376$

29. (a) 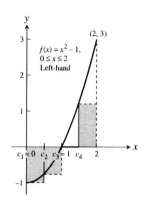   (b) 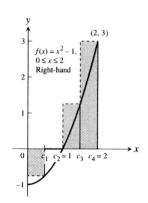   (c)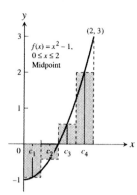

31. (a) 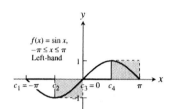   (b) 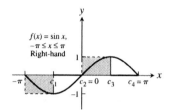   (c)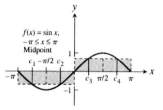

33. $|x_1 - x_0| = |1.2 - 0| = 1.2$, $|x_2 - x_1| = |1.5 - 1.2| = 0.3$, $|x_3 - x_2| = |2.3 - 1.5| = 0.8$, $|x_4 - x_3| = |2.6 - 2.3| = 0.3$, and $|x_5 - x_4| = |3 - 2.6| = 0.4$; the largest is $\|P\| = 1.2$.

35. $f(x) = 1 - x^2$

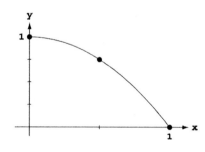

Since f is decreasing on $[0, 1]$ we use left endpoints to obtain upper sums. $\triangle x = \frac{1-0}{n} = \frac{1}{n}$ and $x_i = i\triangle x = \frac{i}{n}$. So an upper sum is $\sum_{i=0}^{n-1}(1 - x_i^2)\frac{1}{n} = \frac{1}{n}\sum_{i=0}^{n-1}\left(1 - \left(\frac{i}{n}\right)^2\right) = \frac{1}{n^3}\sum_{i=0}^{n-1}(n^2 - i^2)$

$= \frac{n^3}{n^3} - \frac{1}{n^3}\sum_{i=0}^{n}i^2 = 1 - \frac{(n-1)n(2(n-1)+1)}{6n^3} = 1 - \frac{2n^3 - 3n^2 + n}{6n^3}$

$= 1 - \frac{2 - \frac{3}{n} + \frac{1}{n^2}}{6}$. Thus,

$\lim_{n\to\infty}\sum_{i=0}^{n-1}(1 - x_i^2)\frac{1}{n} = \lim_{n\to\infty}\left(1 - \frac{2 - \frac{3}{n} + \frac{1}{n^2}}{6}\right) = 1 - \frac{1}{3} = \frac{2}{3}$

37. $f(x) = x^2 + 1$

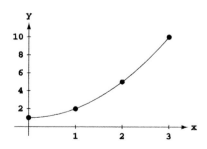

Since f is increasing on $[0, 3]$ we use right endpoints to obtain upper sums. $\triangle x = \frac{3-0}{n} = \frac{3}{n}$ and $x_i = i\triangle x = \frac{3i}{n}$. So an upper sum is $\sum_{i=1}^{n}(x_i^2 + 1)\frac{3}{n} = \sum_{i=1}^{n}\left(\left(\frac{3i}{n}\right)^2 + 1\right)\frac{3}{n} = \frac{3}{n}\sum_{i=1}^{n}\left(\frac{9i^2}{n^2} + 1\right)$

$= \frac{27}{n}\sum_{i=1}^{n}i^2 + \frac{3}{n}\cdot n = \frac{27}{n^3}\left(\frac{n(n+1)(2n+1)}{6}\right) + 3$

$= \frac{9(2n^3 + 3n^2 + n)}{2n^3} + 3 = \frac{18 + \frac{27}{n} + \frac{9}{n^2}}{2} + 3$. Thus,

$\lim_{n\to\infty}\sum_{i=1}^{n}(x_i^2 + 1)\frac{3}{n} = \lim_{n\to\infty}\left(\frac{18 + \frac{27}{n} + \frac{9}{n^2}}{2} + 3\right) = 9 + 3 = 12.$

39. $f(x) = x + x^2 = x(1 + x)$

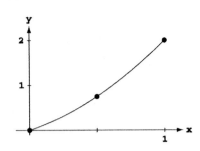

Since f is increasing on $[0, 1]$ we use right endpoints to obtain upper sums. $\triangle x = \frac{1-0}{n} = \frac{1}{n}$ and $x_i = i\triangle x = \frac{i}{n}$. So an upper sum is $\sum_{i=1}^{n}(x_i + x_i^2)\frac{1}{n} = \sum_{i=1}^{n}\left(\frac{i}{n} + \left(\frac{i}{n}\right)^2\right)\frac{1}{n} = \frac{1}{n^2}\sum_{i=1}^{n}i + \frac{1}{n^3}\sum_{i=1}^{n}i^2$

$= \frac{1}{n^2}\left(\frac{n(n+1)}{2}\right) + \frac{1}{n^3}\left(\frac{n(n+1)(2n+1)}{6}\right) = \frac{n^2+n}{2n^2} + \frac{2n^3+3n^2+n}{6n^3}$

$= \frac{1+\frac{1}{n}}{2} + \frac{2+\frac{3}{n}+\frac{1}{n^2}}{6}$. Thus, $\lim_{n\to\infty}\sum_{i=1}^{n}(x_i + x_i^2)\frac{1}{n}$

$= \lim_{n\to\infty}\left[\left(\frac{1+\frac{1}{n}}{2}\right) + \left(\frac{2+\frac{3}{n}+\frac{1}{n^2}}{6}\right)\right] = \frac{1}{2} + \frac{2}{6} = \frac{5}{6}$.

4.4 THE DEFINITE INTEGRAL

1. $\int_{0}^{2} x^2 \, dx$

3. $\int_{-7}^{5}(x^2 - 3x)\, dx$

5. $\int_{2}^{3} \frac{1}{1-x}\, dx$

7. $\int_{-\pi/4}^{0}(\sec x)\, dx$

9. (a) $\int_{2}^{2} g(x)\, dx = 0$

(b) $\int_{5}^{1} g(x)\, dx = -\int_{1}^{5} g(x)\, dx = -8$

(c) $\int_{1}^{2} 3f(x)\, dx = 3\int_{1}^{2} f(x)\, dx = 3(-4) = -12$

(d) $\int_{2}^{5} f(x)\, dx = \int_{1}^{5} f(x)\, dx - \int_{1}^{2} f(x)\, dx = 6 - (-4) = 10$

(e) $\int_{1}^{5}[f(x) - g(x)]\, dx = \int_{1}^{5} f(x)\, dx - \int_{1}^{5} g(x)\, dx = 6 - 8 = -2$

(f) $\int_{1}^{5}[4f(x) - g(x)]\, dx = 4\int_{1}^{5} f(x)\, dx - \int_{1}^{5} g(x)\, dx = 4(6) - 8 = 16$

11. (a) $\int_{1}^{2} f(u)\, du = \int_{1}^{2} f(x)\, dx = 5$

(b) $\int_{1}^{2}\sqrt{3}\, f(z)\, dz = \sqrt{3}\int_{1}^{2} f(z)\, dz = 5\sqrt{3}$

(c) $\int_{2}^{1} f(t)\, dt = -\int_{1}^{2} f(t)\, dt = -5$

(d) $\int_{1}^{2}[-f(x)]\, dx = -\int_{1}^{2} f(x)\, dx = -5$

13. (a) $\int_{3}^{4} f(z)\, dz = \int_{0}^{4} f(z)\, dz - \int_{0}^{3} f(z)\, dz = 7 - 3 = 4$

(b) $\int_{4}^{3} f(t)\, dt = -\int_{3}^{4} f(t)\, dt = -4$

15. The area of the trapezoid is $A = \frac{1}{2}(B + b)h$

$= \frac{1}{2}(5 + 2)(6) = 21 \Rightarrow \int_{-2}^{4}\left(\frac{x}{2} + 3\right)dx$

$= 21$ square units

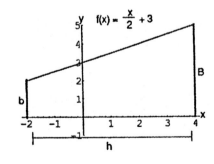

17. The area of the semicircle is A $= \frac{1}{2}\pi r^2 = \frac{1}{2}\pi(3)^2$

$= \frac{9}{2}\pi \Rightarrow \int_{-3}^{3} \sqrt{9-x^2}\, dx = \frac{9}{2}\pi$ square units

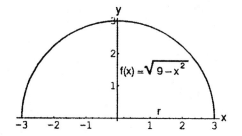

19. The area of the triangle on the left is A $= \frac{1}{2}$ bh $= \frac{1}{2}(2)(2)$

$= 2$. The area of the triangle on the right is A $= \frac{1}{2}$ bh

$= \frac{1}{2}(1)(1) = \frac{1}{2}$. Then, the total area is 2.5

$\Rightarrow \int_{-2}^{1} |x|\, dx = 2.5$ square units

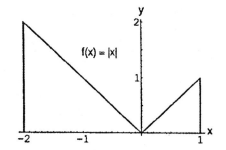

21. The area of the triangular peak is A $= \frac{1}{2}$ bh $= \frac{1}{2}(2)(1) = 1$.

The area of the rectangular base is S $= \ell w = (2)(1) = 2$.

Then the total area is 3 $\Rightarrow \int_{-1}^{1}(2-|x|)\, dx = 3$ square units

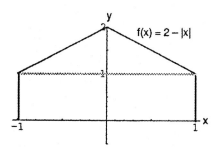

23. $\int_{0}^{b} \frac{x}{2}\, dx = \frac{1}{2}(b)(\frac{b}{2}) = \frac{b^2}{4}$

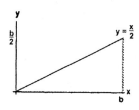

25. $\int_{1}^{\sqrt{2}} x\, dx = \frac{(\sqrt{2})^2}{2} - \frac{(1)^2}{2} = \frac{1}{2}$

27. $\int_{0}^{\sqrt[3]{7}} x^2\, dx = \frac{(\sqrt[3]{7})^3}{3} = \frac{7}{3}$

29. $\int_{a}^{2a} x\, dx = \frac{(2a)^2}{2} - \frac{a^2}{2} = \frac{3a^2}{2}$

31. $\int_{3}^{1} 7\, dx = 7(1-3) = -14$

33. $\int_{0}^{2}(2t-3)\, dt = 2\int_{1}^{1} t\, dt - \int_{0}^{2} 3\, dt = 2\left[\frac{2^2}{2} - \frac{0^2}{2}\right] - 3(2-0) = 4 - 6 = -2$

35. $\int_{1}^{2} 3u^2\, du = 3\int_{1}^{2} u^2\, du = 3\left[\int_{0}^{2} u^2\, du - \int_{0}^{1} u^2\, du\right] = 3\left(\left[\frac{2^3}{3} - \frac{0^3}{3}\right] - \left[\frac{1^3}{3} - \frac{0^3}{3}\right]\right) = 3\left[\frac{2^3}{3} - \frac{1^3}{3}\right] = 3\left(\frac{7}{3}\right) = 7$

37. Let $\Delta x = \frac{b-0}{n} = \frac{b}{n}$ and let $x_0 = 0$, $x_1 = \Delta x$,

$x_2 = 2\Delta x, \dots , x_{n-1} = (n-1)\Delta x$, $x_n = n\Delta x = b$.

Let the c_k's be the right end-points of the subintervals

$\Rightarrow c_1 = x_1$, $c_2 = x_2$, and so on. The rectangles

defined have areas:

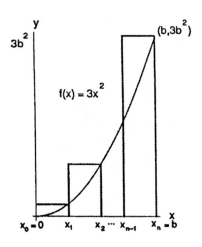

$$f(c_1)\,\Delta x = f(\Delta x)\,\Delta x = 3(\Delta x)^2\,\Delta x = 3(\Delta x)^3$$

$$f(c_2)\,\Delta x = f(2\Delta x)\,\Delta x = 3(2\Delta x)^2\,\Delta x = 3(2)^2(\Delta x)^3$$

$$f(c_3)\,\Delta x = f(3\Delta x)\,\Delta x = 3(3\Delta x)^2\,\Delta x = 3(3)^2(\Delta x)^3$$

$$\vdots$$

$$f(c_n)\,\Delta x = f(n\Delta x)\,\Delta x = 3(n\Delta x)^2\,\Delta x = 3(n)^2(\Delta x)^3$$

Then $S_n = \sum\limits_{k=1}^{n} f(c_k)\,\Delta x = \sum\limits_{k=1}^{n} 3k^2(\Delta x)^3$

$= 3(\Delta x)^3 \sum\limits_{k=1}^{n} k^2 = 3\left(\frac{b^3}{n^3}\right)\left(\frac{n(n+1)(2n+1)}{6}\right)$

$= \frac{b^3}{2}\left(2 + \frac{3}{n} + \frac{1}{n^2}\right) \Rightarrow \int_0^b 3x^2\,dx = \lim\limits_{n\to\infty} \frac{b^3}{2}\left(2 + \frac{3}{n} + \frac{1}{n^2}\right) = b^3.$

39. $\operatorname{av}(f) = \left(\frac{1}{\sqrt{3}-0}\right)\int_0^{\sqrt{3}} (x^2 - 1)\,dx$

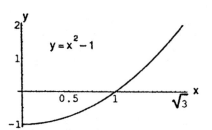

$= \frac{1}{\sqrt{3}}\int_0^{\sqrt{3}} x^2\,dx - \frac{1}{\sqrt{3}}\int_0^{\sqrt{3}} 1\,dx$

$= \frac{1}{\sqrt{3}}\left(\frac{(\sqrt{3})^3}{3}\right) - \frac{1}{\sqrt{3}}\left(\sqrt{3} - 0\right) = 1 - 1 = 0.$

41. $\operatorname{av}(f) = \left(\frac{1}{3-0}\right)\int_0^3 (t-1)^2\,dt$

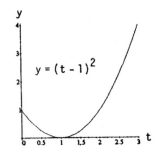

$= \frac{1}{3}\int_0^3 t^2\,dt - \frac{2}{3}\int_0^3 t\,dt + \frac{1}{3}\int_0^3 1\,dt$

$= \frac{1}{3}\left(\frac{3^3}{3}\right) - \frac{2}{3}\left(\frac{3^2}{2} - \frac{0^2}{2}\right) + \frac{1}{3}(3-0) = 1.$

43. (a) $\operatorname{av}(g) = \left(\frac{1}{1-(-1)}\right)\int_{-1}^{1} (|x| - 1)\,dx$

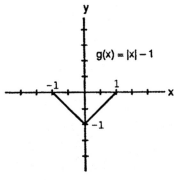

$= \frac{1}{2}\int_{-1}^{0} (-x - 1)\,dx + \frac{1}{2}\int_0^1 (x - 1)\,dx$

$= -\frac{1}{2}\int_{-1}^{0} x\,dx - \frac{1}{2}\int_{-1}^{0} 1\,dx + \frac{1}{2}\int_0^1 x\,dx - \frac{1}{2}\int_0^1 1\,dx$

$= -\frac{1}{2}\left(\frac{0^2}{2} - \frac{(-1)^2}{2}\right) - \frac{1}{2}(0 - (-1)) + \frac{1}{2}\left(\frac{1^2}{2} - \frac{0^2}{2}\right) - \frac{1}{2}(1 - 0)$

$= -\frac{1}{2}.$

(b) $\text{av}(g) = \left(\frac{1}{3-1}\right) \int_1^3 (|x| - 1)\, dx = \frac{1}{2} \int_1^3 (x - 1)\, dx$

$= \frac{1}{2} \int_1^3 x\, dx - \frac{1}{2} \int_1^3 1\, dx = \frac{1}{2} \left(\frac{3^2}{2} - \frac{1^2}{2}\right) - \frac{1}{2}(3 - 1)$

$= 1.$

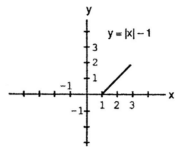

(c) $\text{av}(g) = \left(\frac{1}{3-(-1)}\right) \int_{-1}^3 (|x| - 1)\, dx$

$= \frac{1}{4} \int_{-1}^1 (|x| - 1)\, dx + \frac{1}{4} \int_1^3 (|x| - 1)\, dx$

$= \frac{1}{4}(-1 + 2) = \frac{1}{4}$ (see parts (a) and (b) above).

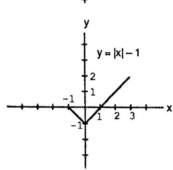

45. To find where $x - x^2 \geq 0$, let $x - x^2 = 0 \Rightarrow x(1 - x) = 0 \Rightarrow x = 0$ or $x = 1$. If $0 < x < 1$, then $0 < x - x^2 \Rightarrow a = 0$ and $b = 1$ maximize the integral.

47. $f(x) = \frac{1}{1+x^2}$ is decreasing on $[0, 1] \Rightarrow$ maximum value of f occurs at $0 \Rightarrow$ max $f = f(0) = 1$; minimum value of f

occurs at $1 \Rightarrow$ min $f = f(1) = \frac{1}{1+1^2} = \frac{1}{2}$. Therefore, $(1 - 0)$ min $f \leq \int_0^1 \frac{1}{1+x^2}\, dx \leq (1 - 0)$ max f

$\Rightarrow \frac{1}{2} \leq \int_0^1 \frac{1}{1+x^2}\, dx \leq 1$. That is, an upper bound $= 1$ and a lower bound $= \frac{1}{2}$.

49. $-1 \leq \sin(x^2) \leq 1$ for all $x \Rightarrow (1 - 0)(-1) \leq \int_0^1 \sin(x^2)\, dx \leq (1 - 0)(1)$ or $\int_0^1 \sin x^2\, dx \leq 1 \Rightarrow \int_0^1 \sin x^2\, dx$ cannot equal 2.

51. If $f(x) \geq 0$ on $[a, b]$, then min $f \geq 0$ and max $f \geq 0$ on $[a, b]$. Now, $(b - a)$ min $f \leq \int_a^b f(x)\, dx \leq (b - a)$ max f.

Then $b \geq a \Rightarrow b - a \geq 0 \Rightarrow (b - a)$ min $f \geq 0 \Rightarrow \int_a^b f(x)\, dx \geq 0.$

53. Yes, for the following reasons: $\text{av}(f) = \frac{1}{b-a} \int_a^b f(x)\, dx$ is a constant K. Thus $\int_a^b \text{av}(f)\, dx = \int_a^b K\, dx$

$= K(b - a) \Rightarrow \int_a^b \text{av}(f)\, dx = (b - a)K = (b - a) \cdot \frac{1}{b-a} \int_a^b f(x)\, dx = \int_a^b f(x)\, dx.$

55. Consider the partition P that subdivides the interval $[a, b]$ into n subintervals of width $\triangle x = \frac{b-a}{n}$ and let c_k be the right

endpoint of each subinterval. So the partition is $P = \{a, a + \frac{b-a}{n}, a + \frac{2(b-a)}{n}, \ldots, a + \frac{n(b-a)}{n}\}$ and $c_k = a + \frac{k(b-a)}{n}$.

We get the Riemann sum $\sum_{k=1}^n f(c_k)\triangle x = \sum_{k=1}^n c \cdot \frac{b-a}{n} = \frac{c(b-a)}{n} \sum_{k=1}^n 1 = \frac{c(b-a)}{n} \cdot n = c(b - a)$. As $n \to \infty$ and $\|P\| \to 0$

this expression remains $c(b - a)$. Thus, $\int_a^b c\, dx = c(b - a).$

57. (a) $U = \max_1 \Delta x + \max_2 \Delta x + \ldots + \max_n \Delta x$ where $\max_1 = f(x_1), \max_2 = f(x_2), \ldots, \max_n = f(x_n)$ since f is increasing on $[a, b]$; $L = \min_1 \Delta x + \min_2 \Delta x + \ldots + \min_n \Delta x$ where $\min_1 = f(x_0), \min_2 = f(x_1), \ldots,$ $\min_n = f(x_{n-1})$ since f is increasing on $[a, b]$. Therefore

$U - L = (\max_1 - \min_1) \Delta x + (\max_2 - \min_2) \Delta x + \ldots + (\max_n - \min_n) \Delta x$

$= (f(x_1) - f(x_0)) \Delta x + (f(x_2) - f(x_1))\Delta x + \ldots + (f(x_n) - f(x_{n-1})) \Delta x = (f(x_n) - f(x_0)) \Delta x = (f(b) - f(a)) \Delta x.$

(b) $U = \max_1 \Delta x_1 + \max_2 \Delta x_2 + \ldots + \max_n \Delta x_n$ where $\max_1 = f(x_1), \max_2 = f(x_2), \ldots, \max_n = f(x_n)$ since f is increasing on $[a, b]$; $L = \min_1 \Delta x_1 + \min_2 \Delta x_2 + \ldots + \min_n \Delta x_n$ where

$\min_1 = f(x_0), \min_2 = f(x_1), \ldots, \min_n = f(x_{n-1})$ since f is increasing on $[a, b]$. Therefore

$U - L = (\max_1 - \min_1) \Delta x_1 + (\max_2 - \min_2) \Delta x_2 + \ldots + (\max_n - \min_n) \Delta x_n$

$= (f(x_1) - f(x_0)) \Delta x_1 + (f(x_2) - f(x_1))\Delta x_2 + \ldots + (f(x_n) - f(x_{n-1})) \Delta x_n$

$\leq (f(x_1) - f(x_0)) \Delta x_{max} + (f(x_2) - f(x_1)) \Delta x_{max} + \ldots + (f(x_n) - f(x_{n-1})) \Delta x_{max}$. Then

$U - L \leq (f(x_n) - f(x_0)) \Delta x_{max} = (f(b) - f(a)) \Delta x_{max} = |f(b) - f(a)| \Delta x_{max}$ since $f(b) \geq f(a)$. Thus

$\lim\limits_{\|P\| \to 0} (U - L) = \lim\limits_{\|P\| \to 0} (f(b) - f(a)) \Delta x_{max} = 0$, since $\Delta x_{max} = \|P\|$.

59. (a) Partition $\left[0, \frac{\pi}{2}\right]$ into n subintervals, each of length $\Delta x = \frac{\pi}{2n}$ with points $x_0 = 0$, $x_1 = \Delta x$,

$x_2 = 2\Delta x, \ldots, x_n = n\Delta x = \frac{\pi}{2}$. Since $\sin x$ is increasing on $\left[0, \frac{\pi}{2}\right]$, the upper sum U is the sum of the areas of the circumscribed rectangles of areas $f(x_1) \Delta x = (\sin \Delta x)\Delta x, f(x_2) \Delta x = (\sin 2\Delta x) \Delta x, \ldots, f(x_n) \Delta x$

$= (\sin n\Delta x) \Delta x$. Then $U = (\sin \Delta x + \sin 2\Delta x + \ldots + \sin n\Delta x) \Delta x = \left[\dfrac{\cos \frac{\Delta x}{2} - \cos\left(\left(n + \frac{1}{2}\right) \Delta x\right)}{2 \sin \frac{\Delta x}{2}}\right] \Delta x$

$= \left[\dfrac{\cos \frac{\pi}{4n} - \cos\left(\left(n + \frac{1}{2}\right) \frac{\pi}{2n}\right)}{2 \sin \frac{\pi}{4n}}\right] \left(\frac{\pi}{2n}\right) = \dfrac{\pi \left(\cos \frac{\pi}{4n} - \cos\left(\frac{\pi}{2} + \frac{\pi}{4n}\right)\right)}{4n \sin \frac{\pi}{4n}} = \dfrac{\cos \frac{\pi}{4n} - \cos\left(\frac{\pi}{2} + \frac{\pi}{4n}\right)}{\left(\frac{\sin \frac{\pi}{4n}}{\frac{\pi}{4n}}\right)}$

(b) The area is $\displaystyle\int_0^{\pi/2} \sin x \, dx = \lim\limits_{n \to \infty} \dfrac{\cos \frac{\pi}{4n} - \cos\left(\frac{\pi}{2} + \frac{\pi}{4n}\right)}{\left(\frac{\sin \frac{\pi}{4n}}{\frac{\pi}{4n}}\right)} = \dfrac{1 - \cos \frac{\pi}{2}}{1} = 1$.

61. By Exercise 60, $U - L = \displaystyle\sum_{i=1}^{n} \triangle x_i \cdot M_i - \sum_{i=1}^{n} \triangle x_i \cdot m_i$ where $M_i = \max\{f(x) \text{ on the ith subinterval}\}$ and

$m_i = \min\{f(x) \text{ on the ith subinterval}\}$. Thus $U - L = \displaystyle\sum_{i=1}^{n} (M_i - m_i)\triangle x_i < \sum_{i=1}^{n} \epsilon \cdot \triangle x_i$ provided $\triangle x_i < \delta$ for each

$i = 1, \ldots, n$. Since $\displaystyle\sum_{i=1}^{n} \epsilon \cdot \triangle x_i = \epsilon \sum_{i=1}^{n} \triangle x_i = \epsilon(b - a)$ the result, $U - L < \epsilon(b - a)$ follows.

4.5 THE FUNDAMENTAL THEOREM OF CALCULUS

1. $\displaystyle\int_{-2}^{0} (2x + 5) \, dx = [x^2 + 5x]_{-2}^{0} = (0^2 + 5(0)) - ((-2)^2 + 5(-2)) = 6$

3. $\displaystyle\int_{0}^{4} \left(3x - \frac{x^3}{4}\right) dx = \left[\frac{3x^2}{2} - \frac{x^4}{16}\right]_{0}^{4} = \left(\frac{3(4)^2}{2} - \frac{4^4}{16}\right) - \left(\frac{3(0)^2}{2} - \frac{(0)^4}{16}\right) = 8$

5. $\displaystyle\int_{0}^{1} \left(x^2 + \sqrt{x}\right) dx = \left[\frac{x^3}{3} + \frac{2}{3} x^{3/2}\right]_{0}^{1} = \left(\frac{1}{3} + \frac{2}{3}\right) - 0 = 1$

7. $\displaystyle\int_{1}^{32} x^{-6/5} \, dx = \left[-5x^{-1/5}\right]_{1}^{32} = \left(-\frac{5}{2}\right) - (-5) = \frac{5}{2}$

9. $\displaystyle\int_{0}^{\pi} \sin x \, dx = [-\cos x]_{0}^{\pi} = (-\cos \pi) - (-\cos 0) = -(-1) - (-1) = 2$

11. $\displaystyle\int_{0}^{\pi/3} 2 \sec^2 x \, dx = [2 \tan x]_{0}^{\pi/3} = \left(2 \tan \left(\frac{\pi}{3}\right)\right) - (2 \tan 0) = 2\sqrt{3} - 0 = 2\sqrt{3}$

13. $\int_{\pi/4}^{3\pi/4} \csc\theta\cot\theta\, d\theta = [-\csc\theta]_{\pi/4}^{3\pi/4} = \left(-\csc\left(\frac{3\pi}{4}\right)\right) - \left(-\csc\left(\frac{\pi}{4}\right)\right) = -\sqrt{2} - \left(-\sqrt{2}\right) = 0$

15. $\int_{\pi/2}^{0} \frac{1+\cos 2t}{2}\, dt = \int_{\pi/2}^{0}\left(\frac{1}{2} + \frac{1}{2}\cos 2t\right) dt = \left[\frac{1}{2}t + \frac{1}{4}\sin 2t\right]_{\pi/2}^{0} = \left(\frac{1}{2}(0) + \frac{1}{4}\sin 2(0)\right) - \left(\frac{1}{2}\left(\frac{\pi}{2}\right) + \frac{1}{4}\sin 2\left(\frac{\pi}{2}\right)\right)$

$= -\frac{\pi}{4}$

17. $\int_{-\pi/2}^{\pi/2}(8y^2 + \sin y)\, dy = \left[\frac{8y^3}{3} - \cos y\right]_{-\pi/2}^{\pi/2} = \left(\frac{8\left(\frac{\pi}{2}\right)^3}{3} - \cos\frac{\pi}{2}\right) - \left(\frac{8\left(-\frac{\pi}{2}\right)^3}{3} - \cos\left(-\frac{\pi}{2}\right)\right) = \frac{2\pi^3}{3}$

19. $\int_{1}^{-1}(r+1)^2\, dr = \int_{1}^{-1}(r^2 + 2r + 1)\, dr = \left[\frac{r^3}{3} + r^2 + r\right]_{1}^{-1} = \left(\frac{(-1)^3}{3} + (-1)^2 + (-1)\right) - \left(\frac{1^3}{3} + 1^2 + 1\right) = -\frac{8}{3}$

21. $\int_{\sqrt{2}}^{1}\left(\frac{u^7}{2} - \frac{1}{u^5}\right) du = \int_{\sqrt{2}}^{1}\left(\frac{u^7}{2} - u^{-5}\right) du = \left[\frac{u^8}{16} + \frac{1}{4u^4}\right]_{\sqrt{2}}^{1} = \left(\frac{1^8}{16} + \frac{1}{4(1)^4}\right) - \left(\frac{\left(\sqrt{2}\right)^8}{16} + \frac{1}{4\left(\sqrt{2}\right)^4}\right) = -\frac{3}{4}$

23. $\int_{1}^{\sqrt{2}} \frac{s^2 + \sqrt{s}}{s^2}\, ds = \int_{1}^{\sqrt{2}}\left(1 + s^{-3/2}\right) ds = \left[s - \frac{2}{\sqrt{s}}\right]_{1}^{\sqrt{2}} = \left(\sqrt{2} - \frac{2}{\sqrt{\sqrt{2}}}\right) - \left(1 - \frac{2}{\sqrt{1}}\right) = \sqrt{2} - 2^{3/4} + 1$

$= \sqrt{2} - \sqrt[4]{8} + 1$

25. $\int_{-4}^{4} |x|\, dx = \int_{-4}^{0} |x|\, dx + \int_{0}^{4} |x|\, dx = -\int_{-4}^{0} x\, dx + \int_{0}^{4} x\, dx = \left[-\frac{x^2}{2}\right]_{-4}^{0} + \left[\frac{x^2}{2}\right]_{0}^{4} = \left(-\frac{0^2}{2} + \frac{(-4)^2}{2}\right) + \left(\frac{4^2}{2} - \frac{0^2}{2}\right)$

$= 16$

27. $\int_{0}^{\ln 2} e^{3x}dx = \frac{1}{3} e^{3x}\Big|_{0}^{\ln 2} = \frac{1}{3} e^{3\ln 2} - \frac{1}{3} e^0 = \frac{1}{3} e^{\ln 8} - \frac{1}{3} = \frac{8}{3} - \frac{1}{3} = \frac{7}{3}$

29. $\int_{0}^{1} \frac{4}{1+x^2}dx = 4\tan^{-1}x\Big|_{0}^{1} = 4\tan^{-1}1 - 4\tan^{-1}0 = 4\left(\frac{\pi}{4}\right) - 4(0) = \pi$

31. $\int_{2}^{4} x^{\pi-1}dx = \frac{x^\pi}{\pi}\Big|_{2}^{4} = \frac{1}{\pi}(4^\pi - 2^\pi)$ 33. $\int_{0}^{1} x e^{x^2}dx = \frac{1}{2}e^{x^2}\Big|_{0}^{1} = \frac{1}{2}e^1 - \frac{1}{2}e^0 = \frac{1}{2}(e-1)$

35. (a) $\int_{0}^{\sqrt{x}} \cos t\, dt = [\sin t]_{0}^{\sqrt{x}} = \sin\sqrt{x} - \sin 0 = \sin\sqrt{x} \Rightarrow \frac{d}{dx}\left(\int_{0}^{\sqrt{x}} \cos t\, dt\right) = \frac{d}{dx}\left(\sin\sqrt{x}\right) = \cos\sqrt{x}\left(\frac{1}{2}x^{-1/2}\right)$

$= \frac{\cos\sqrt{x}}{2\sqrt{x}}$

(b) $\frac{d}{dx}\left(\int_{0}^{\sqrt{x}} \cos t\, dt\right) = \left(\cos\sqrt{x}\right)\left(\frac{d}{dx}\left(\sqrt{x}\right)\right) = \left(\cos\sqrt{x}\right)\left(\frac{1}{2}x^{-1/2}\right) = \frac{\cos\sqrt{x}}{2\sqrt{x}}$

37. (a) $\int_{0}^{t^4} \sqrt{u}\, du = \int_{0}^{t^4} u^{1/2}\, du = \left[\frac{2}{3}u^{3/2}\right]_{0}^{t^4} = \frac{2}{3}(t^4)^{3/2} - 0 = \frac{2}{3}t^6 \Rightarrow \frac{d}{dt}\left(\int_{0}^{t^4} \sqrt{u}\, du\right) = \frac{d}{dt}\left(\frac{2}{3}t^6\right) = 4t^5$

(b) $\frac{d}{dt}\left(\int_{0}^{t^4} \sqrt{u}\, du\right) = \sqrt{t^4}\left(\frac{d}{dt}(t^4)\right) = t^2(4t^3) = 4t^5$

39. (a) $\int_{0}^{x^3} e^{-t}dt = -e^{-t}\Big|_{0}^{x^3} = -e^{-x^3} + 1 \Rightarrow \frac{d}{dx}\left(\int_{0}^{x^3} e^{-t}dt\right) = \frac{d}{dx}\left(-e^{-x^3} + 1\right) = 3x^2e^{-x^3}$

(b) $\frac{d}{dx}\left(\int_{0}^{x^3} e^{-t}dt\right) = e^{-x^3}\cdot\frac{d}{dx}(x^3) = 3x^2e^{-x^3}$

41. $y = \int_0^x \sqrt{1+t^2}\, dt \Rightarrow \frac{dy}{dx} = \sqrt{1+x^2}$

43. $y = \int_{\sqrt{x}}^0 \sin t^2\, dt = -\int_0^{\sqrt{x}} \sin t^2\, dt \Rightarrow \frac{dy}{dx} = -\left(\sin\left(\sqrt{x}\right)^2\right)\left(\frac{d}{dx}\left(\sqrt{x}\right)\right) = -(\sin x)\left(\frac{1}{2}x^{-1/2}\right) = -\frac{\sin x}{2\sqrt{x}}$

45. $y = \int_0^{\sin x} \frac{dt}{\sqrt{1-t^2}}, \ |x| < \frac{\pi}{2} \Rightarrow \frac{dy}{dx} = \frac{1}{\sqrt{1-\sin^2 x}}\left(\frac{d}{dx}(\sin x)\right) = \frac{1}{\sqrt{\cos^2 x}}(\cos x) = \frac{\cos x}{|\cos x|} = \frac{\cos x}{\cos x} = 1$ since $|x| < \frac{\pi}{2}$

47. $y = \int_0^{e^{x^2}} \frac{1}{\sqrt{t}}\, dt \Rightarrow \frac{dy}{dx} = \frac{1}{\sqrt{e^{x^2}}} \cdot \frac{d}{dx}\left(e^{x^2}\right) = \frac{1}{e^{\frac{1}{2}x^2}} \cdot 2x\, e^{x^2} = 2x\, e^{\frac{1}{2}x^2}$

49. $y = \int_0^{\sin^{-1}t} \cos t\, dt \Rightarrow \frac{dy}{dx} = \cos(\sin^{-1}x) \cdot \frac{d}{dx}(\sin^{-1}x) = \sqrt{1-x^2} \cdot \frac{1}{\sqrt{1-x^2}} = 1$

51. $-x^2 - 2x = 0 \Rightarrow -x(x+2) = 0 \Rightarrow x = 0$ or $x = -2$; Area

$= -\int_{-3}^{-2}(-x^2-2x)dx + \int_{-2}^0(-x^2-2x)dx - \int_0^2(-x^2-2x)dx$

$= -\left[-\frac{x^3}{3} - x^2\right]_{-3}^{-2} + \left[-\frac{x^3}{3} - x^2\right]_{-2}^0 - \left[-\frac{x^3}{3} - x^2\right]_0^2$

$= -\left(\left(-\frac{(-2)^3}{3} - (-2)^2\right) - \left(-\frac{(-3)^3}{3} - (-3)^2\right)\right)$

$+ \left(\left(-\frac{0^3}{3} - 0^2\right) - \left(-\frac{(-2)^3}{3} - (-2)^2\right)\right)$

$- \left(\left(-\frac{2^3}{3} - 2^2\right) - \left(-\frac{0^3}{3} - 0^2\right)\right) = \frac{28}{3}$

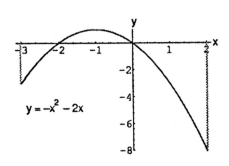

$y = -x^2 - 2x$

53. $x^3 - 3x^2 + 2x = 0 \Rightarrow x(x^2 - 3x + 2) = 0$

$\Rightarrow x(x-2)(x-1) = 0 \Rightarrow x = 0, 1,$ or 2;

Area $= \int_0^1(x^3 - 3x^2 + 2x)dx - \int_1^2(x^3 - 3x^2 + 2x)dx$

$= \left[\frac{x^4}{4} - x^3 + x^2\right]_0^1 - \left[\frac{x^4}{4} - x^3 + x^2\right]_1^2$

$= \left(\frac{1^4}{4} - 1^3 + 1^2\right) - \left(\frac{0^4}{4} - 0^3 + 0^2\right)$

$- \left[\left(\frac{2^4}{4} - 2^3 + 2^2\right) - \left(\frac{1^4}{4} - 1^3 + 1^2\right)\right] = \frac{1}{2}$

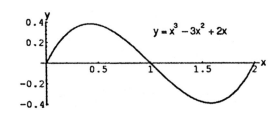

$y = x^3 - 3x^2 + 2x$

55. $x^{1/3} = 0 \Rightarrow x = 0$; Area $= -\int_{-1}^0 x^{1/3}\, dx + \int_0^8 x^{1/3}\, dx$

$= \left[-\frac{3}{4}x^{4/3}\right]_{-1}^0 + \left[\frac{3}{4}x^{4/3}\right]_0^8$

$= \left(-\frac{3}{4}(0)^{4/3}\right) - \left(-\frac{3}{4}(-1)^{4/3}\right) + \left(\frac{3}{4}(8)^{4/3}\right) - \left(\frac{3}{4}(0)^{4/3}\right)$

$= \frac{51}{4}$

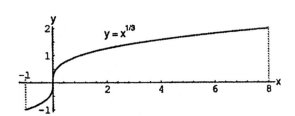

$y = x^{1/3}$

57. The area of the rectangle bounded by the lines $y = 2$, $y = 0$, $x = \pi$, and $x = 0$ is 2π. The area under the curve

$y = 1 + \cos x$ on $[0, \pi]$ is $\int_0^\pi (1 + \cos x)\, dx = [x + \sin x]_0^\pi = (\pi + \sin \pi) - (0 + \sin 0) = \pi$. Therefore the area of

the shaded region is $2\pi - \pi = \pi$.

59. On $\left[-\frac{\pi}{4}, 0\right]$: The area of the rectangle bounded by the lines $y = \sqrt{2}$, $y = 0$, $\theta = 0$, and $\theta = -\frac{\pi}{4}$ is $\sqrt{2}\left(\frac{\pi}{4}\right)$

$= \frac{\pi\sqrt{2}}{4}$. The area between the curve $y = \sec \theta \tan \theta$ and $y = 0$ is $-\int_{-\pi/4}^0 \sec \theta \tan \theta\, d\theta = [-\sec \theta]_{-\pi/4}^0$

$= (-\sec 0) - \left(-\sec\left(-\frac{\pi}{4}\right)\right) = \sqrt{2} - 1$. Therefore the area of the shaded region on $\left[-\frac{\pi}{4}, 0\right]$ is $\frac{\pi\sqrt{2}}{4} + \left(\sqrt{2} - 1\right)$.

On $\left[0, \frac{\pi}{4}\right]$: The area of the rectangle bounded by $\theta = \frac{\pi}{4}$, $\theta = 0$, $y = \sqrt{2}$, and $y = 0$ is $\sqrt{2}\left(\frac{\pi}{4}\right) = \frac{\pi\sqrt{2}}{4}$. The area under the curve $y = \sec\theta \tan\theta$ is $\int_0^{\pi/4} \sec\theta \tan\theta \, d\theta = [\sec\theta]_0^{\pi/4} = \sec\frac{\pi}{4} - \sec 0 = \sqrt{2} - 1$. Therefore the area of the shaded region on $\left[0, \frac{\pi}{4}\right]$ is $\frac{\pi\sqrt{2}}{4} - \left(\sqrt{2} - 1\right)$. Thus, the area of the total shaded region is $\left(\frac{\pi\sqrt{2}}{4} + \sqrt{2} - 1\right) + \left(\frac{\pi\sqrt{2}}{4} - \sqrt{2} + 1\right) = \frac{\pi\sqrt{2}}{2}$.

61. $y = \int_\pi^x \frac{1}{t} \, dt - 3 \Rightarrow \frac{dy}{dx} = \frac{1}{x}$ and $y(\pi) = \int_\pi^\pi \frac{1}{t} \, dt - 3 = 0 - 3 = -3 \Rightarrow$ (d) is a solution to this problem.

63. $y = \int_0^x \sec t \, dt + 4 \Rightarrow \frac{dy}{dx} = \sec x$ and $y(0) = \int_0^0 \sec t \, dt + 4 = 0 + 4 = 4 \Rightarrow$ (b) is a solution to this problem.

65. $y = \int_2^x \sec t \, dt + 3$ 67. $s = \int_{t_0}^t f(x) \, dx + s_0$

69. Area $= \int_{-b/2}^{b/2} \left(h - \left(\frac{4h}{b^2}\right) x^2\right) dx = \left[hx - \frac{4hx^3}{3b^2}\right]_{-b/2}^{b/2}$

$= \left(h\left(\frac{b}{2}\right) - \frac{4h\left(\frac{b}{2}\right)^3}{3b^2}\right) - \left(h\left(-\frac{b}{2}\right) - \frac{4h\left(-\frac{b}{2}\right)^3}{3b^2}\right)$

$= \left(\frac{bh}{2} - \frac{bh}{6}\right) - \left(-\frac{bh}{2} + \frac{bh}{6}\right) = bh - \frac{bh}{3} = \frac{2}{3} bh$

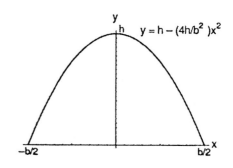

71. $\frac{dc}{dx} = \frac{1}{2\sqrt{x}} = \frac{1}{2} x^{-1/2} \Rightarrow c = \int_0^x \frac{1}{2} t^{-1/2} dt = \left[t^{1/2}\right]_0^x = \sqrt{x}$

$c(100) - c(1) = \sqrt{100} - \sqrt{1} = \$9.00$

73. $\int_1^x f(t) \, dt = x^2 - 2x + 1 \Rightarrow f(x) = \frac{d}{dx} \int_1^x f(t) \, dt = \frac{d}{dx}\left(x^2 - 2x + 1\right) = 2x - 2$

4.6 INDEFINTE INTEGRALS AND THE SUBSTITUTION RULE

1. Let $u = 3x \Rightarrow du = 3 \, dx \Rightarrow \frac{1}{3} du = dx$

$\int \sin 3x \, dx = \int \frac{1}{3} \sin u \, du = -\frac{1}{3} \cos u + C = -\frac{1}{3} \cos 3x + C$

3. Let $u = 2t \Rightarrow du = 2 \, dt \Rightarrow \frac{1}{2} du = dt$

$\int \sec 2t \tan 2t \, dt = \int \frac{1}{2} \sec u \tan u \, du = \frac{1}{2} \sec u + C = \frac{1}{2} \sec 2t + C$

5. Let $u = 7x - 2 \Rightarrow du = 7 \, dx \Rightarrow \frac{1}{7} du = dx$

$\int 28(7x - 2)^{-5} \, dx = \int \frac{1}{7} (28)u^{-5} \, du = \int 4u^{-5} \, du = -u^{-4} + C = -(7x - 2)^{-4} + C$

7. Let $u = 1 - r^3 \Rightarrow du = -3r^2 \, dr \Rightarrow -3 \, du = 9r^2 \, dr$

$\int \frac{9r^2 \, dr}{\sqrt{1 - r^3}} = \int -3u^{-1/2} \, du = -3(2)u^{1/2} + C = -6\left(1 - r^3\right)^{1/2} + C$

9. Let $u = x^{3/2} - 1 \Rightarrow du = \frac{3}{2} x^{1/2} \, dx \Rightarrow \frac{2}{3} du = \sqrt{x} \, dx$

$\int \sqrt{x} \sin^2\left(x^{3/2} - 1\right) dx = \int \frac{2}{3} \sin^2 u \, du = \frac{2}{3}\left(\frac{u}{2} - \frac{1}{4} \sin 2u\right) + C = \frac{1}{3}\left(x^{3/2} - 1\right) - \frac{1}{6} \sin\left(2x^{3/2} - 2\right) + C$

11. (a) Let $u = \cot 2\theta \Rightarrow du = -2 \csc^2 2\theta\, d\theta \Rightarrow -\frac{1}{2}\, du = \csc^2 2\theta\, d\theta$

$$\int \csc^2 2\theta \cot 2\theta\, d\theta = -\int \frac{1}{2} u\, du = -\frac{1}{2}\left(\frac{u^2}{2}\right) + C = -\frac{u^2}{4} + C = -\frac{1}{4}\cot^2 2\theta + C$$

(b) Let $u = \csc 2\theta \Rightarrow du = -2\csc 2\theta \cot 2\theta\, d\theta \Rightarrow -\frac{1}{2}\, du = \csc 2\theta \cot 2\theta\, d\theta$

$$\int \csc^2 2\theta \cot 2\theta\, d\theta = \int -\frac{1}{2} u\, du = -\frac{1}{2}\left(\frac{u^2}{2}\right) + C = -\frac{u^2}{4} + C = -\frac{1}{4}\csc^2 2\theta + C$$

13. Let $u = 3 - 2s \Rightarrow du = -2\, ds \Rightarrow -\frac{1}{2}\, du = ds$

$$\int \sqrt{3 - 2s}\, ds = \int \sqrt{u}\left(-\frac{1}{2}\, du\right) = -\frac{1}{2}\int u^{1/2}\, du = \left(-\frac{1}{2}\right)\left(\frac{2}{3} u^{3/2}\right) + C = -\frac{1}{3}(3 - 2s)^{3/2} + C$$

15. Let $u = 5s + 4 \Rightarrow du = 5\, ds \Rightarrow \frac{1}{5}\, du = ds$

$$\int \frac{1}{\sqrt{5s+4}}\, ds = \int \frac{1}{\sqrt{u}}\left(\frac{1}{5}\, du\right) = \frac{1}{5}\int u^{-1/2}\, du = \left(\frac{1}{5}\right)\left(2u^{1/2}\right) + C = \frac{2}{5}\sqrt{5s+4} + C$$

17. Let $u = 1 - \theta^2 \Rightarrow du = -2\theta\, d\theta \Rightarrow -\frac{1}{2}\, du = \theta\, d\theta$

$$\int \theta \sqrt[4]{1 - \theta^2}\, d\theta = \int \sqrt[4]{u}\left(-\frac{1}{2}\, du\right) = -\frac{1}{2}\int u^{1/4}\, du = \left(-\frac{1}{2}\right)\left(\frac{4}{5} u^{5/4}\right) + C = -\frac{2}{5}\left(1 - \theta^2\right)^{5/4} + C$$

19. Let $u = 1 + \sqrt{x} \Rightarrow du = \frac{1}{2\sqrt{x}}\, dx \Rightarrow 2\, du = \frac{1}{\sqrt{x}}\, dx$

$$\int \frac{1}{\sqrt{x}\left(1 + \sqrt{x}\right)^2}\, dx = \int \frac{2\, du}{u^2} = -\frac{2}{u} + C = \frac{-2}{1 + \sqrt{x}} + C$$

21. Let $u = 3z + 4 \Rightarrow du = 3\, dz \Rightarrow \frac{1}{3}\, du = dz$

$$\int \cos(3z + 4)\, dz = \int (\cos u)\left(\frac{1}{3}\, du\right) = \frac{1}{3}\int \cos u\, du = \frac{1}{3}\sin u + C = \frac{1}{3}\sin(3z + 4) + C$$

23. Let $u = \cos x \Rightarrow du = -\sin x\, dx \Rightarrow -du = \sin x\, dx$

$$\int \tan x\, dx = \int \frac{\sin x}{\cos x}\, dx = -\int \frac{1}{u}\, du = -\ln|u| + C = -\ln|\cos x| + C = \ln|\cos x|^{-1} + C = \ln|\sec x| + C$$

25. Let $u = \frac{r^3}{18} - 1 \Rightarrow du = \frac{r^2}{6}\, dr \Rightarrow 6\, du = r^2\, dr$

$$\int r^2 \left(\frac{r^3}{18} - 1\right)^5\, dr = \int u^5 (6\, du) = 6\int u^5\, du = 6\left(\frac{u^6}{6}\right) + C = \left(\frac{r^3}{18} - 1\right)^6 + C$$

27. Let $u = x^{3/2} + 1 \Rightarrow du = \frac{3}{2} x^{1/2}\, dx \Rightarrow \frac{2}{3}\, du = x^{1/2}\, dx$

$$\int x^{1/2}\sin\left(x^{3/2} + 1\right) dx = \int (\sin u)\left(\frac{2}{3}\, du\right) = \frac{2}{3}\int \sin u\, du = \frac{2}{3}(-\cos u) + C = -\frac{2}{3}\cos\left(x^{3/2} + 1\right) + C$$

29. Let $u = \cos(2t + 1) \Rightarrow du = -2\sin(2t + 1)\, dt \Rightarrow -\frac{1}{2}\, du = \sin(2t + 1)\, dt$

$$\int \frac{\sin(2t + 1)}{\cos^2(2t + 1)}\, dt = \int -\frac{1}{2}\frac{du}{u^2} = \frac{1}{2u} + C = \frac{1}{2\cos(2t + 1)} + C$$

31. Let $u = \sin\frac{1}{\theta} \Rightarrow du = \left(\cos\frac{1}{\theta}\right)\left(-\frac{1}{\theta^2}\right) d\theta \Rightarrow -du = \frac{1}{\theta^2}\cos\frac{1}{\theta}\, d\theta$

$$\int \frac{1}{\theta^2}\sin\frac{1}{\theta}\cos\frac{1}{\theta}\, d\theta = \int -u\, du = -\frac{1}{2} u^2 + C = -\frac{1}{2}\sin^2\frac{1}{\theta} + C$$

33. Let $u = \frac{1}{t} - 1 = t^{-1} - 1 \Rightarrow du = -t^{-2}\, dt \Rightarrow -du = \frac{1}{t^2}\, dt$

$$\int \frac{1}{t^2}\cos\left(\frac{1}{t} - 1\right) dt = \int (\cos u)(-du) = -\int \cos u\, du = -\sin u + C = -\sin\left(\frac{1}{t} - 1\right) + C$$

35. Let $u = s^3 + 2s^2 - 5s + 5 \Rightarrow du = (3s^2 + 4s - 5)\, ds$

$$\int (s^3 + 2s^2 - 5s + 5)(3s^2 + 4s - 5)\, ds = \int u\, du = \frac{u^2}{2} + C = \frac{(s^3 + 2s^2 - 5s + 5)^2}{2} + C$$

37. Let $u = 1 - \frac{1}{x} \Rightarrow du = \frac{1}{x^2} dx$

$\int \sqrt{\frac{x-1}{x^5}} \, dx = \int \frac{1}{x^2} \sqrt{\frac{x-1}{x}} \, dx = \int \frac{1}{x^2} \sqrt{1 - \frac{1}{x}} \, dx = \int \sqrt{u} \, du = \int u^{1/2} \, du = \frac{2}{3} u^{3/2} + C = \frac{2}{3} \left(1 - \frac{1}{x}\right)^{3/2} + C$

39. Let $u = \sin x \Rightarrow du = \cos x \, dx$

$\int (\cos x) \, e^{\sin x} \, dx = \int e^u \, du = e^u + C = e^{\sin x} + C$

41. Let $u = e^{\sqrt{x}} + 1 \Rightarrow du = \frac{1}{2\sqrt{x}} e^{\sqrt{x}} dx \Rightarrow 2 \, du = \frac{1}{\sqrt{x}} e^{\sqrt{x}} dx = \frac{1}{\sqrt{x} e^{-\sqrt{x}}} \, dx$

$\int \frac{1}{\sqrt{x} e^{-\sqrt{x}}} \sec^2 (e^{\sqrt{x}} + 1) dx = 2 \int \sec^2 u \, du = 2 \tan u + C = 2 \tan(e^{\sqrt{x}} + 1) + C$

43. Let $u = \ln x \Rightarrow du = \frac{1}{x} dx$

$\int \frac{1}{x \ln x} dx = \int \frac{1}{u} du = \ln |u| + C = \ln |\ln x| + C$

45. Let $u = e^{-z} + 1 \Rightarrow du = -e^{-z} dz \Rightarrow -du = e^{-z} dz$

$\int \frac{dz}{1 + e^z} = \int \frac{1}{1 + e^z} \cdot \frac{e^{-z}}{e^{-z}} dz = \int \frac{e^{-z}}{e^{-z} + 1} dz = -\int \frac{1}{u} du = -\ln |u| + C = -\ln (e^{-z} + 1) + C = -\ln \left(\frac{1 + e^z}{e^z}\right) + C$

$= -(\ln (1 + e^z) - \ln e^z) + C = z - \ln (1 + e^z) + C$

47. Let $u = \frac{2}{3} r \Rightarrow du = \frac{2}{3} dr \Rightarrow \frac{3}{2} du = dr$

$\int \frac{5}{9 + 4r^2} dr = \frac{5}{9} \int \frac{1}{1 + \left(\frac{2}{3} r\right)^2} dr = \frac{5}{9} \int \frac{\frac{3}{2}}{1 + u^2} du = \frac{5}{6} \tan^{-1} u + C = \frac{5}{6} \tan^{-1} \left(\frac{2}{3} r\right) + C$

49. $\int \frac{e^{\sin^{-1} x}}{\sqrt{1 - x^2}} \, dx = \int e^u \, du$, where $u = \sin^{-1} x$ and $du = \frac{dx}{\sqrt{1 - x^2}}$

$= e^u + C = e^{\sin^{-1} x} + C$

51. $\int \frac{(\sin^{-1} x)^2}{\sqrt{1 - x^2}} \, dx = \int u^2 \, du$, where $u = \sin^{-1} x$ and $du = \frac{dx}{\sqrt{1 - x^2}}$

$= \frac{u^3}{3} + C = \frac{(\sin^{-1} x)^3}{3} + C$

53. $\int \frac{1}{(\tan^{-1} y)(1 + y^2)} \, dy = \int \frac{\left(\frac{1}{1 + y^2}\right)}{\tan^{-1} y} \, dy = \int \frac{1}{u} \, du$, where $u = \tan^{-1} y$ and $du = \frac{dy}{1 + y^2}$

$= \ln |u| + C = \ln |\tan^{-1} y| + C$

55. Let $u = 3t^2 - 1 \Rightarrow du = 6t \, dt \Rightarrow 2 \, du = 12t \, dt$

$s = \int 12t (3t^2 - 1)^3 \, dt = \int u^3 (2 \, du) = 2 \left(\frac{1}{4} u^4\right) + C = \frac{1}{2} u^4 + C = \frac{1}{2} (3t^2 - 1)^4 + C;$

$s = 3$ when $t = 1 \Rightarrow 3 = \frac{1}{2} (3 - 1)^4 + C \Rightarrow 3 = 8 + C \Rightarrow C = -5 \Rightarrow s = \frac{1}{2} (3t^2 - 1)^4 - 5$

57. Let $u = 2t - \frac{\pi}{2} \Rightarrow du = 2 \, dt \Rightarrow -2 \, du = -4 \, dt$

$\frac{ds}{dt} = \int -4 \sin \left(2t - \frac{\pi}{2}\right) dt = \int (\sin u)(-2 \, du) = 2 \cos u + C_1 = 2 \cos \left(2t - \frac{\pi}{2}\right) + C_1;$

at $t = 0$ and $\frac{ds}{dt} = 100$ we have $100 = 2 \cos \left(-\frac{\pi}{2}\right) + C_1 \Rightarrow C_1 = 100 \Rightarrow \frac{ds}{dt} = 2 \cos \left(2t - \frac{\pi}{2}\right) + 100$

$\Rightarrow s = \int \left(2 \cos \left(2t - \frac{\pi}{2}\right) + 100\right) dt = \int (\cos u + 50) \, du = \sin u + 50u + C_2 = \sin \left(2t - \frac{\pi}{2}\right) + 50 \left(2t - \frac{\pi}{2}\right) + C_2;$

at $t = 0$ and $s = 0$ we have $0 = \sin \left(-\frac{\pi}{2}\right) + 50 \left(-\frac{\pi}{2}\right) + C_2 \Rightarrow C_2 = 1 + 25\pi$

$\Rightarrow s = \sin \left(2t - \frac{\pi}{2}\right) + 100t - 25\pi + (1 + 25\pi) \Rightarrow s = \sin \left(2t - \frac{\pi}{2}\right) + 100t + 1$

59. Let $u = 2t \Rightarrow du = 2\,dt \Rightarrow 3\,du = 6\,dt$

$s = \int 6 \sin 2t\,dt = \int (\sin u)(3\,du) = -3 \cos u + C = -3 \cos 2t + C;$

at $t = 0$ and $s = 0$ we have $0 = -3 \cos 0 + C \Rightarrow C = 3 \Rightarrow s = 3 - 3 \cos 2t \Rightarrow s\left(\frac{\pi}{2}\right) = 3 - 3 \cos(\pi) = 6$ m

4.7 SUBSTITUTION AND AREA BETWEEN CURVES

1. (a) Let $u = y + 1 \Rightarrow du = dy; y = 0 \Rightarrow u = 1, y = 3 \Rightarrow u = 4$

$\int_0^3 \sqrt{y+1}\,dy = \int_1^4 u^{1/2}\,du = \left[\frac{2}{3} u^{3/2}\right]_1^4 = \left(\frac{2}{3}\right)(4)^{3/2} - \left(\frac{2}{3}\right)(1)^{3/2} = \left(\frac{2}{3}\right)(8) - \left(\frac{2}{3}\right)(1) = \frac{14}{3}$

   (b) Use the same substitution for u as in part (a); $y = -1 \Rightarrow u = 0, y = 0 \Rightarrow u = 1$

$\int_{-1}^0 \sqrt{y+1}\,dy = \int_0^1 u^{1/2}\,du = \left[\frac{2}{3} u^{3/2}\right]_0^1 = \left(\frac{2}{3}\right)(1)^{3/2} - 0 = \frac{2}{3}$

3. (a) Let $u = \tan x \Rightarrow du = \sec^2 x\,dx; x = 0 \Rightarrow u = 0, x = \frac{\pi}{4} \Rightarrow u = 1$

$\int_0^{\pi/4} \tan x \sec^2 x\,dx = \int_0^1 u\,du = \left[\frac{u^2}{2}\right]_0^1 = \frac{1^2}{2} - 0 = \frac{1}{2}$

   (b) Use the same substitution as in part (a); $x = -\frac{\pi}{4} \Rightarrow u = -1, x = 0 \Rightarrow u = 0$

$\int_{-\pi/4}^0 \tan x \sec^2 x\,dx = \int_{-1}^0 u\,du = \left[\frac{u^2}{2}\right]_{-1}^0 = 0 - \frac{1}{2} = -\frac{1}{2}$

5. (a) $u = 1 + t^4 \Rightarrow du = 4t^3\,dt \Rightarrow \frac{1}{4}\,du = t^3\,dt; t = 0 \Rightarrow u = 1, t = 1 \Rightarrow u = 2$

$\int_0^1 t^3\left(1 + t^4\right)^3\,dt = \int_1^2 \frac{1}{4} u^3\,du = \left[\frac{u^4}{16}\right]_1^2 = \frac{2^4}{16} - \frac{1^4}{16} = \frac{15}{16}$

   (b) Use the same substitution as in part (a); $t = -1 \Rightarrow u = 2, t = 1 \Rightarrow u = 2$

$\int_{-1}^1 t^3\left(1 + t^4\right)^3\,dt = \int_2^2 \frac{1}{4} u^3\,du = 0$

7. (a) Let $u = 4 + r^2 \Rightarrow du = 2r\,dr \Rightarrow \frac{1}{2}\,du = r\,dr; r = -1 \Rightarrow u = 5, r = 1 \Rightarrow u = 5$

$\int_{-1}^1 \frac{5r}{(4+r^2)^2}\,dr = 5\int_5^5 \frac{1}{2} u^{-2}\,du = 0$

   (b) Use the same substitution as in part (a); $r = 0 \Rightarrow u = 4, r = 1 \Rightarrow u = 5$

$\int_0^1 \frac{5r}{(4+r^2)^2}\,dr = 5\int_4^5 \frac{1}{2} u^{-2}\,du = 5\left[-\frac{1}{2} u^{-1}\right]_4^5 = 5\left(-\frac{1}{2}(5)^{-1}\right) - 5\left(-\frac{1}{2}(4)^{-1}\right) = \frac{1}{8}$

9. (a) Let $u = x^2 + 1 \Rightarrow du = 2x\,dx \Rightarrow 2\,du = 4x\,dx; x = 0 \Rightarrow u = 1, x = \sqrt{3} \Rightarrow u = 4$

$\int_0^{\sqrt{3}} \frac{4x}{\sqrt{x^2+1}}\,dx = \int_1^4 \frac{2}{\sqrt{u}}\,du = \int_1^4 2u^{-1/2}\,du = \left[4u^{1/2}\right]_1^4 = 4(4)^{1/2} - 4(1)^{1/2} = 4$

   (b) Use the same substitution as in part (a); $x = -\sqrt{3} \Rightarrow u = 4, x = \sqrt{3} \Rightarrow u = 4$

$\int_{-\sqrt{3}}^{\sqrt{3}} \frac{4x}{\sqrt{x^2+1}}\,dx = \int_4^4 \frac{2}{\sqrt{u}}\,du = 0$

11. (a) Let $u = 1 - \cos 3t \Rightarrow du = 3 \sin 3t\,dt \Rightarrow \frac{1}{3}\,du = \sin 3t\,dt; t = 0 \Rightarrow u = 0, t = \frac{\pi}{6} \Rightarrow u = 1 - \cos\frac{\pi}{2} = 1$

$\int_0^{\pi/6} (1 - \cos 3t) \sin 3t\,dt = \int_0^1 \frac{1}{3} u\,du = \left[\frac{1}{3}\left(\frac{u^2}{2}\right)\right]_0^1 = \frac{1}{6}(1)^2 - \frac{1}{6}(0)^2 = \frac{1}{6}$

   (b) Use the same substitution as in part (a); $t = \frac{\pi}{6} \Rightarrow u = 1, t = \frac{\pi}{3} \Rightarrow u = 1 - \cos \pi = 2$

$\int_{\pi/6}^{\pi/3} (1 - \cos 3t) \sin 3t\,dt = \int_1^2 \frac{1}{3} u\,du = \left[\frac{1}{3}\left(\frac{u^2}{2}\right)\right]_1^2 = \frac{1}{6}(2)^2 - \frac{1}{6}(1)^2 = \frac{1}{2}$

13. (a) Let $u = 4 + 3 \sin z \Rightarrow du = 3 \cos z\,dz \Rightarrow \frac{1}{3}\,du = \cos z\,dz; z = 0 \Rightarrow u = 4, z = 2\pi \Rightarrow u = 4$

$\int_0^{2\pi} \frac{\cos z}{\sqrt{4+3\sin z}}\,dz = \int_4^4 \frac{1}{\sqrt{u}}\left(\frac{1}{3}\,du\right) = 0$

(b) Use the same substitution as in part (a); $z = -\pi \Rightarrow u = 4 + 3\sin(-\pi) = 4, z = \pi \Rightarrow u = 4$

$$\int_{-\pi}^{\pi} \frac{\cos z}{\sqrt{4 + 3\sin z}} \, dz = \int_{4}^{4} \frac{1}{\sqrt{u}} \left(\frac{1}{3} \, du\right) = 0$$

15. Let $u = t^5 + 2t \Rightarrow du = (5t^4 + 2) \, dt; t = 0 \Rightarrow u = 0, t = 1 \Rightarrow u = 3$

$$\int_{0}^{1} \sqrt{t^5 + 2t} \, (5t^4 + 2) \, dt = \int_{0}^{3} u^{1/2} \, du = \left[\frac{2}{3} u^{3/2}\right]_{0}^{3} = \frac{2}{3}(3)^{3/2} - \frac{2}{3}(0)^{3/2} = 2\sqrt{3}$$

17. Let $u = \cos 2\theta \Rightarrow du = -2\sin 2\theta \, d\theta \Rightarrow -\frac{1}{2} \, du = \sin 2\theta \, d\theta; \theta = 0 \Rightarrow u = 1, \theta = \frac{\pi}{6} \Rightarrow u = \cos 2\left(\frac{\pi}{6}\right) = \frac{1}{2}$

$$\int_{0}^{\pi/6} \cos^{-3} 2\theta \sin 2\theta \, d\theta = \int_{1}^{1/2} u^{-3} \left(-\frac{1}{2} \, du\right) = -\frac{1}{2} \int_{1}^{1/2} u^{-3} \, du = \left[-\frac{1}{2}\left(\frac{u^{-2}}{-2}\right)\right]_{1}^{1/2} = \frac{1}{4\left(\frac{1}{2}\right)^2} - \frac{1}{4(1)^2} = \frac{3}{4}$$

19. Let $u = 5 - 4\cos t \Rightarrow du = 4\sin t \, dt \Rightarrow \frac{1}{4} \, du = \sin t \, dt; t = 0 \Rightarrow u = 5 - 4\cos 0 = 1, t = \pi \Rightarrow$
    $u = 5 - 4\cos \pi = 9$

$$\int_{0}^{\pi} 5(5 - 4\cos t)^{1/4} \sin t \, dt = \int_{1}^{9} 5u^{1/4} \left(\frac{1}{4} \, du\right) = \frac{5}{4} \int_{1}^{9} u^{1/4} \, du = \left[\frac{5}{4}\left(\frac{4}{5} u^{5/4}\right)\right]_{1}^{9} = 9^{5/4} - 1 = 3^{5/2} - 1$$

21. Let $u = 4y - y^2 + 4y^3 + 1 \Rightarrow du = (4 - 2y + 12y^2) \, dy; y = 0 \Rightarrow u = 1, y = 1 \Rightarrow u = 4(1) - (1)^2 + 4(1)^3 + 1 = 8$

$$\int_{0}^{1} (4y - y^2 + 4y^3 + 1)^{-2/3} (12y^2 - 2y + 4) \, dy = \int_{1}^{8} u^{-2/3} \, du = \left[3u^{1/3}\right]_{1}^{8} = 3(8)^{1/3} - 3(1)^{1/3} = 3$$

23. Let $u = \theta^{3/2} \Rightarrow du = \frac{3}{2} \theta^{1/2} \, d\theta \Rightarrow \frac{2}{3} \, du = \sqrt{\theta} \, d\theta; \theta = 0 \Rightarrow u = 0, \theta = \sqrt[3]{\pi^2} \Rightarrow u = \pi$

$$\int_{0}^{\sqrt[3]{\pi^2}} \sqrt{\theta} \cos^2\left(\theta^{3/2}\right) \, d\theta = \int_{0}^{\pi} \cos^2 u \left(\frac{2}{3} \, du\right) = \left[\frac{2}{3}\left(\frac{u}{2} + \frac{1}{4} \sin 2u\right)\right]_{0}^{\pi} = \frac{2}{3}\left(\frac{\pi}{2} + \frac{1}{4} \sin 2\pi\right) - \frac{2}{3}(0) = \frac{\pi}{3}$$

25. Let $u = \tan \theta \Rightarrow du = \sec^2 \theta \, d\theta; \theta = 0 \Rightarrow u = 0, \theta = \frac{\pi}{4} \Rightarrow u = 1;$

$$\int_{0}^{\pi/4} \left(1 + e^{\tan \theta}\right) \sec^2 \theta \, d\theta = \int_{0}^{\pi/4} \sec^2 \theta \, d\theta + \int_{0}^{1} e^u \, du = \left[\tan \theta\right]_{0}^{\pi/4} + \left[e^u\right]_{0}^{1} = \left[\tan\left(\frac{\pi}{4}\right) - \tan(0)\right] + \left(e^1 - e^0\right)$$
$$= (1 - 0) + (e - 1) = e$$

27. $\int_{0}^{\pi} \frac{\sin t}{2 - \cos t} \, dt = \left[\ln|2 - \cos t|\right]_{0}^{\pi} = \ln 3 - \ln 1 = \ln 3;$ or let $u = 2 - \cos t \Rightarrow du = \sin t \, dt$ with $t = 0$

$\Rightarrow u = 1$ and $t = \pi \Rightarrow u = 3 \Rightarrow \int_{0}^{\pi} \frac{\sin t}{2 - \cos t} \, dt = \int_{1}^{3} \frac{1}{u} \, du = \left[\ln|u|\right]_{1}^{3} = \ln 3 - \ln 1 = \ln 3$

29. Let $u = \ln x \Rightarrow du = \frac{1}{x} \, dx; x = 1 \Rightarrow u = 0$ and $x = 2 \Rightarrow u = \ln 2;$

$$\int_{1}^{2} \frac{2 \ln x}{x} \, dx = \int_{0}^{\ln 2} 2u \, du = \left[u^2\right]_{0}^{\ln 2} = (\ln 2)^2$$

31. Let $u = \ln x \Rightarrow du = \frac{1}{x} \, dx; x = 2 \Rightarrow u = \ln 2$ and $x = 4 \Rightarrow u = \ln 4;$

$$\int_{2}^{4} \frac{dx}{x(\ln x)^2} = \int_{\ln 2}^{\ln 4} u^{-2} \, du = \left[-\frac{1}{u}\right]_{\ln 2}^{\ln 4} = -\frac{1}{\ln 4} + \frac{1}{\ln 2} = -\frac{1}{\ln 2^2} + \frac{1}{\ln 2} = -\frac{1}{2 \ln 2} + \frac{1}{\ln 2} = \frac{1}{2 \ln 2} = \frac{1}{\ln 4}$$

33. Let $u = \cos \frac{x}{2} \Rightarrow du = -\frac{1}{2} \sin \frac{x}{2} \, dx \Rightarrow -2 \, du = \sin \frac{x}{2} \, dx; x = 0 \Rightarrow u = 1$ and $x = \frac{\pi}{2} \Rightarrow u = \frac{1}{\sqrt{2}};$

$$\int_{0}^{\pi/2} \tan \frac{x}{2} \, dx = \int_{0}^{\pi/2} \frac{\sin \frac{x}{2}}{\cos \frac{x}{2}} \, dx = -2 \int_{1}^{1/\sqrt{2}} \frac{du}{u} = \left[-2 \ln|u|\right]_{1}^{1/\sqrt{2}} = -2 \ln \frac{1}{\sqrt{2}} = 2 \ln \sqrt{2} = \ln 2$$

35. Let $u = \sin \frac{\theta}{3} \Rightarrow du = \frac{1}{3} \cos \frac{\theta}{3} \, d\theta \Rightarrow 6 \, du = 2 \cos \frac{\theta}{3} \, d\theta; \theta = \frac{\pi}{2} \Rightarrow u = \frac{1}{2}$ and $\theta = \pi \Rightarrow u = \frac{\sqrt{3}}{2};$

$$\int_{\pi/2}^{\pi} 2 \cot \frac{\theta}{3} \, d\theta = \int_{\pi/2}^{\pi} \frac{2 \cos \frac{\theta}{3}}{\sin \frac{\theta}{3}} \, d\theta = 6 \int_{1/2}^{\sqrt{3}/2} \frac{du}{u} = 6 \left[\ln|u|\right]_{1/2}^{\sqrt{3}/2} = 6 \left(\ln \frac{\sqrt{3}}{2} - \ln \frac{1}{2}\right) = 6 \ln \sqrt{3} = \ln 27$$

37. $\int_{-\pi/2}^{\pi/2} \frac{2\cos\theta\,d\theta}{1+(\sin\theta)^2} = 2\int_{-1}^{1} \frac{du}{1+u^2}$, where $u = \sin\theta$ and $du = \cos\theta\,d\theta$; $\theta = -\frac{\pi}{2} \Rightarrow u = -1$, $\theta = \frac{\pi}{2} \Rightarrow u = 1$

$= \left[2\tan^{-1} u\right]_{-1}^{1} = 2\left(\tan^{-1} 1 - \tan^{-1}(-1)\right) = 2\left[\frac{\pi}{4} - \left(-\frac{\pi}{4}\right)\right] = \pi$

39. $\int_{0}^{\ln\sqrt{3}} \frac{e^x\,dx}{1+e^{2x}} = \int_{1}^{\sqrt{3}} \frac{du}{1+u^2}$, where $u = e^x$ and $du = e^x\,dx$; $x = 0 \Rightarrow u = 1$, $x = \ln\sqrt{3} \Rightarrow u = \sqrt{3}$

$= \left[\tan^{-1} u\right]_{1}^{\sqrt{3}} = \tan^{-1}\sqrt{3} - \tan^{-1} 1 = \frac{\pi}{3} - \frac{\pi}{4} = \frac{\pi}{12}$

41. $\int_{0}^{1} \frac{4\,ds}{\sqrt{4-s^2}} = \left[4\sin^{-1} \frac{s}{2}\right]_{0}^{1} = 4\left(\sin^{-1} \frac{1}{2} - \sin^{-1} 0\right) = 4\left(\frac{\pi}{6} - 0\right) = \frac{2\pi}{3}$

43. $\int_{\sqrt{2}}^{2} \frac{\sec^2(\sec^{-1} x)}{x\sqrt{x^2-1}}\,dx = \int_{\pi/4}^{\pi/3} \sec^2 u\,du$, where $u = \sec^{-1} x$ and $du = \frac{dx}{x\sqrt{x^2-1}}$; $x = \sqrt{2} \Rightarrow u = \frac{\pi}{4}$, $x = 2 \Rightarrow u = \frac{\pi}{3}$

$= \left[\tan u\right]_{\pi/4}^{\pi/3} = \tan\frac{\pi}{3} - \tan\frac{\pi}{4} = \sqrt{3} - 1$

45. $\int_{-1}^{-\sqrt{2}/2} \frac{dy}{y\sqrt{4y^2-1}} = \int_{-2}^{-\sqrt{2}} \frac{du}{u\sqrt{u^2-1}}$, where $u = 2y$ and $du = 2\,dy$; $y = -1 \Rightarrow u = -2$, $y = -\frac{\sqrt{2}}{2} \Rightarrow u = -\sqrt{2}$

$= \left[\sec^{-1} |u|\right]_{-2}^{-\sqrt{2}} = \sec^{-1}\left|-\sqrt{2}\right| - \sec^{-1} |-2| = \frac{\pi}{4} - \frac{\pi}{3} = -\frac{\pi}{12}$

47. Let $u = 4 - x^2 \Rightarrow du = -2x\,dx \Rightarrow -\frac{1}{2}\,du = x\,dx$; $x = -2 \Rightarrow u = 0$, $x = 0 \Rightarrow u = 4$, $x = 2 \Rightarrow u = 0$

$A = -\int_{-2}^{0} x\sqrt{4-x^2}\,dx + \int_{0}^{2} x\sqrt{4-x^2}\,dx = -\int_{0}^{4} -\frac{1}{2}u^{1/2}\,du + \int_{4}^{0} -\frac{1}{2}u^{1/2}\,du = 2\int_{0}^{4} \frac{1}{2}u^{1/2}\,du = \int_{0}^{4} u^{1/2}\,du$

$= \left[\frac{2}{3}u^{3/2}\right]_{0}^{4} = \frac{2}{3}(4)^{3/2} - \frac{2}{3}(0)^{3/2} = \frac{16}{3}$

49. Let $u = 1 + \cos x \Rightarrow du = -\sin x\,dx \Rightarrow -du = \sin x\,dx$; $x = -\pi \Rightarrow u = 1 + \cos(-\pi) = 0$, $x = 0$

$\Rightarrow u = 1 + \cos 0 = 2$

$A = -\int_{-\pi}^{0} 3(\sin x)\sqrt{1+\cos x}\,dx = -\int_{0}^{2} 3u^{1/2}(-du) = 3\int_{0}^{2} u^{1/2}\,du = \left[2u^{3/2}\right]_{0}^{2} = 2(2)^{3/2} - 2(0)^{3/2} = 2^{5/2}$

51. For the sketch given, $c = 0$, $d = 1$; $f(y) - g(y) = (12y^2 - 12y^3) - (2y^2 - 2y) = 10y^2 - 12y^3 + 2y$;

$A = \int_{0}^{1}(10y^2 - 12y^3 + 2y)\,dy = \int_{0}^{1} 10y^2\,dy - \int_{0}^{1} 12y^3\,dy + \int_{0}^{1} 2y\,dy = \left[\frac{10}{3}y^3\right]_{0}^{1} - \left[\frac{12}{4}y^4\right]_{0}^{1} + \left[\frac{2}{2}y^2\right]_{0}^{1}$

$= \left(\frac{10}{3} - 0\right) - (3 - 0) + (1 - 0) = \frac{4}{3}$

53. We want the area between the line $y = 1$, $0 \le x \le 2$, and the curve $y = \frac{x^2}{4}$, *minus* the area of a triangle

(formed by $y = x$ and $y = 1$) with base 1 and height 1. Thus, $A = \int_{0}^{2}\left(1 - \frac{x^2}{4}\right)dx - \frac{1}{2}(1)(1) = \left[x - \frac{x^3}{12}\right]_{0}^{2} - \frac{1}{2}$

$= \left(2 - \frac{8}{12}\right) - \frac{1}{2} = 2 - \frac{2}{3} - \frac{1}{2} = \frac{5}{6}$

55. $a = -2$, $b = 2$;

$f(x) - g(x) = 2 - (x^2 - 2) = 4 - x^2$

$\Rightarrow A = \int_{-2}^{2}(4 - x^2)dx = \left[4x - \frac{x^3}{3}\right]_{-2}^{2} = \left(8 - \frac{8}{3}\right) - \left(-8 + \frac{8}{3}\right)$

$= 2\cdot\left(\frac{24}{3} - \frac{8}{3}\right) = \frac{32}{3}$

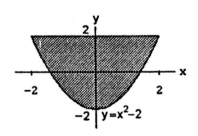

57. Limits of integration: $x^2 = -x^2 + 4x \Rightarrow 2x^2 - 4x = 0$

$\Rightarrow 2x(x - 2) = 0 \Rightarrow a = 0$ and $b = 2$;

$f(x) - g(x) = (-x^2 + 4x) - x^2 = -2x^2 + 4x$

$\Rightarrow A = \int_0^2 (-2x^2 + 4x)\, dx = \left[\frac{-2x^3}{3} + \frac{4x^2}{2} \right]_0^2$

$= -\frac{16}{3} + \frac{16}{2} = \frac{-32 + 48}{6} = \frac{8}{3}$

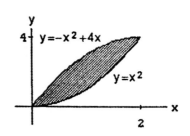

59. Limits of integration: $x^4 - 4x^2 + 4 = x^2$

$\Rightarrow x^4 - 5x^2 + 4 = 0 \Rightarrow (x^2 - 4)(x^2 - 1) = 0$

$\Rightarrow (x + 2)(x - 2)(x + 1)(x - 1) = 0 \Rightarrow x = -2, -1, 1, 2$;

$f(x) - g(x) = (x^4 - 4x^2 + 4) - x^2 = x^4 - 5x^2 + 4$ and

$g(x) - f(x) = x^2 - (x^4 - 4x^2 + 4) = -x^4 + 5x^2 - 4$

$\Rightarrow A = \int_{-2}^{-1} (-x^4 + 5x^2 - 4)dx + \int_{-1}^{1} (x^4 - 5x^2 + 4)dx$

$+ \int_1^2 (-x^4 + 5x^2 - 4)dx$

$= \left[-\frac{x^5}{5} + \frac{5x^3}{3} - 4x \right]_{-2}^{-1} + \left[\frac{x^5}{5} - \frac{5x^3}{3} + 4x \right]_{-1}^{1} + \left[\frac{-x^5}{5} + \frac{5x^3}{3} - 4x \right]_1^2$

$= \left(\frac{1}{5} - \frac{5}{3} + 4 \right) - \left(\frac{32}{5} - \frac{40}{3} + 8 \right) + \left(\frac{1}{5} - \frac{5}{3} + 4 \right) - \left(-\frac{1}{5} + \frac{5}{3} - 4 \right) + \left(-\frac{32}{5} + \frac{40}{3} - 8 \right) - \left(-\frac{1}{5} + \frac{5}{3} - 4 \right)$

$= -\frac{60}{5} + \frac{60}{3} = \frac{300 - 180}{15} = 8$

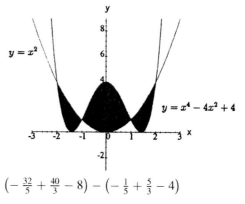

61. Limits of integration: $y = \sqrt{|x|} = \begin{cases} \sqrt{-x}, & x \le 0 \\ \sqrt{x}, & x \ge 0 \end{cases}$ and

$5y = x + 6$ or $y = \frac{x}{5} + \frac{6}{5}$; for $x \le 0$: $\sqrt{-x} = \frac{x}{5} + \frac{6}{5}$

$\Rightarrow 5\sqrt{-x} = x + 6 \Rightarrow 25(-x) = x^2 + 12x + 36$

$\Rightarrow x^2 + 37x + 36 = 0 \Rightarrow (x + 1)(x + 36) = 0$

$\Rightarrow x = -1, -36$ (but $x = -36$ is not a solution);

for $x \ge 0$: $5\sqrt{x} = x + 6 \Rightarrow 25x = x^2 + 12x + 36$

$\Rightarrow x^2 - 13x + 36 = 0 \Rightarrow (x - 4)(x - 9) = 0$

$\Rightarrow x = 4, 9$; there are three intersection points and

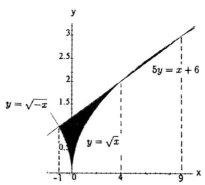

$A = \int_{-1}^0 \left(\frac{x+6}{5} - \sqrt{-x} \right)dx + \int_0^4 \left(\frac{x+6}{5} - \sqrt{x} \right)dx + \int_4^9 \left(\sqrt{x} - \frac{x+6}{5} \right)dx$

$= \left[\frac{(x+6)^2}{10} + \frac{2}{3}(-x)^{3/2} \right]_{-1}^0 + \left[\frac{(x+6)^2}{10} - \frac{2}{3}x^{3/2} \right]_0^4 + \left[\frac{2}{3}x^{3/2} - \frac{(x+6)^2}{10} \right]_4^9$

$= \left(\frac{36}{10} - \frac{25}{10} - \frac{2}{3} \right) + \left(\frac{100}{10} - \frac{2}{3} \cdot 4^{3/2} - \frac{36}{10} + 0 \right) + \left(\frac{2}{3} \cdot 9^{3/2} - \frac{225}{10} - \frac{2}{3} \cdot 4^{3/2} + \frac{100}{10} \right) = -\frac{50}{10} + \frac{20}{3} = \frac{5}{3}$

63. Limits of integration: $c = 0$ and $d = 3$;

$f(y) - g(y) = 2y^2 - 0 = 2y^2$

$\Rightarrow A = \int_0^3 2y^2\, dy = \left[\frac{2y^3}{3} \right]_0^3 = 2 \cdot 9 = 18$

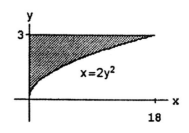

65. Limits of integration: $x = -y^2$ and $x = 2 - 3y^2$

$\Rightarrow -y^2 = 2 - 3y^2 \Rightarrow 2y^2 - 2 = 0$

$\Rightarrow 2(y - 1)(y + 1) = 0 \Rightarrow c = -1 \text{ and } d = 1;$

$f(y) - g(y) = (2 - 3y^2) - (-y^2) = 2 - 2y^2 = 2(1 - y^2)$

$\Rightarrow A = 2 \int_{-1}^{1} (1 - y^2) \, dy = 2 \left[y - \frac{y^3}{3} \right]_{-1}^{1}$

$= 2 \left(1 - \frac{1}{3} \right) - 2 \left(-1 + \frac{1}{3} \right) = 4 \left(\frac{2}{3} \right) = \frac{8}{3}$

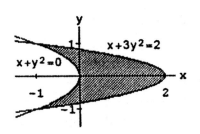

67. Limits of integration: $x = y^2 - 1$ and $x = |y| \sqrt{1 - y^2}$

$\Rightarrow y^2 - 1 = |y| \sqrt{1 - y^2} \Rightarrow y^4 - 2y^2 + 1 = y^2 (1 - y^2)$

$\Rightarrow y^4 - 2y^2 + 1 = y^2 - y^4 \Rightarrow 2y^4 - 3y^2 + 1 = 0$

$\Rightarrow (2y^2 - 1)(y^2 - 1) = 0 \Rightarrow 2y^2 - 1 = 0 \text{ or } y^2 - 1 = 0$

$\Rightarrow y^2 = \frac{1}{2} \text{ or } y^2 = 1 \Rightarrow y = \pm \frac{\sqrt{2}}{2} \text{ or } y = \pm 1.$

Substitution shows that $\frac{\pm\sqrt{2}}{2}$ are not solutions $\Rightarrow y = \pm 1;$

for $-1 \le y \le 0$, $f(x) - g(x) = -y\sqrt{1 - y^2} - (y^2 - 1)$

$= 1 - y^2 - y(1 - y^2)^{1/2}$, and by symmetry of the graph,

$A = 2 \int_{-1}^{0} \left[1 - y^2 - y(1 - y^2)^{1/2} \right] dy$

$= 2 \int_{-1}^{0} (1 - y^2) \, dy - 2 \int_{-1}^{0} y(1 - y^2)^{1/2} \, dy$

$= 2 \left[y - \frac{y^3}{3} \right]_{-1}^{0} + 2 \left(\frac{1}{2} \right) \left[\frac{2(1 - y^2)^{3/2}}{3} \right]_{-1}^{0} = 2 \left[(0 - 0) - \left(-1 + \frac{1}{3} \right) \right] + \left(\frac{2}{3} - 0 \right) = 2$

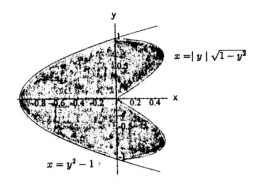

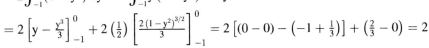

69. Limits of integration: $y = -4x^2 + 4$ and $y = x^4 - 1$

$\Rightarrow x^4 - 1 = -4x^2 + 4 \Rightarrow x^4 + 4x^2 - 5 = 0$

$\Rightarrow (x^2 + 5)(x - 1)(x + 1) = 0 \Rightarrow a = -1 \text{ and } b = 1;$

$f(x) - g(x) = -4x^2 + 4 - x^4 + 1 = -4x^2 - x^4 + 5$

$\Rightarrow A = \int_{-1}^{1} (-4x^2 - x^4 + 5) \, dx = \left[-\frac{4x^3}{3} - \frac{x^5}{5} + 5x \right]_{-1}^{1}$

$= \left(-\frac{4}{3} - \frac{1}{5} + 5 \right) - \left(\frac{4}{3} + \frac{1}{5} - 5 \right) = 2 \left(-\frac{4}{3} - \frac{1}{5} + 5 \right) = \frac{104}{15}$

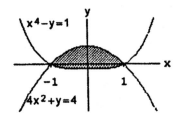

71. $a = 0, b = \pi; f(x) - g(x) = 2 \sin x - \sin 2x$

$\Rightarrow A = \int_{0}^{\pi} (2 \sin x - \sin 2x) \, dx = \left[-2 \cos x + \frac{\cos 2x}{2} \right]_{0}^{\pi}$

$= \left[-2(-1) + \frac{1}{2} \right] - \left(-2 \cdot 1 + \frac{1}{2} \right) = 4$

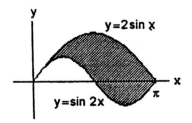

73. $a = -\frac{\pi}{4}, b = \frac{\pi}{4}; f(x) - g(x) = \sec^2 x - \tan^2 x$

$\Rightarrow A = \int_{-\pi/4}^{\pi/4} (\sec^2 x - \tan^2 x) \, dx$

$= \int_{-\pi/4}^{\pi/4} [\sec^2 x - (\sec^2 x - 1)] \, dx$

$= \int_{-\pi/4}^{\pi/4} 1 \cdot dx = [x]_{-\pi/4}^{\pi/4} = \frac{\pi}{4} - \left(-\frac{\pi}{4} \right) = \frac{\pi}{2}$

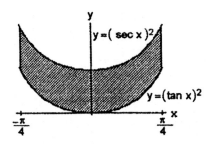

75. $A = A_1 + A_2$

Limits of integration: $x = y^3$ and $x = y \Rightarrow y = y^3$

$\Rightarrow y^3 - y = 0 \Rightarrow y(y - 1)(y + 1) = 0 \Rightarrow c_1 = -1, d_1 = 0$

and $c_2 = 0, d_2 = 1$; $f_1(y) - g_1(y) = y^3 - y$ and

$f_2(y) - g_2(y) = y - y^3 \Rightarrow$ by symmetry about the origin,

$A_1 + A_2 = 2A_2 \Rightarrow A = 2\int_0^1 (y - y^3)\,dy = 2\left[\frac{y^2}{2} - \frac{y^4}{4}\right]_0^1$

$= 2\left(\frac{1}{2} - \frac{1}{4}\right) = \frac{1}{2}$

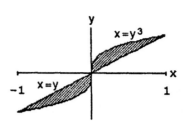

77. $\int_1^5 (\ln 2x - \ln x)\,dx = \int_1^5 (-\ln x + \ln 2 + \ln x)\,dx = (\ln 2)\int_1^5 dx = (\ln 2)(5 - 1) = \ln 2^4 = \ln 16$

79. $\int_0^{\ln 3} (e^{2x} - e^x)\,dx = \left[\frac{e^{2x}}{2} - e^x\right]_0^{\ln 3} = \left(\frac{e^{2\ln 3}}{2} - e^{\ln 3}\right) - \left(\frac{e^0}{2} - e^0\right) = \left(\frac{9}{2} - 3\right) - \left(\frac{1}{2} - 1\right) = \frac{8}{2} - 2 = 2$

81. $A = \int_{-2}^2 \frac{2x}{1+x^2}\,dx = 2\int_0^2 \frac{2x}{1+x^2}\,dx$; $[u = 1 + x^2 \Rightarrow du = 2x\,dx; x = 0 \Rightarrow u = 1, x = 2 \Rightarrow u = 5]$

$\rightarrow A = 2\int_1^5 \frac{1}{u}\,du = 2\,[\ln |u|]_1^5 = 2(\ln 5 - \ln 1) = 2\ln 5$

83. Limits of integration: $y = 1 + \sqrt{x}$ and $y = \frac{2}{\sqrt{x}}$

$\Rightarrow 1 + \sqrt{x} = \frac{2}{\sqrt{x}}, x \neq 0 \Rightarrow \sqrt{x} + x = 2 \Rightarrow x = (2 - x)^2$

$\Rightarrow x = 4 - 4x + x^2 \Rightarrow x^2 - 5x + 4 = 0$

$\Rightarrow (x - 4)(x - 1) = 0 \Rightarrow x = 1, 4$ (but $x = 4$ does not

satisfy the equation); $y = \frac{2}{\sqrt{x}}$ and $y = \frac{x}{4} \Rightarrow \frac{2}{\sqrt{x}} = \frac{x}{4}$

$\Rightarrow 8 = x\sqrt{x} \Rightarrow 64 = x^3 \Rightarrow x = 4$.

Therefore, AREA $= A_1 + A_2$: $f_1(x) - g_1(x)$

$= \left(1 + x^{1/2}\right) - \frac{x}{4} \Rightarrow A_1 = \int_0^1 \left(1 + x^{1/2} - \frac{x}{4}\right)\,dx$

$= \left[x + \frac{2}{3}x^{3/2} - \frac{x^2}{8}\right]_0^1$

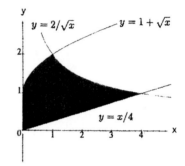

$= \left(1 + \frac{2}{3} - \frac{1}{8}\right) - 0 = \frac{37}{24}$; $f_2(x) - g_2(x) = 2x^{-1/2} - \frac{x}{4} \Rightarrow A_2 = \int_1^4 \left(2x^{-1/2} - \frac{x}{4}\right)\,dx = \left[4x^{1/2} - \frac{x^2}{8}\right]_1^4$

$= \left(4 \cdot 2 - \frac{16}{8}\right) - \left(4 - \frac{1}{8}\right) = 4 - \frac{15}{8} = \frac{17}{8}$; Therefore, AREA $= A_1 + A_2 = \frac{37}{24} + \frac{17}{8} = \frac{37+51}{24} = \frac{88}{24} = \frac{11}{3}$

85. Let $u = 2x \Rightarrow du = 2\,dx \Rightarrow \frac{1}{2}du = dx$; $x = 1 \Rightarrow u = 2, x = 3 \Rightarrow u = 6$

$\int_1^3 \frac{\sin 2x}{x}\,dx = \int_2^6 \frac{\sin u}{\left(\frac{u}{2}\right)}\left(\frac{1}{2}\,du\right) = \int_2^6 \frac{\sin u}{u}\,du = [F(u)]_2^6 = F(6) - F(2)$

87. (a) Let $u = -x \Rightarrow du = -dx$; $x = -1 \Rightarrow u = 1, x = 0 \Rightarrow u = 0$

f odd $\Rightarrow f(-x) = -f(x)$. Then $\int_{-1}^0 f(x)\,dx = \int_1^0 f(-u)\,(-du) = \int_1^0 -f(u)\,(-du) = \int_1^0 f(u)\,du = -\int_0^1 f(u)\,du$

$= -3$

(b) Let $u = -x \Rightarrow du = -dx$; $x = -1 \Rightarrow u = 1, x = 0 \Rightarrow u = 0$

f even $\Rightarrow f(-x) = f(x)$. Then $\int_{-1}^0 f(x)\,dx = \int_1^0 f(-u)\,(-du) = -\int_1^0 f(u)\,du = \int_0^1 f(u)\,du = 3$

89. Let $u = a - x \Rightarrow du = -dx$; $x = 0 \Rightarrow u = a, x = a \Rightarrow u = 0$

$I = \int_0^a \frac{f(x)\,dx}{f(x)+f(a-x)} = \int_a^0 \frac{f(a-u)}{f(a-u)+f(u)}\,(-du) = \int_0^a \frac{f(a-u)\,du}{f(u)+f(a-u)} = \int_0^a \frac{f(a-x)\,dx}{f(x)+f(a-x)}$

$\Rightarrow I + I = \int_0^a \frac{f(x)\,dx}{f(x)+f(a-x)} + \int_0^a \frac{f(a-x)\,dx}{f(x)+f(a-x)} = \int_0^a \frac{f(x)+f(a-x)}{f(x)+f(a-x)}\,dx = \int_0^a dx = [x]_0^a = a - 0 = a$.

Therefore, $2I = a \Rightarrow I = \frac{a}{2}$.

CHAPTER 4 PRACTICE AND ADDITIONAL EXERCISES

1. $\int (x^3 + 5x - 7)\, dx = \frac{x^4}{4} + \frac{5x^2}{2} - 7x + C$

3. Let $u = r + 5 \Rightarrow du = dr$

$\int \frac{dr}{(r+5)^2} = \int \frac{du}{u^2} = \int u^{-2}\, du = \frac{u^{-1}}{-1} + C = -u^{-1} + C = -\frac{1}{(r+5)} + C$

5. Let $u = \theta^2 + 1 \Rightarrow du = 2\theta\, d\theta \Rightarrow \frac{1}{2}\, du = \theta\, d\theta$

$\int 3\theta\sqrt{\theta^2 + 1}\, d\theta = \int \sqrt{u}\left(\frac{3}{2}\, du\right) = \frac{3}{2}\int u^{1/2}\, du = \frac{3}{2}\left(\frac{u^{3/2}}{\frac{3}{2}}\right) + C = u^{3/2} + C = \left(\theta^2 + 1\right)^{3/2} + C$

7. Let $u = 1 + x^4 \Rightarrow du = 4x^3\, dx \Rightarrow \frac{1}{4}\, du = x^3\, dx$

$\int x^3\left(1 + x^4\right)^{-1/4} dx = \int u^{-1/4}\left(\frac{1}{4}\, du\right) = \frac{1}{4}\int u^{-1/4}\, du = \frac{1}{4}\left(\frac{u^{3/4}}{\frac{3}{4}}\right) + C = \frac{1}{3}u^{3/4} + C = \frac{1}{3}\left(1 + x^4\right)^{3/4} + C$

9. Let $u = \frac{s}{10} \Rightarrow du = \frac{1}{10}\, ds \Rightarrow 10\, du = ds$

$\int \sec^2 \frac{s}{10}\, ds = \int \left(\sec^2 u\right)(10\, du) = 10\int \sec^2 u\, du = 10\tan u + C = 10\tan \frac{s}{10} + C$

11. Let $u = \frac{x}{4} \Rightarrow du = \frac{1}{4}\, dx \Rightarrow 4\, du = dx$

$\int \sin^2 \frac{x}{4}\, dx = \int \left(\sin^2 u\right)(4\, du) = \int 4\left(\frac{1 - \cos 2u}{2}\right) du = 2\int (1 - \cos 2u)\, du = 2\left(u - \frac{\sin 2u}{2}\right) + C$
$= 2u - \sin 2u + C = 2\left(\frac{x}{4}\right) - \sin 2\left(\frac{x}{4}\right) + C = \frac{x}{2} - \sin \frac{x}{2} + C$

13. $\int \left(\frac{3}{x} - x\right) dx = 3\ln|x| - \frac{x^2}{2} + C$

15. $\int \left(\theta^{1-\pi}\right) d\theta = \frac{\theta^{2-\pi}}{2-\pi} + C$

17. $y = \int \frac{x^2+1}{x^2}\, dx = \int \left(1 + x^{-2}\right) dx = x - x^{-1} + C = x - \frac{1}{x} + C;\ y = -1$ when $x = 1 \Rightarrow 1 - \frac{1}{1} + C = -1$
$\Rightarrow C = -1 \Rightarrow y = x - \frac{1}{x} - 1$

19. $\frac{dr}{dt} = \int \left(15\sqrt{t} + \frac{3}{\sqrt{t}}\right) dt = \int \left(15t^{1/2} + 3t^{-1/2}\right) dt = 10t^{3/2} + 6t^{1/2} + C;\ \frac{dr}{dt} = 8$ when $t = 1$
$\Rightarrow 10(1)^{3/2} + 6(1)^{1/2} + C = 8 \Rightarrow C = -8.$ Thus $\frac{dr}{dt} = 10t^{3/2} + 6t^{1/2} - 8 \Rightarrow r = \int \left(10t^{3/2} + 6t^{1/2} - 8\right) dt$
$= 4t^{5/2} + 4t^{3/2} - 8t + C;\ r = 0$ when $t = 1 \Rightarrow 4(1)^{5/2} + 4(1)^{3/2} - 8(1) + C_1 = 0 \Rightarrow C_1 = 0.$ Therefore,
$r = 4t^{5/2} + 4t^{3/2} - 8t$

21. (a) Each time subinterval is of length $\Delta t = 0.4$ sec. The distance traveled over each subinterval, using the midpoint rule, is $\Delta h = \frac{1}{2}\left(v_i + v_{i+1}\right)\Delta t$, where v_i is the velocity at the left endpoint and v_{i+1} the velocity at the right endpoint of the subinterval. We then add Δh to the height attained so far at the left endpoint v_i to arrive at the height associated with velocity v_{i+1} at the right endpoint. Using this methodology we build the following table based on the figure in the text:

t (sec)	0	0.4	0.8	1.2	1.6	2.0	2.4	2.8	3.2	3.6	4.0	4.4	4.8	5.2	5.6	6.0
v (fps)	0	10	25	55	100	190	180	165	150	140	130	115	105	90	76	65
h (ft)	0	2	9	25	56	114	188	257	320	378	432	481	525	564	592	620.2

t (sec)	6.4	6.8	7.2	7.6	8.0
v (fps)	50	37	25	12	0
h (ft)	643.2	660.6	672	679.4	681.8

NOTE: Your table values may vary slightly from ours depending on the v-values you read from the graph.

Remember that some shifting of the graph occurs in the printing process.

The total height attained is about 680 ft.

(b) The graph is based on the table in part (a).

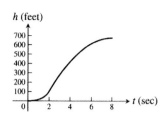

23. (a) $\sum_{k=1}^{10} \frac{a_k}{4} = \frac{1}{4} \sum_{k=1}^{10} a_k = \frac{1}{4}(-2) = -\frac{1}{2}$ (b) $\sum_{k=1}^{10} (b_k - 3a_k) = \sum_{k=1}^{10} b_k - 3 \sum_{k=1}^{10} a_k = 25 - 3(-2) = 31$

(c) $\sum_{k=1}^{10} (a_k + b_k - 1) = \sum_{k=1}^{10} a_k + \sum_{k=1}^{10} b_k - \sum_{k=1}^{10} 1 = -2 + 25 - (1)(10) = 13$

(d) $\sum_{k=1}^{10} \left(\frac{5}{2} - b_k\right) = \sum_{k=1}^{10} \frac{5}{2} - \sum_{k=1}^{10} b_k = \frac{5}{2}(10) - 25 = 0$

25. Let $u = 2x - 1 \Rightarrow du = 2\,dx \Rightarrow \frac{1}{2}\,du = dx; x = 1 \Rightarrow u = 1, x = 5 \Rightarrow u = 9$

$\int_1^5 (2x - 1)^{-1/2}\,dx = \int_1^9 u^{-1/2} \left(\frac{1}{2}\,du\right) = \left[u^{1/2}\right]_1^9 = 3 - 1 = 2$

27. Let $u = \frac{x}{2} \Rightarrow 2\,du = dx; x = -\pi \Rightarrow u = -\frac{\pi}{2}, x = 0 \Rightarrow u = 0$

$\int_{-\pi}^0 \cos\left(\frac{x}{2}\right) dx = \int_{-\pi/2}^0 (\cos u)(2\,du) = [2 \sin u]_{-\pi/2}^0 = 2 \sin 0 - 2 \sin\left(-\frac{\pi}{2}\right) = 2(0 - (-1)) = 2$

29. (a) $\int_{-2}^2 f(x)\,dx = \frac{1}{3}\int_{-2}^2 3\,f(x)\,dx = \frac{1}{3}(12) = 4$ (b) $\int_2^5 f(x)\,dx = \int_{-2}^5 f(x)\,dx - \int_{-2}^2 f(x)\,dx = 6 - 4 = 2$

(c) $\int_5^{-2} g(x)\,dx = -\int_{-2}^5 g(x)\,dx = -2$ (d) $\int_{-2}^5 (-\pi\, g(x))\,dx = -\pi \int_{-2}^5 g(x)\,dx = -\pi(2) = -2\pi$

(e) $\int_{-2}^5 \left(\frac{f(x) + g(x)}{5}\right) dx = \frac{1}{5}\int_{-2}^5 f(x)\,dx + \frac{1}{5}\int_{-2}^5 g(x)\,dx = \frac{1}{5}(6) + \frac{1}{5}(2) = \frac{8}{5}$

31. $x^2 - 4x + 3 = 0 \Rightarrow (x - 3)(x - 1) = 0 \Rightarrow x = 3$ or $x = 1$;

$\text{Area} = \int_0^1 (x^2 - 4x + 3)\,dx - \int_1^3 (x^2 - 4x + 3)\,dx$

$= \left[\frac{x^3}{3} - 2x^2 + 3x\right]_0^1 - \left[\frac{x^3}{3} - 2x^2 + 3x\right]_1^3$

$= \left[\left(\frac{1^3}{3} - 2(1)^2 + 3(1)\right) - 0\right]$

$\quad - \left[\left(\frac{3^3}{3} - 2(3)^2 + 3(3)\right) - \left(\frac{1^3}{3} - 2(1)^2 + 3(1)\right)\right]$

$= \left(\frac{1}{3} + 1\right) - \left[0 - \left(\frac{1}{3} + 1\right)\right] = \frac{8}{3}$

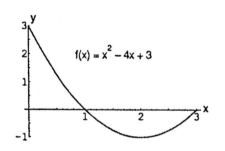

33. $5 - 5x^{2/3} = 0 \Rightarrow 1 - x^{2/3} = 0 \Rightarrow x = \pm 1$;

$\text{Area} = \int_{-1}^1 \left(5 - 5x^{2/3}\right) dx - \int_1^8 \left(5 - 5x^{2/3}\right) dx$

$= \left[5x - 3x^{5/3}\right]_{-1}^1 - \left[5x - 3x^{5/3}\right]_1^8$

$= \left[\left(5(1) - 3(1)^{5/3}\right) - \left(5(-1) - 3(-1)^{5/3}\right)\right]$

$\quad - \left[\left(5(8) - 3(8)^{5/3}\right) - \left(5(1) - 3(1)^{5/3}\right)\right]$

$= [2 - (-2)] - [(40 - 96) - 2] = 62$

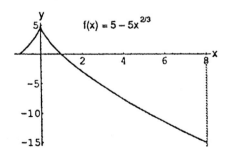

35. $f(x) = x$, $g(x) = \frac{1}{x^2}$, $a = 1$, $b = 2$ $\Rightarrow$ $A = \int_a^b [f(x) - g(x)]\, dx$

$= \int_1^2 \left(x - \frac{1}{x^2} \right) dx = \left[\frac{x^2}{2} + \frac{1}{x} \right]_1^2 = \left(\frac{4}{2} + \frac{1}{2} \right) - \left(\frac{1}{2} + 1 \right) = 1$

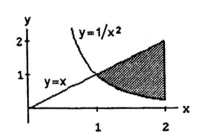

37. $f(y) = 2y^2$, $g(y) = 0$, $c = 0$, $d = 3$

$\Rightarrow A = \int_c^d [f(y) - g(y)]\, dy = \int_0^3 (2y^2 - 0)\, dy$

$= 2 \int_0^3 y^2\, dy = \frac{2}{3} [y^3]_0^3 = 18$

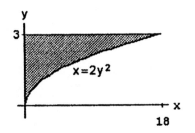

39. Let us find the intersection points: $\frac{y^2}{4} = \frac{y+2}{4}$

$\Rightarrow y^2 - y - 2 = 0 \Rightarrow (y - 2)(y + 1) = 0 \Rightarrow y = -1$

or $y = 2 \Rightarrow c = -1$, $d = 2$; $f(y) = \frac{y+2}{4}$, $g(y) = \frac{y^2}{4}$

$\Rightarrow A = \int_c^d [f(y) - g(y)]\, dy = \int_{-1}^2 \left(\frac{y+2}{4} - \frac{y^2}{4} \right) dy$

$= \frac{1}{4} \int_{-1}^2 (y + 2 - y^2)\, dy = \frac{1}{4} \left[\frac{y^2}{2} + 2y - \frac{y^3}{3} \right]_{-1}^2$

$= \frac{1}{4} \left[\left(\frac{4}{2} + 4 - \frac{8}{3} \right) - \left(\frac{1}{2} - 2 + \frac{1}{3} \right) \right] = \frac{9}{8}$

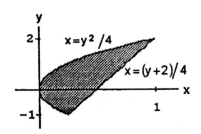

41. $f(x) = x$, $g(x) = \sin x$, $a = 0$, $b = \frac{\pi}{4}$

$\Rightarrow A = \int_a^b [f(x) - g(x)]\, dx = \int_0^{\pi/4} (x - \sin x)\, dx$

$= \left[\frac{x^2}{2} + \cos x \right]_0^{\pi/4} = \left(\frac{\pi^2}{32} + \frac{\sqrt{2}}{2} \right) - 1$

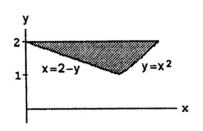

43. $f(y) = \sqrt{y}$, $g(y) = 2 - y$, $c = 1$, $d = 2$

$\Rightarrow A = \int_c^d [f(y) - g(y)]\, dy = \int_1^2 \left[\sqrt{y} - (2 - y) \right] dy$

$= \int_1^2 \left(\sqrt{y} - 2 + y \right) dy = \left[\frac{2}{3} y^{3/2} - 2y + \frac{y^2}{2} \right]_1^2$

$= \left(\frac{4}{3} \sqrt{2} - 4 + 2 \right) - \left(\frac{2}{3} - 2 + \frac{1}{2} \right) = \frac{4}{3} \sqrt{2} - \frac{7}{6} = \frac{8\sqrt{2}-7}{6}$

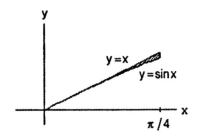

45. $f(x) = x^3 - 3x^2 = x^2(x - 3) \Rightarrow f'(x) = 3x^2 - 6x = 3x(x - 2) \Rightarrow f' = +++ \mid ----- \mid +++$
$ 0 2$

$\Rightarrow f(0) = 0$ is a maximum and $f(2) = -4$ is a minimum. $A = -\int_0^3 (x^3 - 3x^2)\, dx = -\left[\frac{x^4}{4} - x^3 \right]_0^3$

$= -\left(\frac{81}{4} - 27 \right) = \frac{27}{4}$

47. The area above the x-axis is $A_1 = \int_0^1 (y^{2/3} - y)\, dy$

$= \left[\frac{3y^{5/3}}{5} - \frac{y^2}{2}\right]_0^1 = \frac{1}{10}$; the area below the x-axis is

$A_2 = \int_{-1}^0 (y^{2/3} - y)\, dy = \left[\frac{3y^{5/3}}{5} - \frac{y^2}{2}\right]_{-1}^0 = \frac{11}{10}$

$\Rightarrow$ the total area is $A_1 + A_2 = \frac{6}{5}$

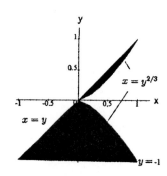

49. $A = \int_1^e \frac{2 \ln x}{x}\, dx = \int_0^1 2u\, du = [u^2]_0^1 = 1$, where

$u = \ln x$ and $du = \frac{1}{x}\, dx$; $x = 1 \Rightarrow u = 0$, $x = e \Rightarrow u = 1$

51. Let $u = \cos x \Rightarrow du = -\sin x\, dx \Rightarrow -du = \sin x\, dx$

$\int 2(\cos x)^{-1/2} \sin x\, dx = \int 2u^{-1/2}(-du) = -2\int u^{-1/2}\, du = -2\left(\frac{u^{1/2}}{\frac{1}{2}}\right) + C = -4u^{1/2} + C$

$= -4(\cos x)^{1/2} + C$

53. Let $u = 2\theta + 1 \Rightarrow du = 2\, d\theta \Rightarrow \frac{1}{2}\, du = d\theta$

$\int [2\theta + 1 + 2\cos(2\theta + 1)]\, d\theta = \int (u + 2\cos u)\left(\frac{1}{2}\, du\right) = \frac{u^2}{4} + \sin u + C_1 = \frac{(2\theta + 1)^2}{4} + \sin(2\theta + 1) + C_1$

$= \theta^2 + \theta + \sin(2\theta + 1) + C$, where $C = C_1 + \frac{1}{4}$ is still an arbitrary constant

55. $\int \left(t - \frac{2}{t}\right)\left(t + \frac{2}{t}\right) dt = \int \left(t^2 - \frac{4}{t^2}\right) dt = \int (t^2 - 4t^{-2})\, dt = \frac{t^3}{3} - 4\left(\frac{t^{-1}}{-1}\right) + C = \frac{t^3}{3} + \frac{4}{t} + C$

57. Let $u = 2t^{3/2} \Rightarrow du = 3\sqrt{t}\, dt \Rightarrow \frac{1}{3}\, du = \sqrt{t}\, dt$

$\int \sqrt{t} \sin(2t^{3/2})\, dt = \frac{1}{3}\int \sin u\, du = -\frac{1}{3}\cos u + C = -\frac{1}{3}\cos(2t^{3/2}) + C$

59. $\int e^x \sec^2(e^x - 7)\, dx = \int \sec^2 u\, du$, where $u = e^x - 7$ and $du = e^x\, dx$

$= \tan u + C = \tan(e^x - 7) + C$

61. $\int (\sec^2 x) e^{\tan x}\, dx = \int e^u\, du$, where $u = \tan x$ and $du = \sec^2 x\, dx$

$= e^u + C = e^{\tan x} + C$

63. $\int_{-1}^1 \frac{1}{3x - 4}\, dx = \frac{1}{3}\int_{-7}^{-1} \frac{1}{u}\, du$, where $u = 3x - 4$, $du = 3\, dx$; $x = -1 \Rightarrow u = -7$, $x = 1 \Rightarrow u = -1$

$= \frac{1}{3}[\ln|u|]_{-7}^{-1} = \frac{1}{3}[\ln|-1| - \ln|-7|] = \frac{1}{3}[0 - \ln 7] = -\frac{\ln 7}{3}$

65. $\int_0^4 \frac{2t}{t^2 - 25}\, dt = \int_{-25}^{-9} \frac{1}{u}\, du$, where $u = t^2 - 25$, $du = 2t\, dt$; $t = 0 \Rightarrow u = -25$, $t = 4 \Rightarrow u = -9$

$= [\ln|u|]_{-25}^{-9} = \ln|-9| - \ln|-25| = \ln 9 - \ln 25 = \ln\frac{9}{25}$

67. $\int \frac{(\ln x)^{-3}}{x}\, dx = \int u^{-3}\, du$, where $u = \ln x$ and $du = \frac{1}{x}\, dx$

$= \frac{u^{-2}}{-2} + C = -\frac{1}{2}(\ln x)^{-2} + C$

69. $\int x3^{x^2}\,dx = \frac{1}{2}\int 3^u\,du$, where $u = x^2$ and $du = 2x\,dx$

    $= \frac{1}{2\ln 3}(3^u) + C = \frac{1}{2\ln 3}\left(3^{x^2}\right) + C$

71. $\int \frac{3\,dr}{\sqrt{1 - 4(r-1)^2}} = \frac{3}{2}\int \frac{du}{\sqrt{1 - u^2}}$, where $u = 2(r-1)$ and $du = 2\,dr$

    $= \frac{3}{2}\sin^{-1}u + C = \frac{3}{2}\sin^{-1}2(r-1) + C$

73. $\int \frac{dx}{(2x-1)\sqrt{(2x-1)^2 - 4}} = \frac{1}{2}\int \frac{du}{u\sqrt{u^2 - 4}}$, where $u = 2x - 1$ and $du = 2\,dx$

    $= \frac{1}{2}\cdot\frac{1}{2}\sec^{-1}\left|\frac{u}{2}\right| + C = \frac{1}{4}\sec^{-1}\left|\frac{2x-1}{2}\right| + C$

75. $\int \frac{1}{\sqrt{\tan^{-1}y}\,(1 + y^2)}\,dy = \int \frac{\left(\frac{1}{1+y^2}\right)}{\sqrt{\tan^{-1}y}}\,dy = \int u^{-1/2}\,du$, where $u = \tan^{-1}y$ and $du = \frac{dy}{1+y^2}$

    $= 2u^{1/2} + C = 2\sqrt{\tan^{-1}y} + C$

77. $\int_{-1}^{1}(3x^2 - 4x + 7)\,dx = [x^3 - 2x^2 + 7x]_{-1}^{1} = [1^3 - 2(1)^2 + 7(1)] - [(-1)^3 - 2(-1)^2 + 7(-1)] = 6 - (-10) = 16$

79. $\int_{1}^{2}\frac{4}{v^2}\,dv = \int_{1}^{2}4v^{-2}\,dv = [-4v^{-1}]_{1}^{2} = \left(\frac{-4}{2}\right) - \left(\frac{-4}{1}\right) = 2$

81. $\int_{1}^{4}\frac{dt}{t\sqrt{t}} = \int_{1}^{4}\frac{dt}{t^{3/2}} = \int_{1}^{4}t^{-3/2}\,dt = [-2t^{-1/2}]_{1}^{4} = \frac{-2}{\sqrt{4}} - \frac{(-2)}{\sqrt{1}} = 1$

83. Let $u = 1 - x^{2/3} \Rightarrow du = -\frac{2}{3}x^{-1/3}\,dx \Rightarrow -\frac{3}{2}\,du = x^{-1/3}\,dx$; $x = \frac{1}{8} \Rightarrow u = 1 - \left(\frac{1}{8}\right)^{2/3} = \frac{3}{4}$,

    $x = 1 \Rightarrow u = 1 - 1^{2/3} = 0$

    $\int_{1/8}^{1}x^{-1/3}\left(1 - x^{2/3}\right)^{3/2}\,dx = \int_{3/4}^{0}u^{3/2}\left(-\frac{3}{2}\,du\right) = \left[\left(-\frac{3}{2}\right)\left(\frac{u^{5/2}}{\frac{5}{2}}\right)\right]_{3/4}^{0} = \left[-\frac{3}{5}u^{5/2}\right]_{3/4}^{0} = -\frac{3}{5}(0)^{5/2} - \left(-\frac{3}{5}\right)\left(\frac{3}{4}\right)^{5/2}$

    $= \frac{27\sqrt{3}}{160}$

85. Let $u = 5r \Rightarrow du = 5\,dr \Rightarrow \frac{1}{5}\,du = dr$; $r = 0 \Rightarrow u = 0$, $r = \pi \Rightarrow u = 5\pi$

    $\int_{0}^{\pi}\sin^2 5r\,dr = \int_{0}^{5\pi}(\sin^2 u)\left(\frac{1}{5}\,du\right) = \frac{1}{5}\left[\frac{u}{2} - \frac{\sin 2u}{4}\right]_{0}^{5\pi} = \left(\frac{\pi}{2} - \frac{\sin 10\pi}{20}\right) - \left(0 - \frac{\sin 0}{20}\right) = \frac{\pi}{2}$

87. Let $u = \frac{x}{6} \Rightarrow du = \frac{1}{6}\,dx \Rightarrow 6\,du = dx$; $x = \pi \Rightarrow u = \frac{\pi}{6}$, $x = 3\pi \Rightarrow u = \frac{\pi}{2}$

    $\int_{\pi}^{3\pi}\cot^2 \frac{x}{6}\,dx = \int_{\pi/6}^{\pi/2}6\cot^2 u\,du = 6\int_{\pi/6}^{\pi/2}(\csc^2 u - 1)\,du = [6(-\cot u - u)]_{\pi/6}^{\pi/2} = 6\left(-\cot\frac{\pi}{2} - \frac{\pi}{2}\right) - 6\left(-\cot\frac{\pi}{6} - \frac{\pi}{6}\right)$

    $= 6\sqrt{3} - 2\pi$

89. $\int_{-\pi/3}^{0}\sec x\tan x\,dx = [\sec x]_{-\pi/3}^{0} = \sec 0 - \sec\left(-\frac{\pi}{3}\right) = 1 - 2 = -1$

91. Let $u = \sin x \Rightarrow du = \cos x\,dx$; $x = 0 \Rightarrow u = 0$, $x = \frac{\pi}{2} \Rightarrow u = 1$

    $\int_{0}^{\pi/2}5(\sin x)^{3/2}\cos x\,dx = \int_{0}^{1}5u^{3/2}\,du = \left[5\left(\frac{2}{5}\right)u^{5/2}\right]_{0}^{1} = [2u^{5/2}]_{0}^{1} = 2(1)^{5/2} - 2(0)^{5/2} = 2$

93. Let $u = 1 + 3\sin^2 x \Rightarrow du = 6\sin x\cos x\,dx \Rightarrow \frac{1}{2}\,du = 3\sin x\cos x\,dx$; $x = 0 \Rightarrow u = 1$, $x = \frac{\pi}{2}$

    $\Rightarrow u = 1 + 3\sin^2\frac{\pi}{2} = 4$

    $\int_{0}^{\pi/2}\frac{3\sin x\cos x}{\sqrt{1 + 3\sin^2 x}}\,dx = \int_{1}^{4}\frac{1}{\sqrt{u}}\left(\frac{1}{2}\,du\right) = \int_{1}^{4}\frac{1}{2}u^{-1/2}\,du = \left[\frac{1}{2}\left(\frac{u^{1/2}}{\frac{1}{2}}\right)\right]_{1}^{4} = [u^{1/2}]_{1}^{4} = 4^{1/2} - 1^{1/2} = 1$

95. $\int_1^4 \left(\frac{x}{8} + \frac{1}{2x}\right) dx = \frac{1}{2}\int_1^4 \left(\frac{1}{4}x + \frac{1}{x}\right) dx = \frac{1}{2}\left[\frac{1}{8}x^2 + \ln|x|\right]_1^4 = \frac{1}{2}\left[\left(\frac{16}{8} + \ln 4\right) - \left(\frac{1}{8} + \ln 1\right)\right] = \frac{15}{16} + \frac{1}{2}\ln 4$

$\quad = \frac{15}{16} + \ln\sqrt{4} = \frac{15}{16} + \ln 2$

97. $\int_{-2}^{-1} e^{-(x+1)} dx = -\int_1^0 e^u \, du$, where $u = -(x+1)$, $du = -dx$; $x = -2 \Rightarrow u = 1$, $x = -1 \Rightarrow u = 0$

$\quad = -[e^u]_1^0 = -(e^0 - e^1) = e - 1$

99. $\int_1^{\ln 5} e^r (3e^r + 1)^{-3/2} dr = \frac{1}{3}\int_4^{16} u^{-3/2} du$, where $u = 3e^r + 1$, $du = 3e^r dr$; $r = 0 \Rightarrow u = 4$, $r = \ln 5 \Rightarrow u = 16$

$\quad = -\frac{2}{3}\left[u^{-1/2}\right]_4^{16} = -\frac{2}{3}\left(16^{-1/2} - 4^{-1/2}\right) = \left(-\frac{2}{3}\right)\left(\frac{1}{4} - \frac{1}{2}\right) = \left(-\frac{2}{3}\right)\left(-\frac{1}{4}\right) = \frac{1}{6}$

101. $\int_1^e \frac{1}{x}(1 + 7\ln x)^{-1/3} dx = \frac{1}{7}\int_1^8 u^{-1/3} du$, where $u = 1 + 7\ln x$, $du = \frac{7}{x} dx$, $x = 1 \Rightarrow u = 1$, $x = e \Rightarrow u = 8$

$\quad = \frac{3}{14}\left[u^{2/3}\right]_1^8 = \frac{3}{14}\left(8^{2/3} - 1^{2/3}\right) = \left(\frac{3}{14}\right)(4 - 1) = \frac{9}{14}$

103. $\int_1^8 \frac{\log_4 \theta}{\theta} d\theta = \frac{1}{\ln 4}\int_1^8 (\ln\theta)\left(\frac{1}{\theta}\right) d\theta = \frac{1}{\ln 4}\int_0^{\ln 8} u \, du$, where $u = \ln\theta$, $du = \frac{1}{\theta} d\theta$, $\theta = 1 \Rightarrow u = 0$, $\theta = 8 \Rightarrow u = \ln 8$

$\quad = \frac{1}{2\ln 4}\left[u^2\right]_0^{\ln 8} = \frac{1}{\ln 16}\left[(\ln 8)^2 - 0^2\right] = \frac{(3\ln 2)^2}{4\ln 2} = \frac{9\ln 2}{4}$

105. $\int_{-3/4}^{3/4} \frac{6}{\sqrt{9 - 4x^2}} dx = 3\int_{-3/4}^{3/4} \frac{2}{\sqrt{3^2 - (2x)^2}} dx = 3\int_{-3/2}^{3/2} \frac{1}{\sqrt{3^2 - u^2}} du$, where $u = 2x$, $du = 2 \, dx$;

$\quad\quad\quad\quad\quad\quad\quad\quad x = -\frac{3}{4} \Rightarrow u = -\frac{3}{2}, x = \frac{3}{4} \Rightarrow u = \frac{3}{2}$

$\quad = 3\left[\sin^{-1}\left(\frac{u}{3}\right)\right]_{-3/2}^{3/2} = 3\left[\sin^{-1}\left(\frac{1}{2}\right) - \sin^{-1}\left(-\frac{1}{2}\right)\right] = 3\left[\frac{\pi}{6} - \left(-\frac{\pi}{6}\right)\right] = 3\left(\frac{\pi}{3}\right) = \pi$

107. $\int_{-2}^2 \frac{3}{4 + 3t^2} dt = \sqrt{3}\int_{-2}^2 \frac{\sqrt{3}}{2^2 + \left(\sqrt{3}t\right)^2} dt = \sqrt{3}\int_{-2\sqrt{3}}^{2\sqrt{3}} \frac{1}{2^2 + u^2} du$, where $u = \sqrt{3}t$, $du = \sqrt{3} \, dt$;

$\quad\quad\quad\quad\quad\quad\quad\quad t = -2 \Rightarrow u = -2\sqrt{3}, t = 2 \Rightarrow u = 2\sqrt{3}$

$\quad = \sqrt{3}\left[\frac{1}{2}\tan^{-1}\left(\frac{u}{2}\right)\right]_{-2\sqrt{3}}^{2\sqrt{3}} = \frac{\sqrt{3}}{2}\left[\tan^{-1}\left(\sqrt{3}\right) - \tan^{-1}\left(-\sqrt{3}\right)\right] = \frac{\sqrt{3}}{2}\left[\frac{\pi}{3} - \left(-\frac{\pi}{3}\right)\right] = \frac{\pi}{\sqrt{3}}$

109. $\int \frac{1}{y\sqrt{4y^2 - 1}} dy = \int \frac{2}{(2y)\sqrt{(2y)^2 - 1}} dy = \int \frac{1}{u\sqrt{u^2 - 1}} du$, where $u = 2y$ and $du = 2 \, dy$

$\quad = \sec^{-1}|u| + C = \sec^{-1}|2y| + C$

111. $\int_{\sqrt{2}/3}^{2/3} \frac{1}{|y|\sqrt{9y^2 - 1}} dy = \int_{\sqrt{2}/3}^{2/3} \frac{3}{|3y|\sqrt{(3y)^2 - 1}} dy = \int_{\sqrt{2}}^2 \frac{1}{|u|\sqrt{u^2 - 1}} du$, where $u = 3y$, $du = 3 \, dy$;

$\quad\quad\quad\quad\quad\quad\quad\quad y = \frac{\sqrt{2}}{3} \Rightarrow u = \sqrt{2}, y = \frac{2}{3} \Rightarrow u = 2$

$\quad = \left[\sec^{-1} u\right]_{\sqrt{2}}^2 = \left[\sec^{-1} 2 - \sec^{-1}\sqrt{2}\right] = \frac{\pi}{3} - \frac{\pi}{4} = \frac{\pi}{12}$

113. (a) $av(f) = \frac{1}{1 - (-1)}\int_{-1}^1 (mx + b) dx = \frac{1}{2}\left[\frac{mx^2}{2} + bx\right]_{-1}^1 = \frac{1}{2}\left[\left(\frac{m(1)^2}{2} + b(1)\right) - \left(\frac{m(-1)^2}{2} + b(-1)\right)\right] = \frac{1}{2}(2b) = b$

$\quad$ (b) $av(f) = \frac{1}{k - (-k)}\int_{-k}^k (mx + b) dx = \frac{1}{2k}\left[\frac{mx^2}{2} + bx\right]_{-k}^k = \frac{1}{2k}\left[\left(\frac{m(k)^2}{2} + b(k)\right) - \left(\frac{m(-k)^2}{2} + b(-k)\right)\right]$

$\quad\quad = \frac{1}{2k}(2bk) = b$

115. (a) $\frac{d}{dx}(x\ln x - x + C) = x \cdot \frac{1}{x} + \ln x - 1 + 0 = \ln x$

$\quad$ (b) average value $= \frac{1}{e - 1}\int_1^e \ln x \, dx = \frac{1}{e - 1}[x\ln x - x]_1^e = \frac{1}{e - 1}[(e\ln e - e) - (1\ln 1 - 1)]$

$\quad\quad = \frac{1}{e - 1}(e - e + 1) = \frac{1}{e - 1}$

117. We want to evaluate

$$\frac{1}{365-0}\int_0^{365} f(x)\,dx = \frac{1}{365}\int_0^{365}\left(37\sin\left[\frac{2\pi}{365}(x-101)\right]+25\right)dx = \frac{37}{365}\int_0^{365}\sin\left[\frac{2\pi}{365}(x-101)\right]dx + \frac{25}{365}\int_0^{365}dx$$

Notice that the period of $y = \sin\left[\frac{2\pi}{365}(x-101)\right]$ is $\frac{2\pi}{\frac{2\pi}{365}} = 365$ and that we are integrating this function over an iterval of

length 365. Thus the value of $\frac{37}{365}\int_0^{365}\sin\left[\frac{2\pi}{365}(x-101)\right]dx + \frac{25}{365}\int_0^{365}dx$ is $\frac{37}{365}\cdot 0 + \frac{25}{365}\cdot 365 = 25°F.$

119. $\frac{dy}{dx} = \sqrt{2+\cos^3 x}$

121. $y = \int_{\ln x^2}^{0} e^{\cos t}\,dt = -\int_0^{\ln x^2} e^{\cos t}\,dt \Rightarrow \frac{dy}{dx} = -e^{\cos(\ln x^2)}\cdot\frac{d}{dx}(\ln x^2) = -\frac{2}{x}e^{\cos(\ln x^2)}$

123. $y = \int_0^{\sin^{-1}x}\frac{dt}{\sqrt{1-2t^2}} \Rightarrow \frac{dy}{dx} = \frac{1}{\sqrt{1-2(\sin^{-1}x)^2}}\cdot\frac{d}{dx}(\sin^{-1}x) = \frac{1}{\sqrt{1-2(\sin^{-1}x)^2}}\cdot\frac{1}{\sqrt{1-x^2}}$

125. Yes. The function f, being differentiable on [a, b], is then continuous on [a, b]. The Fundamental Theorem of Calculus says that every continuous function on [a, b] is the derivative of a function on [a, b].

127. $y = \ln x \Rightarrow \frac{dy}{dx} = \frac{1}{x}; \frac{dy}{dt} = \frac{dy}{dx}\frac{dx}{dt} \Rightarrow \frac{dy}{dt} = \left(\frac{1}{x}\right)\sqrt{x} = \frac{1}{\sqrt{x}} \Rightarrow \frac{dy}{dt}\Big|_{e^2} = \frac{1}{e}$ m/sec

129. $\int\sinh u\,du = \int\frac{e^u-e^{-u}}{2}\,du = \int\left(\frac{e^u}{2}-\frac{e^{-u}}{2}\right)du = \frac{1}{2}\int e^u du - \frac{1}{2}\int e^{-u}du = \frac{1}{2}e^u + \frac{1}{2}e^{-u} + C = \frac{e^u+e^{-u}}{2} + C = \cosh u + C$

$\int\cosh u\,du = \int\frac{e^u+e^{-u}}{2}\,du = \int\left(\frac{e^u}{2}+\frac{e^{-u}}{2}\right)du = \frac{1}{2}\int e^u du + \frac{1}{2}\int e^{-u}du = \frac{1}{2}e^u - \frac{1}{2}e^{-u} + C = \frac{e^u-e^{-u}}{2} + C = \sinh u + C$

$\int\text{sech}^2 u\,du = \int\left(\frac{2}{e^u+e^{-u}}\right)^2 du = \int\frac{4}{(e^u+e^{-u})^2}\,du = \int\frac{2+2}{(e^u+e^{-u})^2}\,du = \int\left(\frac{2}{(e^u+e^{-u})^2}+\frac{2}{(e^u+e^{-u})^2}\right)du$

$= \int\left(\frac{2}{(e^u+e^{-u})^2}\cdot\frac{e^{-2u}}{e^{-2u}}+\frac{2}{(e^u+e^{-u})^2}\cdot\frac{e^{2u}}{e^{2u}}\right)du = \int\left(\frac{2e^{-2u}}{(e^u+e^{-u})^2(e^{-u})^2}+\frac{2e^{2u}}{(e^u+e^{-u})^2(e^u)^2}\right)du$

$= \int\left(\frac{2e^{-2u}}{(1+e^{-2u})^2}+\frac{2e^{2u}}{(e^{2u}+1)^2}\right)du = \int(1+e^{-2u})^{-2}(2e^{-2u})du + \int(e^{2u}+1)^{-2}(2e^{2u})du$

$= -\frac{(1+e^{-2u})^{-1}}{-1}+\frac{(e^{2u}+1)^{-1}}{-1}+C = \frac{1}{1+e^{-2u}}-\frac{1}{e^{2u}+1}+C = \frac{1}{1+e^{-2u}}\cdot\frac{e^u}{e^u}-\frac{1}{e^{2u}+1}\cdot\frac{e^{-u}}{e^{-u}}+C$

$= \frac{e^u}{e^u+e^{-u}}-\frac{e^{-u}}{e^u+e^{-u}}+C = \frac{e^u-e^{-u}}{e^u+e^{-u}}+C = \tanh u + C$

$\int\text{csch}^2 u\,du = \int\left(\frac{2}{e^u-e^{-u}}\right)^2 du = \int\frac{4}{(e^u-e^{-u})^2}\,du = \int\frac{2+2}{(e^u-e^{-u})^2}\,du = \int\left(\frac{2}{(e^u-e^{-u})^2}+\frac{2}{(e^u-e^{-u})^2}\right)du$

$= \int\left(\frac{2}{(e^u-e^{-u})^2}\cdot\frac{e^{-2u}}{e^{-2u}}+\frac{2}{(e^u-e^{-u})^2}\cdot\frac{e^{2u}}{e^{2u}}\right)du = \int\left(\frac{2e^{-2u}}{(e^u-e^{-u})^2(e^{-u})^2}+\frac{2e^{2u}}{(e^u-e^{-u})^2(e^u)^2}\right)du$

$= \int\left(\frac{2e^{-2u}}{(1-e^{-2u})^2}+\frac{2e^{2u}}{(e^{2u}-1)^2}\right)du = \int(1-e^{-2u})^{-2}(2e^{-2u})du + \int(e^{2u}-1)^{-2}(2e^{2u})du$

$= \frac{(1-e^{-2u})^{-1}}{-1}+\frac{(e^{2u}-1)^{-1}}{-1}+C = -\frac{1}{1-e^{-2u}}-\frac{1}{e^{2u}-1}+C = -\frac{1}{1-e^{-2u}}\cdot\frac{e^u}{e^u}-\frac{1}{e^{2u}-1}\cdot\frac{e^{-u}}{e^{-u}}+C$

$= -\frac{e^u}{e^u-e^{-u}}-\frac{e^{-u}}{e^u-e^{-u}}+C = \frac{-e^u-e^{-u}}{e^u-e^{-u}}+C = -\frac{e^u+e^{-u}}{e^u-e^{-u}}+C = -\coth u + C$

$\int\text{sech}\,u\tanh u\,du = \int\frac{2}{e^u+e^{-u}}\cdot\frac{e^u-e^{-u}}{e^u+e^{-u}}\,du = 2\int\frac{e^u-e^{-u}}{(e^u+e^{-u})^2}\,du \quad [w = e^u+e^{-u}, dw = (e^u-e^{-u})du]$

$= 2\int\frac{1}{w^2}\,dw = -2w^{-1}+C = -\frac{2}{e^u+e^{-u}}+C = -\text{sech}\,u + C$

$\int\text{csch}\,u\coth u\,du = \int\frac{2}{e^u-e^{-u}}\cdot\frac{e^u+e^{-u}}{e^u-e^{-u}}\,du = 2\int\frac{e^u+e^{-u}}{(e^u-e^{-u})^2}\,du \quad [w = e^u-e^{-u}, dw = (e^u+e^{-u})du]$

$= 2\int\frac{1}{w^2}\,dw = -2w^{-1}+C = -\frac{2}{e^u-e^{-u}}+C = -\text{csch}\,u + C$

131. (a) $\int_0^{2\sqrt{3}} \frac{dx}{\sqrt{4+x^2}} = \left[\sinh^{-1} \frac{x}{2}\right]_0^{2\sqrt{3}} = \sinh^{-1} \sqrt{3} - \sinh 0 = \sinh^{-1} \sqrt{3}$

   (b) $\sinh^{-1} \sqrt{3} = \ln\left(\sqrt{3} + \sqrt{3+1}\right) = \ln\left(\sqrt{3}+2\right)$

133. (a) $\int_{5/4}^{2} \frac{1}{1-x^2}\, dx = \left[\coth^{-1} x\right]_{5/4}^{2} = \coth^{-1} 2 - \coth^{-1} \frac{5}{4}$

   (b) $\coth^{-1} 2 - \coth^{-1} \frac{5}{4} = \frac{1}{2}\left[\ln 3 - \ln\left(\frac{9/4}{1/4}\right)\right] = \frac{1}{2} \ln \frac{1}{3}$

135. (a) $\int_{1/5}^{3/13} \frac{dx}{x\sqrt{1-16x^2}} = \int_{4/5}^{12/13} \frac{du}{u\sqrt{a^2-u^2}}$, where u = 4x, du = 4 dx, a = 1

   $= \left[-\operatorname{sech}^{-1} u\right]_{4/5}^{12/13} = -\operatorname{sech}^{-1} \frac{12}{13} + \operatorname{sech}^{-1} \frac{4}{5}$

   (b) $-\operatorname{sech}^{-1} \frac{12}{13} + \operatorname{sech}^{-1} \frac{4}{5} = -\ln\left(\frac{1+\sqrt{1-(12/13)^2}}{(12/13)}\right) + \ln\left(\frac{1+\sqrt{1-(4/5)^2}}{(4/5)}\right)$

   $= -\ln\left(\frac{13+\sqrt{169-144}}{12}\right) + \ln\left(\frac{5+\sqrt{25-16}}{4}\right) = \ln\left(\frac{5+3}{4}\right) - \ln\left(\frac{13+5}{12}\right) = \ln 2 - \ln \frac{3}{2}$

   $= \ln\left(2 \cdot \frac{2}{3}\right) = \ln \frac{4}{3}$

137. (a) $\int_0^{\pi} \frac{\cos x}{\sqrt{1+\sin^2 x}}\, dx = \int_0^0 \frac{1}{\sqrt{1+u^2}}\, du = \left[\sinh^{-1} u\right]_0^0 = \sinh^{-1} 0 - \sinh^{-1} 0 = 0$, where u = sin x, du = cos x dx

   (b) $\sinh^{-1} 0 - \sinh^{-1} 0 = \ln\left(0 + \sqrt{0+1}\right) - \ln\left(0 + \sqrt{0+1}\right) = 0$

139. $\int_{-8}^{3} f(x)\, dx = \int_{-8}^{0} x^{2/3}\, dx + \int_{0}^{3} -4\, dx$

   $= \left[\frac{3}{5} x^{5/3}\right]_{-8}^{0} + \left[-4x\right]_0^3$

   $= \left(0 - \frac{3}{5}(-8)^{5/3}\right) + (-4(3) - 0) = \frac{96}{5} - 12$

   $= \frac{36}{5}$

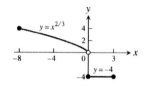

141. $\int_0^2 g(t)\, dt = \int_0^1 t\, dt + \int_1^2 \sin \pi t\, dt$

   $= \left[\frac{t^2}{2}\right]_0^1 + \left[-\frac{1}{\pi} \cos \pi t\right]_1^2$

   $= \left(\frac{1}{2} - 0\right) + \left[-\frac{1}{\pi} \cos 2\pi - \left(-\frac{1}{\pi} \cos \pi\right)\right]$

   $= \frac{1}{2} - \frac{2}{\pi}$

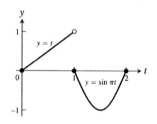

143. $\int_{-2}^{2} f(x)\, dx = \int_{-2}^{-1} dx + \int_{-1}^{1} (1-x^2)\, dx + \int_{1}^{2} 2\, dx$

   $= \left[x\right]_{-2}^{-1} + \left[x - \frac{x^3}{3}\right]_{-1}^{1} + \left[2x\right]_1^2$

   $= (-1 - (-2)) + \left[\left(1 - \frac{1^3}{3}\right) - \left(-1 - \frac{(-1)^3}{3}\right)\right]$

   $+ \left[2(2) - 2(1)\right]$

   $= 1 + \frac{2}{3} - \left(-\frac{2}{3}\right) + 4 - 2 = \frac{13}{3}$

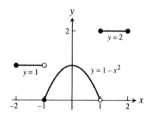

NOTES:

CHAPTER 5 TECHNIQUES OF INTEGRATION

5.1 INTEGRATION BY PARTS

1. $u = x$, $du = dx$; $dv = \sin \frac{x}{2}\, dx$, $v = -2 \cos \frac{x}{2}$;

$\int x \sin \frac{x}{2}\, dx = -2x \cos \frac{x}{2} - \int \left(-2 \cos \frac{x}{2}\right) dx = -2x \cos\left(\frac{x}{2}\right) + 4 \sin\left(\frac{x}{2}\right) + C$

3.

		$\cos t$
t^2	$\xrightarrow{(+)}$	$\sin t$
$2t$	$\xrightarrow{(-)}$	$-\cos t$
2	$\xrightarrow{(+)}$	$-\sin t$
0		

$\int t^2 \cos t\, dt = t^2 \sin t + 2t \cos t - 2 \sin t + C$

5. $u = \ln x$, $du = \frac{dx}{x}$; $dv = x\, dx$, $v = \frac{x^2}{2}$;

$\int_1^2 x \ln x\, dx = \left[\frac{x^2}{2} \ln x\right]_1^2 - \int_1^2 \frac{x^2}{2} \frac{dx}{x} = 2 \ln 2 - \left[\frac{x^2}{4}\right]_1^2 = 2 \ln 2 - \frac{3}{4} = \ln 4 - \frac{3}{4}$

7. $u = \tan^{-1} y$, $du = \frac{dy}{1+y^2}$; $dv = dy$, $v = y$;

$\int \tan^{-1} y\, dy = y \tan^{-1} y - \int \frac{y\, dy}{(1+y^2)} = y \tan^{-1} y - \frac{1}{2} \ln(1+y^2) + C = y \tan^{-1} y - \ln \sqrt{1+y^2} + C$

9. $u = x$, $du = dx$; $dv = \sec^2 x\, dx$, $v = \tan x$;

$\int x \sec^2 x\, dx = x \tan x - \int \tan x\, dx = x \tan x + \ln |\cos x| + C$

11.

		e^x
x^3	$\xrightarrow{(+)}$	e^x
$3x^2$	$\xrightarrow{(-)}$	e^x
$6x$	$\xrightarrow{(+)}$	e^x
6	$\xrightarrow{(-)}$	e^x
0		

$\int x^3 e^x\, dx = x^3 e^x - 3x^2 e^x + 6x e^x - 6e^x + C = (x^3 - 3x^2 + 6x - 6)\, e^x + C$

13.

		e^x
$x^2 - 5x$	$\xrightarrow{(+)}$	e^x
$2x - 5$	$\xrightarrow{(-)}$	e^x
2	$\xrightarrow{(+)}$	e^x
0		

$\int (x^2 - 5x) e^x\, dx = (x^2 - 5x)\, e^x - (2x - 5)e^x + 2e^x + C = x^2 e^x - 7x e^x + 7e^x + C$
$= (x^2 - 7x + 7)\, e^x + C$

15.

$$\begin{array}{ll} & e^x \\ x^5 \xrightarrow{(+)} & e^x \\ 5x^4 \xrightarrow{(-)} & e^x \\ 20x^3 \xrightarrow{(+)} & e^x \\ 60x^2 \xrightarrow{(-)} & e^x \\ 120x \xrightarrow{(+)} & e^x \\ 120 \xrightarrow{(-)} & e^x \\ 0 \end{array}$$

$$\int x^5 e^x \, dx = x^5 e^x - 5x^4 e^x + 20x^3 e^x - 60x^2 e^x + 120x e^x - 120 e^x + C$$
$$= \left(x^5 - 5x^4 + 20x^3 - 60x^2 + 120x - 120 \right) e^x + C$$

17.

$$\begin{array}{ll} & \sin 2\theta \\ \theta^2 \xrightarrow{(+)} & -\tfrac{1}{2} \cos 2\theta \\ 2\theta \xrightarrow{(-)} & -\tfrac{1}{4} \sin 2\theta \\ 2 \xrightarrow{(+)} & \tfrac{1}{8} \cos 2\theta \\ 0 \end{array}$$

$$\int_0^{\pi/2} \theta^2 \sin 2\theta \, d\theta = \left[-\tfrac{\theta^2}{2} \cos 2\theta + \tfrac{\theta}{2} \sin 2\theta + \tfrac{1}{4} \cos 2\theta \right]_0^{\pi/2}$$
$$= \left[-\tfrac{\pi^2}{8} \cdot (-1) + \tfrac{\pi}{4} \cdot 0 + \tfrac{1}{4} \cdot (-1) \right] - \left[0 + 0 + \tfrac{1}{4} \cdot 1 \right] = \tfrac{\pi^2}{8} - \tfrac{1}{2} = \tfrac{\pi^2 - 4}{8}$$

19. $u = \sec^{-1} t, \, du = \tfrac{dt}{t\sqrt{t^2-1}}$; $dv = t \, dt, \, v = \tfrac{t^2}{2}$;

$$\int_{2/\sqrt{3}}^2 t \sec^{-1} t \, dt = \left[\tfrac{t^2}{2} \sec^{-1} t \right]_{2/\sqrt{3}}^2 - \int_{2/\sqrt{3}}^2 \left(\tfrac{t^2}{2} \right) \tfrac{dt}{t\sqrt{t^2-1}} = \left(2 \cdot \tfrac{\pi}{3} - \tfrac{2}{3} \cdot \tfrac{\pi}{6} \right) - \int_{2/\sqrt{3}}^2 \tfrac{t \, dt}{2\sqrt{t^2-1}}$$
$$= \tfrac{5\pi}{9} - \left[\tfrac{1}{2} \sqrt{t^2-1} \right]_{2/\sqrt{3}}^2 = \tfrac{5\pi}{9} - \tfrac{1}{2} \left(\sqrt{3} - \sqrt{\tfrac{4}{3} - 1} \right) = \tfrac{5\pi}{9} - \tfrac{1}{2} \left(\sqrt{3} - \tfrac{\sqrt{3}}{3} \right) = \tfrac{5\pi}{9} - \tfrac{\sqrt{3}}{3} = \tfrac{5\pi - 3\sqrt{3}}{9}$$

21. $I = \int e^\theta \sin \theta \, d\theta$; $[u = \sin \theta, \, du = \cos \theta \, d\theta; \, dv = e^\theta \, d\theta, \, v = e^\theta] \Rightarrow I = e^\theta \sin \theta - \int e^\theta \cos \theta \, d\theta$;

$[u = \cos \theta, \, du = -\sin \theta \, d\theta; \, dv = e^\theta \, d\theta, \, v = e^\theta] \Rightarrow I = e^\theta \sin \theta - \left(e^\theta \cos \theta + \int e^\theta \sin \theta \, d\theta \right)$

$= e^\theta \sin \theta - e^\theta \cos \theta - I + C' \Rightarrow 2I = \left(e^\theta \sin \theta - e^\theta \cos \theta \right) + C' \Rightarrow I = \tfrac{1}{2} \left(e^\theta \sin \theta - e^\theta \cos \theta \right) + C$, where $C = \tfrac{C'}{2}$ is another arbitrary constant

23. $I = \int e^{2x} \cos 3x \, dx$; $\left[u = \cos 3x; \, du = -3 \sin 3x \, dx, \, dv = e^{2x} \, dx; \, v = \tfrac{1}{2} e^{2x} \right]$

$\Rightarrow I = \tfrac{1}{2} e^{2x} \cos 3x + \tfrac{3}{2} \int e^{2x} \sin 3x \, dx$; $\left[u = \sin 3x, \, du = 3 \cos 3x, \, dv = e^{2x} \, dx, \, v = \tfrac{1}{2} e^{2x} \right]$

$\Rightarrow I = \tfrac{1}{2} e^{2x} \cos 3x + \tfrac{3}{2} \left(\tfrac{1}{2} e^{2x} \sin 3x - \tfrac{3}{2} \int e^{2x} \cos 3x \, dx \right) = \tfrac{1}{2} e^{2x} \cos 3x + \tfrac{3}{4} e^{2x} \sin 3x - \tfrac{9}{4} I + C'$

$\Rightarrow \tfrac{13}{4} I = \tfrac{1}{2} e^{2x} \cos 3x + \tfrac{3}{4} e^{2x} \sin 3x + C' \Rightarrow \tfrac{e^{2x}}{13} \left(3 \sin 3x + 2 \cos 3x \right) + C$, where $C = \tfrac{4}{13} C'$

25. $\int e^{\sqrt{3s+9}} \, ds$; $\begin{bmatrix} 3s + 9 = x^2 \\ ds = \tfrac{2}{3} x \, dx \end{bmatrix} \rightarrow \int e^x \cdot \tfrac{2}{3} x \, dx = \tfrac{2}{3} \int x e^x \, dx$; $[u = x, \, du = dx; \, dv = e^x \, dx, \, v = e^x]$;

$\tfrac{2}{3} \int x e^x \, dx = \tfrac{2}{3} \left(x e^x - \int e^x \, dx \right) = \tfrac{2}{3} \left(x e^x - e^x \right) + C = \tfrac{2}{3} \left(\sqrt{3s+9} \, e^{\sqrt{3s+9}} - e^{\sqrt{3s+9}} \right) + C$

27. $u = x$, $du = dx$; $dv = \tan^2 x\, dx$, $v = \int \tan^2 x\, dx = \int \frac{\sin^2 x}{\cos^2 x}\, dx = \int \frac{1 - \cos^2 x}{\cos^2 x}\, dx = \int \frac{dx}{\cos^2 x} - \int dx$

$= \tan x - x; \int_0^{\pi/3} x \tan^2 x\, dx = [x(\tan x - x)]_0^{\pi/3} - \int_0^{\pi/3} (\tan x - x)\, dx = \frac{\pi}{3}\left(\sqrt{3} - \frac{\pi}{3}\right) + \left[\ln |\cos x| + \frac{x^2}{2}\right]_0^{\pi/3}$

$= \frac{\pi}{3}\left(\sqrt{3} - \frac{\pi}{3}\right) + \ln \frac{1}{2} + \frac{\pi^2}{18} = \frac{\pi\sqrt{3}}{3} - \ln 2 - \frac{\pi^2}{18}$

29. $\int \sin(\ln x)\, dx$; $\begin{bmatrix} u = \ln x \\ du = \frac{1}{x}\, dx \\ dx = e^u\, du \end{bmatrix} \rightarrow \int (\sin u)\, e^u\, du$. From Exercise 21, $\int (\sin u)\, e^u\, du = e^u \left(\frac{\sin u - \cos u}{2}\right) + C$

$= \frac{1}{2}[-x \cos(\ln x) + x \sin(\ln x)] + C$

31. (a) $u = x$, $du = dx$; $dv = \sin x\, dx$, $v = -\cos x$;

$S_1 = \int_0^\pi x \sin x\, dx = [-x \cos x]_0^\pi + \int_0^\pi \cos x\, dx = \pi + [\sin x]_0^\pi = \pi$

(b) $S_2 = -\int_\pi^{2\pi} x \sin x\, dx = -\left[[-x \cos x]_\pi^{2\pi} + \int_\pi^{2\pi} \cos x\, dx\right] = -[-3\pi + [\sin x]_\pi^{2\pi}] = 3\pi$

(c) $S_3 = \int_{2\pi}^{3\pi} x \sin x\, dx = [-x \cos x]_{2\pi}^{3\pi} + \int_{2\pi}^{3\pi} \cos x\, dx = 5\pi + [\sin x]_{2\pi}^{3\pi} = 5\pi$

(d) $S_{n+1} = (-1)^{n+1} \int_{n\pi}^{(n+1)\pi} x \sin x\, dx = (-1)^{n+1}\left[[-x \cos x]_{n\pi}^{(n+1)\pi} + [\sin x]_{n\pi}^{(n+1)\pi}\right]$

$= (-1)^{n+1}[-(n+1)\pi(-1)^n + n\pi(-1)^{n+1}] + 0 = (2n+1)\pi$

33. $av(y) = \frac{1}{2\pi} \int_0^{2\pi} 2e^{-t} \cos t\, dt$

$= \frac{1}{\pi}\left[e^{-t}\left(\frac{\sin t - \cos t}{2}\right)\right]_0^{2\pi}$

(see Exercise 22) $\Rightarrow av(y) = \frac{1}{2\pi}(1 - e^{-2\pi})$

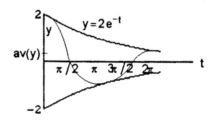

35. $I = \int x^n \cos x\, dx$; $[u = x^n, du = nx^{n-1}\, dx; dv = \cos x\, dx, v = \sin x]$

$\Rightarrow I = x^n \sin x - \int nx^{n-1} \sin x\, dx$

37. $I = \int x^n e^{ax}\, dx$; $\left[u = x^n, du = nx^{n-1}\, dx; dv = e^{ax}\, dx, v = \frac{1}{a}e^{ax}\right]$

$\Rightarrow I = \frac{x^n e^{ax}}{a} e^{ax} - \frac{n}{a} \int x^{n-1} e^{ax}\, dx$, $a \neq 0$

39. $\int \sin^{-1} x\, dx = x \sin^{-1} x - \int \sin y\, dy = x \sin^{-1} x + \cos y + C = x \sin^{-1} x + \cos(\sin^{-1} x) + C$

41. $\int \sec^{-1} x\, dx = x \sec^{-1} x - \int \sec y\, dy = x \sec^{-1} x - \ln |\sec y + \tan y| + C$

$= x \sec^{-1} x - \ln |\sec(\sec^{-1} x) + \tan(\sec^{-1} x)| + C = x \sec^{-1} x - \ln \left|x + \sqrt{x^2 - 1}\right| + C$

43. Yes, $\cos^{-1} x$ is the angle whose cosine is x which implies $\sin(\cos^{-1} x) = \sqrt{1 - x^2}$.

45. (a) $\int \sinh^{-1} x\, dx = x \sinh^{-1} x - \int \sinh y\, dy = x \sinh^{-1} x - \cosh y + C = x \sinh^{-1} x - \cosh(\sinh^{-1} x) + C$;

check: $d[x \sinh^{-1} x - \cosh(\sinh^{-1} x) + C] = \left[\sinh^{-1} x + \frac{x}{\sqrt{1 + x^2}} - \sinh(\sinh^{-1} x)\frac{1}{\sqrt{1 + x^2}}\right] dx = \sinh^{-1} x\, dx$

(b) $\int \sinh^{-1} x \, dx = x \sinh^{-1} x - \int x \left(\frac{1}{\sqrt{1+x^2}} \right) dx = x \sinh^{-1} x - \frac{1}{2} \int (1+x^2)^{-1/2} 2x \, dx$

$\qquad = x \sinh^{-1} x - (1+x^2)^{1/2} + C$

check: $d \left[x \sinh^{-1} x - (1+x^2)^{1/2} + C \right] = \left[\sinh^{-1} x + \frac{x}{\sqrt{1+x^2}} - \frac{x}{\sqrt{1+x^2}} \right] dx = \sinh^{-1} x \, dx$

5.2 TRIGONOMETRIC INTEGRALS

1. $\int \sin^5 x \, dx = \int (\sin^2 x)^2 \sin x \, dx = \int (1 - \cos^2 x)^2 \sin x \, dx = \int (1 - 2\cos^2 x + \cos^4 x) \sin x \, dx$

$\qquad = \int \sin x \, dx - \int 2\cos^2 x \sin x \, dx + \int \cos^4 x \sin x \, dx = -\cos x + 2\frac{\cos^3 x}{3} - \frac{\cos^5 x}{5} + C$

3. $\int \cos^3 x \, dx = \int (\cos^2 x) \cos x \, dx = \int (1 - \sin^2 x) \cos x \, dx = \int \cos x \, dx - \int \sin^2 x \cos x \, dx = \sin x - \frac{\sin^3 x}{3} + C$

5. $\int_0^{\pi/2} \sin^7 y \, dy = \int_0^{\pi/2} \sin^6 y \sin y \, dy = \int_0^{\pi/2} (1 - \cos^2 y)^3 \sin y \, dy = \int_0^{\pi/2} \sin y \, dy - 3 \int_0^{\pi/2} \cos^2 y \sin y \, dy$

$\qquad + 3 \int_0^{\pi/2} \cos^4 y \sin y \, dy - \int_0^{\pi/2} \cos^6 y \sin y \, dy = \left[-\cos y + 3\frac{\cos^3 y}{3} - 3\frac{\cos^5 y}{5} + \frac{\cos^7 y}{7} \right]_0^{\pi/2} = (0) - \left(-1 + 1 - \frac{3}{5} + \frac{1}{7} \right) = \frac{16}{35}$

7. $\int_0^{\pi} 8 \sin^4 x \, dx = 8 \int_0^{\pi} \left(\frac{1 - \cos 2x}{2} \right)^2 dx = 2 \int_0^{\pi} (1 - 2\cos 2x + \cos^2 2x) dx = 2 \int_0^{\pi} dx - 2 \int_0^{\pi} \cos 2x \cdot 2 dx + 2 \int_0^{\pi} \frac{1 + \cos 4x}{2} dx$

$\qquad = [2x - 2\sin 2x]_0^{\pi} + \int_0^{\pi} dx + \int_0^{\pi} \cos 4x \, dx = 2\pi + \left[x + \frac{1}{2}\sin 4x \right]_0^{\pi} = 2\pi + \pi = 3\pi$

9. $\int 16 \sin^2 x \cos^2 x \, dx = 16 \int \left(\frac{1 - \cos 2x}{2} \right) \left(\frac{1 + \cos 2x}{2} \right) dx = 4 \int (1 - \cos^2 2x) dx = 4 \int dx - 4 \int \left(\frac{1 + \cos 4x}{2} \right) dx$

$\qquad = 4x - 2 \int dx - 2 \int \cos 4x \, dx = 4x - 2x - \frac{\sin 4x}{2} + C = 2x - \frac{\sin 4x}{2} + C$

11. $\int 35 \sin^4 x \cos^3 x \, dx = \int 35 \sin^4 x (1 - \sin^2 x) \cos x \, dx = 35 \int \sin^4 x \cos x \, dx - 35 \int \sin^6 x \cos x \, dx$

$\qquad = 35\frac{\sin^5 x}{5} - 35\frac{\sin^7 x}{7} + C = 7\sin^5 x - 5\sin^7 x + C$

13. $\int 8\cos^3 2\theta \sin 2\theta \, d\theta = 8 \left(-\frac{1}{2} \right) \frac{\cos^4 2\theta}{4} + C = -\cos^4 2\theta + C$

15. $\int_0^{2\pi} \sqrt{\frac{1 - \cos x}{2}} \, dx = \int_0^{2\pi} \left| \sin \frac{x}{2} \right| dx = \int_0^{2\pi} \sin \frac{x}{2} \, dx = \left[-2\cos \frac{x}{2} \right]_0^{2\pi} = 2 + 2 = 4$

17. $\int_0^{\pi} \sqrt{1 - \sin^2 t} \, dt = \int_0^{\pi} |\cos t| \, dt = \int_0^{\pi/2} \cos t \, dt - \int_{\pi/2}^{\pi} \cos t \, dt = [\sin t]_0^{\pi/2} - [\sin t]_{\pi/2}^{\pi} = 1 - 0 - 0 + 1 = 2$

19. $\int_{-\pi/4}^{\pi/4} \sqrt{1 + \tan^2 x} \, dx = \int_{-\pi/4}^{\pi/4} |\sec x| \, dx = \int_{-\pi/4}^{\pi/4} \sec x \, dx = [\ln|\sec x + \tan x|]_{-\pi/4}^{\pi/4} = \ln\left(\sqrt{2} + 1 \right) - \ln\left(\sqrt{2} - 1 \right)$

$\qquad = \ln\left(\frac{\sqrt{2}+1}{\sqrt{2}-1} \right) = 2\ln\left(1 + \sqrt{2} \right)$

21. $\int_0^{\pi/2} \theta \sqrt{1 - \cos 2\theta} \, d\theta = \int_0^{\pi/2} \theta \sqrt{2} |\sin \theta| \, d\theta = \sqrt{2} \int_0^{\pi/2} \theta \sin \theta \, d\theta = \sqrt{2} [-\theta\cos\theta + \sin\theta]_0^{\pi/2} = \sqrt{2}(1) = \sqrt{2}$

23. $\int_{-\pi/3}^{0} 2 \sec^3 x \, dx$; $u = \sec x$, $du = \sec x \tan x \, dx$, $dv = \sec^2 x \, dx$, $v = \tan x$;

$\qquad \int_{-\pi/3}^{0} 2 \sec^3 x \, dx = [2 \sec x \tan x]_{-\pi/3}^{0} - 2 \int_{-\pi/3}^{0} \sec x \tan^2 x \, dx = 2 \cdot 1 \cdot 0 - 2 \cdot 2 \cdot \sqrt{3} - 2 \int_{-\pi/3}^{0} \sec x (\sec^2 x - 1) dx$

$\qquad = 4\sqrt{3} - 2 \int_{-\pi/3}^{0} \sec^3 x \, dx + 2 \int_{-\pi/3}^{0} \sec x \, dx$; $2 \int_{-\pi/3}^{0} 2 \sec^3 x \, dx = 4\sqrt{3} + [2\ln|\sec x + \tan x|]_{-\pi/3}^{0}$

$$2\int_{-\pi/3}^{0} 2\sec^3 x\, dx = 4\sqrt{3} + 2\ln|1+0| - 2\ln|2 - \sqrt{3}| = 4\sqrt{3} - 2\ln\left(2 - \sqrt{3}\right)$$

$$\int_{-\pi/3}^{0} 2\sec^3 x\, dx = 2\sqrt{3} - \ln\left(2 - \sqrt{3}\right)$$

25. $\int \sec^4\theta\, d\theta = \int (1 + \tan^2\theta)\sec^2\theta\, d\theta = \int \sec^2\theta\, d\theta + \int \tan^2\theta\sec^2\theta\, d\theta = \tan\theta + \frac{\tan^3\theta}{3} + C$

27. $\int_{\pi/4}^{\pi/2} \csc^4\theta\, d\theta = \int_{\pi/4}^{\pi/2} (1 + \cot^2\theta)\csc^2\theta\, d\theta = \int_{\pi/4}^{\pi/2} \csc^2\theta\, d\theta + \int_{\pi/4}^{\pi/2} \cot^2\theta\csc^2\theta\, d\theta = \left[-\cot\theta - \frac{\cot^3\theta}{3}\right]_{\pi/4}^{\pi/2}$

$= (0) - \left(-1 - \frac{1}{3}\right) = \frac{4}{3}$

29. $\int 4\tan^3 x\, dx = 4\int (\sec^2 x - 1)\tan x\, dx = 4\int \sec^2 x \tan x\, dx - 4\int \tan x\, dx = 4\frac{\tan^2 x}{2} - 4\ln|\sec x| + C$

$= 2\tan^2 x - 4\ln|\sec x| + C$

31. $\int_{\pi/6}^{\pi/3} \cot^3 x\, dx = \int_{\pi/6}^{\pi/3} (\csc^2 x - 1)\cot x\, dx = \int_{\pi/6}^{\pi/3} \csc^2 x \cot x\, dx - \int_{\pi/6}^{\pi/3} \cot x\, dx = \left[-\frac{\cot^2 x}{2} + \ln|\csc x|\right]_{\pi/6}^{\pi/3}$

$= -\frac{1}{2}\left(\frac{1}{3} - 3\right) + \left(\ln\frac{2}{\sqrt{3}} - \ln 2\right) = \frac{4}{3} - \ln\sqrt{3}$

33. $\int \sin 3x \cos 2x\, dx = \frac{1}{2}\int (\sin x + \sin 5x)\, dx = \frac{1}{2}\left(-\cos x - \frac{1}{5}\cos 5x\right) + C = -\frac{1}{2}\cos x - \frac{1}{10}\cos 5x + C$

35. $\int_{-\pi}^{\pi} \sin 3x \sin 3x\, dx = \frac{1}{2}\int_{-\pi}^{\pi} (\cos 0 - \cos 6x)\, dx = \frac{1}{2}\int_{-\pi}^{\pi} dx - \frac{1}{2}\int_{-\pi}^{\pi} \cos 6x\, dx = \frac{1}{2}\left[x - \frac{1}{12}\sin 6x\right]_{-\pi}^{\pi} = \frac{\pi}{2} + \frac{\pi}{2} - 0 = \pi$

37. $\int_{0}^{\pi} \cos 3x \cos 4x\, dx = \frac{1}{2}\int_{0}^{\pi} (\cos(-x) + \cos 7x)\, dx = \frac{1}{2}\left[-\sin(-x) + \frac{1}{7}\sin 7x\right]_{0}^{\pi} = \frac{1}{2}(0) = 0$

39. $A = \int_{0}^{\pi} \sqrt{1 + \cos 4x}\, dx = \int_{0}^{\pi} \sqrt{2}\,|\cos 2x|dx = \sqrt{2}\int_{0}^{\pi/4} \cos 2x\, dx - \sqrt{2}\int_{\pi/4}^{3\pi/4} \cos 2x\, dx + \sqrt{2}\int_{3\pi/4}^{\pi} \cos 2x\, dx$

$= \frac{\sqrt{2}}{2}[\sin 2x]_{0}^{\pi/4} - \frac{\sqrt{2}}{2}[\sin 2x]_{\pi/4}^{3\pi/4} + \frac{\sqrt{2}}{2}[\sin 2x]_{3\pi/4}^{\pi} = \frac{\sqrt{2}}{2}(1 - 0) - \frac{\sqrt{2}}{2}(-1 - 1) + \frac{\sqrt{2}}{2}(0 + 1) = \sqrt{2} + \sqrt{2} = 2\sqrt{2}$

5.3 TRIGONOMETRIC SUBSTITUTIONS

1. $y = 3\tan\theta, -\frac{\pi}{2} < \theta < \frac{\pi}{2}, dy = \frac{3\, d\theta}{\cos^2\theta}, 9 + y^2 = 9(1 + \tan^2\theta) = \frac{9}{\cos^2\theta} \Rightarrow \frac{1}{\sqrt{9+y^2}} = \frac{|\cos\theta|}{3} = \frac{\cos\theta}{3}$

(because $\cos\theta > 0$ when $-\frac{\pi}{2} < \theta < \frac{\pi}{2}$);

$\int \frac{dy}{\sqrt{9+y^2}} = 3\int \frac{\cos\theta\, d\theta}{3\cos^2\theta} = \int \frac{d\theta}{\cos\theta} = \ln|\sec\theta + \tan\theta| + C' = \ln\left|\frac{\sqrt{9+y^2}}{3} + \frac{y}{3}\right| + C' = \ln\left|\sqrt{9+y^2} + y\right| + C$

3. $\int_{-2}^{2} \frac{dx}{4+x^2} = \left[\frac{1}{2}\tan^{-1}\frac{x}{2}\right]_{-2}^{2} = \frac{1}{2}\tan^{-1} 1 - \frac{1}{2}\tan^{-1}(-1) = \left(\frac{1}{2}\right)\left(\frac{\pi}{4}\right) - \left(\frac{1}{2}\right)\left(-\frac{\pi}{4}\right) = \frac{\pi}{4}$

5. $\int_{0}^{3/2} \frac{dx}{\sqrt{9-x^2}} = \left[\sin^{-1}\frac{x}{3}\right]_{0}^{3/2} = \sin^{-1}\frac{1}{2} - \sin^{-1} 0 = \frac{\pi}{6} - 0 = \frac{\pi}{6}$

7. $t = 5\sin\theta, -\frac{\pi}{2} < \theta < \frac{\pi}{2}, dt = 5\cos\theta\, d\theta, \sqrt{25 - t^2} = 5\cos\theta$;

$\int \sqrt{25 - t^2}\, dt = \int (5\cos\theta)(5\cos\theta)\, d\theta = 25\int \cos^2\theta\, d\theta = 25\int \frac{1 + \cos 2\theta}{2}\, d\theta = 25\left(\frac{\theta}{2} + \frac{\sin 2\theta}{4}\right) + C$

$= \frac{25}{2}(\theta + \sin\theta\cos\theta) + C = \frac{25}{2}\left[\sin^{-1}\left(\frac{t}{5}\right) + \left(\frac{t}{5}\right)\left(\frac{\sqrt{25-t^2}}{5}\right)\right] + C = \frac{25}{2}\sin^{-1}\left(\frac{t}{5}\right) + \frac{t\sqrt{25-t^2}}{2} + C$

9. $x = \frac{7}{2}\sec\theta, 0 < \theta < \frac{\pi}{2}, dx = \frac{7}{2}\sec\theta\tan\theta\, d\theta, \sqrt{4x^2 - 49} = \sqrt{49\sec^2\theta - 49} = 7\tan\theta$;

$\int \frac{dx}{\sqrt{4x^2 - 49}} = \int \frac{\left(\frac{7}{2}\sec\theta\tan\theta\right) d\theta}{7\tan\theta} = \frac{1}{2}\int \sec\theta\, d\theta = \frac{1}{2}\ln|\sec\theta + \tan\theta| + C = \frac{1}{2}\ln\left|\frac{2x}{7} + \frac{\sqrt{4x^2-49}}{7}\right| + C$

11. $y = 7 \sec \theta, 0 < \theta < \frac{\pi}{2}, dy = 7 \sec \theta \tan \theta \, d\theta, \sqrt{y^2 - 49} = 7 \tan \theta$;

$\int \frac{\sqrt{y^2-49}}{y} \, dy = \int \frac{(7 \tan \theta)(7 \sec \theta \tan \theta) \, d\theta}{7 \sec \theta} = 7 \int \tan^2 \theta \, d\theta = 7 \int (\sec^2 \theta - 1) \, d\theta = 7(\tan \theta - \theta) + C$

$= 7 \left[\frac{\sqrt{y^2-49}}{7} - \sec^{-1} \left(\frac{y}{7} \right) \right] + C$

13. $x = \sec \theta, 0 < \theta < \frac{\pi}{2}, dx = \sec \theta \tan \theta \, d\theta, \sqrt{x^2 - 1} = \tan \theta$;

$\int \frac{dx}{x^2 \sqrt{x^2-1}} = \int \frac{\sec \theta \tan \theta \, d\theta}{\sec^2 \theta \tan \theta} = \int \frac{d\theta}{\sec \theta} = \sin \theta + C = \frac{\sqrt{x^2-1}}{x} + C$

15. $x = 2 \tan \theta, -\frac{\pi}{2} < \theta < \frac{\pi}{2}, dx = \frac{2 \, d\theta}{\cos^2 \theta}, \sqrt{x^2 + 4} = \frac{2}{\cos \theta}$;

$\int \frac{x^3 \, dx}{\sqrt{x^2+4}} = \int \frac{(8 \tan^3 \theta)(\cos \theta) \, d\theta}{\cos^2 \theta} = 8 \int \frac{\sin^3 \theta \, d\theta}{\cos^4 \theta} = 8 \int \frac{(\cos^2 \theta - 1)(-\sin \theta) \, d\theta}{\cos^4 \theta}$;

$[t = \cos \theta] \rightarrow 8 \int \frac{t^2-1}{t^4} \, dt = 8 \int \left(\frac{1}{t^2} - \frac{1}{t^4} \right) dt = 8 \left(-\frac{1}{t} + \frac{1}{3t^3} \right) + C = 8 \left(-\sec \theta + \frac{\sec^3 \theta}{3} \right) + C$

$= 8 \left(-\frac{\sqrt{x^2+4}}{2} + \frac{(x^2+4)^{3/2}}{8 \cdot 3} \right) + C = \frac{1}{3} (x^2 + 4)^{3/2} - 4\sqrt{x^2 + 4} + C$

17. $w = 2 \sin \theta, -\frac{\pi}{2} < \theta < \frac{\pi}{2}, dw = 2 \cos \theta \, d\theta, \sqrt{4 - w^2} = 2 \cos \theta$;

$\int \frac{8 \, dw}{w^2 \sqrt{4-w^2}} = \int \frac{8 \cdot 2 \cos \theta \, d\theta}{4 \sin^2 \theta \cdot 2 \cos \theta} = 2 \int \frac{d\theta}{\sin^2 \theta} = -2 \cot \theta + C = \frac{-2\sqrt{4-w^2}}{w} + C$

19. $x = \sin \theta, 0 \le \theta \le \frac{\pi}{3}, dx = \cos \theta \, d\theta, (1 - x^2)^{3/2} = \cos^3 \theta$;

$\int_0^{\sqrt{3}/2} \frac{4x^2 \, dx}{(1-x^2)^{3/2}} = \int_0^{\pi/3} \frac{4 \sin^2 \theta \cos \theta \, d\theta}{\cos^3 \theta} = 4 \int_0^{\pi/3} \left(\frac{1 - \cos^2 \theta}{\cos^2 \theta} \right) d\theta = 4 \int_0^{\pi/3} (\sec^2 \theta - 1) \, d\theta = 4 [\tan \theta - \theta]_0^{\pi/3} = 4\sqrt{3} - \frac{4\pi}{3}$

21. $x = \sec \theta, 0 < \theta < \frac{\pi}{2}, dx = \sec \theta \tan \theta \, d\theta, (x^2 - 1)^{3/2} = \tan^3 \theta$;

$\int \frac{dx}{(x^2-1)^{3/2}} = \int \frac{\sec \theta \tan \theta \, d\theta}{\tan^3 \theta} = \int \frac{\cos \theta \, d\theta}{\sin^2 \theta} = -\frac{1}{\sin \theta} + C = -\frac{x}{\sqrt{x^2-1}} + C$

23. $x = \sin \theta, -\frac{\pi}{2} < \theta < \frac{\pi}{2}, dx = \cos \theta \, d\theta, (1 - x^2)^{3/2} = \cos^3 \theta$;

$\int \frac{(1-x^2)^{3/2} \, dx}{x^6} = \int \frac{\cos^3 \theta \cdot \cos \theta \, d\theta}{\sin^6 \theta} = \int \cot^4 \theta \csc^2 \theta \, d\theta = -\frac{\cot^5 \theta}{5} + C = -\frac{1}{5} \left(\frac{\sqrt{1-x^2}}{x} \right)^5 + C$

25. $x = \frac{1}{2} \tan \theta, -\frac{\pi}{2} < \theta < \frac{\pi}{2}, dx = \frac{1}{2} \sec^2 \theta \, d\theta, (4x^2 + 1)^2 = \sec^4 \theta$;

$\int \frac{8 \, dx}{(4x^2+1)^2} = \int \frac{8 \left(\frac{1}{2} \sec^2 \theta \right) d\theta}{\sec^4 \theta} = 4 \int \cos^2 \theta \, d\theta = 2(\theta + \sin \theta \cos \theta) + C = 2 \tan^{-1} 2x + \frac{4x}{(4x^2+1)} + C$

27. $v = \sin \theta, -\frac{\pi}{2} < \theta < \frac{\pi}{2}, dv = \cos \theta \, d\theta, (1 - v^2)^{5/2} = \cos^5 \theta$;

$\int \frac{v^2 \, dv}{(1-v^2)^{5/2}} = \int \frac{\sin^2 \theta \cos \theta \, d\theta}{\cos^5 \theta} = \int \tan^2 \theta \sec^2 \theta \, d\theta = \frac{\tan^3 \theta}{3} + C = \frac{1}{3} \left(\frac{v}{\sqrt{1-v^2}} \right)^3 + C$

29. Let $e^t = 3 \tan \theta, t = \ln (3 \tan \theta), \tan^{-1} \left(\frac{1}{3} \right) \le \theta \le \tan^{-1} \left(\frac{4}{3} \right), dt = \frac{\sec^2 \theta}{\tan \theta} \, d\theta, \sqrt{e^{2t} + 9} = \sqrt{9 \tan^2 \theta + 9} = 3 \sec \theta$;

$\int_0^{\ln 4} \frac{e^t \, dt}{\sqrt{e^{2t}+9}} = \int_{\tan^{-1}(1/3)}^{\tan^{-1}(4/3)} \frac{3 \tan \theta \cdot \sec^2 \theta \, d\theta}{\tan \theta \cdot 3 \sec \theta} = \int_{\tan^{-1}(1/3)}^{\tan^{-1}(4/3)} \sec \theta \, d\theta = [\ln |\sec \theta + \tan \theta|]_{\tan^{-1}(1/3)}^{\tan^{-1}(4/3)}$

$= \ln \left(\frac{5}{3} + \frac{4}{3} \right) - \ln \left(\frac{\sqrt{10}}{3} + \frac{1}{3} \right) = \ln 9 - \ln \left(1 + \sqrt{10} \right)$

31. $\int_{1/12}^{1/4} \frac{2 \, dt}{\sqrt{t} + 4t\sqrt{t}}; \left[u = 2\sqrt{t}, du = \frac{1}{\sqrt{t}} \, dt \right] \rightarrow \int_{1/\sqrt{3}}^{1} \frac{2 \, du}{1+u^2}; u = \tan \theta, \frac{\pi}{6} \le \theta \le \frac{\pi}{4}, du = \sec^2 \theta \, d\theta, 1 + u^2 = \sec^2 \theta$;

$\int_{1/\sqrt{3}}^{1} \frac{2 \, du}{1+u^2} = \int_{\pi/6}^{\pi/4} \frac{2 \sec^2 \theta \, d\theta}{\sec^2 \theta} = [2\theta]_{\pi/6}^{\pi/4} = 2 \left(\frac{\pi}{4} - \frac{\pi}{6} \right) = \frac{\pi}{6}$

33. $x = \sec\theta, 0 < \theta < \frac{\pi}{2}, dx = \sec\theta\tan\theta\,d\theta, \sqrt{x^2-1} = \sqrt{\sec^2\theta-1} = \tan\theta$;

$\int \frac{dx}{x\sqrt{x^2-1}} = \int \frac{\sec\theta\tan\theta\,d\theta}{\sec\theta\tan\theta} = \theta + C = \sec^{-1} x + C$

35. $x = \sec\theta, dx = \sec\theta\tan\theta\,d\theta, \sqrt{x^2-1} = \sqrt{\sec^2\theta-1} = \tan\theta$;

$\int \frac{x\,dx}{\sqrt{x^2-1}} = \int \frac{\sec\theta\cdot\sec\theta\tan\theta\,d\theta}{\tan\theta} = \int \sec^2\theta\,d\theta = \tan\theta + C = \sqrt{x^2-1} + C$

37. $x\frac{dy}{dx} = \sqrt{x^2-4}; dy = \sqrt{x^2-4}\,\frac{dx}{x}; y = \int \frac{\sqrt{x^2-4}}{x}\,dx; \begin{bmatrix} x = 2\sec\theta, 0 < \theta < \frac{\pi}{2} \\ dx = 2\sec\theta\tan\theta\,d\theta \\ \sqrt{x^2-4} = 2\tan\theta \end{bmatrix}$

$\rightarrow y = \int \frac{(2\tan\theta)(2\sec\theta\tan\theta)\,d\theta}{2\sec\theta} = 2\int \tan^2\theta\,d\theta = 2\int (\sec^2\theta - 1)\,d\theta = 2(\tan\theta - \theta) + C$

$= 2\left[\frac{\sqrt{x^2-4}}{2} - \sec^{-1}\left(\frac{x}{2}\right)\right] + C; x = 2 \text{ and } y = 0 \Rightarrow 0 = 0 + C \Rightarrow C = 0 \Rightarrow y = 2\left[\frac{\sqrt{x^2-4}}{2} - \sec^{-1}\frac{x}{2}\right]$

39. $(x^2+4)\frac{dy}{dx} = 3, dy = \frac{3\,dx}{x^2+4}; y = 3\int \frac{dx}{x^2+4} = \frac{3}{2}\tan^{-1}\frac{x}{2} + C; x = 2 \text{ and } y = 0 \Rightarrow 0 = \frac{3}{2}\tan^{-1} 1 + C$

$\Rightarrow C = -\frac{3\pi}{8} \Rightarrow y = \frac{3}{2}\tan^{-1}\left(\frac{x}{2}\right) - \frac{3\pi}{8}$

41. $A = \int_0^3 \frac{\sqrt{9-x^2}}{3}\,dx; x = 3\sin\theta, 0 \le \theta \le \frac{\pi}{2}, dx = 3\cos\theta\,d\theta, \sqrt{9-x^2} = \sqrt{9 - 9\sin^2\theta} = 3\cos\theta$;

$A = \int_0^{\pi/2} \frac{3\cos\theta\cdot 3\cos\theta\,d\theta}{3} = 3\int_0^{\pi/2}\cos^2\theta\,d\theta = \frac{3}{2}[\theta + \sin\theta\cos\theta]_0^{\pi/2} = \frac{3\pi}{4}$

5.4 INTEGRATION OF RATIONAL FUNCTIONS BY PARTIAL FRACTIONS

1. $\frac{5x-13}{(x-3)(x-2)} = \frac{A}{x-3} + \frac{B}{x-2} \Rightarrow 5x - 13 = A(x-2) + B(x-3) = (A+B)x - (2A+3B)$

$\Rightarrow \left.\begin{array}{c} A + B = 5 \\ 2A + 3B = 13 \end{array}\right\} \Rightarrow -B = (10-13) \Rightarrow B = 3 \Rightarrow A = 2; \text{ thus, } \frac{5x-13}{(x-3)(x-2)} = \frac{2}{x-3} + \frac{3}{x-2}$

3. $\frac{x+4}{(x+1)^2} = \frac{A}{x+1} + \frac{B}{(x+1)^2} \Rightarrow x + 4 = A(x+1) + B = Ax + (A+B) \Rightarrow \left.\begin{array}{c} A = 1 \\ A + B = 4 \end{array}\right\} \Rightarrow A = 1 \text{ and } B = 3$;

thus, $\frac{x+4}{(x+1)^2} = \frac{1}{x+1} + \frac{3}{(x+1)^2}$

5. $\frac{z+1}{z^2(z-1)} = \frac{A}{z} + \frac{B}{z^2} + \frac{C}{z-1} \Rightarrow z + 1 = Az(z-1) + B(z-1) + Cz^2 \Rightarrow z + 1 = (A+C)z^2 + (-A+B)z - B$

$\Rightarrow \left.\begin{array}{c} A + C = 0 \\ -A + B = 1 \\ -B = 1 \end{array}\right\} \Rightarrow B = -1 \Rightarrow A = -2 \Rightarrow C = 2; \text{ thus, } \frac{z+1}{z^2(z-1)} = \frac{-2}{z} + \frac{-1}{z^2} + \frac{2}{z-1}$

7. $\frac{t^2+8}{t^2-5t+6} = 1 + \frac{5t+2}{t^2-5t+6}$ (after long division); $\frac{5t+2}{t^2-5t+6} = \frac{5t+2}{(t-3)(t-2)} = \frac{A}{t-3} + \frac{B}{t-2}$

$\Rightarrow 5t + 2 = A(t-2) + B(t-3) = (A+B)t + (-2A-3B) \Rightarrow \left.\begin{array}{c} A + B = 5 \\ -2A - 3B = 2 \end{array}\right\} \Rightarrow -B = (10+2) = 12$

$\Rightarrow B = -12 \Rightarrow A = 17; \text{ thus, } \frac{t^2+8}{t^2-5t+6} = 1 + \frac{17}{t-3} + \frac{-12}{t-2}$

9. $\frac{1}{1-x^2} = \frac{A}{1-x} + \frac{B}{1+x} \Rightarrow 1 = A(1+x) + B(1-x); x = 1 \Rightarrow A = \frac{1}{2}; x = -1 \Rightarrow B = \frac{1}{2}$;

$\int \frac{dx}{1-x^2} = \frac{1}{2}\int \frac{dx}{1-x} + \frac{1}{2}\int \frac{dx}{1+x} = \frac{1}{2}[\ln|1+x| - \ln|1-x|] + C$

11. $\frac{x+4}{x^2+5x-6} = \frac{A}{x+6} + \frac{B}{x-1} \Rightarrow x + 4 = A(x-1) + B(x+6); x = 1 \Rightarrow B = \frac{5}{7}; x = -6 \Rightarrow A = \frac{-2}{-7} = \frac{2}{7}$;

$\int \frac{x+4}{x^2+5x-6}\,dx = \frac{2}{7}\int \frac{dx}{x+6} + \frac{5}{7}\int \frac{dx}{x-1} = \frac{2}{7}\ln|x+6| + \frac{5}{7}\ln|x-1| + C = \frac{1}{7}\ln|(x+6)^2(x-1)^5| + C$

13. $\frac{y}{y^2-2y-3} = \frac{A}{y-3} + \frac{B}{y+1} \Rightarrow y = A(y+1) + B(y-3); y = -1 \Rightarrow B = \frac{-1}{-4} = \frac{1}{4}; y = 3 \Rightarrow A = \frac{3}{4};$

$\int_4^8 \frac{y\,dy}{y^2-2y-3} = \frac{3}{4}\int_4^8 \frac{dy}{y-3} + \frac{1}{4}\int_4^8 \frac{dy}{y+1} = \left[\frac{3}{4}\ln|y-3| + \frac{1}{4}\ln|y+1|\right]_4^8 = \left(\frac{3}{4}\ln 5 + \frac{1}{4}\ln 9\right) - \left(\frac{3}{4}\ln 1 + \frac{1}{4}\ln 5\right)$

$= \frac{1}{2}\ln 5 + \frac{1}{2}\ln 3 = \frac{\ln 15}{2}$

15. $\frac{1}{t^3+t^2-2t} = \frac{A}{t} + \frac{B}{t+2} + \frac{C}{t-1} \Rightarrow 1 = A(t+2)(t-1) + Bt(t-1) + Ct(t+2); t = 0 \Rightarrow A = -\frac{1}{2}; t = -2$

$\Rightarrow B = \frac{1}{6}; t = 1 \Rightarrow C = \frac{1}{3}; \int \frac{dt}{t^3+t^2-2t} = -\frac{1}{2}\int \frac{dt}{t} + \frac{1}{6}\int \frac{dt}{t+2} + \frac{1}{3}\int \frac{dt}{t-1}$

$= -\frac{1}{2}\ln|t| + \frac{1}{6}\ln|t+2| + \frac{1}{3}\ln|t-1| + C$

17. $\frac{x^3}{x^2+2x+1} = (x-2) + \frac{3x+2}{(x+1)^2}$ (after long division); $\frac{3x+2}{(x+1)^2} = \frac{A}{x+1} + \frac{B}{(x+1)^2} \Rightarrow 3x+2 = A(x+1) + B$

$= Ax + (A+B) \Rightarrow A = 3, A+B = 2 \Rightarrow A = 3, B = -1; \int_0^1 \frac{x^3\,dx}{x^2+2x+1}$

$= \int_0^1 (x-2)\,dx + 3\int_0^1 \frac{dx}{x+1} - \int_0^1 \frac{dx}{(x+1)^2} = \left[\frac{x^2}{2} - 2x + 3\ln|x+1| + \frac{1}{x+1}\right]_0^1$

$= \left(\frac{1}{2} - 2 + 3\ln 2 + \frac{1}{2}\right) - (1) = 3\ln 2 - 2$

19. $\frac{1}{(x^2-1)^2} = \frac{A}{x+1} + \frac{B}{x-1} + \frac{C}{(x+1)^2} + \frac{D}{(x-1)^2} \Rightarrow 1 = A(x+1)(x-1)^2 + B(x-1)(x+1)^2 + C(x-1)^2 + D(x+1)^2;$

$x = -1 \Rightarrow C = \frac{1}{4}; x = 1 \Rightarrow D = \frac{1}{4};$ coefficient of $x^3 = A + B \Rightarrow A + B = 0;$ constant $= A - B + C + D$

$\Rightarrow A - B + C + D = 1 \Rightarrow A - B = \frac{1}{2};$ thus, $A = \frac{1}{4} \Rightarrow B = -\frac{1}{4}; \int \frac{dx}{(x^2-1)^2}$

$= \frac{1}{4}\int \frac{dx}{x+1} - \frac{1}{4}\int \frac{dx}{x-1} + \frac{1}{4}\int \frac{dx}{(x+1)^2} + \frac{1}{4}\int \frac{dx}{(x-1)^2} = \frac{1}{4}\ln\left|\frac{x+1}{x-1}\right| - \frac{x}{2(x^2-1)} + C$

21. $\frac{1}{(x+1)(x^2+1)} = \frac{A}{x+1} + \frac{Bx+C}{x^2+1} \Rightarrow 1 = A(x^2+1) + (Bx+C)(x+1); x = -1 \Rightarrow A = \frac{1}{2};$ coefficient of x^2

$= A + B \Rightarrow A + B = 0 \Rightarrow B = -\frac{1}{2};$ constant $= A + C \Rightarrow A + C = 1 \Rightarrow C = \frac{1}{2}; \int_0^1 \frac{dx}{(x+1)(x^2+1)}$

$= \frac{1}{2}\int_0^1 \frac{dx}{x+1} + \frac{1}{2}\int_0^1 \frac{(-x+1)}{x^2+1}\,dx = \left[\frac{1}{2}\ln|x+1| - \frac{1}{4}\ln(x^2+1) + \frac{1}{2}\tan^{-1}x\right]_0^1$

$= \left(\frac{1}{2}\ln 2 - \frac{1}{4}\ln 2 + \frac{1}{2}\tan^{-1}1\right) - \left(\frac{1}{2}\ln 1 - \frac{1}{4}\ln 1 + \frac{1}{2}\tan^{-1}0\right) = \frac{1}{4}\ln 2 + \frac{1}{2}\left(\frac{\pi}{4}\right) = \frac{(\pi + 2\ln 2)}{8}$

23. $\frac{y^2+2y+1}{(y^2+1)^2} = \frac{Ay+B}{y^2+1} + \frac{Cy+D}{(y^2+1)^2} \Rightarrow y^2+2y+1 = (Ay+B)(y^2+1) + Cy + D$

$= Ay^3 + By^2 + (A+C)y + (B+D) \Rightarrow A = 0, B = 1; A + C = 2 \Rightarrow C = 2; B + D = 1 \Rightarrow D = 0;$

$\int \frac{y^2+2y+1}{(y^2+1)^2}\,dy = \int \frac{1}{y^2+1}\,dy + 2\int \frac{y}{(y^2+1)^2}\,dy = \tan^{-1}y - \frac{1}{y^2+1} + C$

25. $\frac{2s+2}{(s^2+1)(s-1)^3} = \frac{As+B}{s^2+1} + \frac{C}{s-1} + \frac{D}{(s-1)^2} + \frac{E}{(s-1)^3} \Rightarrow 2s+2$

$= (As+B)(s-1)^3 + C(s^2+1)(s-1)^2 + D(s^2+1)(s-1) + E(s^2+1)$

$= [As^4 + (-3A+B)s^3 + (3A-3B)s^2 + (-A+3B)s - B] + C(s^4 - 2s^3 + 2s^2 - 2s + 1) + D(s^3 - s^2 + s - 1)$

$\quad + E(s^2+1)$

$= (A+C)s^4 + (-3A+B-2C+D)s^3 + (3A-3B+2C-D+E)s^2 + (-A+3B-2C+D)s + (-B+C-D+E)$

$\Rightarrow \left.\begin{array}{l} A \quad\quad + C \quad\quad\quad\quad\quad = 0 \\ -3A + \ B - 2C + D \quad\quad = 0 \\ 3A - 3B + 2C - D + E = 0 \\ -A + 3B - 2C + D \quad\quad = 2 \\ \quad\ - B + \ C - D + E = 2 \end{array}\right\}$ summing all equations $\Rightarrow 2E = 4 \Rightarrow E = 2;$

summing eqs (2) and (3) $\Rightarrow -2B + 2 = 0 \Rightarrow B = 1;$ summing eqs (3) and (4) $\Rightarrow 2A + 2 = 2 \Rightarrow A = 0; C = 0$

from eq (1); then $-1 + 0 - D + 2 = 2$ from eq (5) $\Rightarrow D = -1;$

$\int \frac{2s+2}{(s^2+1)(s-1)^3}\,ds = \int \frac{ds}{s^2+1} - \int \frac{ds}{(s-1)^2} + 2\int \frac{ds}{(s-1)^3} = -(s-1)^{-2} + (s-1)^{-1} + \tan^{-1}s + C$

27. $\frac{2\theta^3 + 5\theta^2 + 8\theta + 4}{(\theta^2 + 2\theta + 2)^2} = \frac{A\theta + B}{\theta^2 + 2\theta + 2} + \frac{C\theta + D}{(\theta^2 + 2\theta + 2)^2} \Rightarrow 2\theta^3 + 5\theta^2 + 8\theta + 4 = (A\theta + B)(\theta^2 + 2\theta + 2) + C\theta + D$

$= A\theta^3 + (2A + B)\theta^2 + (2A + 2B + C)\theta + (2B + D) \Rightarrow A = 2; 2A + B = 5 \Rightarrow B = 1; 2A + 2B + C = 8 \Rightarrow C = 2;$

$2B + D = 4 \Rightarrow D = 2; \int \frac{2\theta^3 + 5\theta^2 + 8\theta + 4}{(\theta^2 + 2\theta + 2)^2} \, d\theta = \int \frac{2\theta + 1}{(\theta^2 + 2\theta + 2)} \, d\theta + \int \frac{2\theta + 2}{(\theta^2 + 2\theta + 2)^2} \, d\theta$

$= \int \frac{2\theta + 2}{\theta^2 + 2\theta + 2} \, d\theta - \int \frac{d\theta}{\theta^2 + 2\theta + 2} + \int \frac{d(\theta^2 + 2\theta + 2)}{(\theta^2 + 2\theta + 2)^2} = \int \frac{d(\theta^2 + 2\theta + 2)}{\theta^2 + 2\theta + 2} - \int \frac{d\theta}{(\theta + 1)^2 + 1} - \frac{1}{\theta^2 + 2\theta + 2}$

$= \frac{-1}{\theta^2 + 2\theta + 2} + \ln(\theta^2 + 2\theta + 2) - \tan^{-1}(\theta + 1) + C$

29. $\frac{2x^3 - 2x^2 + 1}{x^2 - x} = 2x + \frac{1}{x^2 - x} = 2x + \frac{1}{x(x - 1)}; \frac{1}{x(x - 1)} = \frac{A}{x} + \frac{B}{x - 1} \Rightarrow 1 = A(x - 1) + Bx; x = 0 \Rightarrow A = -1;$

$x = 1 \Rightarrow B = 1; \int \frac{2x^3 - 2x^2 + 1}{x^2 - x} = \int 2x \, dx - \int \frac{dx}{x} + \int \frac{dx}{x - 1} = x^2 - \ln|x| + \ln|x - 1| + C = x^2 + \ln\left|\frac{x - 1}{x}\right| + C$

31. $\frac{9x^3 - 3x + 1}{x^3 - x^2} = 9 + \frac{9x^2 - 3x + 1}{x^2(x - 1)}$ (after long division); $\frac{9x^2 - 3x + 1}{x^2(x - 1)} = \frac{A}{x} + \frac{B}{x^2} + \frac{C}{x - 1}$

$\Rightarrow 9x^2 - 3x + 1 = Ax(x - 1) + B(x - 1) + Cx^2; x = 1 \Rightarrow C = 7; x = 0 \Rightarrow B = -1; A + C = 9 \Rightarrow A = 2;$

$\int \frac{9x^3 - 3x + 1}{x^3 - x^2} \, dx = \int 9 \, dx + 2 \int \frac{dx}{x} - \int \frac{dx}{x^2} + 7 \int \frac{dx}{x - 1} = 9x + 2 \ln|x| + \frac{1}{x} + 7 \ln|x - 1| + C$

33. $\frac{y^4 + y^2 - 1}{y^3 + y} = y - \frac{1}{y(y^2 + 1)}; \frac{1}{y(y^2 + 1)} = \frac{A}{y} + \frac{By + C}{y^2 + 1} \Rightarrow 1 = A(y^2 + 1) + (By + C)y = (A + B)y^2 + Cy + A$

$\Rightarrow A = 1; A + B = 0 \Rightarrow B = -1; C = 0; \int \frac{y^4 + y^2 - 1}{y^3 + y} \, dy = \int y \, dy - \int \frac{dy}{y} + \int \frac{y \, dy}{y^2 + 1}$

$= \frac{y^2}{2} - \ln|y| + \frac{1}{2} \ln(1 + y^2) + C$

35. $\int \frac{e^t \, dt}{e^{2t} + 3e^t + 2} = [e^t = y] \int \frac{dy}{y^2 + 3y + 2} = \int \frac{dy}{y + 1} - \int \frac{dy}{y + 2} = \ln\left|\frac{y + 1}{y + 2}\right| + C = \ln\left(\frac{e^t + 1}{e^t + 2}\right) + C$

37. $\int \frac{\cos y \, dy}{\sin^2 y + \sin y - 6}; [\sin y = t, \cos y \, dy = dt] \rightarrow \int \frac{dy}{t^2 + t - 6} = \frac{1}{5} \int \left(\frac{1}{t - 2} - \frac{1}{t + 3}\right) dt = \frac{1}{5} \ln\left|\frac{t - 2}{t + 3}\right| + C$

$= \frac{1}{5} \ln\left|\frac{\sin y - 2}{\sin y + 3}\right| + C$

39. $\int \frac{(x - 2)^2 \tan^{-1}(2x) - 12x^3 - 3x}{(4x^2 + 1)(x - 2)^2} \, dx = \int \frac{\tan^{-1}(2x)}{4x^2 + 1} \, dx - 3 \int \frac{x}{(x - 2)^2} \, dx$

$= \frac{1}{2} \int \tan^{-1}(2x) \, d(\tan^{-1}(2x)) - 3 \int \frac{dx}{x - 2} - 6 \int \frac{dx}{(x - 2)^2} = \frac{(\tan^{-1} 2x)^2}{4} - 3 \ln|x - 2| + \frac{6}{x - 2} + C$

41. $(t^2 - 3t + 2) \frac{dx}{dt} = 1; x = \int \frac{dt}{t^2 - 3t + 2} = \int \frac{dt}{t - 2} - \int \frac{dt}{t - 1} = \ln\left|\frac{t - 2}{t - 1}\right| + C; \frac{t - 2}{t - 1} = Ce^x; t = 3$ and $x = 0$

$\Rightarrow \frac{1}{2} = C \Rightarrow \frac{t - 2}{t - 1} = \frac{1}{2} e^x \Rightarrow x = \ln\left|2\left(\frac{t - 2}{t - 1}\right)\right| = \ln|t - 2| - \ln|t - 1| + \ln 2$

43. $(t^2 + 2t) \frac{dx}{dt} = 2x + 2; \frac{1}{2} \int \frac{dx}{x + 1} = \int \frac{dt}{t^2 + 2t} \Rightarrow \frac{1}{2} \ln|x + 1| = \frac{1}{2} \int \frac{dt}{t} - \frac{1}{2} \int \frac{dt}{t + 2} \Rightarrow \ln|x + 1| = \ln\left|\frac{t}{t + 2}\right| + C;$

$t = 1$ and $x = 1 \Rightarrow \ln 2 = \ln \frac{1}{3} + C \Rightarrow C = \ln 2 + \ln 3 = \ln 6 \Rightarrow \ln|x + 1| = \ln 6 \left|\frac{t}{t + 2}\right| \Rightarrow x + 1 = \frac{6t}{t + 2}$

$\Rightarrow x = \frac{6t}{t + 2} - 1, t > 0$

45. (a) $\frac{dx}{dt} = kx(N - x) \Rightarrow \int \frac{dx}{x(N - x)} = \int k \, dt \Rightarrow \frac{1}{N} \int \frac{dx}{x} + \frac{1}{N} \int \frac{dx}{N - x} = \int k \, dt \Rightarrow \frac{1}{N} \ln\left|\frac{x}{N - x}\right| = kt + C;$

$k = \frac{1}{250}, N = 1000, t = 0$ and $x = 2 \Rightarrow \frac{1}{1000} \ln\left|\frac{2}{998}\right| = C \Rightarrow \frac{1}{1000} \ln\left|\frac{x}{1000 - x}\right| = \frac{1}{250} + \frac{1}{1000} \ln\left(\frac{1}{499}\right)$

$\Rightarrow \ln\left|\frac{499x}{1000 - x}\right| = 4t \Rightarrow \frac{499x}{1000 - x} = e^{4t} \Rightarrow 499x = e^{4t}(1000 - x) \Rightarrow (499 + e^{4t})x = 1000e^{4t} \Rightarrow x = \frac{1000e^{4t}}{499 + e^{4t}}$

(b) $x = \frac{1}{2} N = 500 \Rightarrow 500 = \frac{1000e^{4t}}{499 + e^{4t}} \Rightarrow 500 \cdot 499 + 500e^{4t} = 1000e^{4t} \Rightarrow e^{4t} = 499 \Rightarrow t = \frac{1}{4} \ln 499 \approx 1.55$ days

5.5 INTEGRAL TABLES AND COMPUTER ALGEBRA SYSTEMS

1. $\int \frac{dx}{x\sqrt{x-3}} = \frac{2}{\sqrt{3}} \tan^{-1} \sqrt{\frac{x-3}{3}} + C$

   (We used FORMULA 13(a) with $a = 1, b = 3$)

3. $\int \frac{x\,dx}{\sqrt{x-2}} = \int \frac{(x-2)\,dx}{\sqrt{x-2}} + 2 \int \frac{dx}{\sqrt{x-2}} = \int \left(\sqrt{x-2}\right)^1 dx + 2 \int \left(\sqrt{x-2}\right)^{-1} dx$

   $= \left(\frac{2}{1}\right) \frac{\left(\sqrt{x-2}\right)^3}{3} + 2 \left(\frac{2}{1}\right) \frac{\left(\sqrt{x-2}\right)^1}{1} = \sqrt{x-2} \left[\frac{2(x-2)}{3} + 4\right] + C$

   (We used FORMULA 11 with $a = 1, b = -2, n = 1$ and $a = 1, b = -2, n = -1$)

5. $\int x\sqrt{2x-3}\,dx = \frac{1}{2} \int (2x-3)\sqrt{2x-3}\,dx + \frac{3}{2} \int \sqrt{2x-3}\,dx = \frac{1}{2} \int \left(\sqrt{2x-3}\right)^3 dx + \frac{3}{2} \int \left(\sqrt{2x-3}\right)^1 dx$

   $= \left(\frac{1}{2}\right)\left(\frac{2}{2}\right) \frac{\left(\sqrt{2x-3}\right)^5}{5} + \left(\frac{3}{2}\right)\left(\frac{2}{2}\right) \frac{\left(\sqrt{2x-3}\right)^3}{3} + C = \frac{(2x-3)^{3/2}}{2}\left[\frac{2x-3}{5} + 1\right] + C = \frac{(2x-3)^{3/2}(x+1)}{5} + C$

   (We used FORMULA 11 with $a = 2, b = -3, n = 3$ and $a = 2, b = -3, n = 1$)

7. $\int \frac{\sqrt{9-4x}}{x^2}\,dx = -\frac{\sqrt{9-4x}}{x} + \frac{(-4)}{2} \int \frac{dx}{x\sqrt{9-4x}} + C$

   (We used FORMULA 14 with $a = -4, b = 9$)

   $= -\frac{\sqrt{9-4x}}{x} - 2\left(\frac{1}{\sqrt{9}}\right) \ln\left|\frac{\sqrt{9-4x}-\sqrt{9}}{\sqrt{9-4x}+\sqrt{9}}\right| + C$

   (We used FORMULA 13(b) with $a = -4, b = 9$)

   $= \frac{-\sqrt{9-4x}}{x} - \frac{2}{3} \ln\left|\frac{\sqrt{9-4x}-3}{\sqrt{9-4x}+3}\right| + C$

9. $\int x\sqrt{4x-x^2}\,dx = \int x\sqrt{2\cdot 2x-x^2}\,dx = \frac{(x+2)(2x-3\cdot2)\sqrt{2\cdot2\cdot x-x^2}}{6} + \frac{2^3}{2} \sin^{-1}\left(\frac{x-2}{2}\right) + C$

   $= \frac{(x+2)(2x-6)\sqrt{4x-x^2}}{6} + 4\sin^{-1}\left(\frac{x-2}{2}\right) + C = \frac{(x+2)(x-3)\sqrt{4x-x^2}}{3} + 4\sin^{-1}\left(\frac{x-2}{2}\right) + C$

   (We used FORMULA 51 with $a = 2$)

11. $\int \frac{dx}{x\sqrt{7+x^2}} = \int \frac{dx}{x\sqrt{\left(\sqrt{7}\right)^2 + x^2}} = -\frac{1}{\sqrt{7}} \ln\left|\frac{\sqrt{7}+\sqrt{\left(\sqrt{7}\right)^2+x^2}}{x}\right| + C = -\frac{1}{\sqrt{7}} \ln\left|\frac{\sqrt{7}+\sqrt{7+x^2}}{x}\right| + C$

   $\left(\text{We used FORMULA 26 with } a = \sqrt{7}\right)$

13. $\int \frac{\sqrt{4-x^2}}{x}\,dx = \int \frac{\sqrt{2^2-x^2}}{x}\,dx = \sqrt{2^2-x^2} - 2\ln\left|\frac{2+\sqrt{2^2-x^2}}{x}\right| + C = \sqrt{4-x^2} - 2\ln\left|\frac{2+\sqrt{4-x^2}}{x}\right| + C$

   (We used FORMULA 31 with $a = 2$)

15. $\int e^{2t} \cos 3t\,dt = \frac{e^{2t}}{2^2+3^2}(2\cos 3t + 3\sin 3t) + C = \frac{e^{2t}}{13}(2\cos 3t + 3\sin 3t) + C$

   (We used FORMULA 108 with $a = 2, b = 3$)

17. $\int x\cos^{-1} x\,dx = \int x^1 \cos^{-1} x\,dx = \frac{x^{1+1}}{1+1}\cos^{-1} x + \frac{1}{1+1}\int \frac{x^{1+1}\,dx}{\sqrt{1-x^2}} = \frac{x^2}{2}\cos^{-1} x + \frac{1}{2}\int \frac{x^2\,dx}{\sqrt{1-x^2}}$

   (We used FORMULA 100 with $a = 1, n = 1$)

   $= \frac{x^2}{2}\cos^{-1} x + \frac{1}{2}\left(\frac{1}{2}\sin^{-1} x\right) - \frac{1}{2}\left(\frac{1}{2}x\sqrt{1-x^2}\right) + C = \frac{x^2}{2}\cos^{-1} x + \frac{1}{4}\sin^{-1} x - \frac{1}{4}x\sqrt{1-x^2} + C$

   (We used FORMULA 33 with $a = 1$)

19. $\int x^2 \tan^{-1} x\, dx = \frac{x^{2+1}}{2+1} \tan^{-1} x - \frac{1}{2+1} \int \frac{x^{2+1}}{1+x^2}\, dx = \frac{x^3}{3} \tan^{-1} x - \frac{1}{3} \int \frac{x^3}{1+x^2}\, dx$

(We used FORMULA 101 with $a = 1$, $n = 2$);

$\int \frac{x^3}{1+x^2}\, dx = \int x\, dx - \int \frac{x\, dx}{1+x^2} = \frac{x^2}{2} - \frac{1}{2} \ln(1 + x^2) + C \Rightarrow \int x^2 \tan^{-1} x\, dx$

$= \frac{x^3}{3} \tan^{-1} x - \frac{x^2}{6} + \frac{1}{6} \ln(1 + x^2) + C$

21. $\int \sin 3x \cos 2x\, dx = -\frac{\cos 5x}{10} - \frac{\cos x}{2} + C$

(We used FORMULA 62(a) with $a = 3$, $b = 2$)

23. $\int 8 \sin 4t \sin \frac{t}{2}\, dt = \frac{8}{7} \sin\left(\frac{7t}{2}\right) - \frac{8}{9} \sin\left(\frac{9t}{2}\right) + C = 8 \left[\frac{\sin\left(\frac{7t}{2}\right)}{7} - \frac{\sin\left(\frac{9t}{2}\right)}{9} \right] + C$

(We used FORMULA 62(b) with $a = 4$, $b = \frac{1}{2}$)

25. $\int \cos \frac{\theta}{3} \cos \frac{\theta}{4}\, d\theta = 6 \sin\left(\frac{\theta}{12}\right) + \frac{6}{7} \sin\left(\frac{7\theta}{12}\right) + C$

(We used FORMULA 62(c) with $a = \frac{1}{3}$, $b = \frac{1}{4}$)

27. $\int \frac{x^3 + x + 1}{(x^2 + 1)^2}\, dx = \int \frac{x\, dx}{x^2 + 1} + \int \frac{dx}{(x^2 + 1)^2} = \frac{1}{2} \int \frac{d(x^2 + 1)}{x^2 + 1} + \int \frac{dx}{(x^2 + 1)^2}$

$= \frac{1}{2} \ln(x^2 + 1) + \frac{x}{2(1 + x^2)} + \frac{1}{2} \tan^{-1} x + C$

(For the second integral we used FORMULA 17 with $a = 1$)

29. $\int \sin^{-1} \sqrt{x}\, dx;\quad \begin{bmatrix} u = \sqrt{x} \\ x = u^2 \\ dx = 2u\, du \end{bmatrix} \rightarrow 2 \int u^1 \sin^{-1} u\, du = 2 \left(\frac{u^{1+1}}{1+1} \sin^{-1} u - \frac{1}{1+1} \int \frac{u^{1+1}}{\sqrt{1 - u^2}}\, du \right)$

$= u^2 \sin^{-1} u - \int \frac{u^2\, du}{\sqrt{1 - u^2}}$

(We used FORMULA 99 with $a = 1$, $n = 1$)

$= u^2 \sin^{-1} u - \left(\frac{1}{2} \sin^{-1} u - \frac{1}{2} u \sqrt{1 - u^2} \right) + C = \left(u^2 - \frac{1}{2} \right) \sin^{-1} u + \frac{1}{2} u \sqrt{1 - u^2} + C$

(We used FORMULA 33 with $a = 1$)

$= \left(x - \frac{1}{2} \right) \sin^{-1} \sqrt{x} + \frac{1}{2} \sqrt{x - x^2} + C$

31. $\int \frac{\sqrt{x}}{\sqrt{1 - x}}\, dx;\quad \begin{bmatrix} u = \sqrt{x} \\ x = u^2 \\ dx = 2u\, du \end{bmatrix} \rightarrow \int \frac{u \cdot 2u}{\sqrt{1 - u^2}}\, du = 2 \int \frac{u^2}{\sqrt{1 - u^2}}\, du = 2 \left(\frac{1}{2} \sin^{-1} u - \frac{1}{2} u \sqrt{1 - u^2} \right) + C$

$= \sin^{-1} u - u \sqrt{1 - u^2} + C$

(We used FORMULA 33 with $a = 1$)

$= \sin^{-1} \sqrt{x} - \sqrt{x} \sqrt{1 - x} + C = \sin^{-1} \sqrt{x} - \sqrt{x - x^2} + C$

33. $\int (\cot t) \sqrt{1 - \sin^2 t}\, dt = \int \frac{\sqrt{1 - \sin^2 t}\, (\cos t)\, dt}{\sin t};\quad \begin{bmatrix} u = \sin t \\ du = \cos t\, dt \end{bmatrix} \rightarrow \int \frac{\sqrt{1 - u^2}\, du}{u}$

$= \sqrt{1 - u^2} - \ln \left| \frac{1 + \sqrt{1 - u^2}}{u} \right| + C$

(We used FORMULA 31 with $a = 1$)

$= \sqrt{1 - \sin^2 t} - \ln \left| \frac{1 + \sqrt{1 - \sin^2 t}}{\sin t} \right| + C$

35. $\int \frac{dy}{y\sqrt{3+(\ln y)^2}}$; $\begin{bmatrix} u = \ln y \\ y = e^u \\ dy = e^u \, du \end{bmatrix}$ $\rightarrow$ $\int \frac{e^u \, du}{e^u\sqrt{3+u^2}} = \int \frac{du}{\sqrt{3+u^2}} = \ln\left|u + \sqrt{3+u^2}\right| + C$

$= \ln\left|\ln y + \sqrt{3+(\ln y)^2}\right| + C$

$\left(\text{We used FORMULA 20 with a} = \sqrt{3}\right)$

37. $\int \sin^5 2x \, dx = -\frac{\sin^4 2x \cos 2x}{5 \cdot 2} + \frac{5-1}{5}\int \sin^3 2x \, dx = -\frac{\sin^4 2x \cos 2x}{10} + \frac{4}{5}\left[-\frac{\sin^2 2x \cos 2x}{3 \cdot 2} + \frac{3-1}{3}\int \sin 2x \, dx\right]$

(We used FORMULA 60 with a = 2, n = 5 and a = 2, n = 3)

$= -\frac{\sin^4 2x \cos 2x}{10} - \frac{2}{15}\sin^2 2x \cos 2x + \frac{8}{15}\left(-\frac{1}{2}\right)\cos 2x + C = -\frac{\sin^4 2x \cos 2x}{10} - \frac{2\sin^2 2x \cos 2x}{15} - \frac{4\cos 2x}{15} + C$

39. $\int \sin^2 2\theta \cos^3 2\theta \, d\theta = \frac{\sin^3 2\theta \cos^2 2\theta}{2(2+3)} + \frac{3-1}{3+2}\int \sin^2 2\theta \cos 2\theta \, d\theta$

(We used FORMULA 69 with a = 2, m = 3, n = 2)

$= \frac{\sin^3 2\theta \cos^2 2\theta}{10} + \frac{2}{5}\int \sin^2 2\theta \cos 2\theta \, d\theta = \frac{\sin^3 2\theta \cos^2 2\theta}{10} + \frac{2}{5}\left[\frac{1}{2}\int \sin^2 2\theta \, d(\sin 2\theta)\right] = \frac{\sin^3 2\theta \cos^2 2\theta}{10} + \frac{\sin^3 2\theta}{15} + C$

41. $\int 4\tan^3 2x \, dx = 4\left(\frac{\tan^2 2x}{2 \cdot 2} - \int \tan 2x \, dx\right) = \tan^2 2x - 4\int \tan 2x \, dx$

(We used FORMULA 86 with n = 3, a = 2)

$= \tan^2 2x - \frac{4}{2}\ln|\sec 2x| + C = \tan^2 2x - 2\ln|\sec 2x| + C$

43. $\int 2\sec^3 \pi x \, dx = 2\left[\frac{\sec \pi x \tan \pi x}{\pi(3-1)} + \frac{3-2}{3-1}\int \sec \pi x \, dx\right]$

(We used FORMULA 92 with n = 3, a = π)

$= \frac{1}{\pi}\sec \pi x \tan \pi x + \frac{1}{\pi}\ln|\sec \pi x + \tan \pi x| + C$

(We used FORMULA 88 with a = π)

45. $\int \csc^5 x \, dx = -\frac{\csc^3 x \cot x}{5-1} + \frac{5-2}{5-1}\int \csc^3 x \, dx = -\frac{\csc^3 x \cot x}{4} + \frac{3}{4}\left(-\frac{\csc x \cot x}{3-1} + \frac{3-2}{3-1}\int \csc x \, dx\right)$

(We used FORMULA 93 with n = 5, a = 1 and n = 3, a = 1)

$= -\frac{1}{4}\csc^3 x \cot x - \frac{3}{8}\csc x \cot x - \frac{3}{8}\ln|\csc x + \cot x| + C$

(We used FORMULA 89 with a = 1)

47. $\int e^t \sec^3 (e^t - 1) \, dt$; $\begin{bmatrix} x = e^t - 1 \\ dx = e^t \, dt \end{bmatrix}$ $\rightarrow$ $\int \sec^3 x \, dx = \frac{\sec x \tan x}{3-1} + \frac{3-2}{3-1}\int \sec x \, dx$

(We used FORMULA 92 with a = 1, n = 3)

$= \frac{\sec x \tan x}{2} + \frac{1}{2}\ln|\sec x + \tan x| + C = \frac{1}{2}\left[\sec(e^t - 1)\tan(e^t - 1) + \ln|\sec(e^t - 1) + \tan(e^t - 1)|\right] + C$

49. $\int_0^1 2\sqrt{x^2 + 1} \, dx$; $[x = \tan t]$ $\rightarrow$ $2\int_0^{\pi/4} \sec t \cdot \sec^2 t \, dt = 2\int_0^{\pi/4} \sec^3 t \, dt = 2\left[\left.\frac{\sec t \cdot \tan t}{3-1}\right]_0^{\pi/4} + \frac{3-2}{3-1}\int_0^{\pi/4} \sec t \, dt\right]$

(We used FORMULA 92 with n = 3, a = 1)

$= [\sec t \cdot \tan t + \ln|\sec t + \tan t|]_0^{\pi/4} = \sqrt{2} + \ln\left(\sqrt{2} + 1\right)$

51. $\int_1^2 \frac{(r^2-1)^{3/2}}{r} \, dr$; $[r = \sec \theta]$ $\rightarrow$ $\int_0^{\pi/3} \frac{\tan^3 \theta}{\sec \theta}(\sec \theta \tan \theta) \, d\theta = \int_0^{\pi/3} \tan^4 \theta \, d\theta = \left[\frac{\tan^3 \theta}{4-1}\right]_0^{\pi/3} - \int_0^{\pi/3} \tan^2 \theta \, d\theta$

$= \left[\frac{\tan^3 \theta}{3} - \tan \theta + \theta\right]_0^{\pi/3} = \frac{3\sqrt{3}}{3} - \sqrt{3} + \frac{\pi}{3} = \frac{\pi}{3}$

(We used FORMULA 86 with a = 1, n = 4 and FORMULA 84 with a = 1)

53. The integrand $f(x) = \sqrt{x - x^2}$ is nonnegative, so the integral is maximized by integrating over the function's entire domain, which runs from $x = 0$ to $x = 1$

$$\Rightarrow \int_0^1 \sqrt{x - x^2}\, dx = \int_0^1 \sqrt{2 \cdot \tfrac{1}{2} x - x^2}\, dx = \left[\frac{\left(x - \frac{1}{2}\right)}{2} \sqrt{2 \cdot \tfrac{1}{2} x - x^2} + \frac{\left(\frac{1}{2}\right)^2}{2} \sin^{-1}\left(\frac{x - \frac{1}{2}}{\frac{1}{2}} \right) \right]_0^1$$

(We used FORMULA 48 with $a = \frac{1}{2}$)

$$= \left[\frac{\left(x - \frac{1}{2}\right)}{2} \sqrt{x - x^2} + \tfrac{1}{8} \sin^{-1}(2x - 1) \right]_0^1 = \tfrac{1}{8} \cdot \tfrac{\pi}{2} - \tfrac{1}{8}\left(-\tfrac{\pi}{2} \right) = \tfrac{\pi}{8}$$

5.6 NUMERICAL INTEGRATION

1. $\int_1^2 x\, dx$

  I. (a) For $n = 4$, $\Delta x = \frac{b-a}{n} = \frac{2-1}{4} = \frac{1}{4} \Rightarrow \frac{\Delta x}{2} = \frac{1}{8}$;

  $\sum mf(x_i) = 12 \Rightarrow T = \frac{1}{8}(12) = \frac{3}{2}$;

  $f(x) = x \Rightarrow f'(x) = 1 \Rightarrow f'' = 0 \Rightarrow M = 0$

  $\Rightarrow |E_T| = 0$

	x_i	$f(x_i)$	m	$mf(x_i)$
x_0	1	1	1	1
x_1	5/4	5/4	2	5/2
x_2	3/2	3/2	2	3
x_3	7/4	7/4	2	7/2
x_4	2	2	1	2

  (b) $\int_1^2 x\, dx = \left[\frac{x^2}{2} \right]_1^2 = 2 - \frac{1}{2} = \frac{3}{2} \Rightarrow |E_T| = \int_1^2 x\, dx - T = 0$

  (c) $\frac{|E_T|}{\text{True Value}} \times 100 = 0\%$

  II. (a) For $n = 4$, $\Delta x = \frac{b-a}{n} = \frac{2-1}{4} = \frac{1}{4} \Rightarrow \frac{\Delta x}{3} = \frac{1}{12}$;

  $\sum mf(x_i) = 18 \Rightarrow S = \frac{1}{12}(18) = \frac{3}{2}$;

  $f^{(4)}(x) = 0 \Rightarrow M = 0 \Rightarrow |E_S| = 0$

	x_i	$f(x_i)$	m	$mf(x_i)$
x_0	1	1	1	1
x_1	5/4	5/4	4	5
x_2	3/2	3/2	2	3
x_3	7/4	7/4	4	7
x_4	2	2	1	2

  (b) $\int_1^2 x\, dx = \frac{3}{2} \Rightarrow |E_S| = \int_1^2 x\, dx - S = \frac{3}{2} - \frac{3}{2} = 0$

  (c) $\frac{|E_S|}{\text{True Value}} \times 100 = 0\%$

3. $\int_{-1}^1 (x^2 + 1)\, dx$

  I. (a) For $n = 4$, $\Delta x = \frac{b-a}{n} = \frac{1-(-1)}{4} = \frac{2}{4} = \frac{1}{2} \Rightarrow \frac{\Delta x}{2} = \frac{1}{4}$;

  $\sum mf(x_i) = 11 \Rightarrow T = \frac{1}{4}(11) = 2.75$;

  $f(x) = x^2 + 1 \Rightarrow f'(x) = 2x \Rightarrow f''(x) = 2 \Rightarrow M = 2$

  $\Rightarrow |E_T| \le \frac{1-(-1)}{12}\left(\frac{1}{2}\right)^2(2) = \frac{1}{12}$ or 0.08333

	x_i	$f(x_i)$	m	$mf(x_i)$
x_0	-1	2	1	2
x_1	$-1/2$	5/4	2	5/2
x_2	0	1	2	2
x_3	1/2	5/4	2	5/2
x_4	1	2	1	2

  (b) $\int_{-1}^1 (x^2 + 1)\, dx = \left[\frac{x^3}{3} + x \right]_{-1}^1 = \left(\frac{1}{3} + 1 \right) - \left(-\frac{1}{3} - 1 \right) = \frac{8}{3} \Rightarrow E_T = \int_{-1}^1 (x^2 + 1)\, dx - T = \frac{8}{3} - \frac{11}{4} = -\frac{1}{12}$

  $\Rightarrow |E_T| = \left| -\frac{1}{12} \right| \approx 0.08333$

  (c) $\frac{|E_T|}{\text{True Value}} \times 100 = \left(\frac{\frac{1}{12}}{\frac{8}{3}} \right) \times 100 \approx 3\%$

  II. (a) For $n = 4$, $\Delta x = \frac{b-a}{n} = \frac{1-(-1)}{4} = \frac{2}{4} = \frac{1}{2} \Rightarrow \frac{\Delta x}{3} = \frac{1}{6}$;

  $\sum mf(x_i) = 16 \Rightarrow S = \frac{1}{6}(16) = \frac{8}{3} = 2.66667$;

  $f^{(3)}(x) = 0 \Rightarrow f^{(4)}(x) = 0 \Rightarrow M = 0 \Rightarrow |E_S| = 0$

	x_i	$f(x_i)$	m	$mf(x_i)$
x_0	-1	2	1	2
x_1	$-1/2$	5/4	4	5
x_2	0	1	2	2
x_3	1/2	5/4	4	5
x_4	1	2	1	2

  (b) $\int_{-1}^1 (x^2 + 1)\, dx = \left[\frac{x^3}{3} + x \right]_{-1}^1 = \frac{8}{3}$

  $\Rightarrow |E_S| = \int_{-1}^1 (x^2 + 1)\, dx - S = \frac{8}{3} - \frac{8}{3} = 0$

  (c) $\frac{|E_S|}{\text{True Value}} \times 100 = 0\%$

5. $\int_0^2 (t^3 + t)\, dt$

I. (a) For $n = 4$, $\Delta x = \frac{b-a}{n} = \frac{2-0}{4} = \frac{2}{4} = \frac{1}{2}$

$\Rightarrow \frac{\Delta x}{2} = \frac{1}{4}$; $\sum mf(t_i) = 25 \Rightarrow T = \frac{1}{4}(25) = \frac{25}{4}$;

$f(t) = t^3 + t \Rightarrow f'(t) = 3t^2 + 1 \Rightarrow f''(t) = 6t$

$\Rightarrow M = 12 = f''(2) \Rightarrow |E_T| \le \frac{2-0}{12}\left(\frac{1}{2}\right)^2(12) = \frac{1}{2}$

(b) $\int_0^2 (t^3 + t)\, dt = \left[\frac{t^4}{4} + \frac{t^2}{2}\right]_0^2 = \left(\frac{2^4}{4} + \frac{2^2}{2}\right) - 0 = 6 \Rightarrow |E_T| = \int_0^2 (t^3 + t)\, dt - T = 6 - \frac{25}{4} = -\frac{1}{4} \Rightarrow |E_T| = \frac{1}{4}$

(c) $\frac{|E_T|}{\text{True Value}} \times 100 = \frac{\left|-\frac{1}{4}\right|}{6} \times 100 \approx 4\%$

	t_i	$f(t_i)$	m	$mf(t_i)$
t_0	0	0	1	0
t_1	1/2	5/8	2	5/4
t_2	1	2	2	4
t_3	3/2	39/8	2	39/4
t_4	2	10	1	10

II. (a) For $n = 4$, $\Delta x = \frac{b-a}{n} = \frac{2-0}{4} = \frac{2}{4} = \frac{1}{2} \Rightarrow \frac{\Delta x}{3} = \frac{1}{6}$;

$\sum mf(t_i) = 36 \Rightarrow S = \frac{1}{6}(36) = 6$;

$f^{(3)}(t) = 6 \Rightarrow f^{(4)}(t) = 0 \Rightarrow M = 0 \Rightarrow |E_S| = 0$

(b) $\int_0^2 (t^3 + t)\, dt = 6 \Rightarrow |E_S| = \int_0^2 (t^3 + t)\, dt - S$

$= 6 - 6 = 0$

(c) $\frac{|E_S|}{\text{True Value}} \times 100 = 0\%$

	t_i	$f(t_i)$	m	$mf(t_i)$
t_0	0	0	1	0
t_1	1/2	5/8	4	5/2
t_2	1	2	2	4
t_3	3/2	39/8	4	39/2
t_4	2	10	1	10

7. $\int_1^2 \frac{1}{s^2}\, ds$

I. (a) For $n = 4$, $\Delta x = \frac{b-a}{n} = \frac{2-1}{4} = \frac{1}{4} \Rightarrow \frac{\Delta x}{2} = \frac{1}{8}$;

$\sum mf(s_i) = \frac{179,573}{44,100} \Rightarrow T = \frac{1}{8}\left(\frac{179,573}{44,100}\right) = \frac{179,573}{352,800}$

≈ 0.50899; $f(s) = \frac{1}{s^2} \Rightarrow f'(s) = -\frac{2}{s^3}$

$\Rightarrow f''(s) = \frac{6}{s^4} \Rightarrow M = 6 = f''(1)$

$\Rightarrow |E_T| \le \frac{2-1}{12}\left(\frac{1}{4}\right)^2(6) = \frac{1}{32} = 0.03125$

(b) $\int_1^2 \frac{1}{s^2}\, ds = \int_1^2 s^{-2}\, ds = \left[-\frac{1}{s}\right]_1^2 = -\frac{1}{2} - \left(-\frac{1}{1}\right) = \frac{1}{2} \Rightarrow E_T = \int_1^2 \frac{1}{s^2}\, ds - T = \frac{1}{2} - 0.50899 = -0.00899$

$\Rightarrow |E_T| = 0.00899$

(c) $\frac{|E_T|}{\text{True Value}} \times 100 = \frac{0.00899}{0.5} \times 100 \approx 2\%$

	s_i	$f(s_i)$	m	$mf(s_i)$
s_0	1	1	1	1
s_1	5/4	16/25	2	32/25
s_2	3/2	4/9	2	8/9
s_3	7/4	16/49	2	32/49
s_4	2	1/4	1	1/4

II. (a) For $n = 4$, $\Delta x = \frac{b-a}{n} = \frac{2-1}{4} = \frac{1}{4} \Rightarrow \frac{\Delta x}{3} = \frac{1}{12}$;

$\sum mf(s_i) = \frac{264,821}{44,100} \Rightarrow S = \frac{1}{12}\left(\frac{264,821}{44,100}\right) = \frac{264,821}{529,200}$

≈ 0.50042; $f^{(3)}(s) = -\frac{24}{s^5} \Rightarrow f^{(4)}(s) = \frac{120}{s^6}$

$\Rightarrow M = 120 \Rightarrow |E_S| \le \left|\frac{2-1}{180}\right|\left(\frac{1}{4}\right)^4(120)$

$= \frac{1}{384} \approx 0.00260$

(b) $\int_1^2 \frac{1}{s^2}\, ds = \frac{1}{2} \Rightarrow E_S = \int_1^2 \frac{1}{s^2}\, ds - S = \frac{1}{2} - 0.50042 = -0.00042 \Rightarrow |E_S| = 0.00042$

(c) $\frac{|E_S|}{\text{True Value}} \times 100 = \frac{0.0004}{0.5} \times 100 \approx 0.08\%$

	s_i	$f(s_i)$	m	$mf(s_i)$
s_0	1	1	1	1
s_1	5/4	16/25	4	64/25
s_2	3/2	4/9	2	8/9
s_3	7/4	16/49	4	64/49
s_4	2	1/4	1	1/4

9. $\int_0^\pi \sin t\, dt$

I. (a) For $n = 4$, $\Delta x = \frac{b-a}{n} = \frac{\pi-0}{4} = \frac{\pi}{4} \Rightarrow \frac{\Delta x}{2} = \frac{\pi}{8}$;

$\sum mf(t_i) = 2 + 2\sqrt{2} \approx 4.8284$

$\Rightarrow T = \frac{\pi}{8}\left(2 + 2\sqrt{2}\right) \approx 1.89612$;

$f(t) = \sin t \Rightarrow f'(t) = \cos t \Rightarrow f''(t) = -\sin t$

$\Rightarrow M = 1 \Rightarrow |E_T| \le \frac{\pi-0}{12}\left(\frac{\pi}{4}\right)^2(1) = \frac{\pi^3}{192}$

≈ 0.16149

(b) $\int_0^\pi \sin t\, dt = [-\cos t]_0^\pi = (-\cos \pi) - (-\cos 0) = 2 \Rightarrow |E_T| = \int_0^\pi \sin t\, dt - T \approx 2 - 1.89612 = 0.10388$

(c) $\frac{|E_T|}{\text{True Value}} \times 100 = \frac{0.10388}{2} \times 100 \approx 5\%$

	t_i	$f(t_i)$	m	$mf(t_i)$
t_0	0	0	1	0
t_1	$\pi/4$	$\sqrt{2}/2$	2	$\sqrt{2}$
t_2	$\pi/2$	1	2	2
t_3	$3\pi/4$	$\sqrt{2}/2$	2	$\sqrt{2}$
t_4	π	0	1	0

II. (a) For $n = 4$, $\Delta x = \frac{b-a}{n} = \frac{\pi - 0}{4} = \frac{\pi}{4} \Rightarrow \frac{\Delta x}{3} = \frac{\pi}{12}$;

$\sum mf(t_i) = 2 + 4\sqrt{2} \approx 7.6569$

$\Rightarrow S = \frac{\pi}{12}\left(2 + 4\sqrt{2}\right) \approx 2.00456$;

$f^{(3)}(t) = -\cos t \Rightarrow f^{(4)}(t) = \sin t$

$\Rightarrow M = 1 \Rightarrow |E_S| \le \frac{\pi - 0}{180}\left(\frac{\pi}{4}\right)^4 (1) \approx 0.00664$

	t_i	$f(t_i)$	m	$mf(t_i)$
t_0	0	0	1	0
t_1	$\pi/4$	$\sqrt{2}/2$	4	$2\sqrt{2}$
t_2	$\pi/2$	1	2	2
t_3	$3\pi/4$	$\sqrt{2}/2$	4	$2\sqrt{2}$
t_4	π	0	1	0

(b) $\int_0^\pi \sin t\, dt = 2 \Rightarrow E_S = \int_0^\pi \sin t\, dt - S \approx 2 - 2.00456 = -0.00456 \Rightarrow |E_S| \approx 0.00456$

(c) $\frac{|E_S|}{\text{True Value}} \times 100 = \frac{0.00456}{2} \times 100 \approx 0\%$

11. (a) $M = 0$ (see Exercise 1): Then $n = 1 \Rightarrow \Delta x = 1 \Rightarrow |E_T| = \frac{1}{12}(1)^2(0) = 0 < 10^{-4}$

(b) $M = 0$ (see Exercise 1): Then $n = 2$ (n must be even) $\Rightarrow \Delta x = \frac{1}{2} \Rightarrow |E_S| = \frac{1}{180}\left(\frac{1}{2}\right)^4(0) = 0 < 10^{-4}$

13. (a) $M = 2$ (see Exercise 3): Then $\Delta x = \frac{2}{n} \Rightarrow |E_T| \le \frac{2}{12}\left(\frac{2}{n}\right)^2(2) = \frac{4}{3n^2} < 10^{-4} \Rightarrow n^2 > \frac{4}{3}(10^4) \Rightarrow n > \sqrt{\frac{4}{3}(10^4)}$

$\Rightarrow n > 115.4$, so let $n = 116$

(b) $M = 0$ (see Exercise 3): Then $n = 2$ (n must be even) $\Rightarrow \Delta x = 1 \Rightarrow |E_S| = \frac{2}{180}(1)^4(0) = 0 < 10^{-4}$

15. (a) $M = 12$ (see Exercise 5): Then $\Delta x = \frac{2}{n} \Rightarrow |E_T| \le \frac{2}{12}\left(\frac{2}{n}\right)^2(12) = \frac{8}{n^2} < 10^{-4} \Rightarrow n^2 > 8(10^4) \Rightarrow n > \sqrt{8(10^4)}$

$\Rightarrow n > 282.8$, so let $n = 283$

(b) $M = 0$ (see Exercise 5): Then $n = 2$ (n must be even) $\Rightarrow \Delta x = 1 \Rightarrow |E_S| = \frac{2}{180}(1)^4(0) = 0 < 10^{-4}$

17. (a) $M = 6$ (see Exercise 7): Then $\Delta x = \frac{1}{n} \Rightarrow |E_T| \le \frac{1}{12}\left(\frac{1}{n}\right)^2(6) = \frac{1}{2n^2} < 10^{-4} \Rightarrow n^2 > \frac{1}{2}(10^4) \Rightarrow n > \sqrt{\frac{1}{2}(10^4)}$

$\Rightarrow n > 70.7$, so let $n = 71$

(b) $M = 120$ (see Exercise 7): Then $\Delta x = \frac{1}{n} \Rightarrow |E_S| = \frac{1}{180}\left(\frac{1}{n}\right)^4(120) = \frac{2}{3n^4} < 10^{-4} \Rightarrow n^4 > \frac{2}{3}(10^4)$

$\Rightarrow n > \sqrt[4]{\frac{2}{3}(10^4)} \Rightarrow n > 9.04$, so let $n = 10$ (n must be even)

19. (a) $f(x) = \sqrt{x+1} \Rightarrow f'(x) = \frac{1}{2}(x+1)^{-1/2} \Rightarrow f''(x) = -\frac{1}{4}(x+1)^{-3/2} = -\frac{1}{4\left(\sqrt{x+1}\right)^3} \Rightarrow M = \frac{1}{4\left(\sqrt{1}\right)^3} = \frac{1}{4}$.

Then $\Delta x = \frac{3}{n} \Rightarrow |E_T| \le \frac{3}{12}\left(\frac{3}{n}\right)^2\left(\frac{1}{4}\right) = \frac{9}{16n^2} < 10^{-4} \Rightarrow n^2 > \frac{9}{16}(10^4) \Rightarrow n > \sqrt{\frac{9}{16}(10^4)} \Rightarrow n > 75$, so let $n = 76$

(b) $f^{(3)}(x) = \frac{3}{8}(x+1)^{-5/2} \Rightarrow f^{(4)}(x) = -\frac{15}{16}(x+1)^{-7/2} = -\frac{15}{16\left(\sqrt{x+1}\right)^7} \Rightarrow M = \frac{15}{16\left(\sqrt{1}\right)^7} = \frac{15}{16}$. Then $\Delta x = \frac{3}{n}$

$\Rightarrow |E_S| \le \frac{3}{180}\left(\frac{3}{n}\right)^4\left(\frac{15}{16}\right) = \frac{3^5(15)}{16(180)n^4} < 10^{-4} \Rightarrow n^4 > \frac{3^5(15)(10^4)}{16(180)} \Rightarrow n > \sqrt[4]{\frac{3^5(15)(10^4)}{16(180)}} \Rightarrow n > 10.6$, so let

$n = 12$ (n must be even)

21. (a) $f(x) = \sin(x+1) \Rightarrow f'(x) = \cos(x+1) \Rightarrow f''(x) = -\sin(x+1) \Rightarrow M = 1$. Then $\Delta x = \frac{2}{n} \Rightarrow |E_T| \le \frac{2}{12}\left(\frac{2}{n}\right)^2(1)$

$= \frac{8}{12n^2} < 10^{-4} \Rightarrow n^2 > \frac{8(10^4)}{12} \Rightarrow n > \sqrt{\frac{8(10^4)}{12}} \Rightarrow n > 81.6$, so let $n = 82$

(b) $f^{(3)}(x) = -\cos(x+1) \Rightarrow f^{(4)}(x) = \sin(x+1) \Rightarrow M = 1$. Then $\Delta x = \frac{2}{n} \Rightarrow |E_S| \le \frac{2}{180}\left(\frac{2}{n}\right)^4(1) = \frac{32}{180n^4} < 10^{-4}$

$\Rightarrow n^4 > \frac{32(10^4)}{180} \Rightarrow n > \sqrt[4]{\frac{32(10^4)}{180}} \Rightarrow n > 6.49$, so let $n = 8$ (n must be even)

23. $\frac{5}{2}(6.0 + 2(8.2) + 2(9.1)\ldots + 2(12.7) + 13.0)(30) = 15{,}990 \text{ ft}^3$.

25. (a) $|E_S| \le \frac{b-a}{180}(\Delta x^4)M$; $n = 4 \Rightarrow \Delta x = \frac{\frac{\pi}{2} - 0}{4} = \frac{\pi}{8}$; $\left|f^{(4)}\right| \le 1 \Rightarrow M = 1 \Rightarrow |E_S| \le \frac{\left(\frac{\pi}{2} - 0\right)}{180}\left(\frac{\pi}{8}\right)^4(1) \approx 0.00021$

(b) $\Delta x = \frac{\pi}{8} \Rightarrow \frac{\Delta x}{3} = \frac{\pi}{24}$;

$\sum mf(x_i) = 10.47208705$

$\Rightarrow S = \frac{\pi}{24}(10.47208705) \approx 1.37079$

	x_i	$f(x_i)$	m	$mf(x_{1i})$
x_0	0	1	1	1
x_1	$\pi/8$	0.974495358	4	3.897981432
x_2	$\pi/4$	0.900316316	2	1.800632632
x_3	$3\pi/8$	0.784213303	4	3.136853212
x_4	$\pi/2$	0.636619772	1	0.636619772

(c) $\approx \left(\frac{0.00021}{1.37079}\right) \times 100 \approx 0.015\%$

27. $T = \frac{\Delta x}{2}(y_0 + 2y_1 + 2y_2 + 2y_3 + \ldots + 2y_{n-1} + y_n)$ where $\Delta x = \frac{b-a}{n}$ and f is continuous on [a, b]. So

$T = \frac{b-a}{n}\frac{(y_0 + y_1 + y_1 + y_2 + y_2 + \ldots + y_{n-1} + y_{n-1} + y_n)}{2} = \frac{b-a}{n}\left(\frac{f(x_0) + f(x_1)}{2} + \frac{f(x_1) + f(x_2)}{2} + \ldots + \frac{f(x_{n-1}) + f(x_n)}{2}\right).$

Since f is continuous on each interval $[x_{k-1}, x_k]$, and $\frac{f(x_{k-1}) + f(x_k)}{2}$ is always between $f(x_{k-1})$ and $f(x_k)$, there is a point c_k in

$[x_{k-1}, x_k]$ with $f(c_k) = \frac{f(x_{k-1}) + f(x_k)}{2}$; this is a consequence of the Intermediate Value Theorem. Thus our sum is

$\sum_{k=1}^{n}\left(\frac{b-a}{n}\right)f(c_k)$ which has the form $\sum_{k=1}^{n}\Delta x_k f(c_k)$ with $\Delta x_k = \frac{b-a}{n}$ for all k. This is a Riemann Sum for f on [a, b].

29. A calculator or computer numerical integrator yields $\sin^{-1} 0.6 \approx 0.643501109$.

5.7 IMPROPER INTEGRALS

1. $\int_0^\infty \frac{dx}{x^2 + 1} = \lim_{b \to \infty} \int_0^b \frac{dx}{x^2 + 1} = \lim_{b \to \infty}[\tan^{-1}x]_0^b = \lim_{b \to \infty}(\tan^{-1}b - \tan^{-1}0) = \frac{\pi}{2} - 0 = \frac{\pi}{2}$

3. $\int_0^1 \frac{dx}{\sqrt{x}} = \lim_{b \to 0^+}\int_b^1 x^{-1/2}dx = \lim_{b \to 0^+}[2x^{1/2}]_b^1 = \lim_{b \to 0^+}(2 - 2\sqrt{b}) = 2 - 0 = 2$

5. $\int_{-1}^1 \frac{dx}{x^{2/3}} = \int_{-1}^0 \frac{dx}{x^{2/3}} + \int_0^1 \frac{dx}{x^{2/3}} = \lim_{b \to 0^-}[3x^{1/3}]_{-1}^b + \lim_{c \to 0^+}[3x^{1/3}]_c^1$

$= \lim_{b \to 0^-}[3b^{1/3} - 3(-1)^{1/3}] + \lim_{c \to 0^+}[3(1)^{1/3} - 3c^{1/3}] = (0 + 3) + (3 - 0) = 6$

7. $\int_0^1 \frac{dx}{\sqrt{1 - x^2}} = \lim_{b \to 1^-}[\sin^{-1}x]_0^b = \lim_{b \to 1^-}(\sin^{-1}b - \sin^{-1}0) = \frac{\pi}{2} - 0 = \frac{\pi}{2}$

9. $\int_{-\infty}^{-2}\frac{2\,dx}{x^2 - 1} = \int_{-\infty}^{-2}\frac{dx}{x - 1} - \int_{-\infty}^{-2}\frac{dx}{x + 1} = \lim_{b \to -\infty}[\ln|x - 1|]_b^{-2} - \lim_{b \to -\infty}[\ln|x + 1|]_b^{-2} = \lim_{b \to -\infty}\left[\ln\left|\frac{x - 1}{x + 1}\right|\right]_b^{-2}$

$= \lim_{b \to -\infty}\left(\ln\left|\frac{-3}{-1}\right| - \ln\left|\frac{b - 1}{b + 1}\right|\right) = \ln 3 - \ln\left(\lim_{b \to -\infty}\frac{b - 1}{b + 1}\right) = \ln 3 - \ln 1 = \ln 3$

11. $\int_2^\infty \frac{2\,dv}{v^2 - v} = \lim_{b \to \infty}\left[2\ln\left|\frac{v - 1}{v}\right|\right]_2^b = \lim_{b \to \infty}\left(2\ln\left|\frac{b - 1}{b}\right| - 2\ln\left|\frac{2 - 1}{2}\right|\right) = 2\ln(1) - 2\ln\left(\frac{1}{2}\right) = 0 + 2\ln 2 = \ln 4$

13. $\int_{-\infty}^\infty \frac{2x\,dx}{(x^2 + 1)^2} = \int_{-\infty}^0 \frac{2x\,dx}{(x^2 + 1)^2} + \int_0^\infty \frac{2x\,dx}{(x^2 + 1)^2}; \begin{bmatrix} u = x^2 + 1 \\ du = 2x\,dx \end{bmatrix} \to \int_\infty^1 \frac{du}{u^2} + \int_1^\infty \frac{du}{u^2} = \lim_{b \to \infty}\left[-\frac{1}{u}\right]_b^1 + \lim_{c \to \infty}\left[-\frac{1}{u}\right]_1^c$

$= \lim_{b \to \infty}\left(-1 + \frac{1}{b}\right) + \lim_{c \to \infty}\left[-\frac{1}{c} - (-1)\right] = (-1 + 0) + (0 + 1) = 0$

15. $\int_0^1 \frac{\theta + 1}{\sqrt{\theta^2 + 2\theta}}d\theta; \begin{bmatrix} u = \theta^2 + 2\theta \\ du = 2(\theta + 1)d\theta \end{bmatrix} \to \int_0^3 \frac{du}{2\sqrt{u}} = \lim_{b \to 0^+}\int_b^3 \frac{du}{2\sqrt{u}} = \lim_{b \to 0^+}[\sqrt{u}]_b^3 = \lim_{b \to 0^+}(\sqrt{3} - \sqrt{b}) = \sqrt{3} - 0$

$= \sqrt{3}$

17. $\int_0^\infty \frac{dx}{(1+x)\sqrt{x}}$; $\begin{bmatrix} u = \sqrt{x} \\ du = \frac{dx}{2\sqrt{x}} \end{bmatrix}$ $\rightarrow \int_0^\infty \frac{2\,du}{u^2+1} = \lim_{b \to \infty} \int_0^b \frac{2\,du}{u^2+1} = \lim_{b \to \infty} \left[2\tan^{-1}u\right]_0^b = \lim_{b \to \infty} \left(2\tan^{-1}b - 2\tan^{-1}0\right)$

$= 2\left(\frac{\pi}{2}\right) - 2(0) = \pi$

19. $\int_0^\infty \frac{dv}{(1+v^2)\left(1+\tan^{-1}v\right)} = \lim_{b \to \infty} \left[\ln\left|1+\tan^{-1}v\right|\right]_0^b = \lim_{b \to \infty} \left[\ln\left|1+\tan^{-1}b\right|\right] - \ln\left|1+\tan^{-1}0\right| = \ln\left(1+\frac{\pi}{2}\right) - \ln(1+0)$

$= \ln\left(1+\frac{\pi}{2}\right)$

21. $\int_{-\infty}^0 \theta e^\theta \, d\theta = \lim_{b \to -\infty} \left[\theta e^\theta - e^\theta\right]_b^0 = (0 \cdot e^0 - e^0) - \lim_{b \to -\infty} \left[be^b - e^b\right] = -1 - \lim_{b \to -\infty} \left(\frac{b-1}{e^{-b}}\right)$

$= -1 - \lim_{b \to -\infty} \left(\frac{1}{-e^{-b}}\right)$ (l'Hôpital's rule for $\frac{\infty}{\infty}$ form)

$= -1 - 0 = -1$

23. $\int_{-\infty}^0 e^{-|x|} \, dx = \int_{-\infty}^0 e^x \, dx = \lim_{b \to -\infty} \left[e^x\right]_b^0 = \lim_{b \to -\infty} \left(1 - e^b\right) = (1-0) = 1$

25. $\int_0^1 x \ln x \, dx = \lim_{b \to 0^+} \left[\frac{x^2}{2}\ln x - \frac{x^2}{4}\right]_b^1 = \left(\frac{1}{2}\ln 1 - \frac{1}{4}\right) - \lim_{b \to 0^+} \left(\frac{b^2}{2}\ln b - \frac{b^2}{4}\right) = -\frac{1}{4} - \lim_{b \to 0^+} \frac{\ln b}{\left(\frac{2}{b^2}\right)} + 0$

$= -\frac{1}{4} - \lim_{b \to 0^+} \frac{\left(\frac{1}{b}\right)}{\left(-\frac{4}{b^3}\right)} = -\frac{1}{4} + \lim_{b \to 0^+} \left(\frac{b^2}{4}\right) = -\frac{1}{4} + 0 = -\frac{1}{4}$

27. $\int_0^2 \frac{ds}{\sqrt{4-s^2}} = \lim_{b \to 2^-} \left[\sin^{-1}\frac{s}{2}\right]_0^b = \lim_{b \to 2^-} \left(\sin^{-1}\frac{b}{2}\right) - \sin^{-1}0 = \frac{\pi}{2} - 0 = \frac{\pi}{2}$

29. $\int_1^2 \frac{ds}{s\sqrt{s^2-1}} = \lim_{b \to 1^+} \left[\sec^{-1}s\right]_b^2 = \sec^{-1}2 - \lim_{b \to 1^+} \sec^{-1}b = \frac{\pi}{3} - 0 = \frac{\pi}{3}$

31. $\int_{-1}^4 \frac{dx}{\sqrt{|x|}} = \lim_{b \to 0^-} \int_{-1}^b \frac{dx}{\sqrt{-x}} + \lim_{c \to 0^+} \int_c^4 \frac{dx}{\sqrt{x}} = \lim_{b \to 0^-} \left[-2\sqrt{-x}\right]_{-1}^b + \lim_{c \to 0^+} \left[2\sqrt{x}\right]_c^4$

$= \lim_{b \to 0^-} \left(-2\sqrt{-b}\right) - \left(-2\sqrt{-(-1)}\right) + 2\sqrt{4} - \lim_{c \to 0^+} 2\sqrt{c} = 0 + 2 + 2 \cdot 2 - 0 = 6$

33. $\int_{-1}^\infty \frac{d\theta}{\theta^2 + 5\theta + 6} = \lim_{b \to \infty} \left[\ln\left|\frac{\theta+2}{\theta+3}\right|\right]_{-1}^b = \lim_{b \to \infty} \left[\ln\left|\frac{b+2}{b+3}\right|\right] - \ln\left|\frac{-1+2}{-1+3}\right| = 0 - \ln\left(\frac{1}{2}\right) = \ln 2$

35. $\int_0^{\pi/2} \tan\theta \, d\theta = \lim_{b \to \frac{\pi}{2}^-} \left[-\ln|\cos\theta|\right]_0^b = \lim_{b \to \frac{\pi}{2}^-} \left[-\ln|\cos b|\right] + \ln 1 = \lim_{b \to \frac{\pi}{2}^-} \left[-\ln|\cos b|\right] = +\infty$, the integral diverges

37. $\int_0^\pi \frac{\sin\theta \, d\theta}{\sqrt{\pi - \theta}}$; $[\pi - \theta = x] \rightarrow -\int_\pi^0 \frac{\sin x \, dx}{\sqrt{x}} = \int_0^\pi \frac{\sin x \, dx}{\sqrt{x}}$. Since $0 \le \frac{\sin x}{\sqrt{x}} \le \frac{1}{\sqrt{x}}$ for all $0 \le x \le \pi$ and $\int_0^\pi \frac{dx}{\sqrt{x}}$ converges, then

$\int_0^\pi \frac{\sin x}{\sqrt{x}} \, dx$ converges by the Direct Comparison Test.

39. $\int_0^{\ln 2} x^{-2} e^{-1/x} \, dx$; $\left[\frac{1}{x} = y\right] \rightarrow \int_\infty^{1/\ln 2} \frac{y^2 e^{-y} \, dy}{-y^2} = \int_{1/\ln 2}^\infty e^{-y} \, dy = \lim_{b \to \infty} \left[-e^{-y}\right]_{1/\ln 2}^b = \lim_{b \to \infty} \left[-e^{-b}\right] - \left[-e^{-1/\ln 2}\right]$

$= 0 + e^{-1/\ln 2} = e^{-1/\ln 2}$, so the integral converges.

41. $\int_0^\pi \frac{dt}{\sqrt{t + \sin t}}$. Since for $0 \le t \le \pi, 0 \le \frac{1}{\sqrt{t+\sin t}} \le \frac{1}{\sqrt{t}}$ and $\int_0^\pi \frac{dt}{\sqrt{t}}$ converges, then the original integral converges as well by the Direct Comparison Test.

43. $\int_0^2 \frac{dx}{1-x^2} = \int_0^1 \frac{dx}{1-x^2} + \int_1^2 \frac{dx}{1-x^2}$ and $\int_0^1 \frac{dx}{1-x^2} = \lim_{b \to 1^-} \left[\frac{1}{2} \ln \left|\frac{1+x}{1-x}\right|\right]_0^b = \lim_{b \to 1^-} \left[\frac{1}{2} \ln \left|\frac{1+b}{1-b}\right|\right] - 0 = \infty$, which diverges

$\Rightarrow \int_0^2 \frac{dx}{1-x^2}$ diverges as well.

45. $\int_{-1}^1 \ln |x| \, dx = \int_{-1}^0 \ln(-x) \, dx + \int_0^1 \ln x \, dx; \int_0^1 \ln x \, dx = \lim_{b \to 0^+} [x \ln x - x]_b^1 = [1 \cdot 0 - 1] - \lim_{b \to 0^+} [b \ln b - b]$

$= -1 - 0 = -1; \int_{-1}^0 \ln(-x) \, dx = -1 \Rightarrow \int_{-1}^1 \ln |x| \, dx = -2$ converges.

47. $\int_1^\infty \frac{dx}{1+x^3}; 0 \le \frac{1}{x^3+1} \le \frac{1}{x^3}$ for $1 \le x < \infty$ and $\int_1^\infty \frac{dx}{x^3}$ converges $\Rightarrow \int_1^\infty \frac{dx}{1+x^3}$ converges by the Direct Comparison Test.

49. $\int_2^\infty \frac{dv}{\sqrt{v-1}}; \lim_{v \to \infty} \frac{\left(\frac{1}{\sqrt{v-1}}\right)}{\left(\frac{1}{\sqrt{v}}\right)} = \lim_{v \to \infty} \frac{\sqrt{v}}{\sqrt{v-1}} = \lim_{v \to \infty} \frac{1}{\sqrt{1-\frac{1}{v}}} = \frac{1}{\sqrt{1-0}} = 1$ and $\int_2^\infty \frac{dv}{\sqrt{v}} = \lim_{b \to \infty} [2\sqrt{v}]_2^b = \infty$,

which diverges $\Rightarrow \int_2^\infty \frac{dv}{\sqrt{v-1}}$ diverges by the Limit Comparison Test.

51. $\int_0^\infty \frac{dx}{\sqrt{x^6+1}} = \int_0^1 \frac{dx}{\sqrt{x^6+1}} + \int_1^\infty \frac{dx}{\sqrt{x^6+1}} < \int_0^1 \frac{dx}{\sqrt{x^6+1}} + \int_1^\infty \frac{dx}{x^3}$ and $\int_1^\infty \frac{dx}{x^3} = \lim_{b \to \infty} \left[-\frac{1}{2x^2}\right]_1^b$

$= \lim_{b \to \infty} \left(-\frac{1}{2b^2} + \frac{1}{2}\right) = \frac{1}{2} \Rightarrow \int_0^\infty \frac{dx}{\sqrt{x^6+1}}$ converges by the Direct Comparison Test.

53. $\int_1^\infty \frac{\sqrt{x+1}}{x^2} \, dx; \lim_{x \to \infty} \frac{\left(\frac{\sqrt{x}}{x^2}\right)}{\left(\frac{\sqrt{x+1}}{x^2}\right)} = \lim_{x \to \infty} \frac{\sqrt{x}}{\sqrt{x+1}} = \lim_{x \to \infty} \frac{1}{\sqrt{1+\frac{1}{x}}} = 1; \int_1^\infty \frac{\sqrt{x}}{x^2} \, dx = \int_1^\infty \frac{dx}{x^{3/2}}$

$= \lim_{b \to \infty} [-2x^{-1/2}]_1^b = \lim_{b \to \infty} \left(\frac{-2}{\sqrt{b}} + 2\right) = 2 \Rightarrow \int_1^\infty \frac{\sqrt{x+1}}{x^2} \, dx$ converges by the Limit Comparison Test.

55. $\int_\pi^\infty \frac{2+\cos x}{x} \, dx; 0 < \frac{1}{x} \le \frac{2+\cos x}{x}$ for $x \ge \pi$ and $\int_\pi^\infty \frac{dx}{x} = \lim_{b \to \infty} [\ln x]_\pi^b = \infty$, which diverges

$\Rightarrow \int_\pi^\infty \frac{2+\cos x}{x} \, dx$ diverges by the Direct Comparison Test.

57. $\int_4^\infty \frac{2 \, dt}{t^{3/2}-1}; \lim_{t \to \infty} \frac{t^{3/2}}{t^{3/2}-1} = 1$ and $\int_4^\infty \frac{2 \, dt}{t^{3/2}} = \lim_{b \to \infty} [-4t^{-1/2}]_4^b = \lim_{b \to \infty} \left(\frac{-4}{\sqrt{b}} + 2\right) = 2 \Rightarrow \int_4^\infty \frac{2 \, dt}{t^{3/2}}$ converges

$\Rightarrow \int_4^\infty \frac{2 \, dt}{t^{3/2}+1}$ converges by the Limit Comparison Test.

59. $\int_1^\infty \frac{e^x}{x} \, dx; 0 < \frac{1}{x} < \frac{e^x}{x}$ for $x > 1$ and $\int_1^\infty \frac{dx}{x}$ diverges $\Rightarrow \int_1^\infty \frac{e^x \, dx}{x}$ diverges by the Direct Comparison Test.

61. $\int_1^\infty \frac{dx}{\sqrt{e^x-x}}; \lim_{x \to \infty} \frac{\left(\frac{1}{\sqrt{e^x-x}}\right)}{\left(\frac{1}{\sqrt{e^x}}\right)} = \lim_{x \to \infty} \frac{\sqrt{e^x}}{\sqrt{e^x-x}} = \lim_{x \to \infty} \frac{1}{\sqrt{1-\frac{x}{e^x}}} = \frac{1}{\sqrt{1-0}} = 1; \int_1^\infty \frac{dx}{\sqrt{e^x}} = \int_1^\infty e^{-x/2} \, dx$

$= \lim_{b \to \infty} [-2e^{-x/2}]_1^b = \lim_{b \to \infty} (-2e^{-b/2} + 2e^{-1/2}) = \frac{2}{\sqrt{e}} \Rightarrow \int_1^\infty e^{-x/2} \, dx$ converges $\Rightarrow \int_1^\infty \frac{dx}{\sqrt{e^x-x}}$ converges

by the Limit Comparison Test.

63. $\int_{-\infty}^\infty \frac{dx}{\sqrt{x^4+1}} = 2 \int_0^\infty \frac{dx}{\sqrt{x^4+1}}; \int_0^\infty \frac{dx}{\sqrt{x^4+1}} = \int_0^1 \frac{dx}{\sqrt{x^4+1}} + \int_1^\infty \frac{dx}{\sqrt{x^4+1}} < \int_0^1 \frac{dx}{\sqrt{x^4+1}} + \int_1^\infty \frac{dx}{x^2}$ and

$\int_1^\infty \frac{dx}{x^2} = \lim_{b \to \infty} \left[-\frac{1}{x}\right]_1^b = \lim_{b \to \infty} \left(-\frac{1}{b} + 1\right) = 1 \Rightarrow \int_{-\infty}^\infty \frac{dx}{\sqrt{x^4+1}}$ converges by the Direct Comparison Test.

65. (a) $\int_1^2 \frac{dx}{x(\ln x)^p}; [t = \ln x] \to \int_0^{\ln 2} \frac{dt}{t^p} = \lim_{b \to 0^+} \left[\frac{1}{-p+1} t^{1-p}\right]_b^{\ln 2} = \lim_{b \to 0^+} \frac{b^{1-p}}{p-1} + \frac{1}{1-p} (\ln 2)^{1-p}$

$\Rightarrow$ the integral converges for $p < 1$ and diverges for $p \ge 1$

(b) $\int_2^{\infty} \frac{dx}{x(\ln x)^p}$; $[t = \ln x] \rightarrow \int_{\ln 2}^{\infty} \frac{dt}{t^p}$ and this integral is essentially the same as in Exercise 65(a): it converges
for $p > 1$ and diverges for $p \le 1$

67. $A = \int_0^{\infty} e^{-x} \, dx = \lim_{b \to \infty} [-e^{-x}]_0^b = \lim_{b \to \infty} (-e^{-b}) - (-e^{-0})$

$= 0 + 1 = 1$

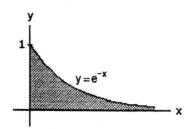

69. $\int_3^{\infty} e^{-3x} \, dx = \lim_{b \to \infty} \left[-\frac{1}{3} e^{-3x}\right]_3^b = \lim_{b \to \infty} \left(-\frac{1}{3} e^{-3b}\right) - \left(-\frac{1}{3} e^{-3 \cdot 3}\right) = 0 + \frac{1}{3} \cdot e^{-9} = \frac{1}{3} e^{-9}$

$\approx 0.0000411 < 0.000042$. Since $e^{-x^2} \le e^{-3x}$ for $x > 3$, then $\int_3^{\infty} e^{-x^2} \, dx < 0.000042$ and therefore

$\int_0^{\infty} e^{-x^2} \, dx$ can be replaced by $\int_0^3 e^{-x^2} \, dx$ without introducing an error greater than 0.000042.

71. (a)

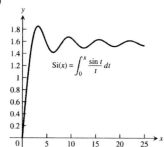

(b) $> \text{int}((\sin(t))/t, t=0..\text{infinity})$; $\left(\text{answer is } \frac{\pi}{2}\right)$

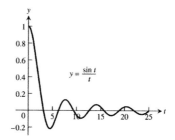

73. (a) $f(x) = \frac{1}{\sqrt{2\pi}} e^{-x^2/2}$

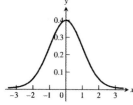

f is increasing on $(-\infty, 0]$. f is decreasing on $[0, \infty)$. f has a local maximum at $(0, f(0)) = \left(0, \frac{1}{\sqrt{2\pi}}\right)$

(b) Maple commands:

$>f: = \exp(-x^2/2)(\text{sqrt}(2*\text{pi}))$;

$>\text{int}(f, x = -1..1)$; ≈ 0.683

$>\text{int}(f, x = -2..2)$; ≈ 0.954

$>\text{int}(f, x = -3..3)$; ≈ 0.997

(c) Part (b) suggests that as n increases, the integral approaches 1. We can take $\int_{-n}^n f(x) \, dx$ as close to 1 as we want by
choosing $n > 1$ large enough. Also, we can make $\int_n^{\infty} f(x) \, dx$ and $\int_{-\infty}^{-n} f(x) \, dx$ as small as we want by choosing n large
enough. This is because $0 < f(x) < e^{-x/2}$ for $x > 1$. (Likewise, $0 < f(x) < e^{x/2}$ for $x < -1$.)

Thus, $\int_n^{\infty} f(x) \, dx < \int_n^{\infty} e^{-x/2} dx$.

$\int_n^{\infty} e^{-x/2} dx = \lim_{c \to \infty} \int_n^c e^{-x/2} dx = \lim_{c \to \infty} \left[-2e^{-x/2}\right]_n^c = \lim_{c \to \infty} \left[-2e^{-c/2} + 2e^{-n/2}\right] = 2e^{-n/2}$

As $n \to \infty$, $2e^{-n/2} \to 0$, for large enough n, $\int_n^\infty f(x)\, dx$ is as small as we want. Likewise for large enough n, $\int_{-\infty}^{-n} f(x)\, dx$ is as small as we want.

CHAPTER 5 PRACTICE AND ADDITIONAL EXERCISES

1. $u = \ln(x+1)$, $du = \frac{dx}{x+1}$; $dv = dx$, $v = x$;

$\int \ln(x+1)\, dx = x\ln(x+1) - \int \frac{x}{x+1}\, dx = x\ln(x+1) - \int dx + \int \frac{dx}{x+1} = x\ln(x+1) - x + \ln(x+1) + C_1$

$= (x+1)\ln(x+1) - x + C_1 = (x+1)\ln(x+1) - (x+1) + C$, where $C = C_1 + 1$

3. $u = \tan^{-1} 3x$, $du = \frac{3\, dx}{1+9x^2}$; $dv = dx$, $v = x$;

$\int \tan^{-1} 3x\, dx = x\tan^{-1} 3x - \int \frac{3x\, dx}{1+9x^2}$; $\begin{bmatrix} y = 1+9x^2 \\ dy = 18x\, dx \end{bmatrix} \to x\tan^{-1} 3x - \frac{1}{6}\int \frac{dy}{y}$

$= x\tan^{-1}(3x) - \frac{1}{6}\ln(1+9x^2) + C$

5.

$(x+1)^2 \xrightarrow{\ \ (+)\ \ } e^x$

$2(x+1) \xrightarrow{\ \ (-)\ \ } e^x$

$2 \xrightarrow{\ \ (+)\ \ } e^x$

$0 \qquad\qquad\qquad \Rightarrow \int (x+1)^2 e^x\, dx = \left[(x+1)^2 - 2(x+1) + 2\right] e^x + C$

7. $u = \cos 2x$, $du = -2\sin 2x\, dx$; $dv = e^x\, dx$, $v = e^x$;

$I = \int e^x \cos 2x\, dx = e^x \cos 2x + 2\int e^x \sin 2x\, dx$;

$u = \sin 2x$, $du = 2\cos 2x\, dx$; $dv = e^x\, dx$, $v = e^x$;

$I = e^x \cos 2x + 2\left[e^x \sin 2x - 2\int e^x \cos 2x\, dx \right] = e^x \cos 2x + 2e^x \sin 2x - 4I \Rightarrow I = \frac{e^x \cos 2x}{5} + \frac{2e^x \sin 2x}{5} + C$

9. $\int \frac{x\, dx}{x^2 - 3x + 2} = \int \frac{2\, dx}{x-2} - \int \frac{dx}{x-1} = 2\ln|x-2| - \ln|x-1| + C$

11. $\int \frac{dx}{x(x+1)^2} = \int \left(\frac{1}{x} - \frac{1}{x+1} + \frac{-1}{(x+1)^2} \right) dx = \ln|x| - \ln|x+1| + \frac{1}{x+1} + C$

13. $\int \frac{\sin\theta\, d\theta}{\cos^2\theta + \cos\theta - 2}$; $[\cos\theta = y] \to -\int \frac{dy}{y^2 + y - 2} = -\frac{1}{3}\int \frac{dy}{y-1} + \frac{1}{3}\int \frac{dy}{y+2} = \frac{1}{3}\ln\left| \frac{y+2}{y-1} \right| + C$

$= \frac{1}{3}\ln\left| \frac{\cos\theta + 2}{\cos\theta - 1} \right| + C = -\frac{1}{3}\ln\left| \frac{\cos\theta - 1}{\cos\theta + 2} \right| + C$

15. $\int \frac{3x^2 + 4x + 4}{x^3 + x}\, dx = \int \frac{4}{x}\, dx - \int \frac{x-4}{x^2+1}\, dx = 4\ln|x| - \frac{1}{2}\ln(x^2 + 1) + 4\tan^{-1} x + C$

17. $\int \frac{(v+3)\, dv}{2v^3 - 8v} = \frac{1}{2}\int \left(-\frac{3}{4v} + \frac{5}{8(v-2)} + \frac{1}{8(v+2)} \right) dv = -\frac{3}{8}\ln|v| + \frac{5}{16}\ln|v-2| + \frac{1}{16}\ln|v+2| + C = \frac{1}{16}\ln\left| \frac{(v-2)^5(v+2)}{v^6} \right| + C$

19. $\int \frac{dt}{t^4 + 4t^2 + 3} = \frac{1}{2}\int \frac{dt}{t^2 + 1} - \frac{1}{2}\int \frac{dt}{t^2 + 3} = \frac{1}{2}\tan^{-1} t - \frac{1}{2\sqrt{3}}\tan^{-1}\left(\frac{t}{\sqrt{3}} \right) + C = \frac{1}{2}\tan^{-1} t - \frac{\sqrt{3}}{6}\tan^{-1}\frac{t}{\sqrt{3}} + C$

21. $\int \frac{x^3 + x^2}{x^2 + x - 2}\, dx = \int \left(x + \frac{2x}{x^2 + x - 2} \right) dx = \int x\, dx + \frac{2}{3}\int \frac{dx}{x-1} + \frac{4}{3}\int \frac{dx}{x+2} = \frac{x^2}{2} + \frac{4}{3}\ln|x+2| + \frac{2}{3}\ln|x-1| + C$

23. $\int \frac{x^3 + 4x^2}{x^2 + 4x + 3}\, dx = \int \left(x - \frac{3x}{x^2 + 4x + 3} \right) dx = \int x\, dx + \frac{3}{2}\int \frac{dx}{x+1} - \frac{9}{2}\int \frac{dx}{x+3} = \frac{x^2}{2} - \frac{9}{2}\ln|x+3| + \frac{3}{2}\ln|x+1| + C$

25. $\int \frac{dx}{x\,(3\sqrt{x+1})}$; $\begin{bmatrix} u = \sqrt{x+1} \\ du = \frac{dx}{2\sqrt{x+1}} \\ dx = 2u\,du \end{bmatrix} \rightarrow \frac{2}{3}\int \frac{u\,du}{(u^2-1)\,u} = \frac{1}{3}\int \frac{du}{u-1} - \frac{1}{3}\int \frac{du}{u+1} = \frac{1}{3}\ln|u-1| - \frac{1}{3}\ln|u+1| + C$

$= \frac{1}{3}\ln\left|\frac{\sqrt{x+1}-1}{\sqrt{x+1}+1}\right| + C$

27. $\int \frac{ds}{e^s-1}$; $\begin{bmatrix} u = e^s - 1 \\ du = e^s\,ds \\ ds = \frac{du}{u+1} \end{bmatrix} \rightarrow \int \frac{du}{u(u+1)} = -\int \frac{du}{u+1} + \int \frac{du}{u} = \ln\left|\frac{u}{u+1}\right| + C = \ln\left|\frac{e^s-1}{e^s}\right| + C = \ln|1 - e^{-s}| + C$

29. (a) $\int \frac{y\,dy}{\sqrt{16-y^2}} = -\frac{1}{2}\int \frac{d(16-y^2)}{\sqrt{16-y^2}} = -\sqrt{16-y^2} + C$

(b) $\int \frac{y\,dy}{\sqrt{16-y^2}}$; $[y = 4\sin x] \rightarrow 4\int \frac{\sin x \cos x\,dx}{\cos x} = -4\cos x + C = -\frac{4\sqrt{16-y^2}}{4} + C = -\sqrt{16-y^2} + C$

31. (a) $\int \frac{x\,dx}{4-x^2} = -\frac{1}{2}\int \frac{d(4-x^2)}{4-x^2} = -\frac{1}{2}\ln|4-x^2| + C$

(b) $\int \frac{x\,dx}{4-x^2}$; $[x = 2\sin\theta] \rightarrow \int \frac{2\sin\theta\cdot 2\cos\theta\,d\theta}{4\cos^2\theta} = \int \tan\theta\,d\theta = -\ln|\cos\theta| + C = -\ln\left(\frac{\sqrt{4-x^2}}{2}\right) + C$

$= -\frac{1}{2}\ln|4-x^2| + C$

33. $\int \frac{x\,dx}{9-x^2}$; $\begin{bmatrix} u = 9 - x^2 \\ du = -2x\,dx \end{bmatrix} \rightarrow -\frac{1}{2}\int \frac{du}{u} = -\frac{1}{2}\ln|u| + C = \ln\frac{1}{\sqrt{u}} + C = \ln\frac{1}{\sqrt{9-x^2}} + C$

35. $\int \frac{dx}{9-x^2} = \frac{1}{6}\int \frac{dx}{3-x} + \frac{1}{6}\int \frac{dx}{3+x} = -\frac{1}{6}\ln|3-x| + \frac{1}{6}\ln|3+x| + C = \frac{1}{6}\ln\left|\frac{x+3}{x-3}\right| + C$

37. $\int \sin^3 x \cos^4 x\,dx = \int \cos^4 x (1 - \cos^2 x)\sin x\,dx = \int \cos^4 x \sin x\,dx - \int \cos^6 x \sin x\,dx = -\frac{\cos^5 x}{5} + \frac{\cos^7 x}{7} + C$

39. $\int \tan^4 x \sec^2 x\,dx = \frac{\tan^5 x}{5} + C$

41. $\int \sin 5\theta \cos 6\theta\,d\theta = \frac{1}{2}\int (\sin(-\theta) + \sin(11\theta))\,d\theta = \frac{1}{2}\int \sin(-\theta)\,d\theta + \frac{1}{2}\int \sin(11\theta)\,d\theta = \frac{1}{2}\cos(-\theta) - \frac{1}{22}\cos 11\theta + C$

$= \frac{1}{2}\cos\theta - \frac{1}{22}\cos 11\theta + C$

43. $\int \sqrt{1 + \cos\left(\frac{t}{2}\right)}\,dt = \int \sqrt{2}\left|\cos\frac{t}{4}\right|\,dt = 4\sqrt{2}\left|\sin\frac{t}{4}\right| + C$

45. $\triangle x = \frac{b-a}{n} = \frac{\pi - 0}{6} = \frac{\pi}{6} \Rightarrow \frac{\triangle x}{2} = \frac{\pi}{12}$;

$\sum_{i=0}^{6} m f(x_i) = 12 \Rightarrow T = \left(\frac{\pi}{12}\right)(12) = \pi$;

	x_i	$f(x_i)$	m	$mf(x_i)$
x_0	0	0	1	0
x_1	$\pi/6$	1/2	2	1
x_2	$\pi/3$	3/2	2	3
x_3	$\pi/2$	2	2	4
x_4	$2\pi/3$	3/2	2	3
x_5	$5\pi/6$	1/2	2	1
x_6	π	0	1	0

$$\sum_{i=0}^{6} mf(x_i) = 18 \text{ and } \frac{\triangle x}{3} = \frac{\pi}{18} \Rightarrow$$

$$S = \left(\frac{\pi}{18}\right)(18) = \pi.$$

	x_i	$f(x_i)$	m	$mf(x_i)$
x_0	0	0	1	0
x_1	$\pi/6$	1/2	4	2
x_2	$\pi/3$	3/2	2	3
x_3	$\pi/2$	2	4	8
x_4	$2\pi/3$	3/2	2	3
x_5	$5\pi/6$	1/2	4	2
x_6	π	0	1	0

47. $y_{av} = \frac{1}{365-0} \int_0^{365} \left[37 \sin\left(\frac{2\pi}{365}(x-101)\right) + 25\right] dx = \frac{1}{365}\left[-37\left(\frac{365}{2\pi}\right)\cos\left(\frac{2\pi}{365}(x-101)\right) + 25x\right]_0^{365}$

$= \frac{1}{365}\left[\left(-37\left(\frac{365}{2\pi}\right)\cos\left[\frac{2\pi}{365}(365-101)\right] + 25(365)\right) - \left(-37\left(\frac{365}{2\pi}\right)\cos\left[\frac{2\pi}{365}(0-101)\right] + 25(0)\right)\right]$

$= -\frac{37}{2\pi}\cos\left(\frac{2\pi}{365}(264)\right) + 25 + \frac{37}{2\pi}\cos\left(\frac{2\pi}{365}(-101)\right) = -\frac{37}{2\pi}\left(\cos\left(\frac{2\pi}{365}(264)\right) - \cos\left(\frac{2\pi}{365}(-101)\right)\right) + 25$

$\approx -\frac{37}{2\pi}(0.16705 - 0.16705) + 25 = 25°\,F$

49. $\int_0^3 \frac{dx}{\sqrt{9-x^2}} = \lim_{b \to 3^-} \int_0^b \frac{dx}{\sqrt{9-x^2}} = \lim_{b \to 3^-}\left[\sin^{-1}\left(\frac{x}{3}\right)\right]_0^b = \lim_{b \to 3^-} \sin^{-1}\left(\frac{b}{3}\right) - \sin^{-1}\left(\frac{0}{3}\right) = \frac{\pi}{2} - 0 = \frac{\pi}{2}$

51. $\int_{-1}^1 \frac{dy}{y^{2/3}} = \int_{-1}^0 \frac{dy}{y^{2/3}} + \int_0^1 \frac{dy}{y^{2/3}} = 2\int_0^1 \frac{dy}{y^{2/3}} = 2 \cdot 3 \lim_{b \to 0^+}\left[y^{1/3}\right]_b^1 = 6\left(1 - \lim_{b \to 0^+} b^{1/3}\right) = 6$

53. $\int_3^\infty \frac{2\,du}{u^2 - 2u} = \int_3^\infty \frac{du}{u-2} - \int_3^\infty \frac{du}{u} = \lim_{b \to \infty}\left[\ln\left|\frac{u-2}{u}\right|\right]_3^b = \lim_{b \to \infty}\left[\ln\left|\frac{b-2}{b}\right|\right] - \ln\left|\frac{3-2}{3}\right| = 0 - \ln\left(\frac{1}{3}\right) = \ln 3$

55. $\int_0^\infty x^2 e^{-x}\,dx = \lim_{b \to \infty}\left[-x^2 e^{-x} - 2xe^{-x} - 2e^{-x}\right]_0^b = \lim_{b \to \infty}\left(-b^2 e^{-b} - 2be^{-b} - 2e^{-b}\right) - (-2) = 0 + 2 = 2$

57. $\int_{-\infty}^\infty \frac{dx}{4x^2 + 9} = 2\int_0^\infty \frac{dx}{4x^2 + 9} = \frac{1}{2}\int_0^\infty \frac{dx}{x^2 + \frac{9}{4}} = \frac{1}{2}\lim_{b \to \infty}\left[\frac{2}{3}\tan^{-1}\left(\frac{2x}{3}\right)\right]_0^b = \frac{1}{2}\lim_{b \to \infty}\left[\frac{2}{3}\tan^{-1}\left(\frac{2b}{3}\right)\right] - \frac{1}{3}\tan^{-1}(0)$

$= \frac{1}{2}\left(\frac{2}{3}\cdot\frac{\pi}{2}\right) - 0 = \frac{\pi}{6}$

59. $\lim_{\theta \to \infty} \frac{\theta}{\sqrt{\theta^2 + 1}} = 1$ and $\int_6^\infty \frac{d\theta}{\theta}$ diverges $\Rightarrow \int_6^\infty \frac{d\theta}{\sqrt{\theta^2 + 1}}$ diverges

61. $\int_1^\infty \frac{\ln z}{z}\,dz = \int_1^e \frac{\ln z}{z}\,dz + \int_e^\infty \frac{\ln z}{z}\,dz = \left[\frac{(\ln z)^2}{2}\right]_1^e + \lim_{b \to \infty}\left[\frac{(\ln z)^2}{2}\right]_e^b = \left(\frac{1^2}{2} - 0\right) + \lim_{b \to \infty}\left[\frac{(\ln b)^2}{2} - \frac{1}{2}\right]$

$= \infty \Rightarrow$ diverges

63. $\int_{-\infty}^\infty \frac{2\,dx}{e^x + e^{-x}} = 2\int_0^\infty \frac{2\,dx}{e^x + e^{-x}} < \int_0^\infty \frac{4\,dx}{e^x}$ converges $\Rightarrow \int_{-\infty}^\infty \frac{2\,dx}{e^x + e^{-x}}$ converges

65. $\int \frac{x\,dx}{1 + \sqrt{x}}$; $\begin{bmatrix} u = \sqrt{x} \\ du = \frac{dx}{2\sqrt{x}} \end{bmatrix} \rightarrow \int \frac{u^2 \cdot 2u\,du}{1+u} = \int \left(2u^2 - 2u + 2 - \frac{2}{1+u}\right)du = \frac{2}{3}u^3 - u^2 + 2u - 2\ln|1+u| + C$

$= \frac{2x^{3/2}}{3} - x + 2\sqrt{x} - 2\ln\left(1 + \sqrt{x}\right) + C$

67. $\int \frac{dx}{x(x^2+1)^2}$; $\begin{bmatrix} x = \tan\theta \\ dx = \sec^2\theta\,d\theta \end{bmatrix} \rightarrow \int \frac{\sec^2\theta\,d\theta}{\tan\theta\sec^4\theta} = \int \frac{\cos^3\theta\,d\theta}{\sin\theta} = \int \left(\frac{1 - \sin^2\theta}{\sin\theta}\right)d(\sin\theta)$

$= \ln|\sin\theta| - \frac{1}{2}\sin^2\theta + C = \ln\left|\frac{x}{\sqrt{x^2+1}}\right| - \frac{1}{2}\left(\frac{x}{\sqrt{x^2+1}}\right)^2 + C$

69. $\int \frac{2 - \cos x + \sin x}{\sin^2 x}\,dx = \int 2\csc^2 x\,dx - \int \frac{\cos x\,dx}{\sin^2 x} + \int \csc x\,dx = -2\cot x + \frac{1}{\sin x} - \ln|\csc x + \cot x| + C$

$= -2\cot x + \csc x - \ln|\csc x + \cot x| + C$

71. $\int \frac{9 \, dv}{81 - v^4} = \frac{1}{2} \int \frac{dv}{v^2 + 9} + \frac{1}{12} \int \frac{dv}{3 - v} + \frac{1}{12} \int \frac{dv}{3 + v} = \frac{1}{12} \ln \left| \frac{3+v}{3-v} \right| + \frac{1}{6} \tan^{-1} \frac{v}{3} + C$

73.
$$\cos(2\theta + 1)$$
$$\theta \xrightarrow{\;(+)\;} \frac{1}{2} \sin(2\theta + 1)$$
$$1 \xrightarrow{\;(-)\;} -\frac{1}{4} \cos(2\theta + 1)$$
$$0 \qquad\qquad \Rightarrow \int \theta \cos(2\theta + 1) \, d\theta = \frac{\theta}{2} \sin(2\theta + 1) + \frac{1}{4} \cos(2\theta + 1) + C$$

75. $\int \frac{\sin 2\theta \, d\theta}{(1 + \cos 2\theta)^2} = -\frac{1}{2} \int \frac{d(1 + \cos 2\theta)}{(1 + \cos 2\theta)^2} = \frac{1}{2(1 + \cos 2\theta)} + C = \frac{1}{4} \sec^2 \theta + C$

77. $\int \frac{x \, dx}{\sqrt{2 - x}} ; \begin{bmatrix} y = 2 - x \\ dy = -dx \end{bmatrix} \rightarrow -\int \frac{(2 - y) \, dy}{\sqrt{y}} = \frac{2}{3} y^{3/2} - 4y^{1/2} + C = \frac{2}{3}(2 - x)^{3/2} - 4(2 - x)^{1/2} + C$

$\qquad = 2\left[\frac{\left(\sqrt{2 - x}\right)^3}{3} - 2\sqrt{2 - x} \right] + C$

79. $\int \frac{dy}{y^2 - 2y + 2} = \int \frac{d(y - 1)}{(y - 1)^2 + 1} = \tan^{-1}(y - 1) + C$

81. $\int \frac{z + 1}{z^2 (z^2 + 4)} \, dz = \frac{1}{4} \int \left(\frac{1}{z} + \frac{1}{z^2} - \frac{z + 1}{z^2 + 4} \right) dz = \frac{1}{4} \ln |z| - \frac{1}{4z} - \frac{1}{8} \ln(z^2 + 4) - \frac{1}{8} \tan^{-1} \frac{z}{2} + C$

83. $\int \frac{t \, dt}{\sqrt{9 - 4t^2}} = -\frac{1}{8} \int \frac{d(9 - 4t^2)}{\sqrt{9 - 4t^2}} = -\frac{1}{4} \sqrt{9 - 4t^2} + C$

85. $\int \frac{e^t \, dt}{e^{2t} + 3e^t + 2} ; [e^t = x] \rightarrow \int \frac{dx}{(x + 1)(x + 2)} = \int \frac{dx}{x + 1} - \int \frac{dx}{x + 2} = \ln |x + 1| - \ln |x + 2| + C$

$\qquad = \ln \left| \frac{x + 1}{x + 2} \right| + C = \ln \left(\frac{e^t + 1}{e^t + 2} \right) + C$

87. $\int_1^\infty \frac{\ln y \, dy}{y^3} ; \begin{bmatrix} x = \ln y \\ dx = \frac{dy}{y} \\ dy = e^x \, dx \end{bmatrix} \rightarrow \int_0^\infty \frac{x \cdot e^x}{e^{3x}} \, dx = \int_0^\infty x e^{-2x} \, dx = \lim_{b \to \infty} \left[-\frac{x}{2} e^{-2x} - \frac{1}{4} e^{-2x} \right]_0^b$

$\qquad = \lim_{b \to \infty} \left(\frac{-b}{2e^{2b}} - \frac{1}{4e^{2b}} \right) - \left(0 - \frac{1}{4} \right) = \frac{1}{4}$

89. $\int e^{\ln \sqrt{x}} \, dx = \int \sqrt{x} \, dx = \frac{2}{3} x^{3/2} + C$

91. $\int \frac{\sin 5t \, dt}{1 + (\cos 5t)^2} ; \begin{bmatrix} u = \cos 5t \\ du = -5 \sin 5t \, dt \end{bmatrix} \rightarrow -\frac{1}{5} \int \frac{du}{1 + u^2} = -\frac{1}{5} \tan^{-1} u + C = -\frac{1}{5} \tan^{-1}(\cos 5t) + C$

93. $\int \frac{dr}{1 + \sqrt{r}} ; \begin{bmatrix} u = \sqrt{r} \\ du = \frac{dr}{2\sqrt{r}} \end{bmatrix} \rightarrow \int \frac{2u \, du}{1 + u} = \int \left(2 - \frac{2}{1 + u} \right) du = 2u - 2 \ln |1 + u| + C = 2\sqrt{r} - 2 \ln \left(1 + \sqrt{r} \right) + C$

NOTES:

CHAPTER 6 APPLICATIONS OF DEFINITE INTEGRALS

6.1 VOLUMES BY SLICING AND ROTATION ABOUT AN AXIS

1. $A(x) = \frac{(\text{diagonal})^2}{2} = \frac{\left(\sqrt{x} - (-\sqrt{x})\right)^2}{2} = 2x$; $a = 0$, $b = 4$;

$V = \int_a^b A(x)\,dx = \int_0^4 2x\,dx = [x^2]_0^4 = 16$

3. $A(x) = (\text{edge})^2 = \left[\sqrt{1-x^2} - \left(-\sqrt{1-x^2}\right)\right]^2 = \left(2\sqrt{1-x^2}\right)^2 = 4\left(1-x^2\right)$; $a = -1$, $b = 1$;

$V = \int_a^b A(x)\,dx = \int_{-1}^1 4(1-x^2)\,dx = 4\left[x - \frac{x^3}{3}\right]_{-1}^1 = 8\left(1 - \frac{1}{3}\right) = \frac{16}{3}$

5. (a) STEP 1) $A(x) = \frac{1}{2}(\text{side}) \cdot (\text{side}) \cdot \left(\sin\frac{\pi}{3}\right) = \frac{1}{2} \cdot \left(2\sqrt{\sin x}\right) \cdot \left(2\sqrt{\sin x}\right)\left(\sin\frac{\pi}{3}\right) = \sqrt{3}\sin x$

   STEP 2) $a = 0$, $b = \pi$

   STEP 3) $V = \int_a^b A(x)\,dx = \sqrt{3}\int_0^\pi \sin x\,dx = \left[-\sqrt{3}\cos x\right]_0^\pi = \sqrt{3}(1+1) = 2\sqrt{3}$

   (b) STEP 1) $A(x) = (\text{side})^2 = \left(2\sqrt{\sin x}\right)\left(2\sqrt{\sin x}\right) = 4\sin x$

   STEP 2) $a = 0$, $b = \pi$

   STEP 3) $V = \int_a^b A(x)\,dx = \int_0^\pi 4\sin x\,dx = [-4\cos x]_0^\pi = 8$

7. $A(y) = \frac{\pi}{4}(\text{diameter})^2 = \frac{\pi}{4}\left(\sqrt{5}y^2 - 0\right)^2 = \frac{5\pi}{4}y^4$;

$c = 0$, $d = 2$; $V = \int_c^d A(y)\,dy = \int_0^2 \frac{5\pi}{4}y^4\,dy$

$= \left[\left(\frac{5\pi}{4}\right)\left(\frac{y^5}{5}\right)\right]_0^2 = \frac{\pi}{4}(2^5 - 0) = 8\pi$

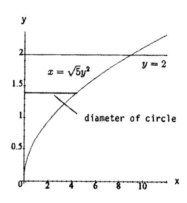

9. (a) $A(x) = \frac{\pi}{4}(\text{diameter})^2 = \frac{\pi}{4}\left[\frac{1}{\sqrt{1+x^2}} - \left(-\frac{1}{\sqrt{1+x^2}}\right)\right]^2 = \frac{\pi}{1+x^2} \Rightarrow V = \int_a^b A(x)\,dx = \int_{-1}^1 \frac{\pi\,dx}{1+x^2}$

   $= \pi\left[\tan^{-1}x\right]_{-1}^1 = (\pi)(2)\left(\frac{\pi}{4}\right) = \frac{\pi^2}{2}$

   (b) $A(x) = (\text{edge})^2 = \left[\frac{1}{\sqrt{1+x^2}} - \left(-\frac{1}{\sqrt{1+x^2}}\right)\right]^2 = \frac{4}{1+x^2} \Rightarrow V = \int_a^b A(x)\,dx = \int_{-1}^1 \frac{4\,dx}{1+x^2}$

   $= 4\left[\tan^{-1}x\right]_{-1}^1 = 4\left[\tan^{-1}(1) - \tan^{-1}(-1)\right] = 4\left[\frac{\pi}{4} - \left(-\frac{\pi}{4}\right)\right] = 2\pi$

11. (a) It follows from Cavalieri's Principle that the volume of a column is the same as the volume of a right prism with a square base of side length s and altitude h. Thus, STEP 1) $A(x) = (\text{side length})^2 = s^2$;

   STEP 2) $a = 0$, $b = h$; STEP 3) $V = \int_a^b A(x)\,dx = \int_0^h s^2\,dx = s^2h$

   (b) From Cavalieri's Principle we conclude that the volume of the column is the same as the volume of the prism described above, regardless of the number of turns $\Rightarrow V = s^2h$

13. $R(x) = y = 1 - \frac{x}{2} \Rightarrow V = \int_0^2 \pi[R(x)]^2 \, dx = \pi \int_0^2 \left(1 - \frac{x}{2}\right)^2 dx = \pi \int_0^2 \left(1 - x + \frac{x^2}{4}\right) dx = \pi \left[x - \frac{x^2}{2} + \frac{x^3}{12}\right]_0^2$

$= \pi \left(2 - \frac{4}{2} + \frac{8}{12}\right) = \frac{2\pi}{3}$

15. $R(x) = \tan\left(\frac{\pi}{4} y\right); u = \frac{\pi}{4} y \Rightarrow du = \frac{\pi}{4} \, dy \Rightarrow 4 \, du = \pi \, dy; y = 0 \Rightarrow u = 0, y = 1 \Rightarrow u = \frac{\pi}{4};$

$V = \int_0^1 \pi[R(y)]^2 \, dy = \pi \int_0^1 \left[\tan\left(\frac{\pi}{4} y\right)\right]^2 dy = 4 \int_0^{\pi/4} \tan^2 u \, du = 4 \int_0^{\pi/4} (-1 + \sec^2 u) \, du = 4[-u + \tan u]_0^{\pi/4}$

$= 4\left(-\frac{\pi}{4} + 1 - 0\right) = 4 - \pi$

17. $R(x) = x^2 \Rightarrow V = \int_0^2 \pi[R(x)]^2 \, dx = \pi \int_0^2 (x^2)^2 \, dx$

$= \pi \int_0^2 x^4 \, dx = \pi \left[\frac{x^5}{5}\right]_0^2 = \frac{32\pi}{5}$

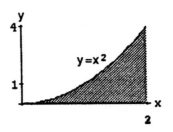

19. $R(x) = \sqrt{9 - x^2} \Rightarrow V = \int_{-3}^3 \pi[R(x)]^2 \, dx = \pi \int_{-3}^3 (9 - x^2) \, dx$

$= \pi \left[9x - \frac{x^3}{3}\right]_{-3}^3 = 2\pi \left[9(3) - \frac{27}{3}\right] = 2 \cdot \pi \cdot 18 = 36\pi$

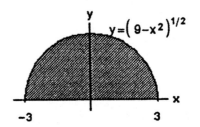

21. $R(x) = \sqrt{\cos x} \Rightarrow V = \int_0^{\pi/2} \pi[R(x)]^2 \, dx = \pi \int_0^{\pi/2} \cos x \, dx$

$= \pi [\sin x]_0^{\pi/2} = \pi(1 - 0) = \pi$

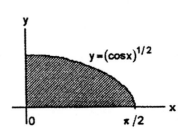

23. $R(x) = e^{-x} \Rightarrow V = \int_0^1 \pi[R(x)]^2 \, dx = \pi \int_0^1 (e^{-x})^2 \, dx$

$= \pi \int_0^1 e^{-2x} \, dx = -\frac{\pi}{2} e^{-2x} \Big|_0^1 = -\frac{\pi}{2}(e^{-2} - 1)$

$= \frac{\pi}{2}\left(1 - \frac{1}{e^2}\right) = \frac{\pi(e^2 - 1)}{2e^2}$

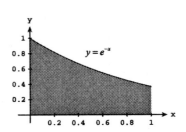

25. $V = \frac{\pi}{4} \int_{1/4}^4 \frac{1}{x} \, dx = \frac{\pi}{4} [\ln x]_{1/4}^4 = \frac{\pi}{4}\left(\ln 4 - \ln \frac{1}{4}\right) = \frac{\pi}{2} \ln 4$

27. $R(x) = \sqrt{2} - \sec x \tan x \;\Rightarrow\; V = \int_0^{\pi/4} \pi [R(x)]^2\, dx$

$= \pi \int_0^{\pi/4} \left(\sqrt{2} - \sec x \tan x\right)^2 dx$

$= \pi \int_0^{\pi/4} \left(2 - 2\sqrt{2}\, \sec x \tan x + \sec^2 x \tan^2 x\right) dx$

$= \pi \left(\int_0^{\pi/4} 2\, dx - 2\sqrt{2} \int_0^{\pi/4} \sec x \tan x\, dx + \int_0^{\pi/4} (\tan x)^2 \sec^2 x\, dx\right)$

$= \pi \left([2x]_0^{\pi/4} - 2\sqrt{2}\,[\sec x]_0^{\pi/4} + \left[\frac{\tan^3 x}{3}\right]_0^{\pi/4}\right)$

$= \pi \left[\left(\frac{\pi}{2} - 0\right) - 2\sqrt{2}\left(\sqrt{2} - 1\right) + \frac{1}{3}(1^3 - 0)\right] = \pi \left(\frac{\pi}{2} + 2\sqrt{2} - \frac{11}{3}\right)$

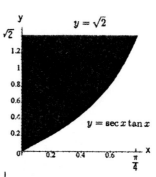

29. $R(y) = \sqrt{5} \cdot y^2 \;\Rightarrow\; V = \int_{-1}^{1} \pi [R(y)]^2\, dy = \pi \int_{-1}^{1} 5y^4\, dy$

$= \pi\, [y^5]_{-1}^{1} = \pi[1 - (-1)] = 2\pi$

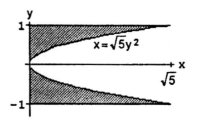

31. $R(y) = \sqrt{2 \sin 2y} \;\Rightarrow\; V = \int_0^{\pi/2} \pi [R(y)]^2\, dy$

$= \pi \int_0^{\pi/2} 2 \sin 2y\, dy = \pi\, [-\cos 2y]_0^{\pi/2}$

$= \pi[1 - (-1)] = 2\pi$

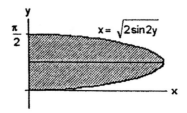

33. $R(y) = \frac{2}{\sqrt{y+1}} \;\Rightarrow\; V = \int_0^3 \pi [R(y)]^2\, dy = 4\pi \int_0^3 \frac{1}{y+1}\, dy$

$= 4\pi\, [\ln|y + 1|]_0^3 = 4\pi[\ln 4 - \ln 1] = 4\pi \ln 4$

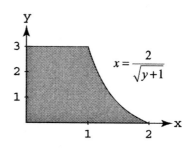

35. For the sketch given, $a = -\frac{\pi}{2}$, $b = \frac{\pi}{2}$; $R(x) = 1$, $r(x) = \sqrt{\cos x}$; $V = \int_a^b \pi \left([R(x)]^2 - [r(x)]^2\right) dx$

$= \int_{-\pi/2}^{\pi/2} \pi(1 - \cos x)\, dx = 2\pi \int_0^{\pi/2} (1 - \cos x)\, dx = 2\pi[x - \sin x]_0^{\pi/2} = 2\pi \left(\frac{\pi}{2} - 1\right) = \pi^2 - 2\pi$

37. $r(x) = x$ and $R(x) = 1 \;\Rightarrow\; V = \int_0^1 \pi \left([R(x)]^2 - [r(x)]^2\right) dx$

$= \int_0^1 \pi (1 - x^2)\, dx = \pi \left[x - \frac{x^3}{3}\right]_0^1 = \pi \left[\left(1 - \frac{1}{3}\right) - 0\right] = \frac{2\pi}{3}$

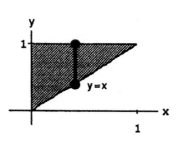

39. $r(x) = x^2 + 1$ and $R(x) = x + 3$

$\Rightarrow V = \int_{-1}^{2} \pi\left([R(x)]^2 - [r(x)]^2\right) dx$

$= \pi \int_{-1}^{2} \left[(x+3)^2 - (x^2+1)^2\right] dx$

$= \pi \int_{-1}^{2} \left[(x^2 + 6x + 9) - (x^4 + 2x^2 + 1)\right] dx$

$= \pi \int_{-1}^{2} \left(-x^4 - x^2 + 6x + 8\right) dx$

$= \pi \left[-\frac{x^5}{5} - \frac{x^3}{3} + \frac{6x^2}{2} + 8x\right]_{-1}^{2}$

$= \pi \left[\left(-\frac{32}{5} - \frac{8}{3} + \frac{24}{2} + 16\right) - \left(\frac{1}{5} + \frac{1}{3} + \frac{6}{2} - 8\right)\right] = \pi\left(-\frac{33}{5} - 3 + 28 - 3 + 8\right) = \pi\left(\frac{5 \cdot 30 - 33}{5}\right) = \frac{117\pi}{5}$

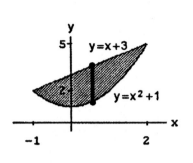

41. $r(x) = \sec x$ and $R(x) = \sqrt{2}$

$\Rightarrow V = \int_{-\pi/4}^{\pi/4} \pi\left([R(x)]^2 - [r(x)]^2\right) dx$

$= \pi \int_{-\pi/4}^{\pi/4} (2 - \sec^2 x) \, dx = \pi[2x - \tan x]_{-\pi/4}^{\pi/4}$

$= \pi\left[\left(\frac{\pi}{2} - 1\right) - \left(-\frac{\pi}{2} + 1\right)\right] = \pi(\pi - 2)$

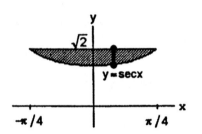

43. $r(y) = 1$ and $R(y) = 1 + y$

$\Rightarrow V = \int_{0}^{1} \pi\left([R(y)]^2 - [r(y)]^2\right) dy$

$= \pi \int_{0}^{1} \left[(1+y)^2 - 1\right] dy = \pi \int_{0}^{1} (1 + 2y + y^2 - 1) \, dy$

$= \pi \int_{0}^{1} (2y + y^2) \, dy = \pi \left[y^2 + \frac{y^3}{3}\right]_{0}^{1} = \pi\left(1 + \frac{1}{3}\right) = \frac{4\pi}{3}$

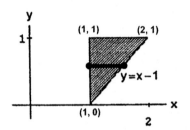

45. $R(y) = 2$ and $r(y) = \sqrt{y}$

$\Rightarrow V = \int_{0}^{4} \pi\left([R(y)]^2 - [r(y)]^2\right) dy$

$= \pi \int_{0}^{4} (4 - y) \, dy = \pi \left[4y - \frac{y^2}{2}\right]_{0}^{4} = \pi(16 - 8) = 8\pi$

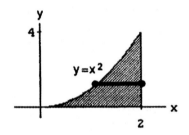

47. $R(y) = 2$ and $r(y) = 1 + \sqrt{y}$

$\Rightarrow V = \int_{0}^{1} \pi\left([R(y)]^2 - [r(y)]^2\right) dy$

$= \pi \int_{0}^{1} \left[4 - \left(1 + \sqrt{y}\right)^2\right] dy$

$= \pi \int_{0}^{1} \left(4 - 1 - 2\sqrt{y} - y\right) dy$

$= \pi \int_{0}^{1} \left(3 - 2\sqrt{y} - y\right) dy$

$= \pi \left[3y - \frac{4}{3} y^{3/2} - \frac{y^2}{2}\right]_{0}^{1}$

$= \pi\left(3 - \frac{4}{3} - \frac{1}{2}\right) = \pi\left(\frac{18 - 8 - 3}{6}\right) = \frac{7\pi}{6}$

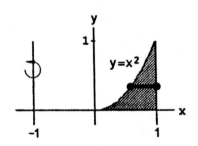

49. (a) $r(x) = \sqrt{x}$ and $R(x) = 2$

$\Rightarrow V = \int_0^4 \pi \left([R(x)]^2 - [r(x)]^2\right) dx$

$= \pi \int_0^4 (4 - x)\, dx = \pi \left[4x - \frac{x^2}{2}\right]_0^4 = \pi(16 - 8) = 8\pi$

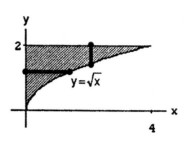

(b) $r(y) = 0$ and $R(y) = y^2$

$\Rightarrow V = \int_0^2 \pi \left([R(y)]^2 - [r(y)]^2\right) dy$

$= \pi \int_0^2 y^4\, dy = \pi \left[\frac{y^5}{5}\right]_0^2 = \frac{32\pi}{5}$

(c) $r(x) = 0$ and $R(x) = 2 - \sqrt{x} \Rightarrow V = \int_0^4 \pi \left([R(x)]^2 - [r(x)]^2\right) dx = \pi \int_0^4 \left(2 - \sqrt{x}\right)^2 dx$

$= \pi \int_0^4 \left(4 - 4\sqrt{x} + x\right) dx = \pi \left[4x - \frac{8x^{3/2}}{3} + \frac{x^2}{2}\right]_0^4 = \pi \left(16 - \frac{64}{3} + \frac{16}{2}\right) = \frac{8\pi}{3}$

(d) $r(y) = 4 - y^2$ and $R(y) = 4 \Rightarrow V = \int_0^2 \pi \left([R(y)]^2 - [r(y)]^2\right) dy = \pi \int_0^2 \left[16 - (4 - y^2)^2\right] dy$

$= \pi \int_0^2 \left(16 - 16 + 8y^2 - y^4\right) dy = \pi \int_0^2 \left(8y^2 - y^4\right) dy = \pi \left[\frac{8}{3} y^3 - \frac{y^5}{5}\right]_0^2 = \pi \left(\frac{64}{3} - \frac{32}{5}\right) = \frac{224\pi}{15}$

51. (a) $r(x) = 0$ and $R(x) = 1 - x^2$

$\Rightarrow V = \int_{-1}^1 \pi \left([R(x)]^2 - [r(x)]^2\right) dx$

$= \pi \int_{-1}^1 (1 - x^2)^2 dx = \pi \int_{-1}^1 (1 - 2x^2 + x^4)\, dx$

$= \pi \left[x - \frac{2x^3}{3} + \frac{x^5}{5}\right]_{-1}^1 = 2\pi \left(1 - \frac{2}{3} + \frac{1}{5}\right)$

$= 2\pi \left(\frac{15 - 10 + 3}{15}\right) = \frac{16\pi}{15}$

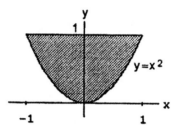

(b) $r(x) = 1$ and $R(x) = 2 - x^2 \Rightarrow V = \int_{-1}^1 \pi \left([R(x)]^2 - [r(x)]^2\right) dx = \pi \int_{-1}^1 \left[(2 - x^2)^2 - 1\right] dx$

$= \pi \int_{-1}^1 (4 - 4x^2 + x^4 - 1)\, dx = \pi \int_{-1}^1 (3 - 4x^2 + x^4)\, dx = \pi \left[3x - \frac{4}{3} x^3 + \frac{x^5}{5}\right]_{-1}^1 = 2\pi \left(3 - \frac{4}{3} + \frac{1}{5}\right)$

$= \frac{2\pi}{15}(45 - 20 + 3) = \frac{56\pi}{15}$

(c) $r(x) = 1 + x^2$ and $R(x) = 2 \Rightarrow V = \int_{-1}^1 \pi \left([R(x)]^2 - [r(x)]^2\right) dx = \pi \int_{-1}^1 \left[4 - (1 + x^2)^2\right] dx$

$= \pi \int_{-1}^1 (4 - 1 - 2x^2 - x^4)\, dx = \pi \int_{-1}^1 (3 - 2x^2 - x^4)\, dx = \pi \left[3x - \frac{2}{3} x^3 - \frac{x^5}{5}\right]_{-1}^1 = 2\pi \left(3 - \frac{2}{3} - \frac{1}{5}\right)$

$= \frac{2\pi}{15}(45 - 10 - 3) = \frac{64\pi}{15}$

6.2 VOLUME BY CYLINDRICAL SHELLS

1. For the sketch given, $a = 0$, $b = 2$;

$V = \int_a^b 2\pi \binom{\text{shell}}{\text{radius}} \binom{\text{shell}}{\text{height}} dx = \int_0^2 2\pi x \left(1 + \frac{x^2}{4}\right) dx = 2\pi \int_0^2 \left(x + \frac{x^3}{4}\right) dx = 2\pi \left[\frac{x^2}{2} + \frac{x^4}{16}\right]_0^2 = 2\pi \left(\frac{4}{2} + \frac{16}{16}\right)$

$= 2\pi \cdot 3 = 6\pi$

3. For the sketch given, $c = 0$, $d = \sqrt{2}$;

$V = \int_c^d 2\pi \binom{\text{shell}}{\text{radius}} \binom{\text{shell}}{\text{height}} dy = \int_0^{\sqrt{2}} 2\pi y \cdot (y^2)\, dy = 2\pi \int_0^{\sqrt{2}} y^3\, dy = 2\pi \left[\frac{y^4}{4}\right]_0^{\sqrt{2}} = 2\pi$

5. For the sketch given, $a = 0$, $b = \sqrt{3}$;

$V = \int_a^b 2\pi \binom{\text{shell}}{\text{radius}} \binom{\text{shell}}{\text{height}} dx = \int_0^{\sqrt{3}} 2\pi x \cdot \left(\sqrt{x^2 + 1}\right) dx$;

$\left[u = x^2 + 1 \Rightarrow du = 2x\, dx; x = 0 \Rightarrow u = 1, x = \sqrt{3} \Rightarrow u = 4\right]$

$\rightarrow V = \pi \int_1^4 u^{1/2}\, du = \pi \left[\frac{2}{3} u^{3/2}\right]_1^4 = \frac{2\pi}{3} \left(4^{3/2} - 1\right) = \left(\frac{2\pi}{3}\right)(8 - 1) = \frac{14\pi}{3}$

7. $a = 0, b = 2$;

$$V = \int_a^b 2\pi \left(\begin{smallmatrix} \text{shell} \\ \text{radius} \end{smallmatrix} \right) \left(\begin{smallmatrix} \text{shell} \\ \text{height} \end{smallmatrix} \right) dx = \int_0^2 2\pi x \left[x - \left(-\frac{x}{2} \right) \right] dx$$

$$= \int_0^2 2\pi x^2 \cdot \frac{3}{2} \, dx = \pi \int_0^2 3x^2 \, dx = \pi \left[x^3 \right]_0^2 = 8\pi$$

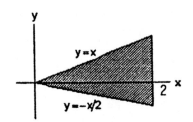

9. $a = 0, b = 1$;

$$V = \int_a^b 2\pi \left(\begin{smallmatrix} \text{shell} \\ \text{radius} \end{smallmatrix} \right) \left(\begin{smallmatrix} \text{shell} \\ \text{height} \end{smallmatrix} \right) dx = \int_0^1 2\pi x \left[(2 - x) - x^2 \right] dx$$

$$= 2\pi \int_0^1 (2x - x^2 - x^3) \, dx = 2\pi \left[x^2 - \frac{x^3}{3} - \frac{x^4}{4} \right]_0^1$$

$$= 2\pi \left(1 - \frac{1}{3} - \frac{1}{4} \right) = 2\pi \left(\frac{12 - 4 - 3}{12} \right) = \frac{10\pi}{12} = \frac{5\pi}{6}$$

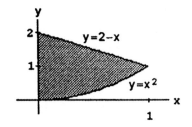

11. $a = 0, b = 1$;

$$V = \int_a^b 2\pi \left(\begin{smallmatrix} \text{shell} \\ \text{radius} \end{smallmatrix} \right) \left(\begin{smallmatrix} \text{shell} \\ \text{height} \end{smallmatrix} \right) dx = \int_0^1 2\pi x \left[\sqrt{x} - (2x - 1) \right] dx$$

$$= 2\pi \int_0^1 \left(x^{3/2} - 2x^2 + x \right) dx = 2\pi \left[\frac{2}{5} x^{5/2} - \frac{2}{3} x^3 + \frac{1}{2} x^2 \right]_0^1$$

$$= 2\pi \left(\frac{2}{5} - \frac{2}{3} + \frac{1}{2} \right) = 2\pi \left(\frac{12 - 20 + 15}{30} \right) = \frac{7\pi}{15}$$

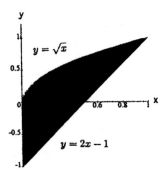

13. (a) $xf(x) = \begin{cases} x \cdot \frac{\sin x}{x}, & 0 < x \le \pi \\ x, & x = 0 \end{cases} \Rightarrow xf(x) = \begin{cases} \sin x, & 0 < x \le \pi \\ 0, & x = 0 \end{cases}$; since $\sin 0 = 0$ we have

$xf(x) = \begin{cases} \sin x, & 0 < x \le \pi \\ \sin x, & x = 0 \end{cases} \Rightarrow xf(x) = \sin x, 0 \le x \le \pi$

(b) $V = \int_a^b 2\pi \left(\begin{smallmatrix} \text{shell} \\ \text{radius} \end{smallmatrix} \right) \left(\begin{smallmatrix} \text{shell} \\ \text{height} \end{smallmatrix} \right) dx = \int_0^\pi 2\pi x \cdot f(x) \, dx$ and $x \cdot f(x) = \sin x, 0 \le x \le \pi$ by part (a)

$\Rightarrow V = 2\pi \int_0^\pi \sin x \, dx = 2\pi [- \cos x]_0^\pi = 2\pi (- \cos \pi + \cos 0) = 4\pi$

15. $c = 0, d = 2$;

$$V = \int_c^d 2\pi \left(\begin{smallmatrix} \text{shell} \\ \text{radius} \end{smallmatrix} \right) \left(\begin{smallmatrix} \text{shell} \\ \text{height} \end{smallmatrix} \right) dy = \int_0^2 2\pi y \left[\sqrt{y} - (-y) \right] dy$$

$$= 2\pi \int_0^2 \left(y^{3/2} + y^2 \right) dy = 2\pi \left[\frac{2y^{5/2}}{5} + \frac{y^3}{3} \right]_0^2$$

$$= 2\pi \left[\frac{2}{5} \left(\sqrt{2} \right)^5 + \frac{2^3}{3} \right] = 2\pi \left(\frac{8\sqrt{2}}{5} + \frac{8}{3} \right) = 16\pi \left(\frac{\sqrt{2}}{5} + \frac{1}{3} \right)$$

$$= \frac{16\pi}{15} \left(3\sqrt{2} + 5 \right)$$

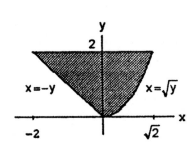

17. $c = 0$, $d = 2$;

$$V = \int_c^d 2\pi \left(\begin{smallmatrix} \text{shell} \\ \text{radius} \end{smallmatrix}\right) \left(\begin{smallmatrix} \text{shell} \\ \text{height} \end{smallmatrix}\right) dy = \int_0^2 2\pi y \,(2y - y^2) dy$$

$$= 2\pi \int_0^2 (2y^2 - y^3)\,dy = 2\pi \left[\frac{2y^3}{3} - \frac{y^4}{4}\right]_0^2 = 2\pi \left(\frac{16}{3} - \frac{16}{4}\right)$$

$$= 32\pi \left(\frac{1}{3} - \frac{1}{4}\right) = \frac{32\pi}{12} = \frac{8\pi}{3}$$

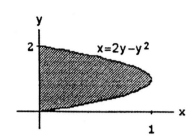

19. $c = 0$, $d = 1$;

$$V = \int_c^d 2\pi \left(\begin{smallmatrix} \text{shell} \\ \text{radius} \end{smallmatrix}\right) \left(\begin{smallmatrix} \text{shell} \\ \text{height} \end{smallmatrix}\right) dy = 2\pi \int_0^1 y[y - (-y)]dy$$

$$= 2\pi \int_0^1 2y^2\,dy = \frac{4\pi}{3}\,[y^3]_0^1 = \frac{4\pi}{3}$$

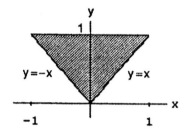

21. $c = 0$, $d = 2$;

$$V = \int_c^d 2\pi \left(\begin{smallmatrix} \text{shell} \\ \text{radius} \end{smallmatrix}\right) \left(\begin{smallmatrix} \text{shell} \\ \text{height} \end{smallmatrix}\right) dy = \int_0^2 2\pi y \left[(2 + y) - y^2\right] dy$$

$$= 2\pi \int_0^2 (2y + y^2 - y^3)\,dy = 2\pi \left[y^2 + \frac{y^3}{3} - \frac{y^4}{4}\right]_0^2$$

$$= 2\pi \left(4 + \frac{8}{3} - \frac{16}{4}\right) = \frac{\pi}{6}(48 + 32 - 48) = \frac{16\pi}{3}$$

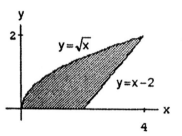

23. (a) $V = \int_c^d 2\pi \left(\begin{smallmatrix} \text{shell} \\ \text{radius} \end{smallmatrix}\right) \left(\begin{smallmatrix} \text{shell} \\ \text{height} \end{smallmatrix}\right) dy = \int_0^1 2\pi y \cdot 12\,(y^2 - y^3)\,dy = 24\pi \int_0^1 (y^3 - y^4)\,dy = 24\pi \left[\frac{y^4}{4} - \frac{y^5}{5}\right]_0^1$

$$= 24\pi \left(\frac{1}{4} - \frac{1}{5}\right) = \frac{24\pi}{20} = \frac{6\pi}{5}$$

(b) $V = \int_c^d 2\pi \left(\begin{smallmatrix} \text{shell} \\ \text{radius} \end{smallmatrix}\right) \left(\begin{smallmatrix} \text{shell} \\ \text{height} \end{smallmatrix}\right) dy = \int_0^1 2\pi(1 - y) \left[12\,(y^2 - y^3)\right] dy = 24\pi \int_0^1 (1 - y)\,(y^2 - y^3)\,dy$

$$= 24\pi \int_0^1 (y^2 - 2y^3 + y^4)\,dy = 24\pi \left[\frac{y^3}{3} - \frac{y^4}{2} + \frac{y^5}{5}\right]_0^1 = 24\pi \left(\frac{1}{3} - \frac{1}{2} + \frac{1}{5}\right) = 24\pi \left(\frac{1}{30}\right) = \frac{4\pi}{5}$$

(c) $V = \int_c^d 2\pi \left(\begin{smallmatrix} \text{shell} \\ \text{radius} \end{smallmatrix}\right) \left(\begin{smallmatrix} \text{shell} \\ \text{height} \end{smallmatrix}\right) dy = \int_0^1 2\pi \left(\frac{8}{5} - y\right) \left[12\,(y^2 - y^3)\right] dy = 24\pi \int_0^1 \left(\frac{8}{5} - y\right)(y^2 - y^3)\,dy$

$$= 24\pi \int_0^1 \left(\frac{8}{5}\,y^2 - \frac{13}{5}\,y^3 + y^4\right) dy = 24\pi \left[\frac{8}{15}\,y^3 - \frac{13}{20}\,y^4 + \frac{y^5}{5}\right]_0^1 = 24\pi \left(\frac{8}{15} - \frac{13}{20} + \frac{1}{5}\right) = \frac{24\pi}{60}(32 - 39 + 12)$$

$$= \frac{24\pi}{12} = 2\pi$$

(d) $V = \int_c^d 2\pi \left(\begin{smallmatrix} \text{shell} \\ \text{radius} \end{smallmatrix}\right) \left(\begin{smallmatrix} \text{shell} \\ \text{height} \end{smallmatrix}\right) dy = \int_0^1 2\pi \left(y + \frac{2}{5}\right) \left[12\,(y^2 - y^3)\right] dy = 24\pi \int_0^1 \left(y + \frac{2}{5}\right)(y^2 - y^3)\,dy$

$$= 24\pi \int_0^1 \left(y^3 - y^4 + \frac{2}{5}\,y^2 - \frac{2}{5}\,y^3\right) dy = 24\pi \int_0^1 \left(\frac{2}{5}\,y^2 + \frac{3}{5}\,y^3 - y^4\right) dy = 24\pi \left[\frac{2}{15}\,y^3 + \frac{3}{20}\,y^4 - \frac{y^5}{5}\right]_0^1$$

$$= 24\pi \left(\frac{2}{15} + \frac{3}{20} - \frac{1}{5}\right) = \frac{24\pi}{60}(8 + 9 - 12) = \frac{24\pi}{12} = 2\pi$$

25. (a) About x-axis: $V = \int_c^d 2\pi \left(\begin{smallmatrix}\text{shell}\\\text{radius}\end{smallmatrix}\right) \left(\begin{smallmatrix}\text{shell}\\\text{height}\end{smallmatrix}\right) dy$

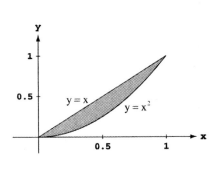

$= \int_0^1 2\pi y \left(\sqrt{y} - y\right) dy = 2\pi \int_0^1 \left(y^{3/2} - y^2\right) dy$

$= 2\pi \left[\frac{2}{5} y^{5/2} - \frac{1}{3} y^3\right]_0^1 = 2\pi \left(\frac{2}{5} - \frac{1}{3}\right) = \frac{2\pi}{15}$

About y-axis: $V = \int_a^b 2\pi \left(\begin{smallmatrix}\text{shell}\\\text{radius}\end{smallmatrix}\right) \left(\begin{smallmatrix}\text{shell}\\\text{height}\end{smallmatrix}\right) dx$

$= \int_0^1 2\pi x (x - x^2) dx = 2\pi \int_0^1 (x^2 - x^3) dx$

$= 2\pi \left[\frac{x^3}{3} - \frac{x^4}{4}\right]_0^1 = 2\pi \left(\frac{1}{3} - \frac{1}{4}\right) = \frac{\pi}{6}$

(b) About x-axis: $R(x) = x$ and $r(x) = x^2 \Rightarrow V = \int_a^b \pi \left[R(x)^2 - r(x)^2\right] dx = \int_0^1 \pi [x^2 - x^4] dx$

$= \pi \left[\frac{x^3}{3} - \frac{x^5}{5}\right]_0^1 = \pi \left(\frac{1}{3} - \frac{1}{5}\right) = \frac{2\pi}{15}$

About y-axis: $R(y) = \sqrt{y}$ and $r(y) = y \Rightarrow V = \int_c^d \pi \left[R(y)^2 - r(y)^2\right] dy = \int_0^1 \pi [y - y^2] dy$

$= \pi \left[\frac{y^2}{2} - \frac{y^3}{3}\right]_0^1 = \pi \left(\frac{1}{2} - \frac{1}{3}\right) = \frac{\pi}{6}$

27. (a) $V = \int_c^d 2\pi \left(\begin{smallmatrix}\text{shell}\\\text{radius}\end{smallmatrix}\right) \left(\begin{smallmatrix}\text{shell}\\\text{height}\end{smallmatrix}\right) dy = \int_1^2 2\pi y (y - 1) \, dy$

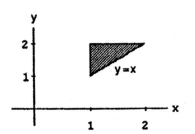

$= 2\pi \int_1^2 (y^2 - y) \, dy = 2\pi \left[\frac{y^3}{3} - \frac{y^2}{2}\right]_1^2$

$= 2\pi \left[\left(\frac{8}{3} - \frac{4}{2}\right) - \left(\frac{1}{3} - \frac{1}{2}\right)\right]$

$= 2\pi \left(\frac{7}{3} - 2 + \frac{1}{2}\right) = \frac{\pi}{3} (14 - 12 + 3) = \frac{5\pi}{3}$

(b) $V = \int_a^b 2\pi \left(\begin{smallmatrix}\text{shell}\\\text{radius}\end{smallmatrix}\right) \left(\begin{smallmatrix}\text{shell}\\\text{height}\end{smallmatrix}\right) dx = \int_1^2 2\pi x (2 - x) \, dx = 2\pi \int_1^2 (2x - x^2) \, dx = 2\pi \left[x^2 - \frac{x^3}{3}\right]_1^2$

$= 2\pi \left[\left(4 - \frac{8}{3}\right) - \left(1 - \frac{1}{3}\right)\right] = 2\pi \left[\left(\frac{12-8}{3}\right) - \left(\frac{3-1}{3}\right)\right] = 2\pi \left(\frac{4}{3} - \frac{2}{3}\right) = \frac{4\pi}{3}$

(c) $V = \int_a^b 2\pi \left(\begin{smallmatrix}\text{shell}\\\text{radius}\end{smallmatrix}\right) \left(\begin{smallmatrix}\text{shell}\\\text{height}\end{smallmatrix}\right) dx = \int_1^2 2\pi \left(\frac{10}{3} - x\right) (2 - x) \, dx = 2\pi \int_1^2 \left(\frac{20}{3} - \frac{16}{3} x + x^2\right) dx$

$= 2\pi \left[\frac{20}{3} x - \frac{8}{3} x^2 + \frac{1}{3} x^3\right]_1^2 = 2\pi \left[\left(\frac{40}{3} - \frac{32}{3} + \frac{8}{3}\right) - \left(\frac{20}{3} - \frac{8}{3} + \frac{1}{3}\right)\right] = 2\pi \left(\frac{3}{3}\right) = 2\pi$

(d) $V = \int_c^d 2\pi \left(\begin{smallmatrix}\text{shell}\\\text{radius}\end{smallmatrix}\right) \left(\begin{smallmatrix}\text{shell}\\\text{height}\end{smallmatrix}\right) dy = \int_1^2 2\pi (y - 1)(y - 1) \, dy = 2\pi \int_1^2 (y - 1)^2 = 2\pi \left[\frac{(y-1)^3}{3}\right]_1^2 = \frac{2\pi}{3}$

29. (a) $V = \int_c^d 2\pi \left(\begin{smallmatrix}\text{shell}\\\text{radius}\end{smallmatrix}\right) \left(\begin{smallmatrix}\text{shell}\\\text{height}\end{smallmatrix}\right) dy = \int_0^1 2\pi y (y - y^3) \, dy$

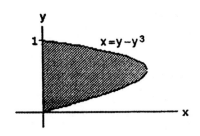

$= \int_0^1 2\pi (y^2 - y^4) \, dy = 2\pi \left[\frac{y^3}{3} - \frac{y^5}{5}\right]_0^1 = 2\pi \left(\frac{1}{3} - \frac{1}{5}\right)$

$= \frac{4\pi}{15}$

(b) $V = \int_c^d 2\pi \left(\begin{smallmatrix}\text{shell}\\\text{radius}\end{smallmatrix}\right) \left(\begin{smallmatrix}\text{shell}\\\text{height}\end{smallmatrix}\right) dy$

$= \int_0^1 2\pi (1 - y)(y - y^3) \, dy$

$= 2\pi \int_0^1 (y - y^2 - y^3 + y^4) \, dy = 2\pi \left[\frac{y^2}{2} - \frac{y^3}{3} - \frac{y^4}{4} + \frac{y^5}{5}\right]_0^1 = 2\pi \left(\frac{1}{2} - \frac{1}{3} - \frac{1}{4} + \frac{1}{5}\right) = \frac{2\pi}{60} (30 - 20 - 15 + 12) = \frac{7\pi}{30}$

31. (a) $V = \int_c^d 2\pi \left(\begin{smallmatrix}\text{shell}\\\text{radius}\end{smallmatrix}\right)\left(\begin{smallmatrix}\text{shell}\\\text{height}\end{smallmatrix}\right) dy = \int_0^2 2\pi y \left(\sqrt{8y} - y^2\right) dy$

$= 2\pi \int_0^2 \left(2\sqrt{2}\, y^{3/2} - y^3\right) dy = 2\pi \left[\frac{4\sqrt{2}}{5} y^{5/2} - \frac{y^4}{4}\right]_0^2$

$= 2\pi \left(\frac{4\sqrt{2}\cdot(\sqrt{2})^5}{5} - \frac{2^4}{4}\right) = 2\pi \left(\frac{4\cdot 2^3}{5} - \frac{4\cdot 4}{4}\right)$

$= 2\pi \cdot 4 \left(\frac{8}{5} - 1\right) = \frac{8\pi}{5}(8-5) = \frac{24\pi}{5}$

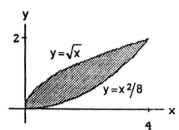

(b) $V = \int_a^b 2\pi \left(\begin{smallmatrix}\text{shell}\\\text{radius}\end{smallmatrix}\right)\left(\begin{smallmatrix}\text{shell}\\\text{height}\end{smallmatrix}\right) dx = \int_0^4 2\pi x \left(\sqrt{x} - \frac{x^2}{8}\right) dx = 2\pi \int_0^4 \left(x^{3/2} - \frac{x^3}{8}\right) dx = 2\pi \left[\frac{2}{5} x^{5/2} - \frac{x^4}{32}\right]_0^4$

$= 2\pi \left(\frac{2\cdot 2^5}{5} - \frac{4^4}{32}\right) = 2\pi \left(\frac{2^6}{5} - \frac{2^8}{32}\right) = \frac{\pi\cdot 2^7}{160}(32-20) = \frac{\pi\cdot 2^9\cdot 3}{160} = \frac{\pi\cdot 2^4\cdot 3}{5} = \frac{48\pi}{5}$

33. (a) $V = \int_0^{\pi/2} 2\pi x \cos x\, dx = 2\pi \left([x \sin x]_0^{\pi/2} - \int_0^{\pi/2} \sin x\, dx\right)$

$= 2\pi \left(\frac{\pi}{2} + [\cos x]_0^{\pi/2}\right) = 2\pi \left(\frac{\pi}{2} + 0 - 1\right) = \pi(\pi - 2)$

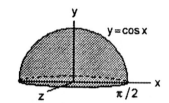

(b) $V = \int_0^{\pi/2} 2\pi \left(\frac{\pi}{2} - x\right)\cos x\, dx$; $u = \frac{\pi}{2} - x$, $du = -dx$; $dv = \cos x\, dx$, $v = \sin x$;

$V = 2\pi \left[\left(\frac{\pi}{2} - x\right)\sin x\right]_0^{\pi/2} + 2\pi \int_0^{\pi/2} \sin x\, dx = 0 + 2\pi[-\cos x]_0^{\pi/2} = 2\pi(0+1) = 2\pi$

35. (a) *Disk:* $V = V_1 - V_2$

$V_1 = \int_{a_1}^{b_1} \pi[R_1(x)]^2\, dx$ and $V_2 = \int_{a_2}^{b_2} \pi[R_2(x)]^2$ with $R_1(x) = \sqrt{\frac{x+2}{3}}$ and $R_2(x) = \sqrt{x}$,

$a_1 = -2$, $b_1 = 1$; $a_2 = 0$, $b_2 = 1$ $\Rightarrow$ two integrals are required

(b) *Washer:* $V = V_1 - V_2$

$V_1 = \int_{a_1}^{b_1} \pi\left([R_1(x)]^2 - [r_1(x)]^2\right) dx$ with $R_1(x) = \sqrt{\frac{x+2}{3}}$ and $r_1(x) = 0$; $a_1 = -2$ and $b_1 = 0$;

$V_2 = \int_{a_2}^{b_2} \pi\left([R_2(x)]^2 - [r_2(x)]^2\right) dx$ with $R_2(x) = \sqrt{\frac{x+2}{3}}$ and $r_2(x) = \sqrt{x}$; $a_2 = 0$ and $b_2 = 1$

$\Rightarrow$ two integrals are required

(c) *Shell:* $V = \int_c^d 2\pi \left(\begin{smallmatrix}\text{shell}\\\text{radius}\end{smallmatrix}\right)\left(\begin{smallmatrix}\text{shell}\\\text{height}\end{smallmatrix}\right) dy = \int_c^d 2\pi y \left(\begin{smallmatrix}\text{shell}\\\text{height}\end{smallmatrix}\right) dy$ where shell height $= y^2 - (3y^2 - 2) = 2 - 2y^2$;

$c = 0$ and $d = 1$. Only *one* integral is required. It is, therefore preferable to use the *shell* method.
However, whichever method you use, you will get $V = \pi$.

37. $W(a) = \int_{f(a)}^{f(a)} \pi\left[\left(f^{-1}(y)\right)^2 - a^2\right] dy = 0 = \int_a^a 2\pi x[f(a) - f(x)]\, dx = S(a)$; $W'(t) = \pi\left[\left(f^{-1}(f(t))\right)^2 - a^2\right] f'(t)$

$= \pi\left(t^2 - a^2\right) f'(t)$; also $S(t) = 2\pi f(t) \int_a^t x\, dx - 2\pi \int_a^t x f(x)\, dx = \left[\pi f(t)t^2 - \pi f(t)a^2\right] - 2\pi \int_a^t x f(x)\, dx$

$\Rightarrow S'(t) = \pi t^2 f'(t) + 2\pi t f(t) - \pi a^2 f'(t) - 2\pi t f(t) = \pi\left(t^2 - a^2\right) f'(t) \Rightarrow W'(t) = S'(t)$. Therefore, $W(t) = S(t)$
for all $t \in [a, b]$.

39. $V = \int_a^b 2\pi \left(\begin{smallmatrix}\text{shell}\\\text{radius}\end{smallmatrix}\right)\left(\begin{smallmatrix}\text{shell}\\\text{height}\end{smallmatrix}\right) dx = \int_0^1 2\pi x\, e^{-x^2} dx = -\pi e^{-x^2}\Big|_0^1 = -\pi(e^{-1} - e^0) = \pi\left(1 - \frac{1}{e}\right)$

6.3 LENGTHS OF PLANE CURVES

1. $\frac{dx}{dt} = -1$ and $\frac{dy}{dt} = 3 \Rightarrow \sqrt{\left(\frac{dx}{dt}\right)^2 + \left(\frac{dy}{dt}\right)^2} = \sqrt{(-1)^2 + (3)^2} = \sqrt{10}$

$\Rightarrow$ Length $= \int_{-2/3}^1 \sqrt{10}\, dt = \sqrt{10}\, [t]_{-2/3}^1 = \sqrt{10} - \left(-\frac{2}{3}\sqrt{10}\right) = \frac{5\sqrt{10}}{3}$

3. $\frac{dx}{dt} = 3t^2$ and $\frac{dy}{dt} = 3t \Rightarrow \sqrt{\left(\frac{dx}{dt}\right)^2 + \left(\frac{dy}{dt}\right)^2} = \sqrt{(3t^2)^2 + (3t)^2} = \sqrt{9t^4 + 9t^2} = 3t\sqrt{t^2 + 1}$ $\left(\text{since } t \geq 0 \text{ on } \left[0, \sqrt{3}\right]\right)$

$\Rightarrow$ Length $= \int_0^{\sqrt{3}} 3t\sqrt{t^2 + 1}\, dt;$ $\left[u = t^2 + 1 \Rightarrow \frac{3}{2} du = 3t\, dt; t = 0 \Rightarrow u = 1, t = \sqrt{3} \Rightarrow u = 4\right]$

$\rightarrow \int_1^4 \frac{3}{2} u^{1/2}\, du = \left[u^{3/2}\right]_1^4 = (8 - 1) = 7$

5. $\frac{dx}{dt} = (2t + 3)^{1/2}$ and $\frac{dy}{dt} = 1 + t \Rightarrow \sqrt{\left(\frac{dx}{dt}\right)^2 + \left(\frac{dy}{dt}\right)^2} = \sqrt{(2t + 3) + (1 + t)^2} = \sqrt{t^2 + 4t + 4} = |t + 2| = t + 2$

since $0 \leq t \leq 3 \Rightarrow$ Length $= \int_0^3 (t + 2)\, dt = \left[\frac{t^2}{2} + 2t\right]_0^3 = \frac{21}{2}$

7. $\frac{dx}{dt} = e^t - 1$ and $\frac{dy}{dt} = 2e^{t/2} \Rightarrow \sqrt{\left(\frac{dx}{dt}\right)^2 + \left(\frac{dy}{dt}\right)^2} = \sqrt{(e^t - 1)^2 + (2e^{t/2})^2} = \sqrt{e^{2t} + 2e^t + 1} = \sqrt{(e^t + 1)^2} = |e^t + 1|$

$= e^t + 1$ (since $e^t + 1 > 0$ for all t) $\Rightarrow L = \int_0^3 (e^t + 1)\, dt = [e^t + t]_0^3 = e^3 + 2$

9. $\frac{dy}{dx} = \frac{1}{3} \cdot \frac{3}{2} (x^2 + 2)^{1/2} \cdot 2x = \sqrt{(x^2 + 2)} \cdot x$

$\Rightarrow L = \int_0^3 \sqrt{1 + (x^2 + 2) x^2}\, dx = \int_0^3 \sqrt{1 + 2x^2 + x^4}\, dx$

$= \int_0^3 \sqrt{(1 + x^2)^2}\, dx = \int_0^3 (1 + x^2)\, dx = \left[x + \frac{x^3}{3}\right]_0^3$

$= 3 + \frac{27}{3} = 12$

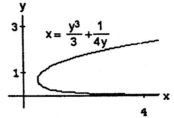

11. $\frac{dx}{dy} = y^2 - \frac{1}{4y^2} \Rightarrow \left(\frac{dx}{dy}\right)^2 = y^4 - \frac{1}{2} + \frac{1}{16y^4}$

$\Rightarrow L = \int_1^3 \sqrt{1 + y^4 - \frac{1}{2} + \frac{1}{16y^4}}\, dy$

$= \int_1^3 \sqrt{y^4 + \frac{1}{2} + \frac{1}{16y^4}}\, dy$

$= \int_1^3 \sqrt{\left(y^2 + \frac{1}{4y^2}\right)^2}\, dy = \int_1^3 \left(y^2 + \frac{1}{4y^2}\right) dy$

$= \left[\frac{y^3}{3} - \frac{y^{-1}}{4}\right]_1^3 = \left(\frac{27}{3} - \frac{1}{12}\right) - \left(\frac{1}{3} - \frac{1}{4}\right) = 9 - \frac{1}{12} - \frac{1}{3} + \frac{1}{4} = 9 + \frac{(-1 - 4 + 3)}{12} = 9 + \frac{(-2)}{12} = \frac{53}{6}$

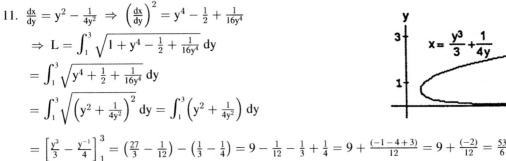

13. $\frac{dx}{dy} = y^3 - \frac{1}{4y^3} \Rightarrow \left(\frac{dx}{dy}\right)^2 = y^6 - \frac{1}{2} + \frac{1}{16y^6}$

$\Rightarrow L = \int_1^2 \sqrt{1 + y^6 - \frac{1}{2} + \frac{1}{16y^6}}\, dy$

$= \int_1^2 \sqrt{y^6 + \frac{1}{2} + \frac{1}{16y^6}}\, dy = \int_1^2 \sqrt{\left(y^3 + \frac{y^{-3}}{4}\right)^2}\, dy$

$= \int_1^2 \left(y^3 + \frac{y^{-3}}{4}\right) dy = \left[\frac{y^4}{4} - \frac{y^{-2}}{8}\right]_1^2$

$= \left(\frac{16}{4} - \frac{1}{(16)(2)}\right) - \left(\frac{1}{4} - \frac{1}{8}\right) = 4 - \frac{1}{32} - \frac{1}{4} + \frac{1}{8} = \frac{128 - 1 - 8 + 4}{32} = \frac{123}{32}$

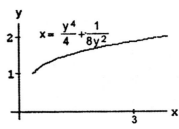

15. $\frac{dy}{dx} = x^{1/3} - \frac{1}{4} x^{-1/3} \Rightarrow \left(\frac{dy}{dx}\right)^2 = x^{2/3} - \frac{1}{2} + \frac{x^{-2/3}}{16}$

$\Rightarrow L = \int_1^8 \sqrt{1 + x^{2/3} - \frac{1}{2} + \frac{x^{-2/3}}{16}}\, dx$

$= \int_1^8 \sqrt{x^{2/3} + \frac{1}{2} + \frac{x^{-2/3}}{16}}\, dx$

$= \int_1^8 \sqrt{\left(x^{1/3} + \frac{1}{4} x^{-1/3}\right)^2}\, dx = \int_1^8 \left(x^{1/3} + \frac{1}{4} x^{-1/3}\right) dx$

$= \left[\frac{3}{4} x^{4/3} + \frac{3}{8} x^{2/3}\right]_1^8 = \frac{3}{8} \left[2x^{4/3} + x^{2/3}\right]_1^8$

$= \frac{3}{8} \left[(2 \cdot 2^4 + 2^2) - (2 + 1)\right] = \frac{3}{8} (32 + 4 - 3) = \frac{99}{8}$

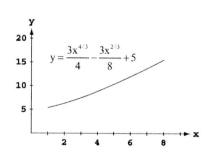

17. $y = \sqrt{1 - x^2} = (1 - x^2)^{1/2} \Rightarrow y' = \left(\frac{1}{2}\right)(1 - x^2)^{-1/2}(-2x) \Rightarrow 1 + (y')^2 = \frac{1}{1 - x^2}\,; L = \int_{-1/2}^{1/2} \sqrt{1 + (y')^2}\, dx$

$= 2 \int_0^{1/2} \frac{1}{\sqrt{1 - x^2}}\, dx = 2 \left[\sin^{-1} x\right]_0^{1/2} = 2 \left(\frac{\pi}{6} - 0\right) = \frac{\pi}{3}$

19. (a) $\frac{dy}{dx} = 2x \Rightarrow \left(\frac{dy}{dx}\right)^2 = 4x^2$

$\Rightarrow L = \int_{-1}^2 \sqrt{1 + \left(\frac{dy}{dx}\right)^2}\, dx$

$= \int_{-1}^2 \sqrt{1 + 4x^2}\, dx$

(c) $L \approx 6.13$

(b)

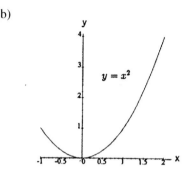

21. (a) $\frac{dx}{dy} = \cos y \Rightarrow \left(\frac{dx}{dy}\right)^2 = \cos^2 y$

$\Rightarrow L = \int_0^\pi \sqrt{1 + \cos^2 y}\, dy$

(c) $L \approx 3.82$

(b)

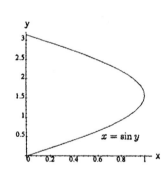

23. (a) $2y + 2 = 2 \frac{dx}{dy} \Rightarrow \left(\frac{dx}{dy}\right)^2 = (y + 1)^2$

$\Rightarrow L = \int_{-1}^3 \sqrt{1 + (y + 1)^2}\, dy$

(c) $L \approx 9.29$

(b)

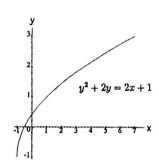

25. (a) $\frac{dy}{dx} = \tan x \Rightarrow \left(\frac{dy}{dx}\right)^2 = \tan^2 x$

   $\Rightarrow L = \int_0^{\pi/6} \sqrt{1 + \tan^2 x}\ dx = \int_0^{\pi/6} \sqrt{\frac{\sin^2 x + \cos^2 x}{\cos^2 x}}\ dx$

   $= \int_0^{\pi/6} \frac{dx}{\cos x} = \int_0^{\pi/6} \sec x\ dx$

   (c) $L \approx 0.55$

(b)

$y = \int_0^x \tan t\,dt$

$= -\ln \cos(x)$

27. $y = \ln(\sec x);\ y' = \frac{\sec x \tan x}{\sec x} = \tan x; (y')^2 = \tan^2 x; \int_0^{\pi/4} \sqrt{1 + \tan^2 x}\ dx = \int_0^{\pi/4} |\sec x|\ dx = [\ln|\sec x + \tan x|]_0^{\pi/4}$

   $= \ln\left(\sqrt{2} + 1\right) - \ln(0 + 1) = \ln\left(\sqrt{2} + 1\right)$

29. (a) $\left(\frac{dy}{dx}\right)^2$ corresponds to $\frac{1}{4x}$ here, so take $\frac{dy}{dx}$ as $\frac{1}{2\sqrt{x}}$. Then $y = \sqrt{x} + C$ and since $(1, 1)$ lies on the curve, $C = 0$.
   So $y = \sqrt{x}$ from $(1, 1)$ to $(4, 2)$.

   (b) Only one. We know the derivative of the function and the value of the function at one value of x.

31. $L = \int_0^1 \sqrt{1 + \frac{e^x}{4}}\ dx \Rightarrow \frac{dy}{dx} = \frac{e^{x/2}}{2} \Rightarrow y = e^{x/2} + C; y(0) = 0 \Rightarrow 0 = e^0 + C \Rightarrow C = -1 \Rightarrow y = e^{x/2} - 1$

33. $\frac{dx}{dt} = \left(\frac{1}{\sec t + \tan t}\right)(\sec t \tan t + \sec^2 t) - \cos t = \sec t - \cos t$ and $\frac{dy}{dt} = -\sin t \Rightarrow \sqrt{\left(\frac{dx}{dt}\right)^2 + \left(\frac{dy}{dt}\right)^2}$

   $= \sqrt{(\sec t - \cos t)^2 + (-\sin t)^2} = \sqrt{\sec^2 t - 1} = \sqrt{\tan^2 t} = |\tan t| = \tan t$ since $0 \le t \le \frac{\pi}{3}$

   $\Rightarrow$ Length $= \int_0^{\pi/3} \tan t\ dt = \int_0^{\pi/3} \frac{\sin t}{\cos t}\ dt = [-\ln|\cos t|]_0^{\pi/3} = -\ln \frac{1}{2} + \ln 1 = \ln 2$

35. $\frac{dx}{dt} = e^t - e^{-t}$ and $\frac{dy}{dt} = -2 \Rightarrow \sqrt{\left(\frac{dx}{dt}\right)^2 + \left(\frac{dy}{dt}\right)^2} = \sqrt{(e^t - e^{-t})^2 + (-2)^2} = \sqrt{e^{2t} + 2 + e^{-2t}} = \sqrt{\frac{e^{4t} + 2e^{2t} + 1}{e^{2t}}}$

   $= \sqrt{\frac{(e^{2t} + 1)^2}{(e^t)^2}} = \frac{e^{2t} + 1}{e^t} = e^t + e^{-t} \Rightarrow$ Length $= \int_0^3 (e^t + e^{-t})dt = (e^t - e^{-t})\Big|_0^3 = (e^3 - e^{-3}) - (e^0 - e^0) = e^3 - \frac{1}{e^3}$

6.4 EXPONENTIAL CHANGE AND SEPARABLE DIFFERENTIAL EQUATIONS

1. (a) $y = e^{-x} \Rightarrow y' = -e^{-x} \Rightarrow 2y' + 3y = 2(-e^{-x}) + 3e^{-x} = e^{-x}$

   (b) $y = e^{-x} + e^{-3x/2} \Rightarrow y' = -e^{-x} - \frac{3}{2}e^{-3x/2} \Rightarrow 2y' + 3y = 2\left(-e^{-x} - \frac{3}{2}e^{-3x/2}\right) + 3\left(e^{-x} + e^{-3x/2}\right) = e^{-x}$

   (c) $y = e^{-x} + Ce^{-3x/2} \Rightarrow y' = -e^{-x} - \frac{3}{2}Ce^{-3x/2} \Rightarrow 2y' + 3y = 2\left(-e^{-x} - \frac{3}{2}Ce^{-3x/2}\right) + 3\left(e^{-x} + Ce^{-3x/2}\right) = e^{-x}$

3. $y = \frac{1}{x}\int_1^x \frac{e^t}{t}\ dt \Rightarrow y' = -\frac{1}{x^2}\int_1^x \frac{e^t}{t}\ dt + \left(\frac{1}{x}\right)\left(\frac{e^x}{x}\right) \Rightarrow x^2 y' = -\int_1^x \frac{e^t}{t}\ dt + e^x = -x\left(\frac{1}{x}\int_1^x \frac{e^t}{t}\ dt\right) + e^x = -xy + e^x$

   $\Rightarrow x^2 y' + xy = e^x$

5. $y = e^{-x}\tan^{-1}(2e^x) \Rightarrow y' = -e^{-x}\tan^{-1}(2e^x) + e^{-x}\left[\frac{1}{1 + (2e^x)^2}\right](2e^x) = -e^{-x}\tan^{-1}(2e^x) + \frac{2}{1 + 4e^{2x}}$

   $\Rightarrow y' = -y + \frac{2}{1 + 4e^{2x}} \Rightarrow y' + y = \frac{2}{1 + 4e^{2x}}; y(-\ln 2) = e^{-(-\ln 2)}\tan^{-1}(2e^{-\ln 2}) = 2\tan^{-1} 1 = 2\left(\frac{\pi}{4}\right) = \frac{\pi}{2}$

7. $y = \frac{\cos x}{x} \Rightarrow y' = \frac{-x\sin x - \cos x}{x^2} \Rightarrow y' = -\frac{\sin x}{x} - \frac{1}{x}\left(\frac{\cos x}{x}\right) \Rightarrow y' = -\frac{\sin x}{x} - \frac{y}{x} \Rightarrow xy' = -\sin x - y$

   $\Rightarrow xy' + y = -\sin x; y\left(\frac{\pi}{2}\right) = \frac{\cos(\pi/2)}{(\pi/2)} = 0$

9. $2\sqrt{xy}\,\frac{dy}{dx} = 1 \Rightarrow 2x^{1/2}y^{1/2}\,dy = dx \Rightarrow 2y^{1/2}\,dy = x^{-1/2}\,dx \Rightarrow \int 2y^{1/2}\,dy = \int x^{-1/2}\,dx \Rightarrow 2\left(\frac{2}{3}y^{3/2}\right)$
$= 2x^{1/2} + C_1 \Rightarrow \frac{2}{3}y^{3/2} - x^{1/2} = C$, where $C = \frac{1}{2}C_1$

11. $\frac{dy}{dx} = e^{x-y} \Rightarrow dy = e^x e^{-y}\,dx \Rightarrow e^y\,dy = e^x\,dx \Rightarrow \int e^y\,dy = \int e^x\,dx \Rightarrow e^y = e^x + C \Rightarrow e^y - e^x = C$

13. $\frac{dy}{dx} = \sqrt{y}\cos^2\sqrt{y} \Rightarrow dy = \left(\sqrt{y}\cos^2\sqrt{y}\right)dx \Rightarrow \frac{\sec^2\sqrt{y}}{\sqrt{y}}\,dy = dx \Rightarrow \int \frac{\sec^2\sqrt{y}}{\sqrt{y}}\,dy = \int dx$. In the integral on the left-hand
side, substitute $u = \sqrt{y} \Rightarrow du = \frac{1}{2\sqrt{y}}\,dy \Rightarrow 2\,du = \frac{1}{\sqrt{y}}\,dy$, and we have $\int \sec^2 u\,du = \int dx \Rightarrow 2\tan u = x + C$
$\Rightarrow -x + 2\tan\sqrt{y} = C$

15. $\sqrt{x}\,\frac{dy}{dx} = e^{y+\sqrt{x}} \Rightarrow \frac{dy}{dx} = \frac{e^y e^{\sqrt{x}}}{\sqrt{x}} \Rightarrow dy = \frac{e^y e^{\sqrt{x}}}{\sqrt{x}}\,dx \Rightarrow e^{-y}\,dy = \frac{e^{\sqrt{x}}}{\sqrt{x}}\,dx \Rightarrow \int e^{-y}\,dy = \int \frac{e^{\sqrt{x}}}{\sqrt{x}}\,dx$. In the integral on the
right-hand side, substitute $u = \sqrt{x} \Rightarrow du = \frac{1}{2\sqrt{x}}\,dx \Rightarrow 2\,du = \frac{1}{\sqrt{x}}\,dx$, and we have $\int e^{-y}\,dy = 2\int e^u\,du$
$\Rightarrow -e^{-y} = 2e^u + C_1 \Rightarrow -e^{-y} = 2e^{\sqrt{x}} + C$, where $C = -C_1$

17. $\frac{dy}{dx} = 2x\sqrt{1-y^2} \Rightarrow dy = 2x\sqrt{1-y^2}\,dx \Rightarrow \frac{dy}{\sqrt{1-y^2}} = 2x\,dx \Rightarrow \int \frac{dy}{\sqrt{1-y^2}} = \int 2x\,dx \Rightarrow \sin^{-1}y = x^2 + C$ since $|y| < 1$
$\Rightarrow y = \sin(x^2 + C)$

19. $\frac{dy}{dt} = -0.6y \Rightarrow y = y_0 e^{-0.6t}; \; y_0 = 100 \Rightarrow y = 100e^{-0.6t} \Rightarrow y = 100e^{-0.6} \approx 54.88$ grams when $t = 1$ hr

21. $L(x) = L_0 e^{-kx} \Rightarrow \frac{L_0}{2} = L_0 e^{-18k} \Rightarrow \ln\frac{1}{2} = -18k \Rightarrow k = \frac{\ln 2}{18} \approx 0.0385 \Rightarrow L(x) = L_0 e^{-0.0385x}$; when the intensity
is one-tenth of the surface value, $\frac{L_0}{10} = L_0 e^{-0.0385x} \Rightarrow \ln 10 = 0.0385x \Rightarrow x \approx 59.8$ ft

23. $y = y_0 e^{kt}$ and $y_0 = 1 \Rightarrow y = e^{kt} \Rightarrow$ at $y = 2$ and $t = 0.5$ we have $2 = e^{0.5k} \Rightarrow \ln 2 = 0.5k \Rightarrow k = \frac{\ln 2}{0.5} = \ln 4$.
Therefore, $y = e^{(\ln 4)t} \Rightarrow y = e^{24\ln 4} = 4^{24} = 2.81474978 \times 10^{14}$ at the end of 24 hrs

25. (a) $10{,}000e^{k(1)} = 7500 \Rightarrow e^k = 0.75 \Rightarrow k = \ln 0.75$ and $y = 10{,}000e^{(\ln 0.75)t}$. Now $1000 = 10{,}000e^{(\ln 0.75)t}$
$\Rightarrow \ln 0.1 = (\ln 0.75)t \Rightarrow t = \frac{\ln 0.1}{\ln 0.75} \approx 8.00$ years (to the nearest hundredth of a year)
(b) $1 = 10{,}000e^{(\ln 0.75)t} \Rightarrow \ln 0.0001 = (\ln 0.75)t \Rightarrow t = \frac{\ln 0.0001}{\ln 0.75} \approx 32.02$ years (to the nearest hundredth of a year)

27. $0.9P_0 = P_0 e^k \Rightarrow k = \ln 0.9$; when the well's output falls to one-fifth of its present value $P = 0.2P_0$
$\Rightarrow 0.2P_0 = P_0 e^{(\ln 0.9)t} \Rightarrow 0.2 = e^{(\ln 0.9)t} \Rightarrow \ln(0.2) = (\ln 0.9)t \Rightarrow t = \frac{\ln 0.2}{\ln 0.9} \approx 15.28$ yr

29. $A = A_0 e^{kt}$ and $A_0 = 10 \Rightarrow A = 10e^{kt}, \; 5 = 10e^{k(24360)} \Rightarrow k = \frac{\ln(0.5)}{24360} \approx -0.000028454 \Rightarrow A = 10e^{-0.000028454t}$,
then $0.2(10) = 10e^{-0.000028254t} \Rightarrow t = \frac{\ln 0.2}{-0.000028454} \approx 56563$ years

31. $y = y_0 e^{-kt} = y_0 e^{-(k)(3/k)} = y_0 e^{-3} = \frac{y_0}{e^3} < \frac{y_0}{20} = (0.05)(y_0) \Rightarrow$ after three mean lifetimes less than 5% remains

33. $T - T_s = (T_0 - T_s)e^{-kt}, \; T_0 = 90°C, \; T_s = 20°C, \; T = 60°C \Rightarrow 60 - 20 = 70e^{-10k} \Rightarrow \frac{4}{7} = e^{-10k}$
$\Rightarrow k = \frac{\ln\left(\frac{7}{4}\right)}{10} \approx 0.05596$
(a) $35 - 20 = 70e^{-0.05596t} \Rightarrow t \approx 27.5$ min is the total time $\Rightarrow$ it will take $27.5 - 10 = 17.5$ minutes longer to reach
$35°C$
(b) $T - T_s = (T_0 - T_s)e^{-kt}, \; T_0 = 90°C, \; T_s = -15°C \Rightarrow 35 + 15 = 105e^{-0.05596t} \Rightarrow t \approx 13.26$ min

35. $T - T_s = (T_0 - T_s) e^{-kt} \Rightarrow 39 - T_s = (46 - T_s) e^{-10k}$ and $33 - T_s = (46 - T_s) e^{-20k} \Rightarrow \frac{39-T_s}{46-T_s} = e^{-10k}$ and

$\frac{33-T_s}{46-T_s} = e^{-20k} = \left(e^{-10k}\right)^2 \Rightarrow \frac{33-T_s}{46-T_s} = \left(\frac{39-T_s}{46-T_s}\right)^2 \Rightarrow (33 - T_s)(46 - T_s) = (39 - T_s)^2 \Rightarrow 1518 - 79T_s + T_s^2$

$= 1521 - 78T_s + T_s^2 \Rightarrow -T_s = 3 \Rightarrow T_s = -3°C$

37. From Example 4, the half-life of carbon-14 is 5700 yr $\Rightarrow \frac{1}{2} c_0 = c_0 e^{-k(5700)} \Rightarrow k = \frac{\ln 2}{5700} \approx 0.0001216$

$\Rightarrow c = c_0 e^{-0.0001216t} \Rightarrow (0.445)c_0 = c_0 e^{-0.0001216t} \Rightarrow t = \frac{\ln(0.445)}{-0.0001216} \approx 6659$ years

39. From Exercise 37, $k \approx 0.0001216$ for carbon-14. Thus, $c = c_0 e^{-0.0001216t} \Rightarrow (0.995)c_0 = c_0 e^{-0.0001216t}$

$\Rightarrow t = \frac{\ln(0.995)}{-0.0001216} \approx 41$ years old

6.5 WORK AND FLUID FORCES

1. The force required to stretch the spring from its natural length of 2 m to a length of 5 m is $F(x) = kx$. The work done

by F is $W = \int_0^3 F(x)\, dx = k \int_0^3 x\, dx = \frac{k}{2} [x^2]_0^3 = \frac{9k}{2}$. This work is equal to 1800 J $\Rightarrow \frac{9}{2} k = 1800 \Rightarrow k = 400$ N/m

3. We find the force constant from Hooke's law: $F = kx$. A force of 2 N stretches the spring to 0.02 m

$\Rightarrow 2 = k \cdot (0.02) \Rightarrow k = 100 \frac{N}{m}$. The force of 4 N will stretch the rubber band y m, where $F = ky \Rightarrow y = \frac{F}{k}$

$\Rightarrow y = \frac{4N}{100 \frac{N}{m}} \Rightarrow y = 0.04$ m = 4 cm. The work done to stretch the rubber band 0.04 m is $W = \int_0^{0.04} kx\, dx$

$= 100 \int_0^{0.04} x\, dx = 100 \left[\frac{x^2}{2}\right]_0^{0.04} = \frac{(100)(0.04)^2}{2} = 0.08$ J

5. (a) We find the spring's constant from Hooke's law: $F = kx \Rightarrow k = \frac{F}{x} = \frac{21,714}{8-5} = \frac{21,714}{3} \Rightarrow k = 7238 \frac{lb}{in}$

(b) The work done to compress the assembly the first half inch is $W = \int_0^{0.5} kx\, dx = 7238 \int_0^{0.5} x\, dx = 7238 \left[\frac{x^2}{2}\right]_0^{0.5}$

$= (7238) \frac{(0.5)^2}{2} = \frac{(7238)(0.25)}{2} \approx 905$ in · lb. The work done to compress the assembly the second half inch is:

$W = \int_{0.5}^{1.0} kx\, dx = 7238 \int_{0.5}^{1.0} x\, dx = 7238 \left[\frac{x^2}{2}\right]_{0.5}^{1.0} = \frac{7238}{2} [1 - (0.5)^2] = \frac{(7238)(0.75)}{2} \approx 2714$ in · lb

7. The force required to haul up the rope is equal to the rope's weight, which varies steadily and is proportional to x,

the length of the rope still hanging: $F(x) = 0.624x$. The work done is: $W = \int_0^{50} F(x)\, dx = \int_0^{50} 0.624x\, dx$

$= 0.624 \left[\frac{x^2}{2}\right]_0^{50} = 780$ J

9. The force required to lift the cable is equal to the weight of the cable paid out: $F(x) = (4.5)(180 - x)$ where x

is the position of the car off the first floor. The work done is: $W = \int_0^{180} F(x)\, dx = 4.5 \int_0^{180} (180 - x)\, dx$

$= 4.5 \left[180x - \frac{x^2}{2}\right]_0^{180} = 4.5 \left(180^2 - \frac{180^2}{2}\right) = \frac{4.5 \cdot 180^2}{2} = 72,900$ ft · lb

11. Let r = the constant rate of leakage. Since the bucket is leaking at a constant rate and the bucket is rising at a constant rate, the amount of water in the bucket is proportional to $(20 - x)$, the distance the bucket is being raised. The leakage rate of the water is 0.8 lb/ft raised and the weight of the water in the bucket is $F = 0.8(20 - x)$. So:

$W = \int_0^{20} 0.8(20 - x)\, dx = 0.8 \left[20x - \frac{x^2}{2}\right]_0^{20} = 160$ ft · lb.

13. We will use the coordinate system given.

    (a) The typical slab between the planes at y and $y + \Delta y$ has a volume of $\Delta V = (10)(12)\,\Delta y = 120\,\Delta y$ ft^3. The force F required to lift the slab is equal to its weight: $F = 62.4\,\Delta V = 62.4 \cdot 120\,\Delta y$ lb. The distance through which F must act is about y ft, so the work done lifting the slab is about $\Delta W = $ force $\times$ distance $= 62.4 \cdot 120 \cdot y \cdot \Delta y$ ft $\cdot$ lb. The work it takes to lift all the water is approximately $W \approx \sum\limits_{0}^{20} \Delta W$

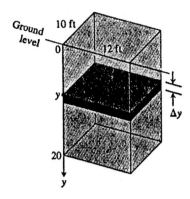

$= \sum\limits_{0}^{20} 62.4 \cdot 120y \cdot \Delta y$ ft $\cdot$ lb. This is a Riemann sum for the function $62.4 \cdot 120y$ over the interval $0 \le y \le 20$. The work of pumping the tank empty is the limit of these sums:

$$W = \int_{0}^{20} 62.4 \cdot 120y \, dy = (62.4)(120)\left[\frac{y^2}{2}\right]_{0}^{20} = (62.4)(120)\left(\frac{400}{2}\right) = (62.4)(120)(200) = 1{,}497{,}600 \text{ ft} \cdot \text{lb}$$

    (b) The time t it takes to empty the full tank with $\left(\frac{5}{11}\right)$–hp motor is $t = \frac{W}{250\,\frac{\text{ft·lb}}{\text{sec}}} = \frac{1{,}497{,}600 \text{ ft·lb}}{250\,\frac{\text{ft·lb}}{\text{sec}}} = 5990.4$ sec

    $= 1.664$ hr $\Rightarrow t \approx 1$ hr and 40 min

    (c) Following all the steps of part (a), we find that the work it takes to lower the water level 10 ft is

    $$W = \int_{0}^{10} 62.4 \cdot 120y \, dy = (62.4)(120)\left[\frac{y^2}{2}\right]_{0}^{10} = (62.4)(120)\left(\frac{100}{2}\right) = 374{,}400 \text{ ft} \cdot \text{lb and the time is } t = \frac{W}{250\,\frac{\text{ft·lb}}{\text{sec}}}$$

    $= 1497.6$ sec $= 0.416$ hr ≈ 25 min

    (d) In a location where water weighs $62.26\,\frac{\text{lb}}{\text{ft}^3}$:

    a) $W = (62.26)(24{,}000) = 1{,}494{,}240$ ft $\cdot$ lb.

    b) $t = \frac{1{,}494{,}240}{250} = 5976.96$ sec ≈ 1.660 hr $\Rightarrow t \approx 1$ hr and 40 min

    In a location where water weighs $62.59\,\frac{\text{lb}}{\text{ft}^3}$

    a) $W = (62.59)(24{,}000) = 1{,}502{,}160$ ft $\cdot$ lb

    b) $t = \frac{1{,}502{,}160}{250} = 6008.64$ sec ≈ 1.669 hr $\Rightarrow t \approx 1$ hr and 40.1 min

15. The slab is a disk of area $\pi x^2 = \pi\left(\frac{y}{2}\right)^2$, thickness $\triangle y$, and height below the top of the tank $(10 - y)$. So the work to pump the oil in this slab, $\triangle W$, is $57(10 - y)\pi\left(\frac{y}{2}\right)^2$. The work to pump all the oil to the top of the tank is

$$W = \int_{0}^{10} \frac{57\pi}{4}(10y^2 - y^3)dy = \frac{57\pi}{4}\left[\frac{10y^3}{3} - \frac{y^4}{4}\right]_{0}^{10} = 11{,}875\pi \text{ ft} \cdot \text{lb} \approx 37{,}306 \text{ ft} \cdot \text{lb}.$$

17. The typical slab between the planes at y and and $y + \Delta y$ has a volume of $\Delta V = \pi(\text{radius})^2(\text{thickness}) = \pi\left(\frac{20}{2}\right)^2\Delta y$ $= \pi \cdot 100\,\Delta y$ ft^3. The force F required to lift the slab is equal to its weight: $F = 51.2\,\Delta V = 51.2 \cdot 100\pi\,\Delta y$ lb $\Rightarrow F = 5120\pi\,\Delta y$ lb. The distance through which F must act is about $(30 - y)$ ft. The work it takes to lift all the kerosene is approximately $W \approx \sum\limits_{0}^{30} \Delta W = \sum\limits_{0}^{30} 5120\pi(30 - y)\,\Delta y$ ft $\cdot$ lb which is a Riemann sum. The work to pump the tank dry is the limit of these sums: $W = \int_{0}^{30} 5120\pi(30 - y)\,dy = 5120\pi\left[30y - \frac{y^2}{2}\right]_{0}^{30} = 5120\pi\left(\frac{900}{2}\right) = (5120)(450\pi)$ $\approx 7{,}238{,}229.48$ ft $\cdot$ lb

19. The typical slab between the planes at y and $y + \Delta y$ has a volume of about $\Delta V = \pi(\text{radius})^2(\text{thickness})$ $= \pi\left(\sqrt{25 - y^2}\right)^2\Delta y$ m^3. The force F(y) required to lift this slab is equal to its weight: $F(y) = 9800 \cdot \Delta V$ $= 9800\pi\left(\sqrt{25 - y^2}\right)^2\Delta y = 9800\pi\left(25 - y^2\right)\Delta y$ N. The distance through which F(y) must act to lift the slab to the level of 4 m above the top of the reservoir is about $(4 - y)$ m, so the work done is approximately $\Delta W \approx 9800\pi\left(25 - y^2\right)(4 - y)\Delta y$ N $\cdot$ m. The work done lifting all the slabs from $y = -5$ m to $y = 0$ m is approximately $W \approx \sum\limits_{-5}^{0} 9800\pi\left(25 - y^2\right)(4 - y)\Delta y$ N $\cdot$ m. Taking the limit of these Riemann sums, we get

$$W = \int_{-5}^{0} 9800\pi \, (25 - y^2)(4 - y) \, dy = 9800\pi \int_{-5}^{0} (100 - 25y - 4y^2 + y^3) \, dy = 9800\pi \left[100y - \tfrac{25}{2} y^2 - \tfrac{4}{3} y^3 + \tfrac{y^4}{4} \right]_{-5}^{0}$$

$$= -9800\pi \left(-500 - \tfrac{25 \cdot 25}{2} + \tfrac{4}{3} \cdot 125 + \tfrac{625}{4} \right) \approx 15{,}073{,}099.75 \text{ J}$$

21. $F = m \frac{dv}{dt} = mv \frac{dv}{dx}$ by the chain rule $\Rightarrow W = \int_{x_1}^{x_2} mv \frac{dv}{dx} \, dx = m \int_{x_1}^{x_2} \left(v \frac{dv}{dx} \right) dx = m \left[\tfrac{1}{2} v^2(x) \right]_{x_1}^{x_2}$

$= \tfrac{1}{2} m \left[v^2(x_2) - v^2(x_1) \right] = \tfrac{1}{2} mv_2^2 - \tfrac{1}{2} mv_1^2$, as claimed.

23. $90 \text{ mph} = \frac{90 \text{ mi}}{1 \text{ hr}} \cdot \frac{1 \text{ hr}}{60 \text{ min}} \cdot \frac{1 \text{ min}}{60 \text{ sec}} \cdot \frac{5280 \text{ ft}}{1 \text{ mi}} = 132 \text{ ft/sec}$; $m = \frac{0.3125 \text{ lb}}{32 \text{ ft/sec}^2} = \frac{0.3125}{32}$ slugs;

$W = \left(\tfrac{1}{2} \right) \left(\frac{0.3125 \text{ lb}}{32 \text{ ft/sec}^2} \right) (132 \text{ ft/sec})^2 \approx 85.1 \text{ ft} \cdot \text{lb}$

25. weight $= 2 \text{ oz} = \tfrac{1}{8} \text{ lb} \Rightarrow m = \frac{\frac{1}{8}}{32}$ slugs $= \frac{1}{256}$ slugs; $124 \text{ mph} = \frac{(124)(5280)}{(60)(60)} \approx 181.87 \text{ ft/sec}$;

$W = \left(\tfrac{1}{2} \right) \left(\frac{1}{256} \text{ slugs} \right) (181.87 \text{ ft/sec})^2 \approx 64.6 \text{ ft} \cdot \text{lb}$

27. We imagine the milkshake divided into thin slabs by planes perpendicular to the y-axis at the points of a partition of the interval $[0, 7]$. The typical slab between the planes at y and $y + \Delta y$ has a volume of about $\Delta V = \pi (\text{radius})^2 (\text{thickness}) = \pi \left(\frac{y + 17.5}{14} \right)^2 \Delta y \text{ in}^3$. The force F(y) required to lift this slab is equal to its weight: $F(y) = \tfrac{4}{9} \Delta V = \frac{4\pi}{9} \left(\frac{y + 17.5}{14} \right)^2 \Delta y \text{ oz}$. The distance through which F(y) must act to lift this slab to the level of 1 inch above the top is about $(8 - y)$ in. The work done lifting the slab is about $\Delta W = \left(\frac{4\pi}{9} \right) \frac{(y + 17.5)^2}{14^2} (8 - y) \Delta y \text{ in} \cdot \text{oz}$. The work done lifting all the slabs from $y = 0$ to $y = 7$ is

approximately $W = \sum_{0}^{7} \frac{4\pi}{9 \cdot 14^2} (y + 17.5)^2 (8 - y) \Delta y \text{ in} \cdot \text{oz}$ which is a Riemann sum. The work is the limit of

these sums as the norm of the partition goes to zero: $W = \int_{0}^{7} \frac{4\pi}{9 \cdot 14^2} (y + 17.5)^2 (8 - y) \, dy$

$$= \frac{4\pi}{9 \cdot 14^2} \int_{0}^{7} (2450 - 26.25y - 27y^2 - y^3) \, dy = \frac{4\pi}{9 \cdot 14^2} \left[-\frac{y^4}{4} - 9y^3 - \frac{26.25}{2} y^2 + 2450y \right]_{0}^{7}$$

$$= \frac{4\pi}{9 \cdot 14^2} \left[-\frac{7^4}{4} - 9 \cdot 7^3 - \frac{26.25}{2} \cdot 7^2 + 2450 \cdot 7 \right] \approx 91.32 \text{ in} \cdot \text{oz}$$

29. To find the width of the plate at a typical depth y, we first find an equation for the line of the plate's right-hand edge: $y = x - 5$. If we let x denote the width of the right-hand half of the triangle at depth y, then $x = 5 + y$ and the total width is $L(y) = 2x = 2(5 + y)$. The depth of the strip is $(-y)$. The force exerted by the water against one side of the plate is therefore $F = \int_{-5}^{-2} w(-y) \cdot L(y) \, dy = \int_{-5}^{-2} 62.4 \cdot (-y) \cdot 2(5 + y) \, dy$

$$= 124.8 \int_{-5}^{-2} (-5y - y^2) \, dy = 124.8 \left[-\tfrac{5}{2} y^2 - \tfrac{1}{3} y^3 \right]_{-5}^{-2} = 124.8 \left[\left(-\tfrac{5}{2} \cdot 4 + \tfrac{1}{3} \cdot 8 \right) - \left(-\tfrac{5}{2} \cdot 25 + \tfrac{1}{3} \cdot 125 \right) \right]$$

$$= (124.8) \left(\tfrac{105}{2} - \tfrac{117}{3} \right) = (124.8) \left(\tfrac{315 - 234}{6} \right) = 1684.8 \text{ lb}$$

31. Using the coordinate system of Exercise 2, we find the equation for the line of the plate's right-hand edge to be $y = 2x - 4 \Rightarrow x = \frac{y + 4}{2}$ and $L(y) = 2x = y + 4$. The depth of the strip is $(1 - y)$.

(a) $F = \int_{-4}^{0} w(1 - y)L(y) \, dy = \int_{-4}^{0} 62.4 \cdot (1 - y)(y + 4) \, dy = 62.4 \int_{-4}^{0} (4 - 3y - y^2) \, dy = 62.4 \left[4y - \frac{3y^2}{2} - \frac{y^3}{3} \right]_{-4}^{0}$

$$= (-62.4) \left[(-4)(4) - \frac{(3)(16)}{2} + \frac{64}{3} \right] = (-62.4) \left(-16 - 24 + \frac{64}{3} \right) = \frac{(-62.4)(-120 + 64)}{3} = 1164.8 \text{ lb}$$

(b) $F = (-64.0) \left[(-4)(4) - \frac{(3)(16)}{2} + \frac{64}{3} \right] = \frac{(-64.0)(-120 + 64)}{3} \approx 1194.7 \text{ lb}$

33. Using the coordinate system given in the accompanying
figure, we see that the total width is $L(y) = 63$ and the depth
of the strip is $(33.5 - y) \Rightarrow F = \int_0^{33} w(33.5 - y)L(y)\, dy$

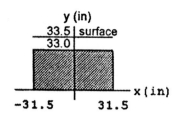

$= \int_0^{33} \frac{64}{12^3} \cdot (33.5 - y) \cdot 63\, dy = \left(\frac{64}{12^3}\right)(63)\int_0^{33} (33.5 - y)\, dy$

$= \left(\frac{64}{12^3}\right)(63)\left[33.5y - \frac{y^2}{2}\right]_0^{33} = \left(\frac{64 \cdot 63}{12^3}\right)\left[(33.5)(33) - \frac{33^2}{2}\right]$

$= \frac{(64)(63)(33)(67 - 33)}{(2)(12^3)} = 1309$ lb

35. The coordinate system is given in the text. The right-hand edge is $x = \sqrt{y}$ and the total width is $L(y) = 2x = 2\sqrt{y}$.

(a) The depth of the strip is $(2 - y)$ so the force exerted by the liquid on the gate is $F = \int_0^1 w(2 - y)L(y)\, dy$

$= \int_0^1 50(2 - y) \cdot 2\sqrt{y}\, dy = 100 \int_0^1 (2 - y)\sqrt{y}\, dy = 100\int_0^1 \left(2y^{1/2} - y^{3/2}\right) dy = 100\left[\frac{4}{3} y^{3/2} - \frac{2}{5} y^{5/2}\right]_0^1$

$= 100\left(\frac{4}{3} - \frac{2}{5}\right) = \left(\frac{100}{15}\right)(20 - 6) = 93.33$ lb

(b) We need to solve $160 = \int_0^1 w(H - y) \cdot 2\sqrt{y}\, dy$ for h. $160 = 100\left(\frac{2H}{3} - \frac{2}{5}\right) \Rightarrow H = 3$ ft.

37. The pressure at level y is $p(y) = w \cdot y \Rightarrow$ the average
pressure is $\bar{p} = \frac{1}{b}\int_0^b p(y)\, dy = \frac{1}{b}\int_0^b w \cdot y\, dy = \frac{1}{b} w \left[\frac{y^2}{2}\right]_0^b$

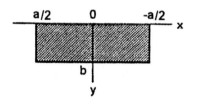

$= \left(\frac{w}{b}\right)\left(\frac{b^2}{2}\right) = \frac{wb}{2}$. This is the pressure at level $\frac{b}{2}$, which
is the pressure at the middle of the plate.

39. When the water reaches the top of the tank the force on the movable side is $\int_{-2}^0 (62.4)\left(2\sqrt{4 - y^2}\right)(-y)\, dy$

$= (62.4)\int_{-2}^0 (4 - y^2)^{1/2}(-2y)\, dy = (62.4)\left[\frac{2}{3}(4 - y^2)^{3/2}\right]_{-2}^0 = (62.4)\left(\frac{2}{3}\right)(4^{3/2}) = 332.8$ ft · lb. The force

compressing the spring is $F = 100x$, so when the tank is full we have $332.8 = 100x \Rightarrow x \approx 3.33$ ft. Therefore the
movable end does not reach the required 5 ft to allow drainage $\Rightarrow$ the tank will overflow.

6.6 MOMENTS AND CENTERS OF MASS

1. Since the plate is symmetric about the y-axis and its density is
constant, the distribution of mass is symmetric about the y-axis
and the center of mass lies on the y-axis. This means that
$\bar{x} = 0$. It remains to find $\bar{y} = \frac{M_x}{M}$. We model the distribution of
mass with *vertical* strips. The typical strip has center of mass:

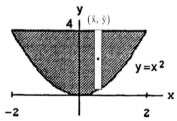

$(\tilde{x}, \tilde{y}) = \left(x, \frac{x^2 + 4}{2}\right)$, length: $4 - x^2$, width: dx, area:

$dA = (4 - x^2)\, dx$, mass: $dm = \delta\, dA = \delta(4 - x^2)\, dx$. The moment of the strip about the x-axis is

$\tilde{y}\, dm = \left(\frac{x^2 + 4}{2}\right)\delta(4 - x^2)\, dx = \frac{\delta}{2}(16 - x^4)\, dx$. The moment of the plate about the x-axis is $M_x = \int \tilde{y}\, dm$

$= \int_{-2}^2 \frac{\delta}{2}(16 - x^4)\, dx = \frac{\delta}{2}\left[16x - \frac{x^5}{5}\right]_{-2}^2 = \frac{\delta}{2}\left[\left(16 \cdot 2 - \frac{2^5}{5}\right) - \left(-16 \cdot 2 + \frac{2^5}{5}\right)\right] = \frac{\delta \cdot 2}{2}\left(32 - \frac{32}{5}\right) = \frac{128\delta}{5}$. The mass of the

plate is $M = \int \delta(4 - x^2)\, dx = \delta\left[4x - \frac{x^3}{3}\right]_{-2}^2 = 2\delta\left(8 - \frac{8}{3}\right) = \frac{32\delta}{3}$. Therefore $\bar{y} = \frac{M_x}{M} = \frac{\left(\frac{128\delta}{5}\right)}{\left(\frac{32\delta}{3}\right)} = \frac{12}{5}$. The plate's center of

mass is the point $(\bar{x}, \bar{y}) = \left(0, \frac{12}{5}\right)$.

3. Intersection points: $x - x^2 = -x \Rightarrow 2x - x^2 = 0$
$\Rightarrow x(2-x) = 0 \Rightarrow x = 0$ or $x = 2$. The typical *vertical*
strip has center of mass: $(\tilde{x}, \tilde{y}) = \left(x, \frac{(x-x^2)+(-x)}{2}\right)$

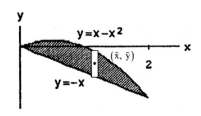

$= \left(x, -\frac{x^2}{2}\right)$, length: $(x-x^2) - (-x) = 2x - x^2$, width: dx,
area: $dA = (2x - x^2)\,dx$, mass: $dm = \delta\,dA = \delta(2x-x^2)\,dx$.
The moment of the strip about the x-axis is

$\tilde{y}\,dm = \left(-\frac{x^2}{2}\right)\delta(2x-x^2)\,dx$; about the y-axis it is $\tilde{x}\,dm = x \cdot \delta(2x-x^2)\,dx$. Thus, $M_x = \int \tilde{y}\,dm$

$= -\int_0^2 \left(\frac{\delta}{2}x^2\right)(2x-x^2)\,dx = -\frac{\delta}{2}\int_0^2 (2x^3-x^4)\,dx = -\frac{\delta}{2}\left[\frac{x^4}{2} - \frac{x^5}{5}\right]_0^2 = -\frac{\delta}{2}\left(2^3 - \frac{2^5}{5}\right) = -\frac{\delta}{2}\cdot 2^3\left(1 - \frac{4}{5}\right)$

$= -\frac{4\delta}{5}$; $M_y = \int \tilde{x}\,dm = \int_0^2 x \cdot \delta(2x-x^2)\,dx = \delta\int_0^2(2x^2-x^3) = \delta\left[\frac{2}{3}x^3 - \frac{x^4}{4}\right]_0^2 = \delta\left(2\cdot\frac{2^3}{3} - \frac{2^4}{4}\right) = \frac{\delta\cdot 2^4}{12} = \frac{4\delta}{3}$;

$M = \int dm = \int_0^2 \delta(2x-x^2)\,dx = \delta\int_0^2(2x-x^2)\,dx = \delta\left[x^2 - \frac{x^3}{3}\right]_0^2 = \delta\left(4 - \frac{8}{3}\right) = \frac{4\delta}{3}$. Therefore, $\bar{x} = \frac{M_y}{M}$

$= \left(\frac{4\delta}{3}\right)\left(\frac{3}{4\delta}\right) = 1$ and $\bar{y} = \frac{M_x}{M} = \left(-\frac{4\delta}{5}\right)\left(\frac{3}{4\delta}\right) = -\frac{3}{5} \Rightarrow (\bar{x},\bar{y}) = \left(1, -\frac{3}{5}\right)$ is the center of mass.

5. The typical *horizontal* strip has center of mass:

$(\tilde{x}, \tilde{y}) = \left(\frac{y-y^3}{2}, y\right)$, length: $y - y^3$, width: dy,

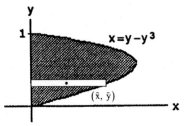

area: $dA = (y-y^3)\,dy$, mass: $dm = \delta\,dA = \delta(y-y^3)\,dy$.
The moment of the strip about the y-axis is

$\tilde{x}\,dm = \delta\left(\frac{y-y^3}{2}\right)(y-y^3)\,dy = \frac{\delta}{2}(y-y^3)^2\,dy$

$= \frac{\delta}{2}(y^2 - 2y^4 + y^6)\,dy$; the moment about the x-axis is

$\tilde{y}\,dm = \delta y(y-y^3)\,dy = \delta(y^2-y^4)\,dy$. Thus, $M_x = \int \tilde{y}\,dm = \delta\int_0^1 (y^2-y^4)\,dy = \delta\left[\frac{y^3}{3} - \frac{y^5}{5}\right]_0^1 = \delta\left(\frac{1}{3} - \frac{1}{5}\right) = \frac{2\delta}{15}$;

$M_y = \int \tilde{x}\,dm = \frac{\delta}{2}\int_0^1 (y^2-2y^4+y^6)\,dy = \frac{\delta}{2}\left[\frac{y^3}{3} - \frac{2y^5}{5} + \frac{y^7}{7}\right]_0^1 = \frac{\delta}{2}\left(\frac{1}{3} - \frac{2}{5} + \frac{1}{7}\right) = \frac{\delta}{2}\left(\frac{35-42+15}{3\cdot5\cdot7}\right) = \frac{4\delta}{105}$; $M = \int dm$

$= \delta\int_0^1 (y-y^3)\,dy = \delta\left[\frac{y^2}{2} - \frac{y^4}{4}\right]_0^1 = \delta\left(\frac{1}{2} - \frac{1}{4}\right) = \frac{\delta}{4}$. Therefore, $\bar{x} = \frac{M_y}{M} = \left(\frac{4\delta}{105}\right)\left(\frac{4}{\delta}\right) = \frac{16}{105}$ and $\bar{y} = \frac{M_x}{M} = \left(\frac{2\delta}{15}\right)\left(\frac{4}{\delta}\right)$

$= \frac{8}{15} \Rightarrow (\bar{x},\bar{y}) = \left(\frac{16}{105}, \frac{8}{15}\right)$ is the center of mass.

7. Applying the symmetry argument analogous to the one used
in Exercise 1, we find $\bar{x} = 0$. The typical *vertical* strip has
center of mass: $(\tilde{x}, \tilde{y}) = \left(x, \frac{\cos x}{2}\right)$, length: $\cos x$, width: dx,

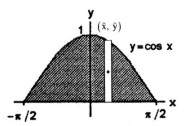

area: $dA = \cos x\,dx$, mass: $dm = \delta\,dA = \delta\cos x\,dx$. The
moment of the strip about the x-axis is $\tilde{y}\,dm = \delta \cdot \frac{\cos x}{2} \cdot \cos x\,dx$

$= \frac{\delta}{2}\cos^2 x\,dx = \frac{\delta}{2}\left(\frac{1+\cos 2x}{2}\right)dx = \frac{\delta}{4}(1+\cos 2x)\,dx$; thus,

$M_x = \int \tilde{y}\,dm = \int_{-\pi/2}^{\pi/2} \frac{\delta}{4}(1+\cos 2x)\,dx = \frac{\delta}{4}\left[x + \frac{\sin 2x}{2}\right]_{-\pi/2}^{\pi/2} = \frac{\delta}{4}\left[\left(\frac{\pi}{2} + 0\right) - \left(-\frac{\pi}{2}\right)\right] = \frac{\delta\pi}{4}$; $M = \int dm = \delta\int_{-\pi/2}^{\pi/2}\cos x\,dx$

$= \delta[\sin x]_{-\pi/2}^{\pi/2} = 2\delta$. Therefore, $\bar{y} = \frac{M_x}{M} = \frac{\delta\pi}{4\cdot 2\delta} = \frac{\pi}{8} \Rightarrow (\bar{x},\bar{y}) = \left(0, \frac{\pi}{8}\right)$ is the center of mass.

9. (a) $M_y = \int_1^2 x\left(\frac{1}{x}\right)dx = 1$, $M_x = \int_1^2 \left(\frac{1}{2x}\right)\left(\frac{1}{x}\right)dx$

(b)

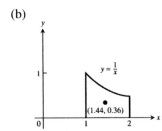

$= \frac{1}{2}\int_1^2 \frac{1}{x^2}\,dx = \left[-\frac{1}{2x}\right]_1^2 = \frac{1}{4}$, $M = \int_1^2 \frac{1}{x}\,dx$

$= [\ln|x|]_1^2 = \ln 2 \Rightarrow \bar{x} = \frac{M_y}{M} = \frac{1}{\ln 2} \approx 1.44$ and

$\bar{y} = \frac{M_x}{M} = \frac{\left(\frac{1}{4}\right)}{\ln 2} \approx 0.36$

11. $M = \int_0^1 \frac{2}{1+x^2} \, dx = 2 \left[\tan^{-1} x \right]_0^1 = \frac{\pi}{2}$ and $M_y = \int_0^1 \frac{2x}{1+x^2} \, dx = \left[\ln (1+x^2) \right]_0^1 = \ln 2 \Rightarrow \bar{x} = \frac{M_y}{M} = \frac{\ln 2}{\left(\frac{\pi}{2}\right)} = \frac{\ln 4}{\pi}$;

$\bar{y} = 0$ by symmetry

13. $M_y = \int_1^{16} x \left(\frac{1}{\sqrt{x}} \right) \, dx = \int_1^{16} x^{1/2} \, dx = \frac{2}{3} \left[x^{3/2} \right]_1^{16} = 42$; $M_x = \int_1^{16} \left(\frac{1}{2\sqrt{x}} \right) \left(\frac{1}{\sqrt{x}} \right) \, dx = \frac{1}{2} \int_1^{16} \frac{1}{x} \, dx$

$= \frac{1}{2} \left[\ln |x| \right]_1^{16} = \ln 4$, $M = \int_1^{16} \frac{1}{\sqrt{x}} \, dx = \left[2x^{1/2} \right]_1^{16} = 6 \Rightarrow \bar{x} = \frac{M_y}{M} = 7$ and $\bar{y} = \frac{M_x}{M} = \frac{\ln 4}{6}$

15. $M_x = \int \tilde{y} \, dm = \int_1^2 \frac{\left(\frac{2}{x^2}\right)}{2} \cdot \delta \cdot \left(\frac{2}{x^2} \right) \, dx$

$= \int_1^2 \left(\frac{1}{x^2} \right) (x^2) \left(\frac{2}{x^2} \right) \, dx = \int_1^2 \frac{2}{x^2} \, dx = 2 \int_1^2 x^{-2} \, dx$

$= 2 \left[-x^{-1} \right]_1^2 = 2 \left[\left(-\frac{1}{2} \right) - (-1) \right] = 2 \left(\frac{1}{2} \right) = 1$;

$M_y = \int \tilde{x} \, dm = \int_1^2 x \cdot \delta \cdot \left(\frac{2}{x^2} \right) \, dx$

$= \int_1^2 x (x^2) \left(\frac{2}{x^2} \right) \, dx = 2 \int_1^2 x \, dx = 2 \left[\frac{x^2}{2} \right]_1^2$

$= 2 \left(2 - \frac{1}{2} \right) = 4 - 1 = 3$; $M = \int dm = \int_1^2 \delta \left(\frac{2}{x^2} \right) \, dx = \int_1^2 x^2 \left(\frac{2}{x^2} \right) \, dx = 2 \int_1^2 \, dx = 2[x]_1^2 = 2(2-1) = 2$. So

$\bar{x} = \frac{M_y}{M} = \frac{3}{2}$ and $\bar{y} = \frac{M_x}{M} = \frac{1}{2} \Rightarrow (\bar{x}, \bar{y}) = \left(\frac{3}{2}, \frac{1}{2} \right)$ is the center of mass.

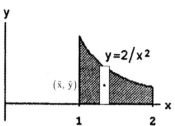

17. $M_y = \int_1^{16} x \left(\frac{1}{\sqrt{x}} \right) \left(\frac{4}{\sqrt{x}} \right) \, dx = 4 \int_1^{16} \, dx = 60$, $M_x = \int_1^{16} \left(\frac{1}{2\sqrt{x}} \right) \left(\frac{1}{\sqrt{x}} \right) \left(\frac{4}{\sqrt{x}} \right) \, dx = 2 \int_1^{16} x^{-3/2} \, dx$

$= -4 \left[x^{-1/2} \right]_1^{16} = 3$, $M = \int_1^{16} \left(\frac{1}{\sqrt{x}} \right) \left(\frac{4}{\sqrt{x}} \right) \, dx = 4 \int_1^{16} \frac{1}{x} \, dx = \left[4 \ln |x| \right]_1^{16} = 4 \ln 16 \Rightarrow \bar{x} = \frac{M_y}{M} = \frac{15}{\ln 16}$ and

$\bar{y} = \frac{M_x}{M} = \frac{3}{4 \ln 16}$

19. (a) We use the shell method: $V = \int_a^b 2\pi \left(\begin{smallmatrix} \text{shell} \\ \text{radius} \end{smallmatrix} \right) \left(\begin{smallmatrix} \text{shell} \\ \text{height} \end{smallmatrix} \right) \, dx = \int_1^4 2\pi x \left[\frac{4}{\sqrt{x}} - \left(-\frac{4}{\sqrt{x}} \right) \right] \, dx$

$= 16\pi \int_1^4 \frac{x}{\sqrt{x}} \, dx = 16\pi \int_1^4 x^{1/2} \, dx = 16\pi \left[\frac{2}{3} x^{3/2} \right]_1^4 = 16\pi \left(\frac{2}{3} \cdot 8 - \frac{2}{3} \right) = \frac{32\pi}{3} (8-1) = \frac{224\pi}{3}$

(b) Since the plate is symmetric about the x-axis and its density $\delta(x) = \frac{1}{x}$ is a function of x alone, the

distribution of its mass is symmetric about the x-axis. This means that $\bar{y} = 0$. We use the vertical strip

approach to find $\bar{x}$: $M_y = \int \tilde{x} \, dm = \int_1^4 x \cdot \left[\frac{4}{\sqrt{x}} - \left(-\frac{4}{\sqrt{x}} \right) \right] \cdot \delta \, dx = \int_1^4 x \cdot \frac{8}{\sqrt{x}} \cdot \frac{1}{x} \, dx = 8 \int_1^4 x^{-1/2} \, dx$

$= 8 \left[2x^{1/2} \right]_1^4 = 8(2 \cdot 2 - 2) = 16$; $M = \int dm = \int_1^4 \left[\frac{4}{\sqrt{x}} - \left(\frac{-4}{\sqrt{x}} \right) \right] \cdot \delta \, dx = 8 \int_1^4 \left(\frac{1}{\sqrt{x}} \right) \left(\frac{1}{x} \right) \, dx = 8 \int_1^4 x^{-3/2} \, dx$

$= 8 \left[-2x^{-1/2} \right]_1^4 = 8[-1 - (-2)] = 8$. So $\bar{x} = \frac{M_y}{M} = \frac{16}{8} = 2 \Rightarrow (\bar{x}, \bar{y}) = (2, 0)$ is the center of mass.

(c)

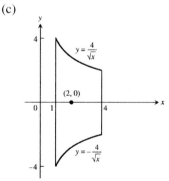

21. The mass of a horizontal strip is $dm = \delta \, dA = \delta L \, dy$, where L is the width of the triangle at a distance of y above

its base on the x-axis as shown in the figure in the text. Also, by similar triangles we have $\frac{L}{b} = \frac{h-y}{h}$

$\Rightarrow L = \frac{b}{h} (h - y)$. Thus, $M_x = \int \tilde{y} \, dm = \int_0^h \delta y \left(\frac{b}{h} \right) (h - y) \, dy = \frac{\delta b}{h} \int_0^h (hy - y^2) \, dy = \frac{\delta b}{h} \left[\frac{hy^2}{2} - \frac{y^3}{3} \right]_0^h$

$= \frac{\delta b}{h}\left(\frac{h^3}{2} - \frac{h^3}{3}\right) = \delta bh^2\left(\frac{1}{2} - \frac{1}{3}\right) = \frac{\delta bh^2}{6}$; $M = \int dm = \int_0^h \delta\left(\frac{b}{h}\right)(h-y)\,dy = \frac{\delta b}{h}\int_0^h (h-y)\,dy = \frac{\delta b}{h}\left[hy - \frac{y^2}{2}\right]_0^h$

$= \frac{\delta b}{h}\left(h^2 - \frac{h^2}{2}\right) = \frac{\delta bh}{2}$. So $\bar{y} = \frac{M_x}{M} = \left(\frac{\delta bh^2}{6}\right)\left(\frac{2}{\delta bh}\right) = \frac{h}{3}$ $\Rightarrow$ the center of mass lies above the base of the

triangle one-third of the way toward the opposite vertex. Similarly the other two sides of the triangle can be placed on the x-axis and the same results will occur. Therefore the centroid does lie at the intersection of the medians, as claimed.

23. From the symmetry about the line x = y it follows that
    $\bar{x} = \bar{y}$. It also follows that the line through the points $(0,0)$
    and $\left(\frac{1}{2}, \frac{1}{2}\right)$ is a median $\Rightarrow \bar{y} = \bar{x} = \frac{2}{3} \cdot \left(\frac{1}{2} - 0\right) = \frac{1}{3}$
    $\Rightarrow (\bar{x}, \bar{y}) = \left(\frac{1}{3}, \frac{1}{3}\right)$.

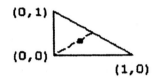

25. The point of intersection of the median from the vertex $(0, b)$
    to the opposite side has coordinates $\left(0, \frac{a}{2}\right)$
    $\Rightarrow \bar{y} = (b - 0) \cdot \frac{1}{3} = \frac{b}{3}$ and $\bar{x} = \left(\frac{a}{2} - 0\right) \cdot \frac{2}{3} = \frac{a}{3}$
    $\Rightarrow (\bar{x}, \bar{y}) = \left(\frac{a}{3}, \frac{b}{3}\right)$.

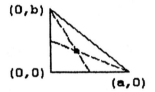

27. $y = x^{1/2} \Rightarrow dy = \frac{1}{2}x^{-1/2}\,dx$
    $\Rightarrow ds = \sqrt{(dx)^2 + (dy)^2} = \sqrt{1 + \frac{1}{4x}}\,dx$;
    $M_x = \delta\int_0^2 \sqrt{x}\sqrt{1 + \frac{1}{4x}}\,dx$
    $= \delta\int_0^2 \sqrt{x + \frac{1}{4}}\,dx = \frac{2\delta}{3}\left[\left(x + \frac{1}{4}\right)^{3/2}\right]_0^2$
    $= \frac{2\delta}{3}\left[\left(2 + \frac{1}{4}\right)^{3/2} - \left(\frac{1}{4}\right)^{3/2}\right]$
    $= \frac{2\delta}{3}\left[\left(\frac{9}{4}\right)^{3/2} - \left(\frac{1}{4}\right)^{3/2}\right] = \frac{2\delta}{3}\left(\frac{27}{8} - \frac{1}{8}\right) = \frac{13\delta}{6}$

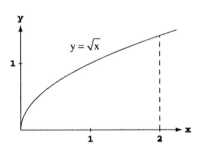

29. From Example 3 we have $M_x = \int_0^\pi a(a\sin\theta)(k\sin\theta)\,d\theta = a^2k\int_0^\pi \sin^2\theta\,d\theta = \frac{a^2k}{2}\int_0^\pi (1 - \cos 2\theta)\,d\theta$
    $= \frac{a^2k}{2}\left[\theta - \frac{\sin 2\theta}{2}\right]_0^\pi = \frac{a^2k\pi}{2}$; $M_y = \int_0^\pi a(a\cos\theta)(k\sin\theta)\,d\theta = a^2k\int_0^\pi \sin\theta\cos\theta\,d\theta = \frac{a^2k}{2}[\sin^2\theta]_0^\pi = 0$;
    $M = \int_0^\pi ak\sin\theta\,d\theta = ak[-\cos\theta]_0^\pi = 2ak$. Therefore, $\bar{x} = \frac{M_y}{M} = 0$ and $\bar{y} = \frac{M_x}{M} = \left(\frac{a^2k\pi}{2}\right)\left(\frac{1}{2ak}\right) = \frac{a\pi}{4} \Rightarrow \left(0, \frac{a\pi}{4}\right)$
    is the center of mass.

31. Consider the curve as an infinite number of line segments joined together. From the derivation of arc
    length we have that the length of a particular segment is $ds = \sqrt{(dx)^2 + (dy)^2}$. This implies that
    $M_x = \int \delta y\,ds$, $M_y = \int \delta x\,ds$ and $M = \int \delta\,ds$. If δ is constant, then $\bar{x} = \frac{M_y}{M} = \frac{\int x\,ds}{\int ds} = \frac{\int x\,ds}{\text{length}}$ and
    $\bar{y} = \frac{M_x}{M} = \frac{\int y\,ds}{\int ds} = \frac{\int y\,ds}{\text{length}}$.

CHAPTER 6 PRACTICE AND ADDITIONAL EXERCISES

1. $A(x) = \frac{\pi}{4} (\text{diameter})^2 = \frac{\pi}{4} \left(\sqrt{x} - x^2 \right)^2$

    $= \frac{\pi}{4} \left(x - 2\sqrt{x} \cdot x^2 + x^4 \right) ; a = 0, b = 1$

    $\Rightarrow V = \int_a^b A(x) \, dx = \frac{\pi}{4} \int_0^1 \left(x - 2x^{5/2} + x^4 \right) dx$

    $= \frac{\pi}{4} \left[\frac{x^2}{2} - \frac{4}{7} x^{7/2} + \frac{x^5}{5} \right]_0^1 = \frac{\pi}{4} \left(\frac{1}{2} - \frac{4}{7} + \frac{1}{5} \right)$

    $= \frac{\pi}{4 \cdot 70} (35 - 40 + 14) = \frac{9\pi}{280}$

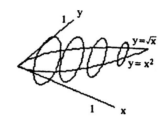

3. $A(x) = \frac{\pi}{4} (\text{diameter})^2 = \frac{\pi}{4} (2 \sin x - 2 \cos x)^2$

    $= \frac{\pi}{4} \cdot 4 \left(\sin^2 x - 2 \sin x \cos x + \cos^2 x \right)$

    $= \pi (1 - \sin 2x); a = \frac{\pi}{4}, b = \frac{5\pi}{4}$

    $\Rightarrow V = \int_a^b A(x) \, dx = \pi \int_{\pi/4}^{5\pi/4} (1 - \sin 2x) \, dx$

    $= \pi \left[x + \frac{\cos 2x}{2} \right]_{\pi/4}^{5\pi/4}$

    $= \pi \left[\left(\frac{5\pi}{4} + \frac{\cos \frac{5\pi}{2}}{2} \right) - \left(\frac{\pi}{4} - \frac{\cos \frac{\pi}{2}}{2} \right) \right] = \pi^2$

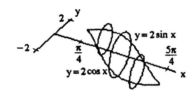

5. $A(x) = \frac{\pi}{4} (\text{diameter})^2 = \frac{\pi}{4} \left(2\sqrt{x} - \frac{x^2}{4} \right)^2 = \frac{\pi}{4} \left(4x - x^{5/2} + \frac{x^4}{16} \right) ; a = 0, b = 4 \Rightarrow V = \int_a^b A(x) \, dx$

    $= \frac{\pi}{4} \int_0^4 \left(4x - x^{5/2} + \frac{x^4}{16} \right) dx = \frac{\pi}{4} \left[2x^2 - \frac{2}{7} x^{7/2} + \frac{x^5}{5 \cdot 16} \right]_0^4 = \frac{\pi}{4} \left(32 - 32 \cdot \frac{8}{7} + \frac{2}{5} \cdot 32 \right)$

    $= \frac{32\pi}{4} \left(1 - \frac{8}{7} + \frac{2}{5} \right) = \frac{8\pi}{35} (35 - 40 + 14) = \frac{72\pi}{35}$

7. (a) *disk method*:

    $V = \int_a^b \pi R^2(x) \, dx = \int_{-1}^1 \pi \left(3x^4 \right)^2 dx = \pi \int_{-1}^1 9x^8 \, dx$

    $= \pi \left[x^9 \right]_{-1}^1 = 2\pi$

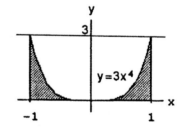

    (b) *shell method*:

    $V = \int_a^b 2\pi \left(\begin{smallmatrix} \text{shell} \\ \text{radius} \end{smallmatrix} \right) \left(\begin{smallmatrix} \text{shell} \\ \text{height} \end{smallmatrix} \right) dx = \int_0^1 2\pi x \left(3x^4 \right) dx = 2\pi \cdot 3 \int_0^1 x^5 \, dx = 2\pi \cdot 3 \left[\frac{x^6}{6} \right]_0^1 = \pi$

    Note: The lower limit of integration is 0 rather than -1.

    (c) *shell method*:

    $V = \int_a^b 2\pi \left(\begin{smallmatrix} \text{shell} \\ \text{radius} \end{smallmatrix} \right) \left(\begin{smallmatrix} \text{shell} \\ \text{height} \end{smallmatrix} \right) dx = 2\pi \int_{-1}^1 (1 - x) \left(3x^4 \right) dx = 2\pi \left[\frac{3x^5}{5} - \frac{x^6}{2} \right]_{-1}^1 = 2\pi \left[\left(\frac{3}{5} - \frac{1}{2} \right) - \left(-\frac{3}{5} - \frac{1}{2} \right) \right] = \frac{12\pi}{5}$

    (d) *washer method*:

    $R(x) = 3, r(x) = 3 - 3x^4 = 3 \left(1 - x^4 \right) \Rightarrow V = \int_a^b \pi \left[R^2(x) - r^2(x) \right] dx = \int_{-1}^1 \pi \left[9 - 9 \left(1 - x^4 \right)^2 \right] dx$

    $= 9\pi \int_{-1}^1 \left[1 - \left(1 - 2x^4 + x^8 \right) \right] dx = 9\pi \int_{-1}^1 \left(2x^4 - x^8 \right) dx = 9\pi \left[\frac{2x^5}{5} - \frac{x^9}{9} \right]_{-1}^1 = 18\pi \left[\frac{2}{5} - \frac{1}{9} \right] = \frac{2\pi \cdot 13}{5} = \frac{26\pi}{5}$

9. (a) *disk method*:

    $V = \pi \int_1^5 \left(\sqrt{x - 1} \right)^2 dx = \pi \int_1^5 (x - 1) \, dx = \pi \left[\frac{x^2}{2} - x \right]_1^5$

    $= \pi \left[\left(\frac{25}{2} - 5 \right) - \left(\frac{1}{2} - 1 \right) \right] = \pi \left(\frac{24}{2} - 4 \right) = 8\pi$

(b) *washer method*:

$R(y) = 5, r(y) = y^2 + 1 \Rightarrow V = \int_c^d \pi [R^2(y) - r^2(y)] \, dy = \pi \int_{-2}^{2} \left[25 - (y^2 + 1)^2\right] dy$

$= \pi \int_{-2}^{2} (25 - y^4 - 2y^2 - 1) \, dy = \pi \int_{-2}^{2} (24 - y^4 - 2y^2) \, dy = \pi \left[24y - \frac{y^5}{5} - \frac{2}{3}y^3\right]_{-2}^{2} = 2\pi \left(24 \cdot 2 - \frac{32}{5} - \frac{2}{3} \cdot 8\right)$

$= 32\pi \left(3 - \frac{2}{5} - \frac{1}{3}\right) = \frac{32\pi}{15} (45 - 6 - 5) = \frac{1088\pi}{15}$

(c) *disk method*:

$R(y) = 5 - (y^2 + 1) = 4 - y^2$

$\Rightarrow V = \int_c^d \pi R^2(y) \, dy = \int_{-2}^{2} \pi (4 - y^2)^2 \, dy$

$= \pi \int_{-2}^{2} (16 - 8y^2 + y^4) \, dy$

$= \pi \left[16y - \frac{8y^3}{3} + \frac{y^5}{5}\right]_{-2}^{2} = 2\pi \left(32 - \frac{64}{3} + \frac{32}{5}\right)$

$= 64\pi \left(1 - \frac{2}{3} + \frac{1}{5}\right) = \frac{64\pi}{15} (15 - 10 + 3) = \frac{512\pi}{15}$

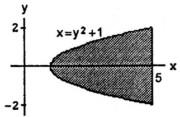

11. *disk method*:

$R(x) = \tan x, a = 0, b = \frac{\pi}{3} \Rightarrow V = \pi \int_0^{\pi/3} \tan^2 x \, dx = \pi \int_0^{\pi/3} (\sec^2 x - 1) \, dx = \pi [\tan x - x]_0^{\pi/3} = \frac{\pi \left(3\sqrt{3} - \pi\right)}{3}$

13. *shell method*:

$V = \int_0^1 2\pi \left(\begin{smallmatrix} \text{shell} \\ \text{radius} \end{smallmatrix}\right) \left(\begin{smallmatrix} \text{shell} \\ \text{height} \end{smallmatrix}\right) dx = \int_0^1 2\pi \, y \, e^{y^2} dy = \pi \, e^{y^2} \Big|_0^1 = \pi(e - 1)$

15. The material removed from the sphere consists of a cylinder and two "caps." From the diagram, the height of the cylinder is 2h, where $h^2 + \left(\sqrt{3}\right)^2 = 2^2$, i.e. $h = 1$. Thus

$V_{cyl} = (2h)\pi \left(\sqrt{3}\right)^2 = 6\pi$ ft^3. To get the volume of a cap,

use the disk method and $x^2 + y^2 = 2^2$: $V_{cap} = \int_1^2 \pi x^2 dy$

$= \int_1^2 \pi(4 - y^2) dy = \pi \left[4y - \frac{y^3}{3}\right]_1^2$

$= \pi \left[\left(8 - \frac{8}{3}\right) - \left(4 - \frac{1}{3}\right)\right] = \frac{5\pi}{3}$ ft^3. Therefore,

$V_{removed} = V_{cyl} + 2V_{cap} = 6\pi + \frac{10\pi}{3} = \frac{28\pi}{3}$ ft^3.

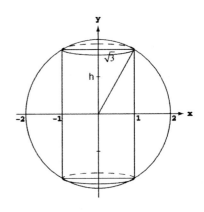

17. $y = x^{1/2} - \frac{x^{3/2}}{3} \Rightarrow \frac{dy}{dx} = \frac{1}{2}x^{-1/2} - \frac{1}{2}x^{1/2} \Rightarrow \left(\frac{dy}{dx}\right)^2 = \frac{1}{4}\left(\frac{1}{x} - 2 + x\right) \Rightarrow L = \int_1^4 \sqrt{1 + \frac{1}{4}\left(\frac{1}{x} - 2 + x\right)} \, dx$

$\Rightarrow L = \int_1^4 \sqrt{\frac{1}{4}\left(\frac{1}{x} + 2 + x\right)} \, dx = \int_1^4 \sqrt{\frac{1}{4}\left(x^{-1/2} + x^{1/2}\right)^2} \, dx = \int_1^4 \frac{1}{2}\left(x^{-1/2} + x^{1/2}\right) dx = \frac{1}{2}\left[2x^{1/2} + \frac{2}{3}x^{3/2}\right]_1^4$

$= \frac{1}{2}\left[\left(4 + \frac{2}{3} \cdot 8\right) - \left(2 + \frac{2}{3}\right)\right] = \frac{1}{2}\left(2 + \frac{14}{3}\right) = \frac{10}{3}$

19. $y = x^2 - \frac{\ln x}{8} \Rightarrow y' = 2x - \frac{1}{8x} \Rightarrow \sqrt{1 + (y')^2} = \sqrt{1 + \left(2x - \frac{1}{8x}\right)^2} = \sqrt{\frac{256x^4 + 32x^2 + 1}{64x^2}} = \sqrt{\frac{(16x^2 + 1)^2}{(8x)^2}} = \frac{16x^2 + 1}{8x}$

$= 2x + \frac{1}{8x} \Rightarrow \text{Length} = \int_1^2 \sqrt{1 + (y')^2} \, dx = \int_1^2 \left(2x + \frac{1}{8x}\right) dx = \left(x^2 + \frac{1}{8}\ln x\right)\Big|_1^2 = \left(4 + \frac{1}{8}\ln 2\right) - \left(1 + \frac{1}{8}\ln 1\right)$

$= 3 + \frac{1}{8}\ln 2$

21. $\frac{dx}{dt} = -5\sin t + 5\sin 5t$ and $\frac{dy}{dt} = 5\cos t - 5\cos 5t \Rightarrow \sqrt{\left(\frac{dx}{dt}\right)^2 + \left(\frac{dy}{dt}\right)^2}$

$= \sqrt{(-5\sin t + 5\sin 5t)^2 + (5\cos t - 5\cos 5t)^2}$

$= 5\sqrt{\sin^2 5t - 2\sin t \sin 5t + \sin^2 t + \cos^2 t - 2\cos t \cos 5t + \cos^2 5t} = 5\sqrt{2 - 2(\sin t \sin 5t + \cos t \cos 5t)}$

$$= 5\sqrt{2(1 - \cos 4t)} = 5\sqrt{4\left(\tfrac{1}{2}\right)(1 - \cos 4t)} = 10\sqrt{\sin^2 2t} = 10|\sin 2t| = 10\sin 2t \text{ (since } 0 \le t \le \tfrac{\pi}{2})$$

$$\Rightarrow \text{Length} = \int_0^{\pi/2} 10\sin 2t \, dt = [-5\cos 2t]_0^{\pi/2} = (-5)(-1) - (-5)(1) = 10$$

23. $\frac{dx}{d\theta} = -3\sin\theta$ and $\frac{dy}{d\theta} = 3\cos\theta \Rightarrow \sqrt{\left(\frac{dx}{d\theta}\right)^2 + \left(\frac{dy}{d\theta}\right)^2} = \sqrt{(-3\sin\theta)^2 + (3\cos\theta)^2} = \sqrt{3(\sin^2\theta + \cos^2\theta)} = 3$

$\Rightarrow \text{Length} = \int_0^{3\pi/2} 3 \, d\theta = 3\int_0^{3\pi/2} d\theta = 3\left(\frac{3\pi}{2} - 0\right) = \frac{9\pi}{2}$

25. $S = \int_a^b 2\pi y \sqrt{1 + \left(\frac{dy}{dx}\right)^2} \, dx; \quad \frac{dy}{dx} = \frac{1}{\sqrt{2x+1}} \Rightarrow \left(\frac{dy}{dx}\right)^2 = \frac{1}{2x+1} \Rightarrow S = \int_0^3 2\pi\sqrt{2x+1} \sqrt{1 + \frac{1}{2x+1}} \, dx$

$= 2\pi\int_0^3 \sqrt{2x+1} \sqrt{\frac{2x+2}{2x+1}} \, dx = 2\sqrt{2}\pi\int_0^3 \sqrt{x+1} \, dx = 2\sqrt{2}\pi \left[\frac{2}{3}(x+1)^{3/2}\right]_0^3 = 2\sqrt{2}\pi \cdot \frac{2}{3}(8-1) = \frac{28\pi\sqrt{2}}{3}$

27. $S = \int_c^d 2\pi x \sqrt{1 + \left(\frac{dx}{dy}\right)^2} \, dy; \quad \frac{dx}{dy} = \frac{\left(\frac{1}{2}\right)(4-2y)}{\sqrt{4y-y^2}} = \frac{2-y}{\sqrt{4y-y^2}} \Rightarrow 1 + \left(\frac{dx}{dy}\right)^2 = \frac{4y - y^2 + 4 - 4y + y^2}{4y - y^2} = \frac{4}{4y - y^2}$

$\Rightarrow S = \int_1^2 2\pi \sqrt{4y - y^2} \sqrt{\frac{4}{4y - y^2}} \, dy = 4\pi\int_1^2 dx = 4\pi$

29. $x = \frac{t^2}{2}$ and $y = 2t, \, 0 \le t \le \sqrt{5} \Rightarrow \frac{dx}{dt} = t$ and $\frac{dy}{dt} = 2 \Rightarrow \text{Surface Area} = \int_0^{\sqrt{5}} 2\pi(2t)\sqrt{t^2 + 4} \, dt = \int_4^9 2\pi u^{1/2} \, du$

$= 2\pi \left[\frac{2}{3} u^{3/2}\right]_4^9 = \frac{76\pi}{3}$, where $u = t^2 + 4 \Rightarrow du = 2t \, dt; \, t = 0 \Rightarrow u = 4, \, t = \sqrt{5} \Rightarrow u = 9$

31. $\frac{dy}{dx} = \sqrt{y}\cos^2\sqrt{y} \Rightarrow \frac{dy}{\sqrt{y}\cos^2\sqrt{y}} = dx \Rightarrow 2\tan\sqrt{y} = x + C \Rightarrow y = \left(\tan^{-1}\left(\frac{x+C}{2}\right)\right)^2$

33. $yy' = \sec(y^2)\sec^2 x \Rightarrow \frac{y \, dy}{\sec(y^2)} = \sec^2 x \, dx \Rightarrow \frac{\sin(y^2)}{2} = \tan x + C \Rightarrow \sin(y^2) = 2\tan x + C_1$

35. $\frac{dy}{dx} = e^{-x-y-2} \Rightarrow e^y dy = e^{-(x+2)} dx \Rightarrow e^y = -e^{-(x+2)} + C$. We have $y(0) = -2$, so $e^{-2} = -e^{-2} + C \Rightarrow C = 2e^{-2}$ and

$e^y = -e^{-(x+2)} + 2e^{-2} \Rightarrow y = \ln\left(-e^{-(x+2)} + 2e^{-2}\right)$

37. $x \, dy - (y + \sqrt{y}) dx = 0 \Rightarrow \frac{dy}{(y+\sqrt{y})} = \frac{dx}{x} \Rightarrow 2\ln\left(\sqrt{y} + 1\right) = \ln x + C$. We have $y(1) = 1 \Rightarrow 2\ln\left(\sqrt{1} + 1\right) = \ln 1 + C$

$\Rightarrow 2\ln 2 = C = \ln 2^2 = \ln 4$. So $2\ln\left(\sqrt{y} + 1\right) = \ln x + \ln 4 = \ln(4x) \Rightarrow \ln\left(\sqrt{y} + 1\right) = \frac{1}{2}\ln(4x) = \ln(4x)^{1/2}$

$\Rightarrow e^{\ln(\sqrt{y}+1)} = e^{\ln(4x)^{1/2}} \Rightarrow \sqrt{y} + 1 = 2\sqrt{x} \Rightarrow y = \left(2\sqrt{x} - 1\right)^2$

39. Since the half life is 5700 years and $A(t) = A_0 e^{kt}$ we have $\frac{A_0}{2} = A_0 e^{5700k} \Rightarrow \frac{1}{2} = e^{5700k} \Rightarrow \ln(0.5) = 5700k$

$\Rightarrow k = \frac{\ln(0.5)}{5700}$. With 10% of the original carbon-14 remaining we have $0.1A_0 = A_0 e^{\frac{\ln(0.5)}{5700} t} \Rightarrow 0.1 = e^{\frac{\ln(0.5)}{5700} t}$

$\Rightarrow \ln(0.1) = \frac{\ln(0.5)}{5700} t \Rightarrow t = \frac{(5700)\ln(0.1)}{\ln(0.5)} \approx 18{,}935$ years (rounded to the nearest year).

41. The equipment alone: the force required to lift the equipment is equal to its weight $\Rightarrow F_1(x) = 100$ N.

The work done is $W_1 = \int_a^b F_1(x) \, dx = \int_0^{40} 100 \, dx = [100x]_0^{40} = 4000$ J; the rope alone: the force required

to lift the rope is equal to the weight of the rope paid out at elevation $x \Rightarrow F_2(x) = 0.8(40 - x)$. The work

done is $W_2 = \int_a^b F_2(x) \, dx = \int_0^{40} 0.8(40 - x) \, dx = 0.8\left[40x - \frac{x^2}{2}\right]_0^{40} = 0.8\left(40^2 - \frac{40^2}{2}\right) = \frac{(0.8)(1600)}{2} = 640$ J;

the total work is $W = W_1 + W_2 = 4000 + 640 = 4640$ J

43. Force constant: $F = kx \Rightarrow 20 = k \cdot 1 \Rightarrow k = 20$ lb/ft; the work to stretch the spring 1 ft is

$$W = \int_0^1 kx \, dx = k \int_0^1 x \, dx = \left[20 \frac{x^2}{2} \right]_0^1 = 10 \text{ ft} \cdot \text{lb}; \text{ the work to stretch the spring an additional foot is}$$

$$W = \int_1^2 kx \, dx = k \int_1^2 x \, dx = 20 \left[\frac{x^2}{2} \right]_1^2 = 20 \left(\frac{4}{2} - \frac{1}{2} \right) = 20 \left(\frac{3}{2} \right) = 30 \text{ ft} \cdot \text{lb}$$

45. We imagine the water divided into thin slabs by planes perpendicular to the y-axis at the points of a partition of the interval [0, 8]. The typical slab between the planes at y and $y + \Delta y$ has a volume of about $\Delta V = \pi(\text{radius})^2(\text{thickness})$ $= \pi \left(\frac{5}{4} y \right)^2 \Delta y = \frac{25\pi}{16} y^2 \, \Delta y \text{ ft}^3$. The force F(y) required to lift this slab is equal to its weight: $F(y) = 62.4 \, \Delta V$ $= \frac{(62.4)(25)}{16} \pi y^2 \, \Delta y$ lb. The distance through which F(y) must act to lift this slab to the level 6 ft above the top is

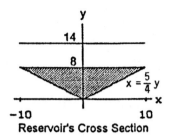

Reservoir's Cross Section

about $(6 + 8 - y)$ ft, so the work done lifting the slab is about $\Delta W = \frac{(62.4)(25)}{16} \pi y^2 (14 - y) \, \Delta y \text{ ft} \cdot \text{lb}$. The work done lifting all the slabs from $y = 0$ to $y = 8$ to the level 6 ft above the top is approximately

$$W \approx \sum_0^8 \frac{(62.4)(25)}{16} \pi y^2 (14 - y) \, \Delta y \text{ ft} \cdot \text{lb so the work to pump the water is the limit of these Riemann sums as the norm of}$$

the partition goes to zero: $W = \int_0^8 \frac{(62.4)(25)}{(16)} \pi y^2 (14 - y) \, dy = \frac{(62.4)(25)\pi}{16} \int_0^8 (14y^2 - y^3) \, dy = (62.4) \left(\frac{25\pi}{16} \right) \left[\frac{14}{3} y^3 - \frac{y^4}{4} \right]_0^8$

$$= (62.4) \left(\frac{25\pi}{16} \right) \left(\frac{14}{3} \cdot 8^3 - \frac{8^4}{4} \right) \approx 418{,}208.81 \text{ ft} \cdot \text{lb}$$

47. The tank's cross section looks like the figure in Exercise 45 with right edge given by $x = \frac{5}{10} y = \frac{y}{2}$. A typical horizontal slab has volume $\Delta V = \pi(\text{radius})^2(\text{thickness}) = \pi \left(\frac{y}{2} \right)^2 \Delta y = \frac{\pi}{4} y^2 \, \Delta y$. The force required to lift this slab is its weight: $F(y) = 60 \cdot \frac{\pi}{4} y^2 \, \Delta y$. The distance through which F(y) must act is $(2 + 10 - y)$ ft, so the work to pump the liquid is $W = 60 \int_0^{10} \pi (12 - y) \left(\frac{y^2}{4} \right) dy = 15\pi \left[\frac{12y^3}{3} - \frac{y^4}{4} \right]_0^{10} = 22{,}500\pi \text{ ft} \cdot \text{lb}$; the time needed to empty the tank is $\frac{22{,}500 \text{ ft·lb}}{275 \text{ ft·lb/sec}} \approx 257 \text{ sec}$

49. Intersection points: $3 - x^2 = 2x^2 \Rightarrow 3x^2 - 3 = 0$ $\Rightarrow 3(x - 1)(x + 1) = 0 \Rightarrow x = -1$ or $x = 1$. Symmetry suggests that $\bar{x} = 0$. The typical *vertical* strip has center of mass: $(\tilde{x}, \tilde{y}) = \left(x, \frac{2x^2 + (3 - x^2)}{2} \right) = \left(x, \frac{x^2 + 3}{2} \right)$, length: $(3 - x^2) - 2x^2 = 3(1 - x^2)$, width: dx, area: $dA = 3(1 - x^2) \, dx$, and mass: $dm = \delta \cdot dA$ $= 3\delta (1 - x^2) \, dx \Rightarrow$ the moment about the x-axis is

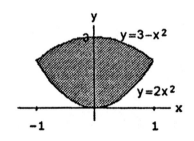

$\tilde{y} \, dm = \frac{3}{2} \delta (x^2 + 3)(1 - x^2) \, dx = \frac{3}{2} \delta (-x^4 - 2x^2 + 3) \, dx \Rightarrow M_x = \int \tilde{y} \, dm = \frac{3}{2} \delta \int_{-1}^1 (-x^4 - 2x^2 + 3) \, dx$

$= \frac{3}{2} \delta \left[-\frac{x^5}{5} - \frac{2x^3}{3} + 3x \right]_{-1}^1 = 3\delta \left(-\frac{1}{5} - \frac{2}{3} + 3 \right) = \frac{3\delta}{15} (-3 - 10 + 45) = \frac{32\delta}{5}$; $M = \int dm = 3\delta \int_{-1}^1 (1 - x^2) \, dx$

$= 3\delta \left[x - \frac{x^3}{3} \right]_{-1}^1 = 6\delta \left(1 - \frac{1}{3} \right) = 4\delta \Rightarrow \bar{y} = \frac{M_x}{M} = \frac{32\delta}{5 \cdot 4\delta} = \frac{8}{5}$. Therefore, the centroid is $(\bar{x}, \bar{y}) = \left(0, \frac{8}{5} \right)$.

51. The typical *vertical* strip has: center of mass: $(\tilde{x}, \tilde{y})$

$= \left(x, \frac{4+\frac{x^2}{4}}{2}\right)$, length: $4 - \frac{x^2}{4}$, width: dx,

area: $dA = \left(4 - \frac{x^2}{4}\right)dx$, mass: $dm = \delta \cdot dA$

$= \delta\left(4 - \frac{x^2}{4}\right)dx \Rightarrow$ the moment about the x-axis is

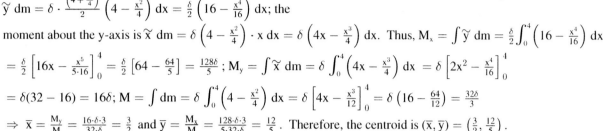

$y = (1/4)x^2$

$\tilde{y}\,dm = \delta \cdot \frac{\left(4 + \frac{x^2}{4}\right)}{2}\left(4 - \frac{x^2}{4}\right)dx = \frac{\delta}{2}\left(16 - \frac{x^4}{16}\right)dx$; the

moment about the y-axis is $\tilde{x}\,dm = \delta\left(4 - \frac{x^2}{4}\right)\cdot x\,dx = \delta\left(4x - \frac{x^3}{4}\right)dx$. Thus, $M_x = \int \tilde{y}\,dm = \frac{\delta}{2}\int_0^4\left(16 - \frac{x^4}{16}\right)dx$

$= \frac{\delta}{2}\left[16x - \frac{x^5}{5\cdot 16}\right]_0^4 = \frac{\delta}{2}\left[64 - \frac{64}{5}\right] = \frac{128\delta}{5}$; $M_y = \int \tilde{x}\,dm = \delta\int_0^4\left(4x - \frac{x^3}{4}\right)dx = \delta\left[2x^2 - \frac{x^4}{16}\right]_0^4$

$= \delta(32 - 16) = 16\delta$; $M = \int dm = \delta\int_0^4\left(4 - \frac{x^2}{4}\right)dx = \delta\left[4x - \frac{x^3}{12}\right]_0^4 = \delta\left(16 - \frac{64}{12}\right) = \frac{32\delta}{3}$

$\Rightarrow \bar{x} = \frac{M_y}{M} = \frac{16\cdot\delta\cdot 3}{32\cdot\delta} = \frac{3}{2}$ and $\bar{y} = \frac{M_x}{M} = \frac{128\cdot\delta\cdot 3}{5\cdot 32\cdot\delta} = \frac{12}{5}$. Therefore, the centroid is $(\bar{x}, \bar{y}) = \left(\frac{3}{2}, \frac{12}{5}\right)$.

53. A typical horizontal strip has: center of mass: $(\tilde{x}, \tilde{y})$

$= \left(\frac{y^2+2y}{2}, y\right)$, length: $2y - y^2$, width: dy,

area: $dA = (2y - y^2)\,dy$, mass: $dm = \delta \cdot dA$

$= (1 + y)(2y - y^2)\,dy \Rightarrow$ the moment about the

x-axis is $\tilde{y}\,dm = y(1 + y)(2y - y^2)\,dy$

$= (2y^2 + 2y^3 - y^3 - y^4)\,dy$

$= (2y^2 + y^3 - y^4)\,dy$; the moment about the y-axis is

$x = y^2$

$x = 2y$

$\tilde{x}\,dm = \left(\frac{y^2+2y}{2}\right)(1 + y)(2y - y^2)\,dy = \frac{1}{2}(4y^2 - y^4)(1 + y)\,dy = \frac{1}{2}(4y^2 + 4y^3 - y^4 - y^5)\,dy$

$\Rightarrow M_x = \int \tilde{y}\,dm = \int_0^2(2y^2 + y^3 - y^4)\,dy = \left[\frac{2}{3}y^3 + \frac{y^4}{4} - \frac{y^5}{5}\right]_0^2 = \left(\frac{16}{3} + \frac{16}{4} - \frac{32}{5}\right) = 16\left(\frac{1}{3} + \frac{1}{4} - \frac{2}{5}\right)$

$= \frac{16}{60}(20 + 15 - 24) = \frac{4}{15}(11) = \frac{44}{15}$; $M_y = \int \tilde{x}\,dm = \int_0^2\frac{1}{2}(4y^2 + 4y^3 - y^4 - y^5)\,dy = \frac{1}{2}\left[\frac{4}{3}y^3 + y^4 - \frac{y^5}{5} - \frac{y^6}{6}\right]_0^2$

$= \frac{1}{2}\left(\frac{4\cdot 2^3}{3} + 2^4 - \frac{2^5}{5} - \frac{2^6}{6}\right) = 4\left(\frac{4}{3} + 2 - \frac{4}{5} - \frac{8}{6}\right) = 4\left(2 - \frac{4}{5}\right) = \frac{24}{5}$; $M = \int dm = \int_0^2(1 + y)(2y - y^2)\,dy$

$= \int_0^2(2y + y^2 - y^3)\,dy = \left[y^2 + \frac{y^3}{3} - \frac{y^4}{4}\right]_0^2 = \left(4 + \frac{8}{3} - \frac{16}{4}\right) = \frac{8}{3} \Rightarrow \bar{x} = \frac{M_y}{M} = \left(\frac{24}{5}\right)\left(\frac{3}{8}\right) = \frac{9}{5}$ and $\bar{y} = \frac{M_x}{M}$

$= \left(\frac{44}{15}\right)\left(\frac{3}{8}\right) = \frac{44}{40} = \frac{11}{10}$. Therefore, the center of mass is $(\bar{x}, \bar{y}) = \left(\frac{9}{5}, \frac{11}{10}\right)$.

55. $F = \int_a^b W \cdot \left(\begin{smallmatrix}\text{strip}\\\text{depth}\end{smallmatrix}\right) \cdot L(y)\,dy \Rightarrow F = 2\int_0^2(62.4)(2 - y)(2y)\,dy = 249.6\int_0^2(2y - y^2)\,dy = 249.6\left[y^2 - \frac{y^3}{3}\right]_0^2$

$= (249.6)\left(4 - \frac{8}{3}\right) = (249.6)\left(\frac{4}{3}\right) = 332.8$ lb

NOTES:

CHAPTER 7 INFINITE SEQUENCES AND SERIES

7.1 SEQUENCES

1. $a_1 = \frac{1-1}{1^2} = 0, a_2 = \frac{1-2}{2^2} = -\frac{1}{4}, a_3 = \frac{1-3}{3^2} = -\frac{2}{9}, a_4 = \frac{1-4}{4^2} = -\frac{3}{16}$

3. $a_1 = \frac{(-1)^2}{2-1} = 1, a_2 = \frac{(-1)^3}{4-1} = -\frac{1}{3}, a_3 = \frac{(-1)^4}{6-1} = \frac{1}{5}, a_4 = \frac{(-1)^5}{8-1} = -\frac{1}{7}$

5. $a_1 = \frac{2}{2^2} = \frac{1}{2}, a_2 = \frac{2^2}{2^3} = \frac{1}{2}, a_3 = \frac{2^3}{2^4} = \frac{1}{2}, a_4 = \frac{2^4}{2^5} = \frac{1}{2}$

7. $a_1 = 1, a_2 = 1 + \frac{1}{2} = \frac{3}{2}, a_3 = \frac{3}{2} + \frac{1}{2^2} = \frac{7}{4}, a_4 = \frac{7}{4} + \frac{1}{2^3} = \frac{15}{8}, a_5 = \frac{15}{8} + \frac{1}{2^4} = \frac{31}{16}, a_6 = \frac{63}{32},$
$a_7 = \frac{127}{64}, a_8 = \frac{255}{128}, a_9 = \frac{511}{256}, a_{10} = \frac{1023}{512}$

9. $a_1 = 2, a_2 = \frac{(-1)^2(2)}{2} = 1, a_3 = \frac{(-1)^3(1)}{2} = -\frac{1}{2}, a_4 = \frac{(-1)^4\left(-\frac{1}{2}\right)}{2} = -\frac{1}{4}, a_5 = \frac{(-1)^5\left(-\frac{1}{4}\right)}{2} = \frac{1}{8},$
$a_6 = \frac{1}{16}, a_7 = -\frac{1}{32}, a_8 = -\frac{1}{64}, a_9 = \frac{1}{128}, a_{10} = \frac{1}{256}$

11. $a_1 = 1, a_2 = 1, a_3 = 1 + 1 = 2, a_4 = 2 + 1 = 3, a_5 = 3 + 2 = 5, a_6 = 8, a_7 = 13, a_8 = 21, a_9 = 34, a_{10} = 55$

13. $a_n = (-1)^{n+1}, n = 1, 2, \ldots$
   15. $a_n = (-1)^{n+1} n^2, n = 1, 2, \ldots$

17. $a_n = n^2 - 1, n = 1, 2, \ldots$
   19. $a_n = 4n - 3, n = 1, 2, \ldots$

21. $a_n = \frac{1 + (-1)^{n+1}}{2}, n = 1, 2, \ldots$

23. $\lim\limits_{n \to \infty} 2 + (0.1)^n = 2 \Rightarrow$ converges   (Theorem 5, #4)

25. $\lim\limits_{n \to \infty} \frac{1 - 2n}{1 + 2n} = \lim\limits_{n \to \infty} \frac{\left(\frac{1}{n}\right) - 2}{\left(\frac{1}{n}\right) + 2} = \lim\limits_{n \to \infty} \frac{-2}{2} = -1 \Rightarrow$ converges

27. $\lim\limits_{n \to \infty} \frac{1 - 5n^4}{n^4 + 8n^3} = \lim\limits_{n \to \infty} \frac{\left(\frac{1}{n^4}\right) - 5}{1 + \left(\frac{8}{n}\right)} = -5 \Rightarrow$ converges

29. $\lim\limits_{n \to \infty} \frac{n^2 - 2n + 1}{n - 1} = \lim\limits_{n \to \infty} \frac{(n-1)(n-1)}{n-1} = \lim\limits_{n \to \infty} (n - 1) = \infty \Rightarrow$ diverges

31. $\lim\limits_{n \to \infty} (1 + (-1)^n)$ does not exist $\Rightarrow$ diverges

33. $\lim\limits_{n \to \infty} \left(\frac{n+1}{2n}\right)\left(1 - \frac{1}{n}\right) = \lim\limits_{n \to \infty} \left(\frac{1}{2} + \frac{1}{2n}\right)\left(1 - \frac{1}{n}\right) = \frac{1}{2} \Rightarrow$ converges

35. $\lim\limits_{n \to \infty} \frac{(-1)^{n+1}}{2n - 1} = 0 \Rightarrow$ converges

37. $\lim\limits_{n \to \infty} \sqrt{\frac{2n}{n+1}} = \sqrt{\lim\limits_{n \to \infty} \frac{2n}{n+1}} = \sqrt{\lim\limits_{n \to \infty} \left(\frac{2}{1 + \frac{1}{n}}\right)} = \sqrt{2} \Rightarrow$ converges

39. $\lim\limits_{n \to \infty} \sin\left(\frac{\pi}{2} + \frac{1}{n}\right) = \sin\left(\lim\limits_{n \to \infty}\left(\frac{\pi}{2} + \frac{1}{n}\right)\right) = \sin\frac{\pi}{2} = 1 \Rightarrow$ converges

41. $\lim\limits_{n \to \infty} \frac{\sin n}{n} = 0$ because $-\frac{1}{n} \le \frac{\sin n}{n} \le \frac{1}{n} \Rightarrow$ converges by the Sandwich Theorem for sequences

43. $\lim\limits_{n \to \infty} \frac{n}{2^n} = \lim\limits_{n \to \infty} \frac{1}{2^n \ln 2} = 0 \Rightarrow$ converges (using l'Hôpital's rule)

45. $\lim\limits_{n \to \infty} \frac{\ln(n+1)}{\sqrt{n}} = \lim\limits_{n \to \infty} \frac{\left(\frac{1}{n+1}\right)}{\left(\frac{1}{2\sqrt{n}}\right)} = \lim\limits_{n \to \infty} \frac{2\sqrt{n}}{n+1} = \lim\limits_{n \to \infty} \frac{\left(\frac{2}{\sqrt{n}}\right)}{1 + \left(\frac{1}{n}\right)} = 0 \Rightarrow$ converges

47. $\lim\limits_{n \to \infty} 8^{1/n} = 1 \Rightarrow$ converges   (Theorem 5, #3)

49. $\lim\limits_{n \to \infty} \left(1 + \frac{7}{n}\right)^n = e^7 \Rightarrow$ converges   (Theorem 5, #5)

51. $\lim\limits_{n \to \infty} \sqrt[n]{10n} = \lim\limits_{n \to \infty} 10^{1/n} \cdot n^{1/n} = 1 \cdot 1 = 1 \Rightarrow$ converges   (Theorem 5, #3 and #2)

53. $\lim\limits_{n \to \infty} \left(\frac{3}{n}\right)^{1/n} = \frac{\lim\limits_{n \to \infty} 3^{1/n}}{\lim\limits_{n \to \infty} n^{1/n}} = \frac{1}{1} = 1 \Rightarrow$ converges   (Theorem 5, #3 and #2)

55. $\lim\limits_{n \to \infty} \frac{\ln n}{n^{1/n}} = \frac{\lim\limits_{n \to \infty} \ln n}{\lim\limits_{n \to \infty} n^{1/n}} = \frac{\infty}{1} = \infty \Rightarrow$ diverges   (Theorem 5, #2)

57. $\lim\limits_{n \to \infty} \sqrt[n]{4^n\, n} = \lim\limits_{n \to \infty} 4\sqrt[n]{n} = 4 \cdot 1 = 4 \Rightarrow$ converges   (Theorem 5, #2)

59. $\lim\limits_{n \to \infty} \frac{n!}{n^n} = \lim\limits_{n \to \infty} \frac{1 \cdot 2 \cdot 3 \cdots (n-1)(n)}{n \cdot n \cdot n \cdots n \cdot n} \le \lim\limits_{n \to \infty} \left(\frac{1}{n}\right) = 0$ and $\frac{n!}{n^n} \ge 0 \Rightarrow \lim\limits_{n \to \infty} \frac{n!}{n^n} = 0 \Rightarrow$ converges

61. $\lim\limits_{n \to \infty} \frac{n!}{10^{6n}} = \lim\limits_{n \to \infty} \frac{1}{\left(\frac{(10^6)^n}{n!}\right)} = \infty \Rightarrow$ diverges   (Theorem 5, #6)

63. $\lim\limits_{n \to \infty} \left(\frac{1}{n}\right)^{1/(\ln n)} = \lim\limits_{n \to \infty} \exp\left(\frac{1}{\ln n} \ln\left(\frac{1}{n}\right)\right) = \lim\limits_{n \to \infty} \exp\left(\frac{\ln 1 - \ln n}{\ln n}\right) = e^{-1} \Rightarrow$ converges

65. $\lim\limits_{n \to \infty} \left(\frac{3n+1}{3n-1}\right)^n = \lim\limits_{n \to \infty} \exp\left(n \ln\left(\frac{3n+1}{3n-1}\right)\right) = \lim\limits_{n \to \infty} \exp\left(\frac{\ln(3n+1) - \ln(3n-1)}{\frac{1}{n}}\right)$

    $= \lim\limits_{n \to \infty} \exp\left(\frac{\frac{3}{3n+1} - \frac{3}{3n-1}}{\left(-\frac{1}{n^2}\right)}\right) = \lim\limits_{n \to \infty} \exp\left(\frac{6n^2}{(3n+1)(3n-1)}\right) = \exp\left(\frac{6}{9}\right) = e^{2/3} \Rightarrow$ converges

67. $\lim\limits_{n \to \infty} \left(\frac{x^n}{2n+1}\right)^{1/n} = \lim\limits_{n \to \infty} x\left(\frac{1}{2n+1}\right)^{1/n} = x \lim\limits_{n \to \infty} \exp\left(\frac{1}{n} \ln\left(\frac{1}{2n+1}\right)\right) = x \lim\limits_{n \to \infty} \exp\left(\frac{-\ln(2n+1)}{n}\right)$

    $= x \lim\limits_{n \to \infty} \exp\left(\frac{-2}{2n+1}\right) = xe^0 = x,\ x > 0 \Rightarrow$ converges

69. $\lim\limits_{n \to \infty} \frac{3^n \cdot 6^n}{2^{-n} \cdot n!} = \lim\limits_{n \to \infty} \frac{36^n}{n!} = 0 \Rightarrow$ converges   (Theorem 5, #6)

71. $\lim\limits_{n \to \infty} \tanh n = \lim\limits_{n \to \infty} \frac{e^n - e^{-n}}{e^n + e^{-n}} = \lim\limits_{n \to \infty} \frac{e^{2n} - 1}{e^{2n} + 1} = \lim\limits_{n \to \infty} \frac{2e^{2n}}{2e^{2n}} = \lim\limits_{n \to \infty} 1 = 1 \Rightarrow$ converges

73. $\lim\limits_{n \to \infty} \frac{n^2 \sin\left(\frac{1}{n}\right)}{2n-1} = \lim\limits_{n \to \infty} \frac{\sin\left(\frac{1}{n}\right)}{\left(\frac{2}{n} - \frac{1}{n^2}\right)} = \lim\limits_{n \to \infty} \frac{-\left(\cos\left(\frac{1}{n}\right)\right)\left(\frac{1}{n^2}\right)}{\left(-\frac{2}{n^2} + \frac{2}{n^3}\right)} = \lim\limits_{n \to \infty} \frac{-\cos\left(\frac{1}{n}\right)}{-2 + \left(\frac{2}{n}\right)} = \frac{1}{2} \Rightarrow$ converges

75. $\lim\limits_{n \to \infty} \tan^{-1} n = \frac{\pi}{2} \Rightarrow$ converges

77. $\lim\limits_{n \to \infty} \left(\frac{1}{3}\right)^n + \frac{1}{\sqrt{2^n}} = \lim\limits_{n \to \infty} \left(\left(\frac{1}{3}\right)^n + \left(\frac{1}{\sqrt{2}}\right)^n\right) = 0 \Rightarrow$ converges   (Theorem 5, #4)

79. $\lim\limits_{n \to \infty} \frac{(\ln n)^{200}}{n} = \lim\limits_{n \to \infty} \frac{200\,(\ln n)^{199}}{n} = \lim\limits_{n \to \infty} \frac{200 \cdot 199\,(\ln n)^{198}}{n} = \ldots = \lim\limits_{n \to \infty} \frac{200!}{n} = 0 \Rightarrow$ converges

81. $\lim\limits_{n \to \infty} \left(n - \sqrt{n^2 - n}\right) = \lim\limits_{n \to \infty} \left(n - \sqrt{n^2 - n}\right)\left(\frac{n + \sqrt{n^2 - n}}{n + \sqrt{n^2 - n}}\right) = \lim\limits_{n \to \infty} \frac{n}{n + \sqrt{n^2 - n}} = \lim\limits_{n \to \infty} \frac{1}{1 + \sqrt{1 - \frac{1}{n}}}$

$= \frac{1}{2} \Rightarrow$ converges

83. $\lim\limits_{n \to \infty} \frac{1}{n} \int_1^n \frac{1}{x}\, dx = \lim\limits_{n \to \infty} \frac{\ln n}{n} = \lim\limits_{n \to \infty} \frac{1}{n} = 0 \Rightarrow$ converges   (Theorem 5, #1)

85. (a) $f(x) = x^2 - 2$; the sequence converges to $1.414213562 \approx \sqrt{2}$
    (b) $f(x) = \tan(x) - 1$; the sequence converges to $0.7853981635 \approx \frac{\pi}{4}$
    (c) $f(x) = e^x$; the sequence $1, 0, -1, -2, -3, -4, -5, \ldots$ diverges

87. If $a = 2n + 1$, then $b = \left\lfloor \frac{a^2}{2} \right\rfloor = \left\lfloor \frac{4n^2 + 4n + 1}{2} \right\rfloor = \left\lfloor 2n^2 + 2n + \frac{1}{2} \right\rfloor = 2n^2 + 2n,\ c = \left\lceil \frac{a^2}{2} \right\rceil = \left\lceil 2n^2 + 2n + \frac{1}{2} \right\rceil$

$= 2n^2 + 2n + 1$ and $a^2 + b^2 = (2n + 1)^2 + \left(2n^2 + 2n\right)^2 = 4n^2 + 4n + 1 + 4n^4 + 8n^3 + 4n^2$

$= 4n^4 + 8n^3 + 8n^2 + 4n + 1 = \left(2n^2 + 2n + 1\right)^2 = c^2.$

89. (a) $\lim\limits_{n \to \infty} \frac{\ln n}{n^c} = \lim\limits_{n \to \infty} \frac{\left(\frac{1}{n}\right)}{cn^{c-1}} = \lim\limits_{n \to \infty} \frac{1}{cn^c} = 0$
    (b) For all $\epsilon > 0$, there exists an $N = e^{-(\ln \epsilon)/c}$ such that $n > e^{-(\ln \epsilon)/c} \Rightarrow \ln n > -\frac{\ln \epsilon}{c} \Rightarrow \ln n^c > \ln\left(\frac{1}{\epsilon}\right)$

$\Rightarrow n^c > \frac{1}{\epsilon} \Rightarrow \frac{1}{n^c} < \epsilon \Rightarrow \left|\frac{1}{n^c} - 0\right| < \epsilon \Rightarrow \lim\limits_{n \to \infty} \frac{1}{n^c} = 0$

91. $\lim\limits_{n \to \infty} n^{1/n} = \lim\limits_{n \to \infty} \exp\left(\frac{1}{n} \ln n\right) = \lim\limits_{n \to \infty} \exp\left(\frac{1}{n}\right) = e^0 = 1$

93. Assume the hypotheses of the theorem and let ϵ be a positive number. For all ϵ there exists a N_1 such that when $n > N_1$ then $|a_n - L| < \epsilon \Rightarrow -\epsilon < a_n - L < \epsilon \Rightarrow L - \epsilon < a_n$, and there exists a N_2 such that when $n > N_2$ then $|c_n - L| < \epsilon$ $\Rightarrow -\epsilon < c_n - L < \epsilon \Rightarrow c_n < L + \epsilon$. If $n > \max\{N_1, N_2\}$, then $L - \epsilon < a_n \le b_n \le c_n < L + \epsilon \Rightarrow |b_n - L| < \epsilon$ $\Rightarrow \lim\limits_{n \to \infty} b_n = L.$

95. $a_{n+1} \ge a_n \Rightarrow \frac{3(n+1)+1}{(n+1)+1} > \frac{3n+1}{n+1} \Rightarrow \frac{3n+4}{n+2} > \frac{3n+1}{n+1} \Rightarrow 3n^2 + 3n + 4n + 4 > 3n^2 + 6n + n + 2 \Rightarrow 4 > 2$; the steps are reversible so the sequence is nondecreasing; $\frac{3n+1}{n+1} < 3 \Rightarrow 3n + 1 < 3n + 3 \Rightarrow 1 < 3$; the steps are reversible so the sequence is bounded above by 3

97. $a_{n+1} \le a_n \Rightarrow \frac{2^{n+1}3^{n+1}}{(n+1)!} \le \frac{2^n 3^n}{n!} \Rightarrow \frac{2^{n+1}3^{n+1}}{2^n 3^n} \le \frac{(n+1)!}{n!} \Rightarrow 2 \cdot 3 \le n + 1$ which is true for $n \ge 5$; the steps are reversible so the sequence is decreasing after a_5, but it is not nondecreasing for all its terms; $a_1 = 6,\ a_2 = 18,\ a_3 = 36,\ a_4 = 54$, $a_5 = \frac{324}{5} = 64.8 \Rightarrow$ the sequence is bounded from above by 64.8

99. $a_n = 1 - \frac{1}{n}$ converges because $\frac{1}{n} \to 0$ by Example 1; also it is a nondecreasing sequence bounded above by 1

101. $a_n = \frac{2^n - 1}{2^n} = 1 - \frac{1}{2^n}$ and $0 < \frac{1}{2^n} < \frac{1}{n}$; since $\frac{1}{n} \to 0$ (by Example 1) $\Rightarrow \frac{1}{2^n} \to 0$, the sequence converges; also it is a nondecreasing sequence bounded above by 1

103. $a_n = ((-1)^n + 1)\left(\frac{n+1}{n}\right)$ diverges because $a_n = 0$ for n odd, while for n even $a_n = 2\left(1 + \frac{1}{n}\right)$ converges to 2; it diverges by definition of divergence

105. If $\{a_n\}$ is nonincreasing with lower bound M, then $\{-a_n\}$ is a nondecreasing sequence with upper bound $-$M. By Theorem 1, $\{-a_n\}$ converges and hence $\{a_n\}$ converges. If $\{a_n\}$ has no lower bound, then $\{-a_n\}$ has no upper bound and therefore diverges. Hence, $\{a_n\}$ also diverges.

107. $a_n \geq a_{n+1} \Leftrightarrow \frac{1 + \sqrt{2n}}{\sqrt{n}} \geq \frac{1 + \sqrt{2(n+1)}}{\sqrt{n+1}} \Leftrightarrow \sqrt{n+1} + \sqrt{2n^2 + 2n} \geq \sqrt{n} + \sqrt{2n^2 + 2n} \Leftrightarrow \sqrt{n+1} \geq \sqrt{n}$ and $\frac{1 + \sqrt{2n}}{\sqrt{n}} \geq \sqrt{2}$; thus the sequence is nonincreasing and bounded below by $\sqrt{2} \Rightarrow$ it converges

109. $\frac{4^{n+1} + 3^n}{4^n} = 4 + \left(\frac{3}{4}\right)^n$ so $a_n \geq a_{n+1} \Leftrightarrow 4 + \left(\frac{3}{4}\right)^n \geq 4 + \left(\frac{3}{4}\right)^{n+1} \Leftrightarrow \left(\frac{3}{4}\right)^n \geq \left(\frac{3}{4}\right)^{n+1} \Leftrightarrow 1 \geq \frac{3}{4}$ and $4 + \left(\frac{3}{4}\right)^n \geq 4$; thus the sequence is nonincreasing and bounded below by $4 \Rightarrow$ it converges

111. Let $0 < M < 1$ and let N be an integer greater than $\frac{M}{1-M}$. Then $n > N \Rightarrow n > \frac{M}{1-M} \Rightarrow n - nM > M$ $\Rightarrow n > M + nM \Rightarrow n > M(n+1) \Rightarrow \frac{n}{n+1} > M$.

113. The sequence $a_n = 1 + \frac{(-1)^n}{2}$ is the sequence $\frac{1}{2}, \frac{3}{2}, \frac{1}{2}, \frac{3}{2}, \ldots$. This sequence is bounded above by $\frac{3}{2}$, but it clearly does not converge, by definition of convergence.

115. Given an $\epsilon > 0$, by definition of convergence there corresponds an N such that for all $n > N$, $|L_1 - a_n| < \epsilon$ and $|L_2 - a_n| < \epsilon$. Now $|L_2 - L_1| = |L_2 - a_n + a_n - L_1| \leq |L_2 - a_n| + |a_n - L_1| < \epsilon + \epsilon = 2\epsilon$. $|L_2 - L_1| < 2\epsilon$ says that the difference between two fixed values is smaller than any positive number 2ϵ. The only nonnegative number smaller than every positive number is 0, so $|L_1 - L_2| = 0$ or $L_1 = L_2$.

117. $a_{2k} \to L \Leftrightarrow$ given an $\epsilon > 0$ there corresponds an N_1 such that $[2k > N_1 \Rightarrow |a_{2k} - L| < \epsilon]$. Similarly, $a_{2k+1} \to L \Leftrightarrow [2k + 1 > N_2 \Rightarrow |a_{2k+1} - L| < \epsilon]$. Let $N = \max\{N_1, N_2\}$. Then $n > N \Rightarrow |a_n - L| < \epsilon$ whether n is even or odd, and hence $a_n \to L$.

119. (a) $f(x) = x^2 - a \Rightarrow f'(x) = 2x \Rightarrow x_{n+1} = x_n - \frac{x_n^2 - a}{2x_n} \Rightarrow x_{n+1} = \frac{2x_n^2 - (x_n^2 - a)}{2x_n} = \frac{x_n^2 + a}{2x_n} = \frac{\left(x_n + \frac{a}{x_n}\right)}{2}$

 (b) $x_1 = 2$, $x_2 = 1.75$, $x_3 = 1.732142857$, $x_4 = 1.73205081$, $x_5 = 1.732050808$; we are finding the positive number where $x^2 - 3 = 0$; that is, where $x^2 = 3$, $x > 0$, or where $x = \sqrt{3}$.

7.2 INFINITE SERIES

1. $s_n = \frac{a(1 - r^n)}{(1 - r)} = \frac{2\left(1 - \left(\frac{1}{3}\right)^n\right)}{1 - \left(\frac{1}{3}\right)} \Rightarrow \lim_{n \to \infty} s_n = \frac{2}{1 - \left(\frac{1}{3}\right)} = 3$

3. $s_n = \frac{a(1 - r^n)}{(1 - r)} = \frac{1 - \left(-\frac{1}{2}\right)^n}{1 - \left(-\frac{1}{2}\right)} \Rightarrow \lim_{n \to \infty} s_n = \frac{1}{\left(\frac{3}{2}\right)} = \frac{2}{3}$

5. $\frac{1}{(n+1)(n+2)} = \frac{1}{n+1} - \frac{1}{n+2} \Rightarrow s_n = \left(\frac{1}{2} - \frac{1}{3}\right) + \left(\frac{1}{3} - \frac{1}{4}\right) + \ldots + \left(\frac{1}{n+1} - \frac{1}{n+2}\right) = \frac{1}{2} - \frac{1}{n+2} \Rightarrow \lim_{n \to \infty} s_n = \frac{1}{2}$

7. $1 - \frac{1}{4} + \frac{1}{16} - \frac{1}{64} + \ldots$, the sum of this geometric series is $\frac{1}{1 - \left(-\frac{1}{4}\right)} = \frac{1}{1 + \left(\frac{1}{4}\right)} = \frac{4}{5}$

9. $\frac{7}{4} + \frac{7}{16} + \frac{7}{64} + \ldots$, the sum of this geometric series is $\frac{\left(\frac{7}{4}\right)}{1 - \left(\frac{1}{4}\right)} = \frac{7}{3}$

11. $(5+1) + \left(\frac{5}{2} + \frac{1}{3}\right) + \left(\frac{5}{4} + \frac{1}{9}\right) + \left(\frac{5}{8} + \frac{1}{27}\right) + \dots$, is the sum of two geometric series; the sum is

$\frac{5}{1 - \left(\frac{1}{2}\right)} + \frac{1}{1 - \left(\frac{1}{3}\right)} = 10 + \frac{3}{2} = \frac{23}{2}$

13. $(1+1) + \left(\frac{1}{2} - \frac{1}{5}\right) + \left(\frac{1}{4} + \frac{1}{25}\right) + \left(\frac{1}{8} - \frac{1}{125}\right) + \dots$, is the sum of two geometric series; the sum is

$\frac{1}{1 - \left(\frac{1}{2}\right)} + \frac{1}{1 + \left(\frac{1}{5}\right)} = 2 + \frac{5}{6} = \frac{17}{6}$

15. $\frac{4}{(4n-3)(4n+1)} = \frac{1}{4n-3} - \frac{1}{4n+1} \Rightarrow s_n = \left(1 - \frac{1}{5}\right) + \left(\frac{1}{5} - \frac{1}{9}\right) + \left(\frac{1}{9} - \frac{1}{13}\right) + \dots + \left(\frac{1}{4n-7} - \frac{1}{4n-3}\right)$

$+ \left(\frac{1}{4n-3} - \frac{1}{4n+1}\right) = 1 - \frac{1}{4n+1} \Rightarrow \lim\limits_{n \to \infty} s_n = \lim\limits_{n \to \infty} \left(1 - \frac{1}{4n+1}\right) = 1$

17. $\frac{40n}{(2n-1)^2(2n+1)^2} = \frac{A}{(2n-1)} + \frac{B}{(2n-1)^2} + \frac{C}{(2n+1)} + \frac{D}{(2n+1)^2} = \frac{A(2n-1)(2n+1)^2 + B(2n+1)^2 + C(2n+1)(2n-1)^2 + D(2n-1)^2}{(2n-1)^2(2n+1)^2}$

$\Rightarrow A(2n-1)(2n+1)^2 + B(2n+1)^2 + C(2n+1)(2n-1)^2 + D(2n-1)^2 = 40n$

$\Rightarrow A\left(8n^3 + 4n^2 - 2n - 1\right) + B\left(4n^2 + 4n + 1\right) + C\left(8n^3 - 4n^2 - 2n + 1\right) = D\left(4n^2 - 4n + 1\right) = 40n$

$\Rightarrow (8A + 8C)n^3 + (4A + 4B - 4C + 4D)n^2 + (-2A + 4B - 2C - 4D)n + (-A + B + C + D) = 40n$

$\Rightarrow \begin{cases} 8A + 8C = 0 \\ 4A + 4B - 4C + 4D = 0 \\ -2A + 4B - 2C - 4D = 40 \\ -A + B + C + D = 0 \end{cases} \Rightarrow \begin{cases} 8A + 8C = 0 \\ A + B - C + D = 0 \\ -A + 2B - C - 2D = 20 \\ -A + B + C + D = 0 \end{cases} \Rightarrow \begin{cases} B + D = 0 \\ 2B - 2D = 20 \end{cases} \Rightarrow 4B = 20 \Rightarrow B = 5$

and $D = -5 \Rightarrow \begin{cases} A + C = 0 \\ -A + 5 + C - 5 = 0 \end{cases} \Rightarrow C = 0$ and $A = 0$. Hence, $\sum\limits_{n=1}^{k} \left[\frac{40n}{(2n-1)^2(2n+1)^2}\right]$

$= 5 \sum\limits_{n=1}^{k} \left[\frac{1}{(2n-1)^2} - \frac{1}{(2n+1)^2}\right] = 5\left(\frac{1}{1} - \frac{1}{9} + \frac{1}{9} - \frac{1}{25} + \frac{1}{25} - \dots - \frac{1}{(2(k-1)+1)^2} + \frac{1}{(2k-1)^2} - \frac{1}{(2k+1)^2}\right)$

$= 5\left(1 - \frac{1}{(2k+1)^2}\right) \Rightarrow$ the sum is $\lim\limits_{n \to \infty} 5\left(1 - \frac{1}{(2k+1)^2}\right) = 5$

19. $s_n = \left(1 - \frac{1}{\sqrt{2}}\right) + \left(\frac{1}{\sqrt{2}} - \frac{1}{\sqrt{3}}\right) + \left(\frac{1}{\sqrt{3}} - \frac{1}{\sqrt{4}}\right) + \dots + \left(\frac{1}{\sqrt{n-1}} + \frac{1}{\sqrt{n}}\right) + \left(\frac{1}{\sqrt{n}} - \frac{1}{\sqrt{n+1}}\right) = 1 - \frac{1}{\sqrt{n+1}}$

$\Rightarrow \lim\limits_{n \to \infty} s_n = \lim\limits_{n \to \infty} \left(1 - \frac{1}{\sqrt{n+1}}\right) = 1$

21. $s_n = \left(\frac{1}{\ln 3} - \frac{1}{\ln 2}\right) + \left(\frac{1}{\ln 4} - \frac{1}{\ln 3}\right) + \left(\frac{1}{\ln 5} - \frac{1}{\ln 4}\right) + \dots + \left(\frac{1}{\ln(n+1)} - \frac{1}{\ln n}\right) + \left(\frac{1}{\ln(n+2)} - \frac{1}{\ln(n+1)}\right)$

$= -\frac{1}{\ln 2} + \frac{1}{\ln(n+2)} \Rightarrow \lim\limits_{n \to \infty} s_n = -\frac{1}{\ln 2}$

23. convergent geometric series with sum $\frac{1}{1 - \left(\frac{1}{\sqrt{2}}\right)} = \frac{\sqrt{2}}{\sqrt{2}-1} = 2 + \sqrt{2}$

25. convergent geometric series with sum $\frac{\left(\frac{3}{2}\right)}{1 - \left(-\frac{1}{2}\right)} = 1$

27. $\lim\limits_{n \to \infty} \cos(n\pi) = \lim\limits_{n \to \infty} (-1)^n \neq 0 \Rightarrow$ diverges

29. convergent geometric series with sum $\frac{1}{1 - \left(\frac{1}{e^2}\right)} = \frac{e^2}{e^2 - 1}$

31. convergent geometric series with sum $\frac{2}{1 - \left(\frac{1}{10}\right)} - 2 = \frac{20}{9} - \frac{18}{9} = \frac{2}{9}$

33. difference of two geometric series with sum $\frac{1}{1 - \left(\frac{2}{3}\right)} - \frac{1}{1 - \left(\frac{1}{3}\right)} = 3 - \frac{3}{2} = \frac{3}{2}$

35. $\lim\limits_{n \to \infty} \frac{n!}{1000^n} = \infty \neq 0 \Rightarrow$ diverges

37. $\sum\limits_{n=1}^{\infty} \ln\left(\frac{n}{n+1}\right) = \sum\limits_{n=1}^{\infty} [\ln(n) - \ln(n+1)] \Rightarrow s_n = [\ln(1) - \ln(2)] + [\ln(2) - \ln(3)] + [\ln(3) - \ln(4)] + \ldots$

$+ [\ln(n-1) - \ln(n)] + [\ln(n) - \ln(n+1)] = \ln(1) - \ln(n+1) = -\ln(n+1) \Rightarrow \lim\limits_{n \to \infty} s_n = -\infty, \Rightarrow$ diverges

39. convergent geometric series with sum $\frac{1}{1 - \left(\frac{e}{\pi}\right)} = \frac{\pi}{\pi - e}$

41. $\sum\limits_{n=0}^{\infty} (-1)^n x^n = \sum\limits_{n=0}^{\infty} (-x)^n; a = 1, r = -x;$ converges to $\frac{1}{1 - (-x)} = \frac{1}{1 + x}$ for $|x| < 1$

43. $a = 3, r = \frac{x-1}{2};$ converges to $\frac{3}{1 - \left(\frac{x-1}{2}\right)} = \frac{6}{3 - x}$ for $-1 < \frac{x-1}{2} < 1$ or $-1 < x < 3$

45. $a = 1, r = 2x;$ converges to $\frac{1}{1 - 2x}$ for $|2x| < 1$ or $|x| < \frac{1}{2}$

47. $a = 1, r = -(x+1)^n;$ converges to $\frac{1}{1 + (x+1)} = \frac{1}{2+x}$ for $|x+1| < 1$ or $-2 < x < 0$

49. $a = 1, r = \sin x;$ converges to $\frac{1}{1 - \sin x}$ for $x \neq (2k+1)\frac{\pi}{2}$, k an integer

51. $0.\overline{23} = \sum\limits_{n=0}^{\infty} \frac{23}{100}\left(\frac{1}{10^2}\right)^n = \frac{\left(\frac{23}{100}\right)}{1 - \left(\frac{1}{100}\right)} = \frac{23}{99}$　　　　53. $0.\overline{7} = \sum\limits_{n=0}^{\infty} \frac{7}{10}\left(\frac{1}{10}\right)^n = \frac{\left(\frac{7}{10}\right)}{1 - \left(\frac{1}{10}\right)} = \frac{7}{9}$

55. $0.0\overline{6} = \sum\limits_{n=0}^{\infty} \left(\frac{1}{10}\right)\left(\frac{6}{10}\right)\left(\frac{1}{10}\right)^n = \frac{\left(\frac{6}{100}\right)}{1 - \left(\frac{1}{10}\right)} = \frac{6}{90} = \frac{1}{15}$

57. $1.24\overline{123} = \frac{124}{100} + \sum\limits_{n=0}^{\infty} \frac{123}{10^5}\left(\frac{1}{10^3}\right)^n = \frac{124}{100} + \frac{\left(\frac{123}{10^5}\right)}{1 - \left(\frac{1}{10^3}\right)} = \frac{124}{100} + \frac{123}{10^5 - 10^2} = \frac{124}{100} + \frac{123}{99,900} = \frac{123,999}{99,900} = \frac{41,333}{33,300}$

59. Let $a_n = b_n = \left(\frac{1}{2}\right)^n$. Then $\sum\limits_{n=1}^{\infty} a_n = \sum\limits_{n=1}^{\infty} b_n = \sum\limits_{n=1}^{\infty} \left(\frac{1}{2}\right)^n = 1$, while $\sum\limits_{n=1}^{\infty} \left(\frac{a_n}{b_n}\right) = \sum\limits_{n=1}^{\infty} (1)$ diverges.

61. Let $a_n = \left(\frac{1}{4}\right)^n$ and $b_n = \left(\frac{1}{2}\right)^n$. Then $A = \sum\limits_{n=1}^{\infty} a_n = \frac{1}{3}$, $B = \sum\limits_{n=1}^{\infty} b_n = 1$ and $\sum\limits_{n=1}^{\infty} \left(\frac{a_n}{b_n}\right) = \sum\limits_{n=1}^{\infty} \left(\frac{1}{2}\right)^n = 1 \neq \frac{A}{B}$.

63. Since the sum of a finite number of terms is finite, adding or subtracting a finite number of terms from a series that diverges does not change the divergence of the series.

65. $s_n = 1 + 2r + r^2 + 2r^3 + r^4 + 2r^5 + \ldots + r^{2n} + 2r^{2n+1}, n = 0, 1, \ldots$

$\Rightarrow s_n = (1 + r^2 + r^4 + \ldots + r^{2n}) + (2r + 2r^3 + 2r^5 + \ldots + 2r^{2n+1}) \Rightarrow \lim\limits_{n \to \infty} s_n = \frac{1}{1 - r^2} + \frac{2r}{1 - r^2}$

$= \frac{1 + 2r}{1 - r^2}$, if $|r^2| < 1$ or $|r| < 1$

7.3 THE INTEGRAL TEST

1. converges; a geometric series with $r = \frac{1}{10} < 1$

3. diverges; by the nth-Term Test for Divergence, $\lim\limits_{n \to \infty} \frac{n}{n+1} = 1 \neq 0$

5. diverges; $\sum_{n=1}^{\infty} \frac{3}{\sqrt{n}} = 3 \sum_{n=1}^{\infty} \frac{1}{\sqrt{n}}$, which is a divergent p-series $(p = \frac{1}{2})$

7. converges; a geometric series with $r = \frac{1}{8} < 1$

9. diverges by the Integral Test: $\int_{2}^{n} \frac{\ln x}{x} \, dx = \frac{1}{2} \left(\ln^2 n - \ln 2 \right) \Rightarrow \int_{2}^{\infty} \frac{\ln x}{x} \, dx \to \infty$

11. converges; a geometric series with $r = \frac{2}{3} < 1$

13. diverges; $\sum_{n=0}^{\infty} \frac{-2}{n+1} = -2 \sum_{n=0}^{\infty} \frac{1}{n+1}$, which diverges by the Integral Test

15. diverges; $\lim_{n \to \infty} a_n = \lim_{n \to \infty} \frac{2^n}{n+1} = \lim_{n \to \infty} \frac{2^n \ln 2}{1} = \infty \neq 0$

17. diverges; $\lim_{n \to \infty} \frac{\sqrt{n}}{\ln n} = \lim_{n \to \infty} \frac{\left(\frac{1}{2\sqrt{n}} \right)}{\left(\frac{1}{n} \right)} = \lim_{n \to \infty} \frac{\sqrt{n}}{2} = \infty \neq 0$

19. diverges; a geometric series with $r = \frac{1}{\ln 2} \approx 1.44 > 1$

21. converges by the Integral Test: $\int_{3}^{\infty} \frac{\left(\frac{1}{x} \right)}{(\ln x) \sqrt{(\ln x)^2 - 1}} \, dx; \begin{bmatrix} u = \ln x \\ du = \frac{1}{x} \, dx \end{bmatrix} \to \int_{\ln 3}^{\infty} \frac{1}{u \sqrt{u^2 - 1}} \, du$

$= \lim_{b \to \infty} \left[\sec^{-1} |u| \right]_{\ln 3}^{b} = \lim_{b \to \infty} \left[\sec^{-1} b - \sec^{-1} (\ln 3) \right] = \lim_{b \to \infty} \left[\cos^{-1} \left(\frac{1}{b} \right) - \sec^{-1} (\ln 3) \right]$

$= \cos^{-1}(0) - \sec^{-1}(\ln 3) = \frac{\pi}{2} - \sec^{-1}(\ln 3) \approx 1.1439$

23. diverges by the nth-Term Test for divergence; $\lim_{n \to \infty} n \sin \left(\frac{1}{n} \right) = \lim_{n \to \infty} \frac{\sin \left(\frac{1}{n} \right)}{\left(\frac{1}{n} \right)} = \lim_{x \to 0} \frac{\sin x}{x} = 1 \neq 0$

25. converges by the Integral Test: $\int_{1}^{\infty} \frac{e^x}{1 + e^{2x}} \, dx; \begin{bmatrix} u = e^x \\ du = e^x \, dx \end{bmatrix} \to \int_{e}^{\infty} \frac{1}{1 + u^2} \, du = \lim_{n \to \infty} \left[\tan^{-1} u \right]_{e}^{b}$

$= \lim_{b \to \infty} \left(\tan^{-1} b - \tan^{-1} e \right) = \frac{\pi}{2} - \tan^{-1} e \approx 0.35$

27. converges by the Integral Test: $\int_{1}^{\infty} \frac{8 \tan^{-1} x}{1 + x^2} \, dx; \begin{bmatrix} u = \tan^{-1} x \\ du = \frac{dx}{1 + x^2} \end{bmatrix} \to \int_{\pi/4}^{\pi/2} 8u \, du = \left[4u^2 \right]_{\pi/4}^{\pi/2} = 4 \left(\frac{\pi^2}{4} - \frac{\pi^2}{16} \right) = \frac{3\pi^2}{4}$

29. converges by the Integral Test: $\int_{1}^{\infty} \operatorname{sech} x \, dx = 2 \lim_{b \to \infty} \int_{1}^{b} \frac{e^x}{1 + (e^x)^2} \, dx = 2 \lim_{b \to \infty} \left[\tan^{-1} e^x \right]_{1}^{b}$

$= 2 \lim_{b \to \infty} \left(\tan^{-1} e^b - \tan^{-1} e \right) = \pi - 2 \tan^{-1} e \approx 0.71$

31. $\int_{1}^{\infty} \left(\frac{a}{x+2} - \frac{1}{x+4} \right) dx = \lim_{b \to \infty} \left[a \ln |x + 2| - \ln |x + 4| \right]_{1}^{b} = \lim_{b \to \infty} \ln \frac{(b+2)^a}{b+4} - \ln \left(\frac{3^a}{5} \right);$

$\lim_{b \to \infty} \frac{(b+2)^a}{b+4} = a \lim_{b \to \infty} (b+2)^{a-1} = \begin{cases} \infty, a > 1 \\ 1, \quad a = 1 \end{cases} \Rightarrow$ the series converges to $\ln \left(\frac{5}{3} \right)$ if $a = 1$ and diverges to ∞ if

$a > 1$. If $a < 1$, the terms of the series eventually become negative and the Integral Test does not apply. From that point on, however, the series behaves like a negative multiple of the harmonic series, and so it diverges.

33. Let $A_n = \sum_{k=1}^{n} a_k$ and $B_n = \sum_{k=1}^{n} 2^k a_{(2^k)}$, where $\{a_k\}$ is a nonincreasing sequence of positive terms converging to 0. Note that $\{A_n\}$ and $\{B_n\}$ are nondecreasing sequences of positive terms. Now, $B_n = 2a_2 + 4a_4 + 8a_8 + \ldots + 2^n a_{(2^n)}$

$= 2a_2 + (2a_4 + 2a_4) + (2a_8 + 2a_8 + 2a_8 + 2a_8) + \ldots + \underbrace{\left(2a_{(2^n)} + 2a_{(2^n)} + \ldots + 2a_{(2^n)}\right)}_{2^{n-1} \text{ terms}}$

$\leq 2a_1 + 2a_2 + (2a_3 + 2a_4) + (2a_5 + 2a_6 + 2a_7 + 2a_8) + \ldots + \left(2a_{(2^{n-1})} + 2a_{(2^{n-1}+1)} + \ldots + 2a_{(2^n)}\right)$

$= 2A_{(2^n)} \leq 2\sum\limits_{k=1}^{\infty} a_k$. Therefore if $\sum a_k$ converges, then $\{B_n\}$ is bounded above $\Rightarrow \sum 2^k a_{(2^k)}$ converges. Conversely,

$A_n = a_1 + (a_2 + a_3) + (a_4 + a_5 + a_6 + a_7) + \ldots + a_n < a_1 + 2a_2 + 4a_4 + \ldots + 2^n a_{(2^n)} = a_1 + B_n < a_1 + \sum\limits_{k=1}^{\infty} 2^k a_{(2^k)}$.

Therefore, if $\sum\limits_{k=1}^{\infty} 2^k a_{(2^k)}$ converges, then $\{A_n\}$ is bounded above and hence converges.

35. (a) $\int_2^{\infty} \frac{dx}{x(\ln x)^p}$; $\begin{bmatrix} u = \ln x \\ du = \frac{dx}{x} \end{bmatrix} \to \int_{\ln 2}^{\infty} u^{-p}\, du = \lim\limits_{b \to \infty} \left[\frac{u^{-p+1}}{-p+1} \right]_{\ln 2}^{b} = \lim\limits_{b \to \infty} \left(\frac{1}{1-p} \right) [b^{-p+1} - (\ln 2)^{-p+1}]$

$= \begin{cases} \frac{1}{p-1}(\ln 2)^{-p+1}, p > 1 \\ \infty, p < 1 \end{cases} \Rightarrow$ the improper integral converges if $p > 1$ and diverges if $p < 1$.

For $p = 1$: $\int_2^{\infty} \frac{dx}{x \ln x} = \lim\limits_{b \to \infty} [\ln(\ln x)]_2^b = \lim\limits_{b \to \infty} [\ln(\ln b) - \ln(\ln 2)] = \infty$, so the improper integral diverges if $p = 1$.

(b) Since the series and the integral converge or diverge together, $\sum\limits_{n=2}^{\infty} \frac{1}{n(\ln n)^p}$ converges if and only if $p > 1$.

37. (a) From Fig. 7.8 in the text with $f(x) = \frac{1}{x}$ and $a_k = \frac{1}{k}$, we have $\int_1^{n+1} \frac{1}{x}\, dx \leq 1 + \frac{1}{2} + \frac{1}{3} + \ldots + \frac{1}{n}$

$\leq 1 + \int_1^n f(x)\, dx \Rightarrow \ln(n+1) \leq 1 + \frac{1}{2} + \frac{1}{3} + \ldots + \frac{1}{n} \leq 1 + \ln n \Rightarrow 0 \leq \ln(n+1) - \ln n$

$\leq \left(1 + \frac{1}{2} + \frac{1}{3} + \ldots + \frac{1}{n}\right) - \ln n \leq 1$. Therefore the sequence $\left\{ \left(1 + \frac{1}{2} + \frac{1}{3} + \ldots + \frac{1}{n}\right) - \ln n \right\}$ is bounded above by 1 and below by 0.

(b) From the graph in Fig. 7.8(a) with $f(x) = \frac{1}{x}$, $\frac{1}{n+1} < \int_n^{n+1} \frac{1}{x}\, dx = \ln(n+1) - \ln n$

$\Rightarrow 0 > \frac{1}{n+1} - [\ln(n+1) - \ln n] = \left(1 + \frac{1}{2} + \frac{1}{3} + \ldots + \frac{1}{n+1} - \ln(n+1)\right) - \left(1 + \frac{1}{2} + \frac{1}{3} + \ldots + \frac{1}{n} - \ln n\right)$.

If we define $a_n = 1 + \frac{1}{2} = \frac{1}{3} + \frac{1}{n} - \ln n$, then $0 > a_{n+1} - a_n \Rightarrow a_{n+1} < a_n \Rightarrow \{a_n\}$ is a decreasing sequence of nonnegative terms.

39. (a) $s_{10} = \sum\limits_{n=1}^{10} \frac{1}{n^3} = 1.97531986$; $\int_{11}^{\infty} \frac{1}{x^3}\, dx = \lim\limits_{b \to \infty} \int_{11}^b x^{-3}\, dx = \lim\limits_{b \to \infty} \left[-\frac{x^{-2}}{2} \right]_{11}^b = \lim\limits_{b \to \infty} \left(-\frac{1}{2b^2} + \frac{1}{242} \right) = \frac{1}{242}$ and

$\int_{10}^{\infty} \frac{1}{x^3}\, dx = \lim\limits_{b \to \infty} \int_{10}^b x^{-3}\, dx = \lim\limits_{b \to \infty} \left[-\frac{x^{-2}}{2} \right]_{10}^b = \lim\limits_{b \to \infty} \left(-\frac{1}{2b^2} + \frac{1}{200} \right) = \frac{1}{200}$

$\Rightarrow 1.97531986 + \frac{1}{242} < s < 1.97531986 + \frac{1}{200} \Rightarrow 1.20166 < s < 1.20253$

(b) $s = \sum\limits_{n=1}^{\infty} \frac{1}{n^3} \approx \frac{1.20166 + 1.20253}{2} = 1.202095$; error $\leq \frac{1.20253 - 1.20166}{2} = 0.000435$

7.4 COMPARISON TESTS

1. diverges by the Limit Comparison Test (part 1) when compared with $\sum\limits_{n=1}^{\infty} \frac{1}{\sqrt{n}}$, a divergent p-series:

$\lim\limits_{n \to \infty} \frac{\left(\frac{1}{2\sqrt{n} + \sqrt[3]{n}} \right)}{\left(\frac{1}{\sqrt{n}} \right)} = \lim\limits_{n \to \infty} \frac{\sqrt{n}}{2\sqrt{n} + \sqrt[3]{n}} = \lim\limits_{n \to \infty} \left(\frac{1}{2 + n^{-1/6}} \right) = \frac{1}{2}$

3. converges by the Direct Comparison Test; $\frac{\sin^2 n}{2^n} \leq \frac{1}{2^n}$, which is the nth term of a convergent geometric series

5. diverges since $\lim\limits_{n \to \infty} \frac{2n}{3n-1} = \frac{2}{3} \neq 0$

7. converges by the Direct Comparison Test; $\left(\frac{n}{3n+1}\right)^n < \left(\frac{n}{3n}\right)^n = \left(\frac{1}{3}\right)^n$, the nth term of a convergent geometric series

9. diverges by the Direct Comparison Test; $n > \ln n \Rightarrow \ln n > \ln \ln n \Rightarrow \frac{1}{n} < \frac{1}{\ln n} < \frac{1}{\ln(\ln n)}$ and $\sum\limits_{n=3}^{\infty} \frac{1}{n}$ diverges

11. converges by the Limit Comparison Test (part 2) when compared with $\sum\limits_{n=1}^{\infty} \frac{1}{n^2}$, a convergent p-series:

$$\lim\limits_{n \to \infty} \frac{\left[\frac{(\ln n)^2}{n^3}\right]}{\left(\frac{1}{n^2}\right)} = \lim\limits_{n \to \infty} \frac{(\ln n)^2}{n} = \lim\limits_{n \to \infty} \frac{2(\ln n)\left(\frac{1}{n}\right)}{1} = 2\lim\limits_{n \to \infty} \frac{\ln n}{n} = 0$$

13. diverges by the Limit Comparison Test (part 3) with $\frac{1}{n}$, the nth term of the divergent harmonic series:

$$\lim\limits_{n \to \infty} \frac{\left[\frac{1}{\sqrt{n}\ln n}\right]}{\left(\frac{1}{n}\right)} = \lim\limits_{n \to \infty} \frac{\sqrt{n}}{\ln n} = \lim\limits_{n \to \infty} \frac{\left(\frac{1}{2\sqrt{n}}\right)}{\left(\frac{1}{n}\right)} = \lim\limits_{n \to \infty} \frac{\sqrt{n}}{2} = \infty$$

15. diverges by the Limit Comparison Test (part 3) with $\frac{1}{n}$, the nth term of the divergent harmonic series:

$$\lim\limits_{n \to \infty} \frac{\left(\frac{1}{1+\ln n}\right)}{\left(\frac{1}{n}\right)} = \lim\limits_{n \to \infty} \frac{n}{1+\ln n} = \lim\limits_{n \to \infty} \frac{1}{\left(\frac{1}{n}\right)} = \lim\limits_{n \to \infty} n = \infty$$

17. diverges by the Integral Test: $\int_2^{\infty} \frac{\ln(x+1)}{x+1} \, dx = \int_{\ln 3}^{\infty} u \, du = \lim\limits_{b \to \infty} \left[\frac{1}{2}u^2\right]_{\ln 3}^{b} = \lim\limits_{b \to \infty} \frac{1}{2}(b^2 - \ln^2 3) = \infty$

19. converges by the Direct Comparison Test with $\frac{1}{n^{3/2}}$, the nth term of a convergent p-series: $n^2 - 1 > n$ for

$n \geq 2 \Rightarrow n^2(n^2-1) > n^3 \Rightarrow n\sqrt{n^2-1} > n^{3/2} \Rightarrow \frac{1}{n^{3/2}} > \frac{1}{n\sqrt{n^2-1}}$ or use Limit Comparison Test with $\frac{1}{n^2}$.

21. converges because $\sum\limits_{n=1}^{\infty} \frac{1-n}{n2^n} = \sum\limits_{n=1}^{\infty} \frac{1}{n2^n} + \sum\limits_{n=1}^{\infty} \frac{-1}{2^n}$ which is the sum of two convergent series:

$\sum\limits_{n=1}^{\infty} \frac{1}{n2^n}$ converges by the Direct Comparison Test since $\frac{1}{n2^n} < \frac{1}{2^n}$, and $\sum\limits_{n=1}^{\infty} \frac{-1}{2^n}$ is a convergent geometric series

23. converges by the Direct Comparison Test: $\frac{1}{3^{n-1}+1} < \frac{1}{3^{n-1}}$, which is the nth term of a convergent geometric series

25. diverges by the Limit Comparison Test (part 1) with $\frac{1}{n}$, the nth term of the divergent harmonic series:

$$\lim\limits_{n \to \infty} \frac{\left(\sin \frac{1}{n}\right)}{\left(\frac{1}{n}\right)} = \lim\limits_{x \to 0} \frac{\sin x}{x} = 1$$

27. converges by the Limit Comparison Test (part 1) with $\frac{1}{n^2}$, the nth term of a convergent p-series:

$$\lim\limits_{n \to \infty} \frac{\left(\frac{10n+1}{n(n+1)(n+2)}\right)}{\left(\frac{1}{n^2}\right)} = \lim\limits_{n \to \infty} \frac{10n^2+n}{n^2+3n+2} = \lim\limits_{n \to \infty} \frac{20n+1}{2n+3} = \lim\limits_{n \to \infty} \frac{20}{2} = 10$$

29. converges by the Direct Comparison Test: $\frac{\tan^{-1}n}{n^{1.1}} < \frac{\frac{\pi}{2}}{n^{1.1}}$ and $\sum\limits_{n=1}^{\infty} \frac{\frac{\pi}{2}}{n^{1.1}} = \frac{\pi}{2}\sum\limits_{n=1}^{\infty} \frac{1}{n^{1.1}}$ is the product of a

convergent p-series and a nonzero constant

31. converges by the Limit Comparison Test (part 1) with $\frac{1}{n^2}$: $\lim\limits_{n \to \infty} \frac{\left(\frac{\coth n}{n^2}\right)}{\left(\frac{1}{n^2}\right)} = \lim\limits_{n \to \infty} \coth n = \lim\limits_{n \to \infty} \frac{e^n + e^{-n}}{e^n - e^{-n}}$

   $= \lim\limits_{n \to \infty} \frac{1 + e^{-2n}}{1 - e^{-2n}} = 1$

33. diverges by the Limit Comparison Test (part 1) with $\frac{1}{n}$: $\lim\limits_{n \to \infty} \frac{\left(\frac{1}{n \sqrt[n]{n}}\right)}{\left(\frac{1}{n}\right)} = \lim\limits_{n \to \infty} \frac{1}{\sqrt[n]{n}} = 1$.

35. $\frac{1}{1+2+3+\ldots+n} = \frac{1}{\left(\frac{n(n+1)}{2}\right)} = \frac{2}{n(n+1)}$. The series converges by the Limit Comparison Test (part 1) with $\frac{1}{n^2}$:

   $\lim\limits_{n \to \infty} \frac{\left(\frac{2}{n(n+1)}\right)}{\left(\frac{1}{n^2}\right)} = \lim\limits_{n \to \infty} \frac{2n^2}{n^2 + n} = \lim\limits_{n \to \infty} \frac{4n}{2n+1} = \lim\limits_{n \to \infty} \frac{4}{2} = 2$.

37. (a) If $\lim\limits_{n \to \infty} \frac{a_n}{b_n} = 0$, then there exists an integer N such that for all $n > N$, $\left|\frac{a_n}{b_n} - 0\right| < 1 \Rightarrow -1 < \frac{a_n}{b_n} < 1$
    $\Rightarrow a_n < b_n$. Thus, if $\sum b_n$ converges, then $\sum a_n$ converges by the Direct Comparison Test.

    (b) If $\lim\limits_{n \to \infty} \frac{a_n}{b_n} = \infty$, then there exists an integer N such that for all $n > N$, $\frac{a_n}{b_n} > 1 \Rightarrow a_n > b_n$. Thus, if
    $\sum b_n$ diverges, then $\sum a_n$ diverges by the Direct Comparison Test.

39. $\lim\limits_{n \to \infty} \frac{a_n}{b_n} = \infty \Rightarrow$ there exists an integer N such that for all $n > N$, $\frac{a_n}{b_n} > 1 \Rightarrow a_n > b_n$. If $\sum a_n$ converges,
    then $\sum b_n$ converges by the Direct Comparison Test

7.5 THE RATIO AND ROOT TESTS

1. converges by the Ratio Test: $\lim\limits_{n \to \infty} \frac{a_{n+1}}{a_n} = \lim\limits_{n \to \infty} \frac{\left[\frac{(n+1)\sqrt{2}}{2^{n+1}}\right]}{\left[\frac{n\sqrt{2}}{2^n}\right]} = \lim\limits_{n \to \infty} \frac{(n+1)\sqrt{2}}{2^{n+1}} \cdot \frac{2^n}{n\sqrt{2}} = \lim\limits_{n \to \infty} \left(1 + \frac{1}{n}\right)^{\sqrt{2}} \left(\frac{1}{2}\right) = \frac{1}{2} < 1$

3. diverges by the Ratio Test: $\lim\limits_{n \to \infty} \frac{a_{n+1}}{a_n} = \lim\limits_{n \to \infty} \frac{\left(\frac{(n+1)!}{e^{n+1}}\right)}{\left(\frac{n!}{e^n}\right)} = \lim\limits_{n \to \infty} \frac{(n+1)!}{e^{n+1}} \cdot \frac{e^n}{n!} = \lim\limits_{n \to \infty} \frac{n+1}{e} = \infty$

5. converges by the Ratio Test: $\lim\limits_{n \to \infty} \frac{a_{n+1}}{a_n} = \lim\limits_{n \to \infty} \frac{\left(\frac{(n+1)^{10}}{10^{n+1}}\right)}{\left(\frac{n^{10}}{10^n}\right)} = \lim\limits_{n \to \infty} \frac{(n+1)^{10}}{10^{n+1}} \cdot \frac{10^n}{n^{10}} = \lim\limits_{n \to \infty} \left(1 + \frac{1}{n}\right)^{10} \left(\frac{1}{10}\right) = \frac{1}{10} < 1$

7. converges by the Direct Comparison Test: $\frac{2 + (-1)^n}{(1.25)^n} = \left(\frac{4}{5}\right)^n [2 + (-1)^n] \le \left(\frac{4}{5}\right)^n (3)$ which is the nth term of a convergent geometric series

9. diverges; $\lim\limits_{n \to \infty} a_n = \lim\limits_{n \to \infty} \left(1 - \frac{3}{n}\right)^n = \lim\limits_{n \to \infty} \left(1 + \frac{-3}{n}\right)^n = e^{-3} \approx 0.05 \ne 0$

11. converges by the Direct Comparison Test: $\frac{\ln n}{n^3} < \frac{n}{n^3} = \frac{1}{n^2}$ for $n \ge 2$, the nth term of a convergent p-series.

13. diverges by the Direct Comparison Test: $\frac{1}{n} - \frac{1}{n^2} = \frac{n-1}{n^2} > \frac{1}{2}\left(\frac{1}{n}\right)$ for $n > 2$ or by the Limit Comparison Test (part 1) with $\frac{1}{n}$.

15. diverges by the Direct Comparison Test: $\frac{\ln n}{n} > \frac{1}{n}$ for $n \ge 3$

17. converges by the Ratio Test: $\lim\limits_{n \to \infty} \frac{a_{n+1}}{a_n} = \lim\limits_{n \to \infty} \frac{(n+2)(n+3)}{(n+1)!} \cdot \frac{n!}{(n+1)(n+2)} = 0 < 1$

19. converges by the Ratio Test: $\lim\limits_{n \to \infty} \frac{a_{n+1}}{a_n} = \lim\limits_{n \to \infty} \frac{(n+4)!}{3!\,(n+1)!\,3^{n+1}} \cdot \frac{3!\,n!\,3^n}{(n+3)!} = \lim\limits_{n \to \infty} \frac{n+4}{3(n+1)} = \frac{1}{3} < 1$

21. converges by the Ratio Test: $\lim\limits_{n \to \infty} \frac{a_{n+1}}{a_n} = \lim\limits_{n \to \infty} \frac{(n+1)!}{(2n+3)!} \cdot \frac{(2n+1)!}{n!} = \lim\limits_{n \to \infty} \frac{n+1}{(2n+3)(2n+2)} = 0 < 1$

23. converges by the Root Test: $\lim\limits_{n \to \infty} \sqrt[n]{a_n} = \lim\limits_{n \to \infty} \sqrt[n]{\frac{n}{(\ln n)^n}} = \lim\limits_{n \to \infty} \frac{\sqrt[n]{n}}{\ln n} = \lim\limits_{n \to \infty} \frac{1}{\ln n} = 0 < 1$

25. converges by the Direct Comparison Test: $\frac{n!\,\ln n}{n(n+2)!} = \frac{\ln n}{n(n+1)(n+2)} < \frac{n}{n(n+1)(n+2)} = \frac{1}{(n+1)(n+2)} < \frac{1}{n^2}$
    which is the nth-term of a convergent p-series

27. converges by the Ratio Test: $\lim\limits_{n \to \infty} \frac{a_{n+1}}{a_n} = \lim\limits_{n \to \infty} \frac{\left(\frac{1+\sin n}{n}\right)a_n}{a_n} = 0 < 1$

29. diverges by the Ratio Test: $\lim\limits_{n \to \infty} \frac{a_{n+1}}{a_n} = \lim\limits_{n \to \infty} \frac{\left(\frac{3n-1}{2n+5}\right)a_n}{a_n} = \lim\limits_{n \to \infty} \frac{3n-1}{2n+5} = \frac{3}{2} > 1$

31. converges by the Ratio Test: $\lim\limits_{n \to \infty} \frac{a_{n+1}}{a_n} = \lim\limits_{n \to \infty} \frac{\left(\frac{2}{n}\right)a_n}{a_n} = \lim\limits_{n \to \infty} \frac{2}{n} = 0 < 1$

33. converges by the Ratio Test: $\lim\limits_{n \to \infty} \frac{a_{n+1}}{a_n} = \lim\limits_{n \to \infty} \frac{\left(\frac{1+\ln n}{n}\right)a_n}{a_n} = \lim\limits_{n \to \infty} \frac{1+\ln n}{n} = \lim\limits_{n \to \infty} \frac{1}{n} = 0 < 1$

35. diverges by the nth-Term Test: $a_1 = \frac{1}{3}$, $a_2 = \sqrt[2]{\frac{1}{3}}$, $a_3 = \sqrt[3]{\sqrt[2]{\frac{1}{3}}} = \sqrt[6]{\frac{1}{3}}$, $a_4 = \sqrt[4]{\sqrt[3]{\sqrt[2]{\frac{1}{3}}}} = \sqrt[4!]{\frac{1}{3}}$, ... ,
    $a_n = \sqrt[n!]{\frac{1}{3}} \Rightarrow \lim\limits_{n \to \infty} a_n = 1$ because $\left\{\sqrt[n!]{\frac{1}{3}}\right\}$ is a subsequence of $\left\{\sqrt[n]{\frac{1}{3}}\right\}$ whose limit is 1 by Table 8.1

37. converges by the Ratio Test: $\lim\limits_{n \to \infty} \frac{a_{n+1}}{a_n} = \lim\limits_{n \to \infty} \frac{2^{n+1}(n+1)!(n+1)!}{(2n+2)!} \cdot \frac{(2n)!}{2^n n!\,n!} = \lim\limits_{n \to \infty} \frac{2(n+1)(n+1)}{(2n+2)(2n+1)}$
    $= \lim\limits_{n \to \infty} \frac{n+1}{2n+1} = \frac{1}{2} < 1$

39. diverges by the Root Test: $\lim\limits_{n \to \infty} \sqrt[n]{a_n} \equiv \lim\limits_{n \to \infty} \sqrt[n]{\frac{(n!)^n}{(n^n)^2}} = \lim\limits_{n \to \infty} \frac{n!}{n^2} = \infty > 1$

41. converges by the Root Test: $\lim\limits_{n \to \infty} \sqrt[n]{a_n} = \lim\limits_{n \to \infty} \sqrt[n]{\frac{n^n}{2^{n^2}}} = \lim\limits_{n \to \infty} \frac{n}{2^n} = \lim\limits_{n \to \infty} \frac{1}{2^n \ln 2} = 0 < 1$

43. converges by the Ratio Test: $\lim\limits_{n \to \infty} \frac{a_{n+1}}{a_n} = \lim\limits_{n \to \infty} \frac{1 \cdot 3 \cdot \cdots \cdot (2n-1)(2n+1)}{4^{n+1} 2^{n+1}(n+1)!} \cdot \frac{4^n 2^n n!}{1 \cdot 3 \cdot \cdots \cdot (2n-1)} = \lim\limits_{n \to \infty} \frac{2n+1}{(4 \cdot 2)(n+1)} = \frac{1}{4} < 1$

7.6 ALTERNATING SERIES, ABSOLUTE AND CONDITIONAL CONVERGENCE

1. converges absolutely $\Rightarrow$ converges by the Absolute Convergence Test since $\sum\limits_{n=1}^{\infty} |a_n| = \sum\limits_{n=1}^{\infty} \frac{1}{n^2}$ which is a
    convergent p-series

3. diverges by the nth-Term Test since for $n > 10 \Rightarrow \frac{n}{10} > 1 \Rightarrow \lim\limits_{n \to \infty} \left(\frac{n}{10}\right)^n \neq 0 \Rightarrow \sum\limits_{n=1}^{\infty} (-1)^{n+1} \left(\frac{n}{10}\right)^n$ diverges

5. converges by the Alternating Series Test because $f(x) = \ln x$ is an increasing function of $x \Rightarrow \frac{1}{\ln x}$ is decreasing
    $\Rightarrow u_n \geq u_{n+1}$ for $n \geq 1$; also $u_n \geq 0$ for $n \geq 1$ and $\lim\limits_{n \to \infty} \frac{1}{\ln n} = 0$

7. diverges by the nth-Term Test since $\lim\limits_{n \to \infty} \frac{\ln n}{\ln n^2} = \lim\limits_{n \to \infty} \frac{\ln n}{2 \ln n} = \lim\limits_{n \to \infty} \frac{1}{2} = \frac{1}{2} \neq 0$

9. converges by the Alternating Series Test since $f(x) = \frac{\sqrt{x}+1}{x+1} \Rightarrow f'(x) = \frac{1-x-2\sqrt{x}}{2\sqrt{x}\,(x+1)^2} < 0 \Rightarrow f(x)$ is decreasing

   $\Rightarrow u_n \geq u_{n+1}$; also $u_n \geq 0$ for $n \geq 1$ and $\lim\limits_{n \to \infty} u_n = \lim\limits_{n \to \infty} \frac{\sqrt{n}+1}{n+1} = 0$

11. converges absolutely since $\sum\limits_{n=1}^{\infty} |a_n| = \sum\limits_{n=1}^{\infty} \left(\frac{1}{10}\right)^n$ a convergent geometric series

13. converges conditionally since $\frac{1}{\sqrt{n}} > \frac{1}{\sqrt{n+1}} > 0$ and $\lim\limits_{n \to \infty} \frac{1}{\sqrt{n}} = 0 \Rightarrow$ convergence; but $\sum\limits_{n=1}^{\infty} |a_n| = \sum\limits_{n=1}^{\infty} \frac{1}{n^{1/2}}$

   is a divergent p-series

15. converges absolutely since $\sum\limits_{n=1}^{\infty} |a_n| = \sum\limits_{n=1}^{\infty} \frac{n}{n^3+1}$ and $\frac{n}{n^3+1} < \frac{1}{n^2}$ which is the nth-term of a converging p-series

17. converges conditionally since $\frac{1}{n+3} > \frac{1}{(n+1)+3} > 0$ and $\lim\limits_{n \to \infty} \frac{1}{n+3} = 0 \Rightarrow$ convergence; but $\sum\limits_{n=1}^{\infty} |a_n|$

   $= \sum\limits_{n=1}^{\infty} \frac{1}{n+3}$ diverges because $\frac{1}{n+3} \geq \frac{1}{4n}$ and $\sum\limits_{n=1}^{\infty} \frac{1}{n}$ is a divergent series

19. diverges by the nth-Term Test since $\lim\limits_{n \to \infty} \frac{3+n}{5+n} = 1 \neq 0$

21. converges conditionally since $f(x) = \frac{1}{x^2} + \frac{1}{x} \Rightarrow f'(x) = -\left(\frac{2}{x^3} + \frac{1}{x^2}\right) < 0 \Rightarrow f(x)$ is decreasing and hence

   $u_n > u_{n+1} > 0$ for $n \geq 1$ and $\lim\limits_{n \to \infty} \left(\frac{1}{n^2} + \frac{1}{n}\right) = 0 \Rightarrow$ convergence; but $\sum\limits_{n=1}^{\infty} |a_n| = \sum\limits_{n=1}^{\infty} \frac{1+n}{n^2}$

   $= \sum\limits_{n=1}^{\infty} \frac{1}{n^2} + \sum\limits_{n=1}^{\infty} \frac{1}{n}$ is the sum of a convergent and divergent series, and hence diverges

23. converges absolutely by the Ratio Test: $\lim\limits_{n \to \infty} \left(\frac{u_{n+1}}{u_n}\right) = \lim\limits_{n \to \infty} \left[\frac{(n+1)^2 \left(\frac{2}{3}\right)^{n+1}}{n^2 \left(\frac{2}{3}\right)^n}\right] = \frac{2}{3} < 1$

25. converges absolutely by the Integral Test since $\int_1^{\infty} (\tan^{-1} x) \left(\frac{1}{1+x^2}\right) dx = \lim\limits_{b \to \infty} \left[\frac{(\tan^{-1} x)^2}{2}\right]_1^b$

   $= \lim\limits_{b \to \infty} \left[(\tan^{-1} b)^2 - (\tan^{-1} 1)^2\right] = \frac{1}{2}\left[\left(\frac{\pi}{2}\right)^2 - \left(\frac{\pi}{4}\right)^2\right] = \frac{3\pi^2}{32}$

27. diverges by the nth-Term Test since $\lim\limits_{n \to \infty} \frac{n}{n+1} = 1 \neq 0$

29. converges absolutely by the Ratio Test: $\lim\limits_{n \to \infty} \left(\frac{u_{n+1}}{u_n}\right) = \lim\limits_{n \to \infty} \frac{(100)^{n+1}}{(n+1)!} \cdot \frac{n!}{(100)^n} = \lim\limits_{n \to \infty} \frac{100}{n+1} = 0 < 1$

31. converges absolutely by the Direct Comparison Test since $\sum\limits_{n=1}^{\infty} |a_n| = \sum\limits_{n=1}^{\infty} \frac{1}{n^2+2n+1}$ and $\frac{1}{n^2+2n+1} < \frac{1}{n^2}$ which is the

   nth-term of a convergent p-series

33. converges absolutely since $\sum\limits_{n=1}^{\infty} |a_n| = \sum\limits_{n=1}^{\infty} \left|\frac{(-1)^n}{n\sqrt{n}}\right| = \sum\limits_{n=1}^{\infty} \frac{1}{n^{3/2}}$ is a convergent p-series

35. converges absolutely by the Root Test: $\lim\limits_{n \to \infty} \sqrt[n]{|a_n|} = \lim\limits_{n \to \infty} \left(\frac{(n+1)^n}{(2n)^n}\right)^{1/n} = \lim\limits_{n \to \infty} \frac{n+1}{2n} = \frac{1}{2} < 1$

37. diverges by the nth-Term Test since $\lim\limits_{n \to \infty} |a_n| = \lim\limits_{n \to \infty} \frac{(2n)!}{2^n n! \, n} = \lim\limits_{n \to \infty} \frac{(n+1)(n+2) \cdots (2n)}{2^n n}$

    $= \lim\limits_{n \to \infty} \frac{(n+1)(n+2) \cdots (n+(n-1))}{2^{n-1}} > \lim\limits_{n \to \infty} \left(\frac{n+1}{2} \right)^{n-1} = \infty \neq 0$

39. converges conditionally since $\frac{\sqrt{n+1} - \sqrt{n}}{1} \cdot \frac{\sqrt{n+1} + \sqrt{n}}{\sqrt{n+1} + \sqrt{n}} = \frac{1}{\sqrt{n+1} + \sqrt{n}}$ and $\left\{ \frac{1}{\sqrt{n+1} + \sqrt{n}} \right\}$ is a

    decreasing sequence of positive terms which converges to $0 \Rightarrow \sum\limits_{n=1}^{\infty} \frac{(-1)^n}{\sqrt{n+1} + \sqrt{n}}$ converges; but

    $\sum\limits_{n=1}^{\infty} |a_n| = \sum\limits_{n=1}^{\infty} \frac{1}{\sqrt{n+1} + \sqrt{n}}$ diverges by the Limit Comparison Test (part 1) with $\frac{1}{\sqrt{n}}$; a divergent p-series:

    $\lim\limits_{n \to \infty} \left(\frac{\frac{1}{\sqrt{n+1} + \sqrt{n}}}{\frac{1}{\sqrt{n}}} \right) = \lim\limits_{n \to \infty} \frac{\sqrt{n}}{\sqrt{n+1} + \sqrt{n}} = \lim\limits_{n \to \infty} \frac{1}{\sqrt{1 + \frac{1}{n}} + 1} = \frac{1}{2}$

41. diverges by the nth-Term Test since $\lim\limits_{n \to \infty} \left(\sqrt{n + \sqrt{n}} - \sqrt{n} \right) = \lim\limits_{n \to \infty} \left[\left(\sqrt{n + \sqrt{n}} - \sqrt{n} \right) \left(\frac{\sqrt{n + \sqrt{n}} + \sqrt{n}}{\sqrt{n + \sqrt{n}} + \sqrt{n}} \right) \right]$

    $= \lim\limits_{n \to \infty} \frac{\sqrt{n}}{\sqrt{n + \sqrt{n}} + \sqrt{n}} = \lim\limits_{n \to \infty} \frac{1}{\sqrt{1 + \frac{1}{\sqrt{n}}} + 1} = \frac{1}{2} \neq 0$

43. converges absolutely by the Direct Comparison Test since $\text{sech}\,(n) = \frac{2}{e^n + e^{-n}} = \frac{2e^n}{e^{2n} + 1} < \frac{2e^n}{e^{2n}} = \frac{2}{e^n}$ which is the
    nth term of a convergent geometric series

45. $|\text{error}| < \left| (-1)^6 \left(\frac{1}{5} \right) \right| = 0.2$ 47. $|\text{error}| < \left| (-1)^6 \frac{(0.01)^5}{5} \right| = 2 \times 10^{-11}$

49. $\frac{1}{(2n)!} < \frac{5}{10^6} \Rightarrow (2n)! > \frac{10^6}{5} = 200{,}000 \Rightarrow n \geq 5 \Rightarrow 1 - \frac{1}{2!} + \frac{1}{4!} - \frac{1}{6!} + \frac{1}{8!} \approx 0.54030$

51. (a) $a_n \geq a_{n+1}$ fails since $\frac{1}{3} < \frac{1}{2}$

    (b) Since $\sum\limits_{n=1}^{\infty} |a_n| = \sum\limits_{n=1}^{\infty} \left[\left(\frac{1}{3} \right)^n + \left(\frac{1}{2} \right)^n \right] = \sum\limits_{n=1}^{\infty} \left(\frac{1}{3} \right)^n + \sum\limits_{n=1}^{\infty} \left(\frac{1}{2} \right)^n$ is the sum of two absolutely convergent

    series, we can rearrange the terms of the original series to find its sum:

    $\left(\frac{1}{3} + \frac{1}{9} + \frac{1}{27} + \ldots \right) - \left(\frac{1}{2} + \frac{1}{4} + \frac{1}{8} + \ldots \right) = \frac{\left(\frac{1}{3} \right)}{1 - \left(\frac{1}{3} \right)} - \frac{\left(\frac{1}{2} \right)}{1 - \left(\frac{1}{2} \right)} = \frac{1}{2} - 1 = -\frac{1}{2}$

53. The unused terms are $\sum\limits_{j=n+1}^{\infty} (-1)^{j+1} a_j = (-1)^{n+1} \left(a_{n+1} - a_{n+2} \right) + (-1)^{n+3} \left(a_{n+3} - a_{n+4} \right) + \ldots$

    $= (-1)^{n+1} \left[\left(a_{n+1} - a_{n+2} \right) + \left(a_{n+3} - a_{n+4} \right) + \ldots \right]$. Each grouped term is positive, so the remainder
    has the same sign as $(-1)^{n+1}$, which is the sign of the first unused term.

55. Theorem 16 states that $\sum\limits_{n=1}^{\infty} |a_n|$ converges $\Rightarrow \sum\limits_{n=1}^{\infty} a_n$ converges. But this is equivalent to $\sum\limits_{n=1}^{\infty} a_n$ diverges $\Rightarrow \sum\limits_{n=1}^{\infty} |a_n|$ diverges

57. (a) $\sum\limits_{n=1}^{\infty} |a_n + b_n|$ converges by the Direct Comparison Test since $|a_n + b_n| \leq |a_n| + |b_n|$ and hence

    $\sum\limits_{n=1}^{\infty} (a_n + b_n)$ converges absolutely

    (b) $\sum\limits_{n=1}^{\infty} |b_n|$ converges $\Rightarrow \sum\limits_{n=1}^{\infty} -b_n$ converges absolutely; since $\sum\limits_{n=1}^{\infty} a_n$ converges absolutely and

    $\sum\limits_{n=1}^{\infty} -b_n$ converges absolutely, we have $\sum\limits_{n=1}^{\infty} [a_n + (-b_n)] = \sum\limits_{n=1}^{\infty} (a_n - b_n)$ converges absolutely by part (a)

(c) $\sum_{n=1}^{\infty} |a_n|$ converges $\Rightarrow$ $|k| \sum_{n=1}^{\infty} |a_n| = \sum_{n=1}^{\infty} |ka_n|$ converges $\Rightarrow$ $\sum_{n=1}^{\infty} ka_n$ converges absolutely

7.7 POWER SERIES

1. $\lim_{n \to \infty} \left| \frac{u_{n+1}}{u_n} \right| < 1 \Rightarrow \lim_{n \to \infty} \left| \frac{x^{n+1}}{x^n} \right| < 1 \Rightarrow |x| < 1 \Rightarrow -1 < x < 1$; when $x = -1$ we have $\sum_{n=1}^{\infty} (-1)^n$, a divergent series;

   when $x = 1$ we have $\sum_{n=1}^{\infty} 1$, a divergent series

   (a) the radius is 1; the interval of convergence is $-1 < x < 1$

   (b) the interval of absolute convergence is $-1 < x < 1$

   (c) there are no values for which the series converges conditionally

3. $\lim_{n \to \infty} \left| \frac{u_{n+1}}{u_n} \right| < 1 \Rightarrow \lim_{n \to \infty} \left| \frac{(4x+1)^{n+1}}{(4x+1)^n} \right| < 1 \Rightarrow |4x + 1| < 1 \Rightarrow -1 < 4x + 1 < 1 \Rightarrow -\frac{1}{2} < x < 0$; when $x = -\frac{1}{2}$ we

   have $\sum_{n=1}^{\infty} (-1)^n (-1)^n = \sum_{n=1}^{\infty} (-1)^{2n} = \sum_{n=1}^{\infty} 1^n$, a divergent series; when $x = 0$ we have $\sum_{n=1}^{\infty} (-1)^n (1)^n = \sum_{n=1}^{\infty} (-1)^n$, a

   divergent series

   (a) the radius is $\frac{1}{4}$; the interval of convergence is $-\frac{1}{2} < x < 0$

   (b) the interval of absolute convergence is $-\frac{1}{2} < x < 0$

   (c) there are no values for which the series converges conditionally

5. $\lim_{n \to \infty} \left| \frac{u_{n+1}}{u_n} \right| < 1 \Rightarrow \lim_{n \to \infty} \left| \frac{(x-2)^{n+1}}{10^{n+1}} \cdot \frac{10^n}{(x-2)^n} \right| < 1 \Rightarrow \frac{|x-2|}{10} < 1 \Rightarrow |x - 2| < 10 \Rightarrow -10 < x - 2 < 10 \Rightarrow -8 < x < 12$;

   when $x = -8$ we have $\sum_{n=1}^{\infty} (-1)^n$, a divergent series; when $x = 12$ we have $\sum_{n=1}^{\infty} 1$, a divergent series

   (a) the radius is 10; the interval of convergence is $-8 < x < 12$

   (b) the interval of absolute convergence is $-8 < x < 12$

   (c) there are no values for which the series converges conditionally

7. $\lim_{n \to \infty} \left| \frac{u_{n+1}}{u_n} \right| < 1 \Rightarrow \lim_{n \to \infty} \left| \frac{(n+1)x^{n+1}}{(n+3)} \cdot \frac{(n+2)}{nx^n} \right| < 1 \Rightarrow |x| \lim_{n \to \infty} \frac{(n+1)(n+2)}{(n+3)(n)} < 1 \Rightarrow |x| < 1 \Rightarrow -1 < x < 1$; when $x = -1$

   we have $\sum_{n=1}^{\infty} (-1)^n \frac{n}{n+2}$, a divergent series by the nth-term Test; when $x = 1$ we have $\sum_{n=1}^{\infty} \frac{n}{n+2}$, a divergent series

   (a) the radius is 1; the interval of convergence is $-1 < x < 1$

   (b) the interval of absolute convergence is $-1 < x < 1$

   (c) there are no values for which the series converges conditionally

9. $\lim_{n \to \infty} \left| \frac{u_{n+1}}{u_n} \right| < 1 \Rightarrow \lim_{n \to \infty} \left| \frac{x^{n+1}}{(n+1)\sqrt{n+1}\, 3^{n+1}} \cdot \frac{n\sqrt{n}\, 3^n}{x^n} \right| < 1 \Rightarrow \frac{|x|}{3} \left(\lim_{n \to \infty} \frac{n}{n+1} \right) \left(\sqrt{\lim_{n \to \infty} \frac{n}{n+1}} \right) < 1$

   $\Rightarrow \frac{|x|}{3} (1)(1) < 1 \Rightarrow |x| < 3 \Rightarrow -3 < x < 3$; when $x = -3$ we have $\sum_{n=1}^{\infty} \frac{(-1)^n}{n^{3/2}}$, an absolutely convergent series;

   when $x = 3$ we have $\sum_{n=1}^{\infty} \frac{1}{n^{3/2}}$, a convergent p-series

   (a) the radius is 3; the interval of convergence is $-3 \leq x \leq 3$

   (b) the interval of absolute convergence is $-3 \leq x \leq 3$

   (c) there are no values for which the series converges conditionally

11. $\lim_{n \to \infty} \left| \frac{u_{n+1}}{u_n} \right| < 1 \Rightarrow \lim_{n \to \infty} \left| \frac{x^{n+1}}{(n+1)!} \cdot \frac{n!}{x^n} \right| < 1 \Rightarrow |x| \lim_{n \to \infty} \left(\frac{1}{n+1} \right) < 1$ for all x

    (a) the radius is ∞; the series converges for all x

(b) the series converges absolutely for all x

(c) there are no values for which the series converges conditionally

13. $\lim\limits_{n \to \infty} \left| \frac{u_{n+1}}{u_n} \right| < 1 \Rightarrow \lim\limits_{n \to \infty} \left| \frac{x^{2n+3}}{(n+1)!} \cdot \frac{n!}{x^{2n+1}} \right| < 1 \Rightarrow x^2 \lim\limits_{n \to \infty} \left(\frac{1}{n+1} \right) < 1$ for all x

(a) the radius is ∞; the series converges for all x

(b) the series converges absolutely for all x

(c) there are no values for which the series converges conditionally

15. $\lim\limits_{n \to \infty} \left| \frac{u_{n+1}}{u_n} \right| < 1 \Rightarrow \lim\limits_{n \to \infty} \left| \frac{x^{n+1}}{\sqrt{(n+1)^2+3}} \cdot \frac{\sqrt{n^2+3}}{x^n} \right| < 1 \Rightarrow |x| \sqrt{\lim\limits_{n \to \infty} \frac{n^2+3}{n^2+2n+4}} < 1 \Rightarrow |x| < 1 \Rightarrow -1 < x < 1$; when

x $= -1$ we have $\sum\limits_{n=1}^{\infty} \frac{(-1)^n}{\sqrt{n^2+3}}$, a conditionally convergent series; when x $= 1$ we have $\sum\limits_{n=1}^{\infty} \frac{1}{\sqrt{n^2+3}}$, a divergent series

(a) the radius is 1; the interval of convergence is $-1 \le x < 1$

(b) the interval of absolute convergence is $-1 < x < 1$

(c) the series converges conditionally at x $= -1$

17. $\lim\limits_{n \to \infty} \left| \frac{u_{n+1}}{u_n} \right| < 1 \Rightarrow \lim\limits_{n \to \infty} \left| \frac{(n+1)(x+3)^{n+1}}{5^{n+1}} \cdot \frac{5^n}{n(x+3)^n} \right| < 1 \Rightarrow \frac{|x+3|}{5} \lim\limits_{n \to \infty} \left(\frac{n+1}{n} \right) < 1 \Rightarrow \frac{|x+3|}{5} < 1 \Rightarrow |x+3| < 5$

$\Rightarrow -5 < x+3 < 5 \Rightarrow -8 < x < 2$; when x $= -8$ we have $\sum\limits_{n=1}^{\infty} \frac{n(-5)^n}{5^n} = \sum\limits_{n=1}^{\infty} (-1)^n\, n$, a divergent series; when x $= 2$ we

have $\sum\limits_{n=1}^{\infty} \frac{n5^n}{5^n} = \sum\limits_{n=1}^{\infty} n$, a divergent series

(a) the radius is 5; the interval of convergence is $-8 < x < 2$

(b) the interval of absolute convergence is $-8 < x < 2$

(c) there are no values for which the series converges conditionally

19. $\lim\limits_{n \to \infty} \left| \frac{u_{n+1}}{u_n} \right| < 1 \Rightarrow \lim\limits_{n \to \infty} \left| \frac{\sqrt{n+1}\, x^{n+1}}{3^{n+1}} \cdot \frac{3^n}{\sqrt{n}\, x^n} \right| < 1 \Rightarrow \frac{|x|}{3} \sqrt{\lim\limits_{n \to \infty} \left(\frac{n+1}{n} \right)} < 1 \Rightarrow \frac{|x|}{3} < 1 \Rightarrow |x| < 3 \Rightarrow -3 < x < 3$;

when x $= -3$ we have $\sum\limits_{n=1}^{\infty} (-1)^n \sqrt{n}$, a divergent series; when x $= 3$ we have $\sum\limits_{n=1}^{\infty} \sqrt{n}$, a divergent series

(a) the radius is 3; the interval of convergence is $-3 < x < 3$

(b) the interval of absolute convergence is $-3 < x < 3$

(c) there are no values for which the series converges conditionally

21. $\lim\limits_{n \to \infty} \left| \frac{u_{n+1}}{u_n} \right| < 1 \Rightarrow \lim\limits_{n \to \infty} \left| \frac{\left(1 + \frac{1}{n+1}\right)^{n+1} x^{n+1}}{\left(1 + \frac{1}{n}\right)^n x^n} \right| < 1 \Rightarrow |x| \left(\frac{\lim\limits_{t \to \infty} \left(1 + \frac{1}{t}\right)^t}{\lim\limits_{n \to \infty} \left(1 + \frac{1}{n}\right)^n} \right) < 1 \Rightarrow |x| \left(\frac{e}{e} \right) < 1 \Rightarrow |x| < 1 \Rightarrow -1 < x < 1$;

when x $= -1$ we have $\sum\limits_{n=1}^{\infty} (-1)^n \left(1 + \frac{1}{n}\right)^n$, a divergent series by the nth-Term Test since $\lim\limits_{n \to \infty} \left(1 + \frac{1}{n}\right)^n = e \ne 0$; when

x $= 1$ we have $\sum\limits_{n=1}^{\infty} \left(1 + \frac{1}{n}\right)^n$, a divergent series

(a) the radius is 1; the interval of convergence is $-1 < x < 1$

(b) the interval of absolute convergence is $-1 < x < 1$

(c) there are no values for which the series converges conditionally

23. $\lim\limits_{n \to \infty} \left| \frac{u_{n+1}}{u_n} \right| < 1 \Rightarrow \lim\limits_{n \to \infty} \left| \frac{(n+1)^{n+1} x^{n+1}}{n^n x^n} \right| < 1 \Rightarrow |x| \left(\lim\limits_{n \to \infty} \left(1 + \frac{1}{n}\right)^n \right) \left(\lim\limits_{n \to \infty} (n+1) \right) < 1 \Rightarrow e\,|x| \lim\limits_{n \to \infty} (n+1) < 1$

$\Rightarrow$ only x $= 0$ satisfies this inequality

(a) the radius is 0; the series converges only for x $= 0$

(b) the series converges absolutely only for x $= 0$

(c) there are no values for which the series converges conditionally

25. $\lim\limits_{n \to \infty} \left| \frac{u_{n+1}}{u_n} \right| < 1 \Rightarrow \lim\limits_{n \to \infty} \left| \frac{(x+2)^{n+1}}{(n+1)\,2^{n+1}} \cdot \frac{n2^n}{(x+2)^n} \right| < 1 \Rightarrow \frac{|x+2|}{2} \lim\limits_{n \to \infty} \left(\frac{n}{n+1} \right) < 1 \Rightarrow \frac{|x+2|}{2} < 1 \Rightarrow |x+2| < 2$

$\Rightarrow -2 < x+2 < 2 \Rightarrow -4 < x < 0$; when $x = -4$ we have $\sum\limits_{n=1}^{\infty} \frac{-1}{n}$, a divergent series; when $x = 0$ we have $\sum\limits_{n=1}^{\infty} \frac{(-1)^{n+1}}{n}$,

the alternating harmonic series which converges conditionally

(a) the radius is 2; the interval of convergence is $-4 < x \le 0$

(b) the interval of absolute convergence is $-4 < x < 0$

(c) the series converges conditionally at $x = 0$

27. $\lim\limits_{n \to \infty} \left| \frac{u_{n+1}}{u_n} \right| < 1 \Rightarrow \lim\limits_{n \to \infty} \left| \frac{x^{n+1}}{(n+1)\,(\ln(n+1))^2} \cdot \frac{n(\ln n)^2}{x^n} \right| < 1 \Rightarrow |x| \left(\lim\limits_{n \to \infty} \frac{n}{n+1} \right) \left(\lim\limits_{n \to \infty} \frac{\ln n}{\ln(n+1)} \right)^2 < 1$

$\Rightarrow |x|\,(1) \left(\lim\limits_{n \to \infty} \frac{\left(\frac{1}{n}\right)}{\left(\frac{1}{n+1}\right)} \right)^2 < 1 \Rightarrow |x| \left(\lim\limits_{n \to \infty} \frac{n+1}{n} \right)^2 < 1 \Rightarrow |x| < 1 \Rightarrow -1 < x < 1$; when $x = -1$ we have

$\sum\limits_{n=1}^{\infty} \frac{(-1)^n}{n(\ln n)^2}$ which converges absolutely; when $x = 1$ we have $\sum\limits_{n=1}^{\infty} \frac{1}{n(\ln n)^2}$ which converges

(a) the radius is 1; the interval of convergence is $-1 \le x \le 1$

(b) the interval of absolute convergence is $-1 \le x \le 1$

(c) there are no values for which the series converges conditionally

29. $\lim\limits_{n \to \infty} \left| \frac{u_{n+1}}{u_n} \right| < 1 \Rightarrow \lim\limits_{n \to \infty} \left| \frac{(4x-5)^{2n+3}}{(n+1)^{3/2}} \cdot \frac{n^{3/2}}{(4x-5)^{2n+1}} \right| < 1 \Rightarrow (4x-5)^2 \left(\lim\limits_{n \to \infty} \frac{n}{n+1} \right)^{3/2} < 1 \Rightarrow (4x-5)^2 < 1$

$\Rightarrow |4x-5| < 1 \Rightarrow -1 < 4x-5 < 1 \Rightarrow 1 < x < \frac{3}{2}$; when $x = 1$ we have $\sum\limits_{n=1}^{\infty} \frac{(-1)^{2n+1}}{n^{3/2}} = \sum\limits_{n=1}^{\infty} \frac{-1}{n^{3/2}}$ which is

absolutely convergent; when $x = \frac{3}{2}$ we have $\sum\limits_{n=1}^{\infty} \frac{(1)^{2n+1}}{n^{3/2}}$, a convergent p-series

(a) the radius is $\frac{1}{4}$; the interval of convergence is $1 \le x \le \frac{3}{2}$

(b) the interval of absolute convergence is $1 \le x \le \frac{3}{2}$

(c) there are no values for which the series converges conditionally

31. $\lim\limits_{n \to \infty} \left| \frac{u_{n+1}}{u_n} \right| < 1 \Rightarrow \lim\limits_{n \to \infty} \left| \frac{(x+\pi)^{n+1}}{\sqrt{n+1}} \cdot \frac{\sqrt{n}}{(x+\pi)^n} \right| < 1 \Rightarrow |x+\pi| \lim\limits_{n \to \infty} \left| \sqrt{\frac{n}{n+1}} \right| < 1$

$\Rightarrow |x+\pi| \sqrt{\lim\limits_{n \to \infty} \left(\frac{n}{n+1} \right)} < 1 \Rightarrow |x+\pi| < 1 \Rightarrow -1 < x+\pi < 1 \Rightarrow -1-\pi < x < 1-\pi$;

when $x = -1-\pi$ we have $\sum\limits_{n=1}^{\infty} \frac{(-1)^n}{\sqrt{n}} = \sum\limits_{n=1}^{\infty} \frac{(-1)^n}{n^{1/2}}$, a conditionally convergent series; when $x = 1-\pi$ we have

$\sum\limits_{n=1}^{\infty} \frac{1^n}{\sqrt{n}} = \sum\limits_{n=1}^{\infty} \frac{1}{n^{1/2}}$, a divergent p-series

(a) the radius is 1; the interval of convergence is $(-1-\pi) \le x < (1-\pi)$

(b) the interval of absolute convergence is $-1-\pi < x < 1-\pi$

(c) the series converges conditionally at $x = -1-\pi$

33. $\lim\limits_{n \to \infty} \left| \frac{u_{n+1}}{u_n} \right| < 1 \Rightarrow \lim\limits_{n \to \infty} \left| \frac{(x-1)^{2n+2}}{4^{n+1}} \cdot \frac{4^n}{(x-1)^{2n}} \right| < 1 \Rightarrow \frac{(x-1)^2}{4} \lim\limits_{n \to \infty} |1| < 1 \Rightarrow (x-1)^2 < 4 \Rightarrow |x-1| < 2$

$\Rightarrow -2 < x-1 < 2 \Rightarrow -1 < x < 3$; at $x = -1$ we have $\sum\limits_{n=0}^{\infty} \frac{(-2)^{2n}}{4^n} = \sum\limits_{n=0}^{\infty} \frac{4^n}{4^n} = \sum\limits_{n=0}^{\infty} 1$, which diverges; at $x = 3$

we have $\sum\limits_{n=0}^{\infty} \frac{2^{2n}}{4^n} = \sum\limits_{n=0}^{\infty} \frac{4^n}{4^n} = \sum\limits_{n=0}^{\infty} 1$, a divergent series; the interval of convergence is $-1 < x < 3$; the series

$\sum\limits_{n=0}^{\infty} \frac{(x-1)^{2n}}{4^n} = \sum\limits_{n=0}^{\infty} \left(\left(\frac{x-1}{2} \right)^2 \right)^n$ is a convergent geometric series when $-1 < x < 3$ and the sum is

$\frac{1}{1 - \left(\frac{x-1}{2} \right)^2} = \frac{1}{\left[\frac{4-(x-1)^2}{4} \right]} = \frac{4}{4 - x^2 + 2x - 1} = \frac{4}{3 + 2x - x^2}$

35. $\lim\limits_{n \to \infty} \left| \frac{u_{n+1}}{u_n} \right| < 1 \Rightarrow \lim\limits_{n \to \infty} \left| \frac{(\sqrt{x} - 2)^{n+1}}{2^{n+1}} \cdot \frac{2^n}{(\sqrt{x} - 2)^n} \right| < 1 \Rightarrow |\sqrt{x} - 2| < 2 \Rightarrow -2 < \sqrt{x} - 2 < 2 \Rightarrow 0 < \sqrt{x} < 4$

$\Rightarrow 0 < x < 16$; when $x = 0$ we have $\sum\limits_{n=0}^{\infty} (-1)^n$, a divergent series; when $x = 16$ we have $\sum\limits_{n=0}^{\infty} (1)^n$, a divergent

series; the interval of convergence is $0 < x < 16$; the series $\sum\limits_{n=0}^{\infty} \left(\frac{\sqrt{x} - 2}{2} \right)^n$ is a convergent geometric series when

$0 < x < 16$ and its sum is $\frac{1}{1 - \left(\frac{\sqrt{x} - 2}{2} \right)} = \frac{1}{\left(\frac{2 - \sqrt{x} + 2}{2} \right)} = \frac{2}{4 - \sqrt{x}}$

37. $\lim\limits_{n \to \infty} \left| \frac{u_{n+1}}{u_n} \right| < 1 \Rightarrow \lim\limits_{n \to \infty} \left| \left(\frac{x^2 + 1}{3} \right)^{n+1} \cdot \left(\frac{3}{x^2 + 1} \right)^n \right| < 1 \Rightarrow \frac{(x^2 + 1)}{3} \lim\limits_{n \to \infty} |1| < 1 \Rightarrow \frac{x^2 + 1}{3} < 1 \Rightarrow x^2 < 2$

$\Rightarrow |x| < \sqrt{2} \Rightarrow -\sqrt{2} < x < \sqrt{2}$; at $x = \pm\sqrt{2}$ we have $\sum\limits_{n=0}^{\infty} (1)^n$ which diverges; the interval of convergence is

$-\sqrt{2} < x < \sqrt{2}$; the series $\sum\limits_{n=0}^{\infty} \left(\frac{x^2 + 1}{3} \right)^n$ is a convergent geometric series when $-\sqrt{2} < x < \sqrt{2}$ and its sum is

$\frac{1}{1 - \left(\frac{x^2 + 1}{3} \right)} = \frac{1}{\left(\frac{3 - x^2 - 1}{3} \right)} = \frac{3}{2 - x^2}$

39. $\lim\limits_{n \to \infty} \left| \frac{(x - 3)^{n+1}}{2^{n+1}} \cdot \frac{2^n}{(x - 3)^n} \right| < 1 \Rightarrow |x - 3| < 2 \Rightarrow 1 < x < 5$; when $x = 1$ we have $\sum\limits_{n=1}^{\infty} (1)^n$ which diverges;

when $x = 5$ we have $\sum\limits_{n=1}^{\infty} (-1)^n$ which also diverges; the interval of convergence is $1 < x < 5$; the sum of this

convergent geometric series is $\frac{1}{1 + \left(\frac{x - 3}{2} \right)} = \frac{2}{x - 1}$. If $f(x) = 1 - \frac{1}{2}(x - 3) + \frac{1}{4}(x - 3)^2 + \ldots + \left(-\frac{1}{2} \right)^n (x - 3)^n + \ldots$

$= \frac{2}{x - 1}$ then $f'(x) = -\frac{1}{2} + \frac{1}{2}(x - 3) + \ldots + \left(-\frac{1}{2} \right)^n n(x - 3)^{n-1} + \ldots$ is convergent when $1 < x < 5$, and diverges

when $x = 1$ or 5. The sum for $f'(x)$ is $\frac{-2}{(x - 1)^2}$, the derivative of $\frac{2}{x - 1}$.

41. (a) Differentiate the series for $\sin x$ to get $\cos x = 1 - \frac{3x^2}{3!} + \frac{5x^4}{5!} - \frac{7x^6}{7!} + \frac{9x^8}{9!} - \frac{11x^{10}}{11!} + \ldots$

$= 1 - \frac{x^2}{2!} + \frac{x^4}{4!} - \frac{x^6}{6!} + \frac{x^8}{8!} - \frac{x^{10}}{10!} + \ldots$. The series converges for all values of x since

$\lim\limits_{n \to \infty} \left| \frac{x^{2n+2}}{(2n + 2)!} \cdot \frac{(2n)!}{x^{2n}} \right| = x^2 \lim\limits_{n \to \infty} \left(\frac{1}{(2n + 1)(2n + 2)} \right) = 0 < 1$ for all x.

(b) $\sin 2x = 2x - \frac{2^3 x^3}{3!} + \frac{2^5 x^5}{5!} - \frac{2^7 x^7}{7!} + \frac{2^9 x^9}{9!} - \frac{2^{11} x^{11}}{11!} + \ldots = 2x - \frac{8x^3}{3!} + \frac{32x^5}{5!} - \frac{128x^7}{7!} + \frac{512x^9}{9!} - \frac{2048x^{11}}{11!} + \ldots$

(c) $2 \sin x \cos x = 2 \left[(0 \cdot 1) + (0 \cdot 0 + 1 \cdot 1)x + \left(0 \cdot \frac{-1}{2} + 1 \cdot 0 + 0 \cdot 1 \right) x^2 + \left(0 \cdot 0 - 1 \cdot \frac{1}{2} + 0 \cdot 0 - 1 \cdot \frac{1}{3!} \right) x^3 \right.$

$+ \left(0 \cdot \frac{1}{4!} + 1 \cdot 0 - 0 \cdot \frac{1}{2} - 0 \cdot \frac{1}{3!} + 0 \cdot 1 \right) x^4 + \left(0 \cdot 0 + 1 \cdot \frac{1}{4!} + 0 \cdot 0 + \frac{1}{2} \cdot \frac{1}{3!} + 0 \cdot 0 + 1 \cdot \frac{1}{5!} \right) x^5$

$+ \left. \left(0 \cdot \frac{1}{6!} + 1 \cdot 0 + 0 \cdot \frac{1}{4!} + 0 \cdot \frac{1}{3!} + 0 \cdot \frac{1}{2} + 0 \cdot \frac{1}{5!} + 0 \cdot 1 \right) x^6 + \ldots \right] = 2 \left[x - \frac{4x^3}{3!} + \frac{16x^5}{5!} - \ldots \right]$

$= 2x - \frac{2^3 x^3}{3!} + \frac{2^5 x^5}{5!} - \frac{2^7 x^7}{7!} + \frac{2^9 x^9}{9!} - \frac{2^{11} x^{11}}{11!} + \ldots$

43. (a) $\ln |\sec x| + C = \int \tan x \, dx = \int \left(x + \frac{x^3}{3} + \frac{2x^5}{15} + \frac{17x^7}{315} + \frac{62x^9}{2835} + \ldots \right) dx$

$= \frac{x^2}{2} + \frac{x^4}{12} + \frac{x^6}{45} + \frac{17x^8}{2520} + \frac{31x^{10}}{14,175} + \ldots + C$; $x = 0 \Rightarrow C = 0 \Rightarrow \ln |\sec x| = \frac{x^2}{2} + \frac{x^4}{12} + \frac{x^6}{45} + \frac{17x^8}{2520} + \frac{31x^{10}}{14,175} + \ldots$,

converges when $-\frac{\pi}{2} < x < \frac{\pi}{2}$

(b) $\sec^2 x = \frac{d(\tan x)}{dx} = \frac{d}{dx} \left(x + \frac{x^3}{3} + \frac{2x^5}{15} + \frac{17x^7}{315} + \frac{62x^9}{2835} + \ldots \right) = 1 + x^2 + \frac{2x^4}{3} + \frac{17x^6}{45} + \frac{62x^8}{315} + \ldots$, converges

when $-\frac{\pi}{2} < x < \frac{\pi}{2}$

(c) $\sec^2 x = (\sec x)(\sec x) = \left(1 + \frac{x^2}{2} + \frac{5x^4}{24} + \frac{61x^6}{720} + \ldots \right) \left(1 + \frac{x^2}{2} + \frac{5x^4}{24} + \frac{61x^6}{720} + \ldots \right)$

$= 1 + \left(\frac{1}{2} + \frac{1}{2} \right) x^2 + \left(\frac{5}{24} + \frac{1}{4} + \frac{5}{24} \right) x^4 + \left(\frac{61}{720} + \frac{5}{48} + \frac{5}{48} + \frac{61}{720} \right) x^6 + \ldots$

$= 1 + x^2 + \frac{2x^4}{3} + \frac{17x^6}{45} + \frac{62x^8}{315} + \ldots, -\frac{\pi}{2} < x < \frac{\pi}{2}$

45. (a) If $f(x) = \sum\limits_{n=0}^{\infty} a_n x^n$, then $f^{(k)}(x) = \sum\limits_{n=k}^{\infty} n(n-1)(n-2)\cdots(n-(k-1))\, a_n x^{n-k}$ and $f^{(k)}(0) = k!a_k$

$\Rightarrow a_k = \frac{f^{(k)}(0)}{k!}$; likewise if $f(x) = \sum\limits_{n=0}^{\infty} b_n x^n$, then $b_k = \frac{f^{(k)}(0)}{k!}$ $\Rightarrow a_k = b_k$ for every nonnegative integer k

(b) If $f(x) = \sum\limits_{n=0}^{\infty} a_n x^n = 0$ for all x, then $f^{(k)}(x) = 0$ for all x $\Rightarrow$ from part (a) that $a_k = 0$ for every nonnegative integer k

47. The series $\sum\limits_{n=1}^{\infty} \frac{x^n}{n}$ converges conditionally at the left-hand endpoint of its interval of convergence $[-1, 1)$; the

series $\sum\limits_{n=1}^{\infty} \frac{x^n}{(n^2)}$ converges absolutely at the left-hand endpoint of its interval of convergence $[-1, 1]$

7.8 TAYLOR AND MACLAURIN SERIES

1. $f(x) = \ln x, f'(x) = \frac{1}{x}, f''(x) = -\frac{1}{x^2}, f'''(x) = \frac{2}{x^3}; f(1) = \ln 1 = 0, f'(1) = 1, f''(1) = -1, f'''(1) = 2 \Rightarrow P_0(x) = 0,$
$P_1(x) = (x-1), P_2(x) = (x-1) - \frac{1}{2}(x-1)^2, P_3(x) = (x-1) - \frac{1}{2}(x-1)^2 + \frac{1}{3}(x-1)^3$

3. $f(x) = \frac{1}{x} = x^{-1}, f'(x) = -x^{-2}, f''(x) = 2x^{-3}, f'''(x) = -6x^{-4}; f(2) = \frac{1}{2}, f'(2) = -\frac{1}{4}, f''(2) = \frac{1}{4}, f'''(x) = -\frac{3}{8}$
$\Rightarrow P_0(x) = \frac{1}{2}, P_1(x) = \frac{1}{2} - \frac{1}{4}(x-2), P_2(x) = \frac{1}{2} - \frac{1}{4}(x-2) + \frac{1}{8}(x-2)^2,$
$P_3(x) = \frac{1}{2} - \frac{1}{4}(x-2) + \frac{1}{8}(x-2)^2 - \frac{1}{16}(x-2)^3$

5. $f(x) = \sin x, f'(x) = \cos x, f''(x) = -\sin x, f'''(x) = -\cos x; f\left(\frac{\pi}{4}\right) = \sin\frac{\pi}{4} = \frac{\sqrt{2}}{2}, f'\left(\frac{\pi}{4}\right) = \cos\frac{\pi}{4} = \frac{\sqrt{2}}{2},$
$f''\left(\frac{\pi}{4}\right) = -\sin\frac{\pi}{4} = -\frac{\sqrt{2}}{2}, f'''\left(\frac{\pi}{4}\right) = -\cos\frac{\pi}{4} = -\frac{\sqrt{2}}{2} \Rightarrow P_0 = \frac{\sqrt{2}}{2}, P_1(x) = \frac{\sqrt{2}}{2} + \frac{\sqrt{2}}{2}\left(x - \frac{\pi}{4}\right),$
$P_2(x) = \frac{\sqrt{2}}{2} + \frac{\sqrt{2}}{2}\left(x - \frac{\pi}{4}\right) - \frac{\sqrt{2}}{4}\left(x - \frac{\pi}{4}\right)^2, P_3(x) = \frac{\sqrt{2}}{2} + \frac{\sqrt{2}}{2}\left(x - \frac{\pi}{4}\right) - \frac{\sqrt{2}}{4}\left(x - \frac{\pi}{4}\right)^2 - \frac{\sqrt{2}}{12}\left(x - \frac{\pi}{4}\right)^3$

7. $f(x) = \sqrt{x} = x^{1/2}, f'(x) = \left(\frac{1}{2}\right)x^{-1/2}, f''(x) = \left(-\frac{1}{4}\right)x^{-3/2}, f'''(x) = \left(\frac{3}{8}\right)x^{-5/2}; f(4) = \sqrt{4} = 2,$
$f'(4) = \left(\frac{1}{2}\right)4^{-1/2} = \frac{1}{4}, f''(4) = \left(-\frac{1}{4}\right)4^{-3/2} = -\frac{1}{32}, f'''(4) = \left(\frac{3}{8}\right)4^{-5/2} = \frac{3}{256} \Rightarrow P_0(x) = 2, P_1(x) = 2 + \frac{1}{4}(x-4),$
$P_2(x) = 2 + \frac{1}{4}(x-4) - \frac{1}{64}(x-4)^2, P_3(x) = 2 + \frac{1}{4}(x-4) - \frac{1}{64}(x-4)^2 + \frac{1}{512}(x-4)^3$

9. $e^x = \sum\limits_{n=0}^{\infty} \frac{x^n}{n!} \Rightarrow e^{-x} = \sum\limits_{n=0}^{\infty} \frac{(-x)^n}{n!} = 1 - x + \frac{x^2}{2!} - \frac{x^3}{3!} + \frac{x^4}{4!} - \cdots$

11. $f(x) = (1+x)^{-1} \Rightarrow f'(x) = -(1+x)^{-2}, f''(x) = 2(1+x)^{-3}, f'''(x) = -3!(1+x)^{-4} \Rightarrow \ldots f^{(k)}(x)$
$= (-1)^k k!(1+x)^{-k-1}; f(0) = 1, f'(0) = -1, f''(0) = 2, f'''(0) = -3!, \ldots, f^{(k)}(0) = (-1)^k k!$
$\Rightarrow \frac{1}{1+x} = 1 - x + x^2 - x^3 + \ldots = \sum\limits_{n=0}^{\infty} (-x)^n = \sum\limits_{n=0}^{\infty} (-1)^n x^n$

13. $\sin x = \sum\limits_{n=0}^{\infty} \frac{(-1)^n x^{2n+1}}{(2n+1)!} \Rightarrow \sin 3x = \sum\limits_{n=0}^{\infty} \frac{(-1)^n (3x)^{2n+1}}{(2n+1)!} = \sum\limits_{n=0}^{\infty} \frac{(-1)^n 3^{2n+1} x^{2n+1}}{(2n+1)!} = 3x - \frac{3^3 x^3}{3!} + \frac{3^5 x^5}{5!} - \cdots$

15. $7\cos(-x) = 7\cos x = 7\sum\limits_{n=0}^{\infty} \frac{(-1)^n x^{2n}}{(2n)!} = 7 - \frac{7x^2}{2!} + \frac{7x^4}{4!} - \frac{7x^6}{6!} + \ldots$, since the cosine is an even function

17. $\cosh x = \frac{e^x + e^{-x}}{2} = \frac{1}{2}\left[\left(1 + x^2 + \frac{x^2}{2!} + \frac{x^3}{3!} + \frac{x^4}{4!} + \ldots\right) + \left(1 - x + \frac{x^2}{2!} - \frac{x^3}{3!} + \frac{x^4}{4!} - \ldots\right)\right] = 1 + \frac{x^2}{2!} + \frac{x^4}{4!} + \frac{x^6}{6!} + \cdots$
$= \sum\limits_{n=0}^{\infty} \frac{x^{2n}}{(2n)!}$

19. $f(x) = x^4 - 2x^3 - 5x + 4 \Rightarrow f'(x) = 4x^3 - 6x^2 - 5, f''(x) = 12x^2 - 12x, f'''(x) = 24x - 12, f^{(4)}(x) = 24$

    $\Rightarrow f^{(n)}(x) = 0$ if $n \geq 5; f(0) = 4, f'(0) = -5, f''(0) = 0, f'''(0) = -12, f^{(4)}(0) = 24, f^{(n)}(0) = 0$ if $n \geq 5$

    $\Rightarrow x^4 - 2x^3 - 5x + 4 = 4 - 5x - \frac{12}{3!}x^3 + \frac{24}{4!}x^4 = x^4 - 2x^3 - 5x + 4$ itself

21. $f(x) = x^3 - 2x + 4 \Rightarrow f'(x) = 3x^2 - 2, f''(x) = 6x, f'''(x) = 6 \Rightarrow f^{(n)}(x) = 0$ if $n \geq 4; f(2) = 8, f'(2) = 10,$

    $f''(2) = 12, f'''(2) = 6, f^{(n)}(2) = 0$ if $n \geq 4 \Rightarrow x^3 - 2x + 4 = 8 + 10(x - 2) + \frac{12}{2!}(x - 2)^2 + \frac{6}{3!}(x - 2)^3$

    $= 8 + 10(x - 2) + 6(x - 2)^2 + (x - 2)^3$

23. $f(x) = x^4 + x^2 + 1 \Rightarrow f'(x) = 4x^3 + 2x, f''(x) = 12x^2 + 2, f'''(x) = 24x, f^{(4)}(x) = 24, f^{(n)}(x) = 0$ if $n \geq 5;$

    $f(-2) = 21, f'(-2) = -36, f''(-2) = 50, f'''(-2) = -48, f^{(4)}(-2) = 24, f^{(n)}(-2) = 0$ if $n \geq 5 \Rightarrow x^4 + x^2 + 1$

    $= 21 - 36(x + 2) + \frac{50}{2!}(x + 2)^2 - \frac{48}{3!}(x + 2)^3 + \frac{24}{4!}(x + 2)^4 = 21 - 36(x + 2) + 25(x + 2)^2 - 8(x + 2)^3 + (x + 2)^4$

25. $f(x) = x^{-2} \Rightarrow f'(x) = -2x^{-3}, f''(x) = 3! x^{-4}, f'''(x) = -4! x^{-5} \Rightarrow f^{(n)}(x) = (-1)^n(n + 1)! x^{-n-2};$

    $f(1) = 1, f'(1) = -2, f''(1) = 3!, f'''(1) = -4!, f^{(n)}(1) = (-1)^n(n + 1)! \Rightarrow \frac{1}{x^2}$

    $= 1 - 2(x - 1) + 3(x - 1)^2 - 4(x - 1)^3 + \ldots = \sum_{n=0}^{\infty} (-1)^n(n + 1)(x - 1)^n$

27. $f(x) = e^x \Rightarrow f'(x) = e^x, f''(x) = e^x \Rightarrow f^{(n)}(x) = e^x; f(2) = e^2, f'(2) = e^2, \ldots f^{(n)}(2) = e^2$

    $\Rightarrow e^x = e^2 + e^2(x - 2) + \frac{e^2}{2}(x - 2)^2 + \frac{e^3}{3!}(x - 2)^3 + \ldots = \sum_{n=0}^{\infty} \frac{e^2}{n!}(x - 2)^n$

29. If $e^x = \sum_{n=0}^{\infty} \frac{f^{(n)}(a)}{n!}(x - a)^n$ and $f(x) = e^x$, we have $f^{(n)}(a) = e^a$ f or all $n = 0, 1, 2, 3, \ldots$

    $\Rightarrow e^x = e^a \left[\frac{(x - a)^0}{0!} + \frac{(x - a)^1}{1!} + \frac{(x - a)^2}{2!} + \ldots\right] = e^a \left[1 + (x - a) + \frac{(x - a)^2}{2!} + \ldots\right]$ at $x = a$

31. $f(x) = f(a) + f'(a)(x - a) + \frac{f''(a)}{2}(x - a)^2 + \frac{f'''(a)}{3!}(x - a)^3 + \ldots \Rightarrow f'(x) = f'(a) + f''(a)(x - a) + \frac{f'''(a)}{3!} 3(x - a)^2 + \ldots$

    $\Rightarrow f''(x) = f''(a) + f'''(a)(x - a) + \frac{f^{(4)}(a)}{4!} 4 \cdot 3(x - a)^2 + \ldots \Rightarrow f^{(n)}(x) = f^{(n)}(a) + f^{(n+1)}(a)(x - a) + \frac{f^{(n+2)}(a)}{2}(x - a)^2 + \ldots$

    $\Rightarrow f(a) = f(a) + 0, f'(a) = f'(a) + 0, \ldots, f^{(n)}(a) = f^{(n)}(a) + 0$

33. $f(x) = \ln(\cos x) \Rightarrow f'(x) = -\tan x$ and $f''(x) = -\sec^2 x; f(0) = 0, f'(0) = 0, f''(0) = -1 \Rightarrow L(x) = 0$ and $Q(x) = -\frac{x^2}{2}$

35. $f(x) = (1 - x^2)^{-1/2} \Rightarrow f'(x) = x(1 - x^2)^{-3/2}$ and $f''(x) = (1 - x^2)^{-3/2} + 3x^2(1 - x^2)^{-5/2}; f(0) = 1, f'(0) = 0,$

    $f''(0) = 1 \Rightarrow L(x) = 1$ and $Q(x) = 1 + \frac{x^2}{2}$

37. $f(x) = \sin x \Rightarrow f'(x) = \cos x$ and $f''(x) = -\sin x; f(0) = 0, f'(0) = 1, f''(0) = 0 \Rightarrow L(x) = x$ and $Q(x) = x$

7.9 CONVERGENCE OF TAYLOR SERIES

1. $e^x = 1 + x + \frac{x^2}{2!} + \ldots = \sum_{n=0}^{\infty} \frac{x^n}{n!} \Rightarrow e^{-5x} = 1 + (-5x) + \frac{(-5x)^2}{2!} + \ldots = 1 - 5x + \frac{5^2 x^2}{2!} - \frac{5^3 x^3}{3!} + \ldots = \sum_{n=0}^{\infty} \frac{(-1)^n 5^n x^n}{n!}$

3. $\sin x = x - \frac{x^3}{3!} + \frac{x^5}{5!} - \ldots = \sum_{n=0}^{\infty} \frac{(-1)^n x^{2n+1}}{(2n+1)!} \Rightarrow 5 \sin(-x) = 5\left[(-x) - \frac{(-x)^3}{3!} + \frac{(-x)^5}{5!} - \ldots\right] = \sum_{n=0}^{\infty} \frac{5(-1)^{n+1} x^{2n+1}}{(2n+1)!}$

5. $\cos x = \sum_{n=0}^{\infty} \frac{(-1)^n x^{2n}}{(2n)!} \Rightarrow \cos\sqrt{x + 1} = \sum_{n=0}^{\infty} \frac{(-1)^n \left[(x+1)^{1/2}\right]^{2n}}{(2n)!} = \sum_{n=0}^{\infty} \frac{(-1)^n (x+1)^n}{(2n)!} = 1 - \frac{x+1}{2!} + \frac{(x+1)^2}{4!} - \frac{(x+1)^3}{6!} + \ldots$

7. $e^x = \sum\limits_{n=0}^{\infty} \frac{x^n}{n!} \Rightarrow xe^x = x\left(\sum\limits_{n=0}^{\infty} \frac{x^n}{n!}\right) = \sum\limits_{n=0}^{\infty} \frac{x^{n+1}}{n!} = x + x^2 + \frac{x^3}{2!} + \frac{x^4}{3!} + \frac{x^5}{4!} + \dots$

9. $\cos x = \sum\limits_{n=0}^{\infty} \frac{(-1)^n x^{2n}}{(2n)!} \Rightarrow \frac{x^2}{2} - 1 + \cos x = \frac{x^2}{2} - 1 + \sum\limits_{n=0}^{\infty} \frac{(-1)^n x^{2n}}{(2n)!} = \frac{x^2}{2} - 1 + 1 - \frac{x^2}{2} + \frac{x^4}{4!} - \frac{x^6}{6!} + \frac{x^8}{8!} - \frac{x^{10}}{10!} + \dots$

$= \frac{x^4}{4!} - \frac{x^6}{6!} + \frac{x^8}{8!} - \frac{x^{10}}{10!} + \dots = \sum\limits_{n=2}^{\infty} \frac{(-1)^n x^{2n}}{(2n)!}$

11. $\cos x = \sum\limits_{n=0}^{\infty} \frac{(-1)^n x^{2n}}{(2n)!} \Rightarrow x\cos \pi x = x\sum\limits_{n=0}^{\infty} \frac{(-1)^n (\pi x)^{2n}}{(2n)!} = \sum\limits_{n=0}^{\infty} \frac{(-1)^n \pi^{2n} x^{2n+1}}{(2n)!} = x - \frac{\pi^2 x^3}{2!} + \frac{\pi^4 x^5}{4!} - \frac{\pi^6 x^7}{6!} + \dots$

13. $\cos^2 x = \frac{1}{2} + \frac{\cos 2x}{2} = \frac{1}{2} + \frac{1}{2}\sum\limits_{n=0}^{\infty} \frac{(-1)^n (2x)^{2n}}{(2n)!} = \frac{1}{2} + \frac{1}{2}\left[1 - \frac{(2x)^2}{2!} + \frac{(2x)^4}{4!} - \frac{(2x)^6}{6!} + \frac{(2x)^8}{8!} - \dots\right]$

$= 1 - \frac{(2x)^2}{2 \cdot 2!} + \frac{(2x)^4}{2 \cdot 4!} - \frac{(2x)^6}{2 \cdot 6!} + \frac{(2x)^8}{2 \cdot 8!} - \dots = 1 + \sum\limits_{n=1}^{\infty} \frac{(-1)^n (2x)^{2n}}{2 \cdot (2n)!} = 1 + \sum\limits_{n=1}^{\infty} \frac{(-1)^n 2^{2n-1} x^{2n}}{(2n)!}$

15. $\frac{x^2}{1-2x} = x^2\left(\frac{1}{1-2x}\right) = x^2\sum\limits_{n=0}^{\infty} (2x)^n = \sum\limits_{n=0}^{\infty} 2^n x^{n+2} = x^2 + 2x^3 + 2^2 x^4 + 2^3 x^5 + \dots$

17. $\frac{1}{1-x} = \sum\limits_{n=0}^{\infty} x^n = 1 + x + x^2 + x^3 + \dots \Rightarrow \frac{d}{dx}\left(\frac{1}{1-x}\right) = \frac{1}{(1-x)^2} = 1 + 2x + 3x^2 + \dots = \sum\limits_{n=1}^{\infty} nx^{n-1} = \sum\limits_{n=0}^{\infty} (n+1)x^n$

19. By the Alternating Series Estimation Theorem, the error is less than $\frac{|x|^5}{5!} \Rightarrow |x|^5 < (5!)\,(5 \times 10^{-4}) \Rightarrow |x|^5 < 600 \times 10^{-4}$

$\Rightarrow |x| < \sqrt[5]{6 \times 10^{-2}} \approx 0.56968$

21. If $\sin x = x$ and $|x| < 10^{-3}$, then the error is less than $\frac{(10^{-3})^3}{3!} \approx 1.67 \times 10^{-10}$, by Alternating Series Estimation Theorem;

The Alternating Series Estimation Theorem says $R_2(x)$ has the same sign as $-\frac{x^3}{3!}$. Moreover, $x < \sin x$

$\Rightarrow 0 < \sin x - x = R_2(x) \Rightarrow x < 0 \Rightarrow -10^{-3} < x < 0.$

23. $|R_2(x)| = \left|\frac{e^c x^3}{3!}\right| < \frac{3^{(0.1)}(0.1)^3}{3!} < 1.87 \times 10^{-4}$, where c is between 0 and x

25. $F(x) = \int_0^x \left(t^2 - \frac{t^6}{3!} + \frac{t^{10}}{5!} - \frac{t^{14}}{7!} + \dots\right) dt = \left[\frac{t^3}{3} - \frac{t^7}{7 \cdot 3!} + \frac{t^{11}}{11 \cdot 5!} - \frac{t^{15}}{15 \cdot 7!} + \dots\right]_0^x \approx \frac{x^3}{3} - \frac{x^7}{7 \cdot 3!} + \frac{x^{11}}{11 \cdot 5!}$

$\Rightarrow |\text{error}| < \frac{1}{15 \cdot 7!} \approx 0.000013$

27. (a) $F(x) = \int_0^x \left(t - \frac{t^3}{3} + \frac{t^5}{5} - \frac{t^7}{7} + \dots\right) dt = \left[\frac{t^2}{2} - \frac{t^4}{12} + \frac{t^6}{30} - \dots\right]_0^x \approx \frac{x^2}{2} - \frac{x^4}{12} \Rightarrow |\text{error}| < \frac{(0.5)^6}{30} \approx .00052$

(b) $|\text{error}| < \frac{1}{33 \cdot 34} \approx .00089$ when $F(x) \approx \frac{x^2}{2} - \frac{x^4}{3 \cdot 4} + \frac{x^6}{5 \cdot 6} - \frac{x^8}{7 \cdot 8} + \dots + (-1)^{15} \frac{x^{32}}{31 \cdot 32}$

29. $e^x \sin x = 0 + x + x^2 + x^3\left(-\frac{1}{3!} + \frac{1}{2!}\right) + x^4\left(-\frac{1}{3!} + \frac{1}{3!}\right) + x^5\left(\frac{1}{5!} - \frac{1}{2!}\frac{1}{3!} + \frac{1}{4!}\right) + x^6\left(\frac{1}{5!} - \frac{1}{3!}\frac{1}{3!} + \frac{1}{5!}\right) + \dots$

$= x + x^2 + \frac{1}{3}x^3 - \frac{1}{30}x^5 - \frac{1}{90}x^6 + \dots$

31. A special case of Taylor's Theorem is $f(b) = f(a) + f'(c)(b - a)$, where c is between a and $b \Rightarrow f(b) - f(a) = f'(c)(b - a)$, the Mean Value Theorem.

33. (a) $f'' \leq 0$, $f'(a) = 0$ and $x = a$ interior to the interval $I \Rightarrow f(x) - f(a) = \frac{f''(c_2)}{2}(x - a)^2 \leq 0$ throughout I

$\Rightarrow f(x) \leq f(a)$ throughout $I \Rightarrow f$ has a local maximum at $x = a$

(b) similar reasoning gives $f(x) - f(a) = \frac{f''(c_2)}{2}(x - a)^2 \geq 0$ throughout I $\Rightarrow$ $f(x) \geq f(a)$ throughout I $\Rightarrow$ f has a local minimum at x = a

35. (a) $f(x) = (1 + x)^k \Rightarrow f'(x) = k(1 + x)^{k-1} \Rightarrow f''(x) = k(k - 1)(1 + x)^{k-2}; f(0) = 1, f'(0) = k,$ and $f''(0) = k(k - 1)$

$\Rightarrow Q(x) = 1 + kx + \frac{k(k-1)}{2}x^2$

(b) $|R_2(x)| = \left|\frac{3 \cdot 2 \cdot 1}{3!}x^3\right| < \frac{1}{100} \Rightarrow |x^3| < \frac{1}{100} \Rightarrow 0 < x < \frac{1}{100^{1/3}}$ or $0 < x < .21544$

37. If $f(x) = \sum_{n=0}^{\infty} a_n x^n$, then $f^{(k)}(x) = \sum_{n=k}^{\infty} n(n - 1)(n - 2) \cdots (n - k + 1)a_n x^{n-k}$ and $f^{(k)}(0) = k! \, a_k$

$\Rightarrow a_k = \frac{f^{(k)}(0)}{k!}$ for k a nonnegative integer. Therefore, the coefficients of f(x) are identical with the corresponding coefficients in the Maclaurin series of f(x) and the statement follows.

39. (a) $e^{-i\pi} = \cos(-\pi) + i\sin(-\pi) = -1 + i(0) = -1$

(b) $e^{i\pi/4} = \cos\left(\frac{\pi}{4}\right) + i\sin\left(\frac{\pi}{4}\right) = \frac{1}{\sqrt{2}} + \frac{i}{\sqrt{2}} = \left(\frac{1}{\sqrt{2}}\right)(1 + i)$

(c) $e^{-i\pi/2} = \cos\left(-\frac{\pi}{2}\right) + i\sin\left(-\frac{\pi}{2}\right) = 0 + i(-1) = -i$

41. $e^x = 1 + x + \frac{x^2}{2!} + \frac{x^3}{3!} + \frac{x^4}{4!} + \ldots \Rightarrow e^{i\theta} = 1 + i\theta + \frac{(i\theta)^2}{2!} + \frac{(i\theta)^3}{3!} + \frac{(i\theta)^4}{4!} + \ldots$ and

$e^{-i\theta} = 1 - i\theta + \frac{(-i\theta)^2}{2!} + \frac{(-i\theta)^3}{3!} + \frac{(-i\theta)^4}{4!} + \ldots = 1 - i\theta + \frac{(i\theta)^2}{2!} - \frac{(i\theta)^3}{3!} + \frac{(i\theta)^4}{4!} - \ldots$

$\Rightarrow \frac{e^{i\theta} + e^{-i\theta}}{2} = \frac{\left(1 + i\theta + \frac{(i\theta)^2}{2!} + \frac{(i\theta)^3}{3!} + \frac{(i\theta)^4}{4!} + \ldots\right) + \left(1 - i\theta + \frac{(i\theta)^2}{2!} - \frac{(i\theta)^3}{3!} + \frac{(i\theta)^4}{4!} - \ldots\right)}{2} = 1 - \frac{\theta^2}{2!} + \frac{\theta^4}{4!} - \frac{\theta^6}{6!} + \ldots = \cos\theta;$

$\frac{e^{i\theta} - e^{-i\theta}}{2i} = \frac{\left(1 + i\theta + \frac{(i\theta)^2}{2!} + \frac{(i\theta)^3}{3!} + \frac{(i\theta)^4}{4!} + \ldots\right) - \left(1 - i\theta + \frac{(i\theta)^2}{2!} - \frac{(i\theta)^3}{3!} + \frac{(i\theta)^4}{4!} - \ldots\right)}{2i} = \theta - \frac{\theta^3}{3!} + \frac{\theta^5}{5!} - \frac{\theta^7}{7!} + \ldots = \sin\theta$

43. (a) $e^{i\theta_1}e^{i\theta_2} = (\cos\theta_1 + i\sin\theta_1)(\cos\theta_2 + i\sin\theta_2) = (\cos\theta_1\cos\theta_2 - \sin\theta_1\sin\theta_2) + i(\sin\theta_1\cos\theta_2 + \sin\theta_2\cos\theta_1)$

$= \cos(\theta_1 + \theta_2) + i\sin(\theta_1 + \theta_2) = e^{i(\theta_1 + \theta_2)}$

(b) $e^{-i\theta} = \cos(-\theta) + i\sin(-\theta) = \cos\theta - i\sin\theta = (\cos\theta - i\sin\theta)\left(\frac{\cos\theta + i\sin\theta}{\cos\theta + i\sin\theta}\right) = \frac{1}{\cos\theta + i\sin\theta} = \frac{1}{e^{i\theta}}$

7.10 THE BINOMIAL SERIES

1. $(1 + x)^{1/2} = 1 + \frac{1}{2}x + \frac{\left(\frac{1}{2}\right)\left(-\frac{1}{2}\right)x^2}{2!} + \frac{\left(\frac{1}{2}\right)\left(-\frac{1}{2}\right)\left(-\frac{3}{2}\right)x^3}{3!} + \ldots = 1 + \frac{1}{2}x - \frac{1}{8}x^2 + \frac{1}{16}x^3 - \ldots$

3. $(1 - x)^{-1/2} = 1 - \frac{1}{2}(-x) + \frac{\left(-\frac{1}{2}\right)\left(-\frac{3}{2}\right)(-x)^2}{2!} + \frac{\left(-\frac{1}{2}\right)\left(-\frac{3}{2}\right)\left(-\frac{5}{2}\right)(-x)^3}{3!} + \ldots = 1 + \frac{1}{2}x + \frac{3}{8}x^2 + \frac{5}{16}x^3 + \ldots$

5. $\left(1 + \frac{x}{2}\right)^{-2} = 1 - 2\left(\frac{x}{2}\right) + \frac{(-2)(-3)\left(\frac{x}{2}\right)^2}{2!} + \frac{(-2)(-3)(-4)\left(\frac{x}{2}\right)^3}{3!} + \ldots = 1 - x + \frac{3}{4}x^2 - \frac{1}{2}x^3$

7. $(1 + x^3)^{-1/2} = 1 - \frac{1}{2}x^3 + \frac{\left(-\frac{1}{2}\right)\left(-\frac{3}{2}\right)(x^3)^2}{2!} + \frac{\left(-\frac{1}{2}\right)\left(-\frac{3}{2}\right)\left(-\frac{5}{2}\right)(x^3)^3}{3!} + \ldots = 1 - \frac{1}{2}x^3 + \frac{3}{8}x^6 - \frac{5}{16}x^9 + \ldots$

9. $\left(1 + \frac{1}{x}\right)^{1/2} = 1 + \frac{1}{2}\left(\frac{1}{x}\right) + \frac{\left(\frac{1}{2}\right)\left(-\frac{1}{2}\right)\left(\frac{1}{x}\right)^2}{2!} + \frac{\left(\frac{1}{2}\right)\left(-\frac{1}{2}\right)\left(-\frac{3}{2}\right)\left(\frac{1}{x}\right)^3}{3!} + \ldots = 1 + \frac{1}{2x} - \frac{1}{8x^2} + \frac{1}{16x^3} + \ldots$

11. $(1 + x)^4 = 1 + 4x + \frac{(4)(3)x^2}{2!} + \frac{(4)(3)(2)x^3}{3!} + \frac{(4)(3)(2)x^4}{4!} = 1 + 4x + 6x^2 + 4x^3 + x^4$

13. $(1 - 2x)^3 = 1 + 3(-2x) + \frac{(3)(2)(-2x)^2}{2!} + \frac{(3)(2)(1)(-2x)^3}{3!} = 1 - 6x + 12x^2 - 8x^3$

15. (a) $(1 - x^2)^{-1/2} \approx 1 + \frac{x^2}{2} + \frac{3x^4}{8} + \frac{5x^6}{16} \Rightarrow \sin^{-1} x \approx x + \frac{x^3}{6} + \frac{3x^5}{40} + \frac{5x^7}{112}$; Using the Ratio Test:

$\lim\limits_{n \to \infty} \left| \frac{1 \cdot 3 \cdot 5 \cdots (2n-1)(2n+1)x^{2n+3}}{2 \cdot 4 \cdot 6 \cdots (2n)(2n+2)(2n+3)} \cdot \frac{2 \cdot 4 \cdot 6 \cdots (2n)(2n+1)}{1 \cdot 3 \cdot 5 \cdots (2n-1)x^{2n+1}} \right| < 1 \Rightarrow x^2 \lim\limits_{n \to \infty} \left| \frac{(2n+1)(2n+1)}{(2n+2)(2n+3)} \right| < 1$

$\Rightarrow |x| < 1 \Rightarrow$ the radius of convergence is 1. See Exercise 19.

(b) $\frac{d}{dx} (\cos^{-1} x) = -(1 - x^2)^{-1/2} \Rightarrow \cos^{-1} x = \frac{\pi}{2} - \sin^{-1} x \approx \frac{\pi}{2} - \left(x + \frac{x^3}{6} + \frac{3x^5}{40} + \frac{5x^7}{112} \right) \approx \frac{\pi}{2} - x - \frac{x^3}{6} - \frac{3x^5}{40} - \frac{5x^7}{112}$

17. $\frac{-1}{1+x} = -\frac{1}{1-(-x)} = -1 + x - x^2 + x^3 - \ldots \Rightarrow \frac{d}{dx} \left(\frac{-1}{1+x} \right) = \frac{1}{1+x^2} = \frac{d}{dx} (-1 + x - x^2 + x^3 - \ldots)$

$= 1 - 2x + 3x^2 - 4x^3 + \ldots$

19. $(1 - x^2)^{-1/2} = (1 + (-x^2))^{-1/2} = (1)^{-1/2} + \left(-\frac{1}{2} \right)(1)^{-3/2} (-x^2) + \frac{\left(-\frac{1}{2} \right) \left(-\frac{3}{2} \right)(1)^{-5/2} (-x^2)^2}{2!}$

$+ \frac{\left(-\frac{1}{2} \right) \left(-\frac{3}{2} \right) \left(-\frac{5}{2} \right)(1)^{-7/2} (-x^2)^3}{3!} + \ldots = 1 + \frac{x^2}{2} + \frac{1 \cdot 3 x^4}{2^2 \cdot 2!} + \frac{1 \cdot 3 \cdot 5 x^6}{2^3 \cdot 3!} + \ldots = 1 + \sum\limits_{n=1}^{\infty} \frac{1 \cdot 3 \cdot 5 \cdots (2n-1)x^{2n}}{2^n \cdot n!}$

$\Rightarrow \sin^{-1} x = \int_0^x (1 - t^2)^{-1/2} \, dt = \int_0^x \left(1 + \sum\limits_{n=1}^{\infty} \frac{1 \cdot 3 \cdot 5 \cdots (2n-1)x^{2n}}{2^n \cdot n!} \right) dt = x + \sum\limits_{n=1}^{\infty} \frac{1 \cdot 3 \cdot 5 \cdots (2n-1)x^{2n+1}}{2 \cdot 4 \cdots (2n)(2n+1)}$,

where $|x| < 1$

CHAPTER 7 PRACTICE AND ADDITIONAL EXERCISES

1. converges to 1, since $\lim\limits_{n \to \infty} a_n = \lim\limits_{n \to \infty} \left(1 + \frac{(-1)^n}{n} \right) = 1$

3. converges to -1, since $\lim\limits_{n \to \infty} a_n = \lim\limits_{n \to \infty} \left(\frac{1 - 2^n}{2^n} \right) = \lim\limits_{n \to \infty} \left(\frac{1}{2^n} - 1 \right) = -1$

5. diverges, since $\left\{ \sin \frac{n\pi}{2} \right\} = \{0, 1, 0, -1, 0, 1, \ldots\}$

7. converges to 0, since $\lim\limits_{n \to \infty} a_n = \lim\limits_{n \to \infty} \frac{\ln n^2}{n} = 2 \lim\limits_{n \to \infty} \frac{\left(\frac{1}{n} \right)}{1} = 0$

9. converges to 1, since $\lim\limits_{n \to \infty} a_n = \lim\limits_{n \to \infty} \left(\frac{n + \ln n}{n} \right) = \lim\limits_{n \to \infty} \frac{1 + \left(\frac{1}{n} \right)}{1} = 1$

11. converges to e^{-5}, since $\lim\limits_{n \to \infty} a_n = \lim\limits_{n \to \infty} \left(\frac{n-5}{n} \right)^n = \lim\limits_{n \to \infty} \left(1 + \frac{(-5)}{n} \right)^n = e^{-5}$ by Theorem 5

13. converges to 3, since $\lim\limits_{n \to \infty} a_n = \lim\limits_{n \to \infty} \left(\frac{3^n}{n} \right)^{1/n} = \lim\limits_{n \to \infty} \frac{3}{n^{1/n}} = \frac{3}{1} = 3$ by Theorem 5

15. converges to $\ln 2$, since $\lim\limits_{n \to \infty} a_n = \lim\limits_{n \to \infty} n \left(2^{1/n} - 1 \right) = \lim\limits_{n \to \infty} \frac{2^{1/n} - 1}{\left(\frac{1}{n} \right)} = \lim\limits_{n \to \infty} \frac{\left[\frac{\left(-2^{1/n} \ln 2 \right)}{n^2} \right]}{\left(\frac{-1}{n^2} \right)} = \lim\limits_{n \to \infty} 2^{1/n} \ln 2$

$= 2^0 \cdot \ln 2 = \ln 2$

17. diverges, since $\lim\limits_{n \to \infty} a_n = \lim\limits_{n \to \infty} \frac{(n+1)!}{n!} = \lim\limits_{n \to \infty} (n+1) = \infty$

19. $\frac{1}{(2n-3)(2n-1)} = \frac{\left(\frac{1}{2} \right)}{2n-3} - \frac{\left(\frac{1}{2} \right)}{2n-1} \Rightarrow s_n = \left[\frac{\left(\frac{1}{2} \right)}{3} - \frac{\left(\frac{1}{2} \right)}{5} \right] + \left[\frac{\left(\frac{1}{2} \right)}{5} - \frac{\left(\frac{1}{2} \right)}{7} \right] + \ldots + \left[\frac{\left(\frac{1}{2} \right)}{2n-3} - \frac{\left(\frac{1}{2} \right)}{2n-1} \right] = \frac{\left(\frac{1}{2} \right)}{3} - \frac{\left(\frac{1}{2} \right)}{2n-1}$

$\Rightarrow \lim\limits_{n \to \infty} s_n = \lim\limits_{n \to \infty} \left[\frac{1}{6} - \frac{\left(\frac{1}{2} \right)}{2n-1} \right] = \frac{1}{6}$

21. $\frac{9}{(3n-1)(3n+2)} = \frac{3}{3n-1} - \frac{3}{3n+2} \Rightarrow s_n = \left(\frac{3}{2} - \frac{3}{5}\right) + \left(\frac{3}{5} - \frac{3}{8}\right) + \left(\frac{3}{8} - \frac{3}{11}\right) + \dots + \left(\frac{3}{3n-1} - \frac{3}{3n+2}\right)$

$= \frac{3}{2} - \frac{3}{3n+2} \Rightarrow \lim_{n \to \infty} s_n = \lim_{n \to \infty} \left(\frac{3}{2} - \frac{3}{3n+2}\right) = \frac{3}{2}$

23. $\sum_{n=0}^{\infty} e^{-n} = \sum_{n=0}^{\infty} \frac{1}{e^n}$, a convergent geometric series with $r = \frac{1}{e}$ and $a = 1 \Rightarrow$ the sum is $\frac{1}{1 - \left(\frac{1}{e}\right)} = \frac{e}{e-1}$

25. diverges, a p-series with $p = \frac{1}{2}$

27. Since $f(x) = \frac{1}{x^{1/2}} \Rightarrow f'(x) = -\frac{1}{2x^{3/2}} < 0 \Rightarrow f(x)$ is decreasing $\Rightarrow a_{n+1} < a_n$, and $\lim_{n \to \infty} a_n = \lim_{n \to \infty} \frac{1}{\sqrt{n}} = 0$, the

series $\sum_{n=1}^{\infty} \frac{(-1)^n}{\sqrt{n}}$ converges by the Alternating Series Test. Since $\sum_{n=1}^{\infty} \frac{1}{\sqrt{n}}$ diverges, the given series converges conditionally.

29. The given series does not converge absolutely by the Direct Comparison Test since $\frac{1}{\ln(n+1)} > \frac{1}{n+1}$, which is the nth

term of a divergent series. Since $f(x) = \frac{1}{\ln(x+1)} \Rightarrow f'(x) = -\frac{1}{(\ln(x+1))^2(x+1)} < 0 \Rightarrow f(x)$ is decreasing $\Rightarrow a_{n+1} < a_n$, and

$\lim_{n \to \infty} a_n = \lim_{n \to \infty} \frac{1}{\ln(n+1)} = 0$, the given series converges conditionally by the Alternating Series Test.

31. converges absolutely by the Direct Comparison Test since $\frac{\ln n}{n^3} < \frac{n}{n^3} = \frac{1}{n^2}$, the nth term of a convergent p-series

33. $\lim_{n \to \infty} \frac{\left(\frac{1}{n\sqrt{n^2+1}}\right)}{\left(\frac{1}{n^2}\right)} = \sqrt{\lim_{n \to \infty} \frac{n^2}{n^2+1}} = \sqrt{1} = 1 \Rightarrow$ converges absolutely by the Limit Comparison Test

35. converges absolutely by the Ratio Test since $\lim_{n \to \infty} \left[\frac{n+2}{(n+1)!} \cdot \frac{n!}{n+1}\right] = \lim_{n \to \infty} \frac{n+2}{(n+1)^2} = 0 < 1$

37. converges absolutely by the Ratio Test since $\lim_{n \to \infty} \left[\frac{3^{n+1}}{(n+1)!} \cdot \frac{n!}{3^n}\right] = \lim_{n \to \infty} \frac{3}{n+1} = 0 < 1$

39. converges absolutely by the Limit Comparison Test since $\lim_{n \to \infty} \frac{\left(\frac{1}{n^{3/2}}\right)}{\left(\frac{1}{\sqrt{n(n+1)(n+2)}}\right)} = \sqrt{\lim_{n \to \infty} \frac{n(n+1)(n+2)}{n^3}} = 1$

41. $\lim_{n \to \infty} \left|\frac{u_{n+1}}{u_n}\right| < 1 \Rightarrow \lim_{n \to \infty} \left|\frac{(x+4)^{n+1}}{(n+1)3^{n+1}} \cdot \frac{n3^n}{(x+4)^n}\right| < 1 \Rightarrow \frac{|x+4|}{3} \lim_{n \to \infty} \left(\frac{n}{n+1}\right) < 1 \Rightarrow \frac{|x+4|}{3} < 1$

$\Rightarrow |x+4| < 3 \Rightarrow -3 < x+4 < 3 \Rightarrow -7 < x < -1$; at $x = -7$ we have $\sum_{n=1}^{\infty} \frac{(-1)^n 3^n}{n3^n} = \sum_{n=1}^{\infty} \frac{(-1)^n}{n}$, the alternating

harmonic series, which converges conditionally; at $x = -1$ we have $\sum_{n=1}^{\infty} \frac{3^n}{n3^n} = \sum_{n=1}^{\infty} \frac{1}{n}$, the divergent harmonic series

   (a) the radius is 3; the interval of convergence is $-7 \le x < -1$
   (b) the interval of absolute convergence is $-7 < x < -1$
   (c) the series converges conditionally at $x = -7$

43. $\lim_{n \to \infty} \left|\frac{u_{n+1}}{u_n}\right| < 1 \Rightarrow \lim_{n \to \infty} \left|\frac{(3x-1)^{n+1}}{(n+1)^2} \cdot \frac{n^2}{(3x-1)^n}\right| < 1 \Rightarrow |3x-1| \lim_{n \to \infty} \frac{n^2}{(n+1)^2} < 1 \Rightarrow |3x-1| < 1$

$\Rightarrow -1 < 3x-1 < 1 \Rightarrow 0 < 3x < 2 \Rightarrow 0 < x < \frac{2}{3}$; at $x = 0$ we have $\sum_{n=1}^{\infty} \frac{(-1)^{n-1}(-1)^n}{n^2} = \sum_{n=1}^{\infty} \frac{(-1)^{2n-1}}{n^2}$

$= -\sum_{n=1}^{\infty} \frac{1}{n^2}$, a nonzero constant multiple of a convergent p-series, which is absolutely convergent; at $x = \frac{2}{3}$ we

have $\sum_{n=1}^{\infty} \frac{(-1)^{n-1}(1)^n}{n^2} = \sum_{n=1}^{\infty} \frac{(-1)^{n-1}}{n^2}$, which converges absolutely

   (a) the radius is $\frac{1}{3}$; the interval of convergence is $0 \le x \le \frac{2}{3}$

(b) the interval of absolute convergence is $0 \le x \le \frac{2}{3}$

(c) there are no values for which the series converges conditionally

45. $\lim\limits_{n \to \infty} \left| \frac{u_{n+1}}{u_n} \right| < 1 \Rightarrow \lim\limits_{n \to \infty} \left| \frac{x^{n+1}}{(n+1)^{n+1}} \cdot \frac{n^n}{x^n} \right| < 1 \Rightarrow |x| \lim\limits_{n \to \infty} \left| \left(\frac{n}{n+1} \right)^n \left(\frac{1}{n+1} \right) \right| < 1 \Rightarrow \frac{|x|}{e} \lim\limits_{n \to \infty} \left(\frac{1}{n+1} \right) < 1$

$\Rightarrow \frac{|x|}{e} \cdot 0 < 1$, which holds for all x

(a) the radius is ∞; the series converges for all x

(b) the series converges absolutely for all x

(c) there are no values for which the series converges conditionally

47. $\lim\limits_{n \to \infty} \left| \frac{u_{n+1}}{u_n} \right| < 1 \Rightarrow \lim\limits_{n \to \infty} \left| \frac{(n+2)x^{2n+1}}{3^{n+1}} \cdot \frac{3^n}{(n+1)x^{2n-1}} \right| < 1 \Rightarrow \frac{x^2}{3} \lim\limits_{n \to \infty} \left(\frac{n+2}{n+1} \right) < 1 \Rightarrow -\sqrt{3} < x < \sqrt{3};$

the series $\sum\limits_{n=1}^{\infty} -\frac{n+1}{\sqrt{3}}$ and $\sum\limits_{n=1}^{\infty} \frac{n+1}{\sqrt{3}}$, obtained with $x = \pm\sqrt{3}$, both diverge

(a) the radius is $\sqrt{3}$; the interval of convergence is $-\sqrt{3} < x < \sqrt{3}$

(b) the interval of absolute convergence is $-\sqrt{3} < x < \sqrt{3}$

(c) there are no values for which the series converges conditionally

49. $\lim\limits_{n \to \infty} \left| \frac{u_{n+1}}{u_n} \right| < 1 \Rightarrow \lim\limits_{n \to \infty} \left| \frac{\operatorname{csch}(n+1)x^{n+1}}{\operatorname{csch}(n)x^n} \right| < 1 \Rightarrow |x| \lim\limits_{n \to \infty} \left| \frac{\left(\frac{2}{e^{n+1} - e^{-n-1}} \right)}{\left(\frac{2}{e^n - e^{-n}} \right)} \right| < 1$

$\Rightarrow |x| \lim\limits_{n \to \infty} \left| \frac{e^{-1} - e^{-2n-1}}{1 - e^{-2n-2}} \right| < 1 \Rightarrow \frac{|x|}{e} < 1 \Rightarrow -e < x < e;$ the series $\sum\limits_{n=1}^{\infty} (\pm e)^n \operatorname{csch} n$, obtained with $x = \pm e$,

both diverge since $\lim\limits_{n \to \infty} (\pm e)^n \operatorname{csch} n \neq 0$

(a) the radius is e; the interval of convergence is $-e < x < e$

(b) the interval of absolute convergence is $-e < x < e$

(c) there are no values for which the series converges conditionally

51. The given series has the form $1 - x + x^2 - x^3 + \ldots + (-x)^n + \ldots = \frac{1}{1+x}$, where $x = \frac{1}{4}$; the sum is $\frac{1}{1 + \left(\frac{1}{4} \right)} = \frac{4}{5}$

53. The given series has the form $x - \frac{x^3}{3!} + \frac{x^5}{5!} - \ldots + (-1)^n \frac{x^{2n+1}}{(2n+1)!} + \ldots = \sin x$, where $x = \pi$; the sum is $\sin \pi = 0$

55. The given series has the form $1 + x + \frac{x^2}{2!} + \frac{x^2}{3!} + \ldots + \frac{x^n}{n!} + \ldots = e^x$, where $x = \ln 2$; the sum is $e^{\ln(2)} = 2$

57. Consider $\frac{1}{1-2x}$ as the sum of a convergent geometric series with $a = 1$ and $r = 2x \Rightarrow \frac{1}{1-2x}$

$= 1 + (2x) + (2x)^2 + (2x)^3 + \ldots = \sum\limits_{n=0}^{\infty} (2x)^n = \sum\limits_{n=0}^{\infty} 2^n x^n$ where $|2x| < 1 \Rightarrow |x| < \frac{1}{2}$

59. $\sin x = \sum\limits_{n=0}^{\infty} \frac{(-1)^n x^{2n+1}}{(2n+1)!} \Rightarrow \sin \pi x = \sum\limits_{n=0}^{\infty} \frac{(-1)^n (\pi x)^{2n+1}}{(2n+1)!} = \sum\limits_{n=0}^{\infty} \frac{(-1)^n \pi^{2n+1} x^{2n+1}}{(2n+1)!}$

61. $\cos x = \sum\limits_{n=0}^{\infty} \frac{(-1)^n x^{2n}}{(2n)!} \Rightarrow \cos \left(x^{5/2} \right) = \sum\limits_{n=0}^{\infty} \frac{(-1)^n \left(x^{5/2} \right)^{2n}}{(2n)!} = \sum\limits_{n=0}^{\infty} \frac{(-1)^n x^{5n}}{(2n)!}$

63. $e^x = \sum\limits_{n=0}^{\infty} \frac{x^n}{n!} \Rightarrow e^{(\pi x/2)} = \sum\limits_{n=0}^{\infty} \frac{\left(\frac{\pi x}{2} \right)^n}{n!} = \sum\limits_{n=0}^{\infty} \frac{\pi^n x^n}{2^n n!}$

65. $f(x) = \sqrt{3+x^2} = (3+x^2)^{1/2} \Rightarrow f'(x) = x(3+x^2)^{-1/2} \Rightarrow f''(x) = -x^2(3+x^2)^{-3/2} + (3+x^2)^{-1/2}$

$\Rightarrow f'''(x) = 3x^3(3+x^2)^{-5/2} - 3x(3+x^2)^{-3/2}; f(-1) = 2, f'(-1) = -\frac{1}{2}, f''(-1) = -\frac{1}{8} + \frac{1}{2} = \frac{3}{8},$

$f'''(-1) = -\frac{3}{32} + \frac{3}{8} = \frac{9}{32} \Rightarrow \sqrt{3+x^2} = 2 - \frac{(x+1)}{2\cdot1!} + \frac{3(x+1)^2}{2^3\cdot2!} + \frac{9(x+1)^3}{2^5\cdot3!} + \cdots$

67. $f(x) = \frac{1}{x+1} = (x+1)^{-1} \Rightarrow f'(x) = -(x+1)^{-2} \Rightarrow f''(x) = 2(x+1)^{-3} \Rightarrow f'''(x) = -6(x+1)^{-4}; f(3) = \frac{1}{4},$

$f'(3) = -\frac{1}{4^2}, f''(3) = \frac{2}{4^3}, f'''(2) = \frac{-6}{4^4} \Rightarrow \frac{1}{x+1} = \frac{1}{4} - \frac{1}{4^2}(x-3) + \frac{1}{4^3}(x-3)^2 - \frac{1}{4^4}(x-3)^3 + \cdots$

69. $\int_0^{1/2} e^{-x^3} dx = \int_0^{1/2} \left(1 - x^3 + \frac{x^6}{2!} - \frac{x^9}{3!} + \frac{x^{12}}{4!} + \cdots\right) dx = \left[x - \frac{x^4}{4} + \frac{x^7}{7\cdot2!} - \frac{x^{10}}{10\cdot3!} + \frac{x^{13}}{13\cdot4!} - \cdots\right]_0^{1/2}$

$\approx \frac{1}{2} - \frac{1}{2^4\cdot4} + \frac{1}{2^7\cdot7\cdot2!} - \frac{1}{2^{10}\cdot10\cdot3!} + \frac{1}{2^{13}\cdot13\cdot4!} - \frac{1}{2^{16}\cdot16\cdot5!} \approx 0.484917143$

71. $\lim_{x\to0}\left(\frac{\sin 3x}{x^3} + \frac{r}{x^2} + s\right) = \lim_{x\to0}\left[\frac{\left(3x - \frac{(3x)^3}{6} + \frac{(3x)^5}{120} - \cdots\right)}{x^3} + \frac{r}{x^2} + s\right] = \lim_{x\to0}\left(\frac{3}{x^2} - \frac{9}{2} + \frac{81x^2}{40} + \cdots + \frac{r}{x^2} + s\right) = 0$

$\Rightarrow \frac{r}{x^2} + \frac{3}{x^2} = 0$ and $s - \frac{9}{2} = 0 \Rightarrow r = -3$ and $s = \frac{9}{2}$

73. $\lim_{n\to\infty}\left|\frac{2\cdot5\cdot8\cdots(3n-1)(3n+2)x^{n+1}}{2\cdot4\cdot6\cdots(2n)(2n+2)} \cdot \frac{2\cdot4\cdot6\cdots(2n)}{2\cdot5\cdot8\cdots(3n-1)x^n}\right| < 1 \Rightarrow |x|\lim_{n\to\infty}\left|\frac{3n+2}{2n+2}\right| < 1 \Rightarrow |x| < \frac{2}{3}$

$\Rightarrow$ the radius of convergence is $\frac{2}{3}$

75. (a) $\lim_{n\to\infty}\left|\frac{1\cdot4\cdot7\cdots(3n-2)(3n+1)x^{3n+3}}{(3n+3)!} \cdot \frac{(3n)!}{1\cdot4\cdot7\cdots(3n-2)x^{3n}}\right| < 1 \Rightarrow |x^3|\lim_{n\to\infty}\frac{(3n+1)}{(3n+1)(3n+2)(3n+3)}$

$= |x^3| \cdot 0 < 1 \Rightarrow$ the radius of convergence is ∞

(b) $y = 1 + \sum_{n=1}^{\infty}\frac{1\cdot4\cdot7\cdots(3n-2)}{(3n)!}x^{3n} \Rightarrow \frac{dy}{dx} = \sum_{n=1}^{\infty}\frac{1\cdot4\cdot7\cdots(3n-2)}{(3n-1)!}x^{3n-1}$

$\Rightarrow \frac{d^2y}{dx^2} = \sum_{n=1}^{\infty}\frac{1\cdot4\cdot7\cdots(3n-2)}{(3n-2)!}x^{3n-2} = x + \sum_{n=2}^{\infty}\frac{1\cdot4\cdot7\cdots(3n-5)}{(3n-3)!}x^{3n-2}$

$= x\left(1 + \sum_{n=1}^{\infty}\frac{1\cdot4\cdot7\cdots(3n-2)}{(3n)!}x^{3n}\right) = xy + 0 \Rightarrow a = 1$ and $b = 0$

77. $\sum_{n=1}^{\infty}(x_{n+1} - x_n) = \lim_{n\to\infty}\sum_{k=1}^{\infty}(x_{k+1} - x_k) = \lim_{n\to\infty}(x_{n+1} - x_1) = \lim_{n\to\infty}(x_{n+1}) - x_1 \Rightarrow$ both the series and

sequence must either converge or diverge.

NOTES:

CHAPTER 8 POLAR COORDINATES AND CONICS

8.1 POLAR COORDINATES

1. a, e; b, g; c, h; d, f

3. (a) $\left(2, \frac{\pi}{2} + 2n\pi\right)$ and $\left(-2, \frac{\pi}{2} + (2n + 1)\pi\right)$, n an integer
    (b) $(2, 2n\pi)$ and $(-2, (2n + 1)\pi)$, n an integer
    (c) $\left(2, \frac{3\pi}{2} + 2n\pi\right)$ and $\left(-2, \frac{3\pi}{2} + (2n + 1)\pi\right)$, n an integer
    (d) $(2, (2n + 1)\pi)$ and $(-2, 2n\pi)$, n an integer

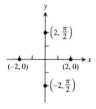

5. (a) $x = r \cos \theta = 3 \cos 0 = 3$, $y = r \sin \theta = 3 \sin 0 = 0 \Rightarrow$ Cartesian coordinates are $(3, 0)$
    (b) $x = r \cos \theta = -3 \cos 0 = -3$, $y = r \sin \theta = -3 \sin 0 = 0 \Rightarrow$ Cartesian coordinates are $(-3, 0)$
    (c) $x = r \cos \theta = 2 \cos \frac{2\pi}{3} = -1$, $y = r \sin \theta = 2 \sin \frac{2\pi}{3} = \sqrt{3} \Rightarrow$ Cartesian coordinates are $\left(-1, \sqrt{3}\right)$
    (d) $x = r \cos \theta = 2 \cos \frac{7\pi}{3} = 1$, $y = r \sin \theta = 2 \sin \frac{7\pi}{3} = \sqrt{3} \Rightarrow$ Cartesian coordinates are $\left(1, \sqrt{3}\right)$
    (e) $x = r \cos \theta = -3 \cos \pi = 3$, $y = r \sin \theta = -3 \sin \pi = 0 \Rightarrow$ Cartesian coordinates are $(3, 0)$
    (f) $x = r \cos \theta = 2 \cos \frac{\pi}{3} = 1$, $y = r \sin \theta = 2 \sin \frac{\pi}{3} = \sqrt{3} \Rightarrow$ Cartesian coordinates are $\left(1, \sqrt{3}\right)$
    (g) $x = r \cos \theta = -3 \cos 2\pi = -3$, $y = r \sin \theta = -3 \sin 2\pi = 0 \Rightarrow$ Cartesian coordinates are $(-3, 0)$
    (h) $x = r \cos \theta = -2 \cos \left(-\frac{\pi}{3}\right) = -1$, $y = r \sin \theta = -2 \sin \left(-\frac{\pi}{3}\right) = \sqrt{3} \Rightarrow$ Cartesian coordinates are $\left(-1, \sqrt{3}\right)$

7.

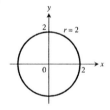

9.

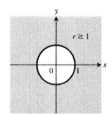

11.

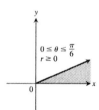

13.

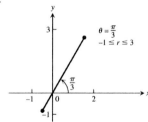

15.

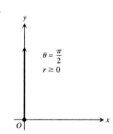

17.

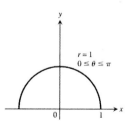

19.

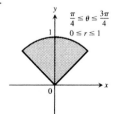

21.

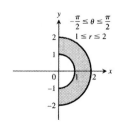

23. $r \cos \theta = 2 \Rightarrow x = 2$, vertical line through $(2, 0)$ 25. $r \sin \theta = 0 \Rightarrow y = 0$, the x-axis

27. $r = 4 \csc \theta \Rightarrow r = \frac{4}{\sin \theta} \Rightarrow r \sin \theta = 4 \Rightarrow y = 4$, a horizontal line through $(0, 4)$

29. $r \cos \theta + r \sin \theta = 1 \Rightarrow x + y = 1$, line with slope $m = -1$ and intercept $b = 1$

31. $r^2 = 1 \Rightarrow x^2 + y^2 = 1$, circle with center $C = (0, 0)$ and radius 1

33. $r = \frac{5}{\sin \theta - 2 \cos \theta} \Rightarrow r \sin \theta - 2r \cos \theta = 5 \Rightarrow y - 2x = 5$, line with slope $m = 2$ and intercept $b = 5$

35. $r = \cot \theta \csc \theta = \left(\frac{\cos \theta}{\sin \theta} \right) \left(\frac{1}{\sin \theta} \right) \Rightarrow r \sin^2 \theta = \cos \theta \Rightarrow r^2 \sin^2 \theta = r \cos \theta \Rightarrow y^2 = x$, parabola with vertex $(0, 0)$
    which opens to the right

37. $r = (\csc \theta) e^{r \cos \theta} \Rightarrow r \sin \theta = e^{r \cos \theta} \Rightarrow y = e^x$, graph of the natural exponential function

39. $r^2 + 2r^2 \cos \theta \sin \theta = 1 \Rightarrow x^2 + y^2 + 2xy = 1 \Rightarrow x^2 + 2xy + y^2 = 1 \Rightarrow (x + y)^2 = 1 \Rightarrow x + y = \pm 1$, two parallel
    straight lines of slope -1 and y-intercepts $b = \pm 1$

41. $r^2 = -4r \cos \theta \Rightarrow x^2 + y^2 = -4x \Rightarrow x^2 + 4x + y^2 = 0 \Rightarrow x^2 + 4x + 4 + y^2 = 4 \Rightarrow (x + 2)^2 + y^2 = 4$, a circle with
    center $C(-2, 0)$ and radius 2

43. $r = 8 \sin \theta \Rightarrow r^2 = 8r \sin \theta \Rightarrow x^2 + y^2 = 8y \Rightarrow x^2 + y^2 - 8y = 0 \Rightarrow x^2 + y^2 - 8y + 16 = 16$
    $\Rightarrow x^2 + (y - 4)^2 = 16$, a circle with center $C(0, 4)$ and radius 4

45. $r = 2 \cos \theta + 2 \sin \theta \Rightarrow r^2 = 2r \cos \theta + 2r \sin \theta \Rightarrow x^2 + y^2 = 2x + 2y \Rightarrow x^2 - 2x + y^2 - 2y = 0$
    $\Rightarrow (x - 1)^2 + (y - 1)^2 = 2$, a circle with center $C(1, 1)$ and radius $\sqrt{2}$

47. $r \sin \left(\theta + \frac{\pi}{6} \right) = 2 \Rightarrow r \left(\sin \theta \cos \frac{\pi}{6} + \cos \theta \sin \frac{\pi}{6} \right) = 2 \Rightarrow \frac{\sqrt{3}}{2} r \sin \theta + \frac{1}{2} r \cos \theta = 2 \Rightarrow \frac{\sqrt{3}}{2} y + \frac{1}{2} x = 2$
    $\Rightarrow \sqrt{3} y + x = 4$, line with slope $m = -\frac{1}{\sqrt{3}}$ and intercept $b = \frac{4}{\sqrt{3}}$

49. $x = 7 \Rightarrow r \cos \theta = 7$ 51. $x = y \Rightarrow r \cos \theta = r \sin \theta \Rightarrow \theta = \frac{\pi}{4}$

53. $x^2 + y^2 = 4 \Rightarrow r^2 = 4 \Rightarrow r = 2$ or $r = -2$

55. $\frac{x^2}{9} + \frac{y^2}{4} = 1 \Rightarrow 4x^2 + 9y^2 = 36 \Rightarrow 4r^2 \cos^2 \theta + 9r^2 \sin^2 \theta = 36$

57. $y^2 = 4x \Rightarrow r^2 \sin^2 \theta = 4r \cos \theta \Rightarrow r \sin^2 \theta = 4 \cos \theta$

59. $x^2 + (y - 2)^2 = 4 \Rightarrow x^2 + y^2 - 4y + 4 = 4 \Rightarrow x^2 + y^2 = 4y \Rightarrow r^2 = 4r \sin \theta \Rightarrow r = 4 \sin \theta$

61. $(x - 3)^2 + (y + 1)^2 = 4 \Rightarrow x^2 - 6x + 9 + y^2 + 2y + 1 = 4 \Rightarrow x^2 + y^2 = 6x - 2y - 6 \Rightarrow r^2 = 6r \cos \theta - 2r \sin \theta - 6$

63. $(0, \theta)$ where θ is any angle

8.2 GRAPHING IN POLAR COORDINATES

1. $1 + \cos(-\theta) = 1 + \cos\theta = r \Rightarrow$ symmetric about the
    x-axis; $1 + \cos(-\theta) \neq -r$ and $1 + \cos(\pi - \theta)$
    $= 1 - \cos\theta \neq r \Rightarrow$ not symmetric about the y-axis;
    therefore not symmetric about the origin

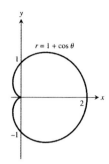

3. $1 - \sin(-\theta) = 1 + \sin\theta \neq r$ and $1 - \sin(\pi - \theta)$
    $= 1 - \sin\theta \neq -r \Rightarrow$ not symmetric about the x-axis;
    $1 - \sin(\pi - \theta) = 1 - \sin\theta = r \Rightarrow$ symmetric about
    the y-axis; therefore not symmetric about the origin

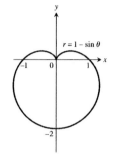

5. $2 + \sin(-\theta) = 2 - \sin\theta \neq r$ and $2 + \sin(\pi - \theta)$
    $= 2 + \sin\theta \neq -r \Rightarrow$ not symmetric about the x-axis;
    $2 + \sin(\pi - \theta) = 2 + \sin\theta = r \Rightarrow$ symmetric about the
    y-axis; therefore not symmetric about the origin

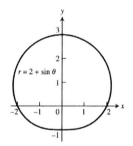

7. $\sin\left(-\frac{\theta}{2}\right) = -\sin\left(\frac{\theta}{2}\right) = -r \Rightarrow$ symmetric about the y-axis;
    $\sin\left(\frac{2\pi - \theta}{2}\right) = \sin\left(\frac{\theta}{2}\right)$, so the graph is symmetric about the
    x-axis, and hence the origin.

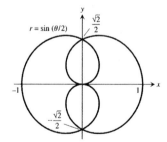

9. $\cos(-\theta) = \cos\theta = r^2 \Rightarrow (r, -\theta)$ and $(-r, -\theta)$ are on the
    graph when (r, θ) is on the graph $\Rightarrow$ symmetric about the
    x-axis and the y-axis; therefore symmetric about the origin

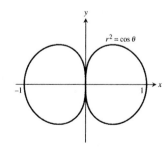

11. $-\sin(\pi-\theta)=-\sin\theta=r^2 \Rightarrow (r,\pi-\theta)$ and $(-r,\pi-\theta)$ are on the graph when (r,θ) is on the graph $\Rightarrow$ symmetric about the y-axis and the x-axis; therefore symmetric about the origin

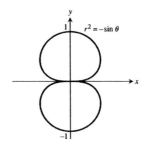

13. Since $(\pm r, -\theta)$ are on the graph when (r,θ) is on the graph $\left((\pm r)^2 = 4\cos 2(-\theta) \Rightarrow r^2 = 4\cos 2\theta\right)$, the graph is symmetric about the x-axis and the y-axis $\Rightarrow$ the graph is symmetric about the origin

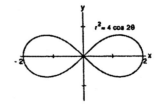

15. Since (r,θ) on the graph $\Rightarrow (-r,\theta)$ is on the graph $\left((\pm r)^2 = -\sin 2\theta \Rightarrow r^2 = -\sin 2\theta\right)$, the graph is symmetric about the origin. But $-\sin 2(-\theta) = -(-\sin 2\theta)$ $\sin 2\theta \neq r^2$ and $-\sin 2(\pi-\theta) = -\sin(2\pi-2\theta)$ $= -\sin(-2\theta) = -(-\sin 2\theta) = \sin 2\theta \neq r^2 \Rightarrow$ the graph is not symmetric about the x-axis; therefore the graph is not symmetric about the y-axis

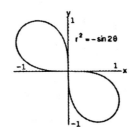

17. $\theta = \frac{\pi}{2} \Rightarrow r = -1 \Rightarrow \left(-1, \frac{\pi}{2}\right)$, and $\theta = -\frac{\pi}{2} \Rightarrow r = -1$ $\Rightarrow \left(-1, -\frac{\pi}{2}\right)$; $r' = \frac{dr}{d\theta} = -\sin\theta$; Slope $= \frac{r'\sin\theta + r\cos\theta}{r'\cos\theta - r\sin\theta}$ $= \frac{-\sin^2\theta + r\cos\theta}{-\sin\theta\cos\theta - r\sin\theta} \Rightarrow$ Slope at $\left(-1, \frac{\pi}{2}\right)$ is $\frac{-\sin^2\left(\frac{\pi}{2}\right)+(-1)\cos\frac{\pi}{2}}{-\sin\frac{\pi}{2}\cos\frac{\pi}{2}-(-1)\sin\frac{\pi}{2}} = -1$; Slope at $\left(-1, -\frac{\pi}{2}\right)$ is $\frac{-\sin^2\left(-\frac{\pi}{2}\right)+(-1)\cos\left(-\frac{\pi}{2}\right)}{-\sin\left(-\frac{\pi}{2}\right)\cos\left(-\frac{\pi}{2}\right)-(-1)\sin\left(-\frac{\pi}{2}\right)} = 1$

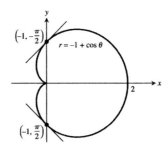

19. $\theta = \frac{\pi}{4} \Rightarrow r = 1 \Rightarrow \left(1, \frac{\pi}{4}\right)$; $\theta = -\frac{\pi}{4} \Rightarrow r = -1$ $\Rightarrow \left(-1, -\frac{\pi}{4}\right)$; $\theta = \frac{3\pi}{4} \Rightarrow r = -1 \Rightarrow \left(-1, \frac{3\pi}{4}\right)$; $\theta = -\frac{3\pi}{4} \Rightarrow r = 1 \Rightarrow \left(1, -\frac{3\pi}{4}\right)$; $r' = \frac{dr}{d\theta} = 2\cos 2\theta$; Slope $= \frac{r'\sin\theta + r\cos\theta}{r'\cos\theta - r\sin\theta} = \frac{2\cos 2\theta\sin\theta + r\cos\theta}{2\cos 2\theta\cos\theta - r\sin\theta}$ $\Rightarrow$ Slope at $\left(1, \frac{\pi}{4}\right)$ is $\frac{2\cos\left(\frac{\pi}{2}\right)\sin\left(\frac{\pi}{4}\right)+(1)\cos\left(\frac{\pi}{4}\right)}{2\cos\left(\frac{\pi}{2}\right)\cos\left(\frac{\pi}{4}\right)-(1)\sin\left(\frac{\pi}{4}\right)} = -1$;

Slope at $\left(-1, -\frac{\pi}{4}\right)$ is $\frac{2\cos\left(-\frac{\pi}{2}\right)\sin\left(-\frac{\pi}{4}\right)+(-1)\cos\left(-\frac{\pi}{4}\right)}{2\cos\left(-\frac{\pi}{2}\right)\cos\left(-\frac{\pi}{4}\right)-(-1)\sin\left(-\frac{\pi}{4}\right)} = 1$;

Slope at $\left(-1, \frac{3\pi}{4}\right)$ is $\frac{2\cos\left(\frac{3\pi}{2}\right)\sin\left(\frac{3\pi}{4}\right)+(-1)\cos\left(\frac{3\pi}{4}\right)}{2\cos\left(\frac{3\pi}{2}\right)\cos\left(\frac{3\pi}{4}\right)-(-1)\sin\left(\frac{3\pi}{4}\right)} = 1$;

Slope at $\left(1, -\frac{3\pi}{4}\right)$ is $\frac{2\cos\left(-\frac{3\pi}{2}\right)\sin\left(-\frac{3\pi}{4}\right)+(1)\cos\left(-\frac{3\pi}{4}\right)}{2\cos\left(-\frac{3\pi}{2}\right)\cos\left(-\frac{3\pi}{4}\right)-(1)\sin\left(-\frac{3\pi}{4}\right)} = -1$

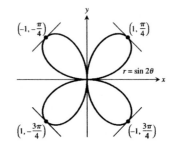

21. (a)

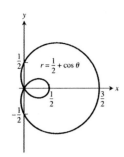

(b)

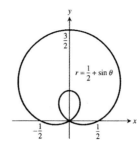

23. (a)

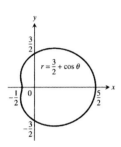

(b)

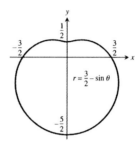

25.

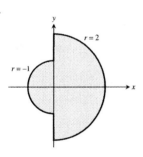

27.

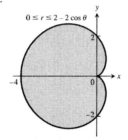

29. Note that (r, θ) and $(-r, \theta + \pi)$ describe the same point in the plane. Then $r = 1 - \cos \theta \iff -1 - \cos(\theta + \pi)$
$= -1 - (\cos \theta \cos \pi - \sin \theta \sin \pi) = -1 + \cos \theta = -(1 - \cos \theta) = -r$; therefore (r, θ) is on the graph of
$r = 1 - \cos \theta \iff (-r, \theta + \pi)$ is on the graph of $r = -1 - \cos \theta \implies$ the answer is (a).

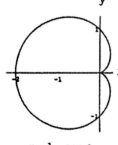

31.

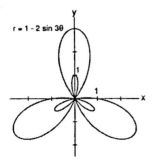

33. (a) 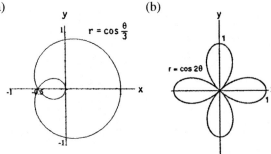   (b) 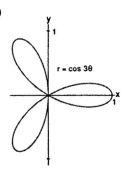   (c)   (d)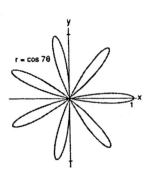

8.3 AREA AND LENGTHS IN POLAR COORDINATES

1. $A = \int_0^{2\pi} \frac{1}{2}(4 + 2\cos\theta)^2 \, d\theta = \int_0^{2\pi} \frac{1}{2}(16 + 16\cos\theta + 4\cos^2\theta) \, d\theta = \int_0^{2\pi} \left[8 + 8\cos\theta + 2\left(\frac{1+\cos 2\theta}{2}\right) \right] d\theta$

    $= \int_0^{2\pi} (9 + 8\cos\theta + \cos 2\theta) \, d\theta = \left[9\theta + 8\sin\theta + \frac{1}{2}\sin 2\theta \right]_0^{2\pi} = 18\pi$

3. $A = 2\int_0^{\pi/4} \frac{1}{2}\cos^2 2\theta \, d\theta = \int_0^{\pi/4} \frac{1+\cos 4\theta}{2} \, d\theta = \frac{1}{2}\left[\theta + \frac{\sin 4\theta}{4} \right]_0^{\pi/4} = \frac{\pi}{8}$

5. $A = \int_0^{\pi/2} \frac{1}{2}(4\sin 2\theta) \, d\theta = \int_0^{\pi/2} 2\sin 2\theta \, d\theta = \left[-\cos 2\theta \right]_0^{\pi/2} = 2$

7. $r = 2\cos\theta$ and $r = 2\sin\theta \Rightarrow 2\cos\theta = 2\sin\theta$

    $\Rightarrow \cos\theta = \sin\theta \Rightarrow \theta = \frac{\pi}{4}$; therefore

    $A = 2\int_0^{\pi/4} \frac{1}{2}(2\sin\theta)^2 \, d\theta = \int_0^{\pi/4} 4\sin^2\theta \, d\theta$

    $= \int_0^{\pi/4} 4\left(\frac{1-\cos 2\theta}{2}\right) d\theta = \int_0^{\pi/4} (2 - 2\cos 2\theta) \, d\theta$

    $= \left[2\theta - \sin 2\theta \right]_0^{\pi/4} = \frac{\pi}{2} - 1$

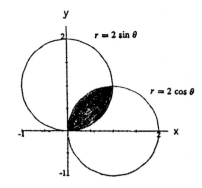

9. $r = 2$ and $r = 2(1 - \cos\theta) \Rightarrow 2 = 2(1 - \cos\theta)$

    $\Rightarrow \cos\theta = 0 \Rightarrow \theta = \pm\frac{\pi}{2}$; therefore

    $A = 2\int_0^{\pi/2} \frac{1}{2}[2(1 - \cos\theta)]^2 \, d\theta + \frac{1}{2}$ area of the circle

    $= \int_0^{\pi/2} 4\left(1 - 2\cos\theta + \cos^2\theta\right) d\theta + \left(\frac{1}{2}\pi\right)(2)^2$

    $= \int_0^{\pi/2} 4\left(1 - 2\cos\theta + \frac{1+\cos 2\theta}{2}\right) d\theta + 2\pi$

    $= \int_0^{\pi/2} (4 - 8\cos\theta + 2 + 2\cos 2\theta) \, d\theta + 2\pi$

    $= \left[6\theta - 8\sin\theta + \sin 2\theta \right]_0^{\pi/2} + 2\pi = 5\pi - 8$

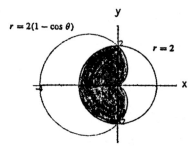

11. $r = \sqrt{3}$ and $r^2 = 6\cos 2\theta \Rightarrow 3 = 6\cos 2\theta \Rightarrow \cos 2\theta = \frac{1}{2}$

    $\Rightarrow \theta = \frac{\pi}{6}$ (in the 1st quadrant); we use symmetry of the

    graph to find the area, so

    $A = 4\int_0^{\pi/6} \left[\frac{1}{2}(6\cos 2\theta) - \frac{1}{2}\left(\sqrt{3}\right)^2 \right] d\theta$

    $= 2\int_0^{\pi/6} (6\cos 2\theta - 3) \, d\theta = 2\left[3\sin 2\theta - 3\theta \right]_0^{\pi/6}$

    $= 3\sqrt{3} - \pi$

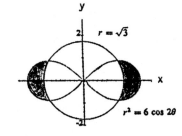

13. $r = 1$ and $r = -2 \cos \theta \Rightarrow 1 = -2 \cos \theta \Rightarrow \cos \theta = -\frac{1}{2}$

$\Rightarrow \theta = \frac{2\pi}{3}$ in quadrant II; therefore

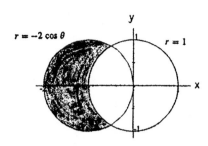

$A = 2 \int_{2\pi/3}^{\pi} \frac{1}{2} \left[(-2\cos\theta)^2 - 1^2 \right] d\theta = \int_{2\pi/3}^{\pi} (4\cos^2\theta - 1) \, d\theta$

$= \int_{2\pi/3}^{\pi} [2(1 + \cos 2\theta) - 1] \, d\theta = \int_{2\pi/3}^{\pi} (1 + 2\cos 2\theta) \, d\theta$

$= [\theta + \sin 2\theta]_{2\pi/3}^{\pi} = \frac{\pi}{3} + \frac{\sqrt{3}}{2}$

15. (a) $r = \tan\theta$ and $r = \left(\frac{\sqrt{2}}{2}\right) \csc\theta \Rightarrow \tan\theta = \left(\frac{\sqrt{2}}{2}\right)\csc\theta$

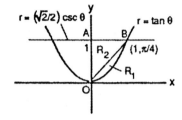

$\Rightarrow \sin^2\theta = \left(\frac{\sqrt{2}}{2}\right)\cos\theta \Rightarrow 1 - \cos^2\theta = \left(\frac{\sqrt{2}}{2}\right)\cos\theta$

$\Rightarrow \cos^2\theta + \left(\frac{\sqrt{2}}{2}\right)\cos\theta - 1 = 0 \Rightarrow \cos\theta = -\sqrt{2}$ or

$\frac{\sqrt{2}}{2}$ (use the quadratic formula) $\Rightarrow \theta = \frac{\pi}{4}$ (the solution

in the first quadrant); therefore the area of R_1 is

$A_1 = \int_0^{\pi/4} \frac{1}{2} \tan^2\theta \, d\theta = \frac{1}{2} \int_0^{\pi/4} (\sec^2\theta - 1) \, d\theta = \frac{1}{2} [\tan\theta - \theta]_0^{\pi/4} = \frac{1}{2} \left(\tan\frac{\pi}{4} - \frac{\pi}{4}\right) = \frac{1}{2} - \frac{\pi}{8}$;

$AO = \left(\frac{\sqrt{2}}{2}\right) \csc\frac{\pi}{2} = \frac{\sqrt{2}}{2}$ and $OB = \left(\frac{\sqrt{2}}{2}\right) \csc\frac{\pi}{4} = 1 \Rightarrow AB = \sqrt{1^2 - \left(\frac{\sqrt{2}}{2}\right)^2} = \frac{\sqrt{2}}{2}$

$\Rightarrow$ the area of R_2 is $A_2 = \frac{1}{2} \left(\frac{\sqrt{2}}{2}\right) \left(\frac{\sqrt{2}}{2}\right) = \frac{1}{4}$; therefore the area of the region shaded in the text is

$2 \left(\frac{1}{2} - \frac{\pi}{8} + \frac{1}{4}\right) = \frac{3}{2} - \frac{\pi}{4}$. Note: The area must be found this way since no common interval generates the region. For

example, the interval $0 \le \theta \le \frac{\pi}{4}$ generates the arc OB of $r = \tan\theta$ but does not generate the segment AB of the line

$r = \frac{\sqrt{2}}{2} \csc\theta$. Instead the interval generates the half-line from B to $+\infty$ on the line $r = \frac{\sqrt{2}}{2} \csc\theta$.

(b) $\lim\limits_{\theta \to \pi/2^-} \tan\theta = \infty$ and the line $x = 1$ is $r = \sec\theta$ in polar coordinates; then $\lim\limits_{\theta \to \pi/2^-} (\tan\theta - \sec\theta)$

$= \lim\limits_{\theta \to \pi/2^-} \left(\frac{\sin\theta}{\cos\theta} - \frac{1}{\cos\theta}\right) = \lim\limits_{\theta \to \pi/2^-} \left(\frac{\sin\theta - 1}{\cos\theta}\right) = \lim\limits_{\theta \to \pi/2^-} \left(\frac{\cos\theta}{-\sin\theta}\right) = 0 \Rightarrow r = \tan\theta$ approaches

$r = \sec\theta$ as $\theta \to \frac{\pi}{2} \Rightarrow r = \sec\theta$ (or $x = 1$) is a vertical asymptote of $r = \tan\theta$. Similarly, $r = -\sec\theta$

(or $x = -1$) is a vertical asymptote of $r = \tan\theta$.

17. $r = \theta^2, 0 \le \theta \le \sqrt{5} \Rightarrow \frac{dr}{d\theta} = 2\theta$; therefore Length $= \int_0^{\sqrt{5}} \sqrt{(\theta^2)^2 + (2\theta)^2} \, d\theta = \int_0^{\sqrt{5}} \sqrt{\theta^4 + 4\theta^2} \, d\theta$

$= \int_0^{\sqrt{5}} |\theta| \sqrt{\theta^2 + 4} \, d\theta = \text{(since } \theta \ge 0\text{)} \int_0^{\sqrt{5}} \theta \sqrt{\theta^2 + 4} \, d\theta; \left[u = \theta^2 + 4 \Rightarrow \frac{1}{2} \, du = \theta \, d\theta; \theta = 0 \Rightarrow u = 4,\right.$

$\left. \theta = \sqrt{5} \Rightarrow u = 9 \right] \to \int_4^9 \frac{1}{2} \sqrt{u} \, du = \frac{1}{2} \left[\frac{2}{3} u^{3/2}\right]_4^9 = \frac{19}{3}$

19. $r = 1 + \cos\theta \Rightarrow \frac{dr}{d\theta} = -\sin\theta$; therefore Length $= \int_0^{2\pi} \sqrt{(1 + \cos\theta)^2 + (-\sin\theta)^2} \, d\theta$

$= 2 \int_0^{\pi} \sqrt{2 + 2\cos\theta} \, d\theta = 2 \int_0^{\pi} \sqrt{\frac{4(1 + \cos\theta)}{2}} \, d\theta = 4 \int_0^{\pi} \sqrt{\frac{1 + \cos\theta}{2}} \, d\theta = 4 \int_0^{\pi} \cos\left(\frac{\theta}{2}\right) d\theta = 4 \left[2 \sin\frac{\theta}{2}\right]_0^{\pi} = 8$

21. $r = \frac{6}{1 + \cos\theta}, 0 \le \theta \le \frac{\pi}{2} \Rightarrow \frac{dr}{d\theta} = \frac{6\sin\theta}{(1 + \cos\theta)^2}$; therefore Length $= \int_0^{\pi/2} \sqrt{\left(\frac{6}{1 + \cos\theta}\right)^2 + \left(\frac{6\sin\theta}{(1 + \cos\theta)^2}\right)^2} \, d\theta$

$= \int_0^{\pi/2} \sqrt{\frac{36}{(1 + \cos\theta)^2} + \frac{36\sin^2\theta}{(1 + \cos\theta)^4}} \, d\theta = 6 \int_0^{\pi/2} \left|\frac{1}{1 + \cos\theta}\right| \sqrt{1 + \frac{\sin^2\theta}{(1 + \cos\theta)^2}} \, d\theta$

$= \left(\text{since } \frac{1}{1 + \cos\theta} > 0 \text{ on } 0 \le \theta \le \frac{\pi}{2}\right) 6 \int_0^{\pi/2} \left(\frac{1}{1 + \cos\theta}\right) \sqrt{\frac{1 + 2\cos\theta + \cos^2\theta + \sin^2\theta}{(1 + \cos\theta)^2}} \, d\theta$

$= 6 \int_0^{\pi/2} \left(\frac{1}{1 + \cos\theta}\right) \sqrt{\frac{2 + 2\cos\theta}{(1 + \cos\theta)^2}} \, d\theta = 6\sqrt{2} \int_0^{\pi/2} \frac{d\theta}{(1 + \cos\theta)^{3/2}} = 6\sqrt{2} \int_0^{\pi/2} \frac{d\theta}{\left(2\cos^2\frac{\theta}{2}\right)^{3/2}} = 3 \int_0^{\pi/2} \left|\sec^3\frac{\theta}{2}\right| d\theta$

$$= 3 \int_0^{\pi/2} \sec^3 \tfrac{\theta}{2} \, d\theta = 6 \int_0^{\pi/4} \sec^3 u \, du = \text{(use tables)} \; 6 \left(\left[\tfrac{\sec u \tan u}{2} \right]_0^{\pi/4} + \tfrac{1}{2} \int_0^{\pi/4} \sec u \, du \right)$$

$$= 6 \left(\tfrac{1}{\sqrt{2}} + \left[\tfrac{1}{2} \ln | \sec u + \tan u | \right]_0^{\pi/4} \right) = 3 \left[\sqrt{2} + \ln \left(1 + \sqrt{2} \right) \right]$$

23. $r = \cos^3 \tfrac{\theta}{3} \Rightarrow \tfrac{dr}{d\theta} = - \sin \tfrac{\theta}{3} \cos^2 \tfrac{\theta}{3}$; therefore Length $= \int_0^{\pi/4} \sqrt{\left(\cos^3 \tfrac{\theta}{3} \right)^2 + \left(- \sin \tfrac{\theta}{3} \cos^2 \tfrac{\theta}{3} \right)^2} \, d\theta$

$$= \int_0^{\pi/4} \sqrt{\cos^6 \left(\tfrac{\theta}{3} \right) + \sin^2 \left(\tfrac{\theta}{3} \right) \cos^4 \left(\tfrac{\theta}{3} \right)} \, d\theta = \int_0^{\pi/4} \left(\cos^2 \tfrac{\theta}{3} \right) \sqrt{\cos^2 \left(\tfrac{\theta}{3} \right) + \sin^2 \left(\tfrac{\theta}{3} \right)} \, d\theta = \int_0^{\pi/4} \cos^2 \left(\tfrac{\theta}{3} \right) \, d\theta$$

$$= \int_0^{\pi/4} \tfrac{1 + \cos \left(\tfrac{2\theta}{3} \right)}{2} \, d\theta = \tfrac{1}{2} \left[\theta + \tfrac{3}{2} \sin \tfrac{2\theta}{3} \right]_0^{\pi/4} = \tfrac{\pi}{8} + \tfrac{3}{8}$$

25. $r = \sqrt{1 + \cos 2\theta} \Rightarrow \tfrac{dr}{d\theta} = \tfrac{1}{2} (1 + \cos 2\theta)^{-1/2} (-2 \sin 2\theta)$; therefore Length $= \int_0^{\pi\sqrt{2}} \sqrt{(1 + \cos 2\theta) + \tfrac{\sin^2 2\theta}{(1 + \cos 2\theta)}} \, d\theta$

$$= \int_0^{\pi\sqrt{2}} \sqrt{\tfrac{1 + 2 \cos 2\theta + \cos^2 2\theta + \sin^2 2\theta}{1 + \cos 2\theta}} \, d\theta = \int_0^{\pi\sqrt{2}} \sqrt{\tfrac{2 + 2 \cos 2\theta}{1 + \cos 2\theta}} \, d\theta = \int_0^{\pi\sqrt{2}} \sqrt{2} \, d\theta = \left[\sqrt{2} \theta \right]_0^{\pi\sqrt{2}} = 2\pi$$

27. Let $r = f(\theta)$. Then $x = f(\theta) \cos \theta \Rightarrow \tfrac{dx}{d\theta} = f'(\theta) \cos \theta - f(\theta) \sin \theta \Rightarrow \left(\tfrac{dx}{d\theta} \right)^2 = [f'(\theta) \cos \theta - f(\theta) \sin \theta]^2$

$= [f'(\theta)]^2 \cos^2 \theta - 2 f'(\theta) f(\theta) \sin \theta \cos \theta + [f(\theta)]^2 \sin^2 \theta$; $y = f(\theta) \sin \theta \Rightarrow \tfrac{dy}{d\theta} = f'(\theta) \sin \theta + f(\theta) \cos \theta$

$\Rightarrow \left(\tfrac{dy}{d\theta} \right)^2 = [f'(\theta) \sin \theta + f(\theta) \cos \theta]^2 = [f'(\theta)]^2 \sin^2 \theta + 2 f'(\theta) f(\theta) \sin \theta \cos \theta + [f(\theta)]^2 \cos^2 \theta$. Therefore

$\left(\tfrac{dx}{d\theta} \right)^2 + \left(\tfrac{dy}{d\theta} \right)^2 = [f'(\theta)]^2 (\cos^2 \theta + \sin^2 \theta) + [f(\theta)]^2 (\cos^2 \theta + \sin^2 \theta) = [f'(\theta)]^2 + [f(\theta)]^2 = r^2 + \left(\tfrac{dr}{d\theta} \right)^2$.

Thus, $L = \int_\alpha^\beta \sqrt{\left(\tfrac{dx}{d\theta} \right)^2 + \left(\tfrac{dy}{d\theta} \right)^2} \, d\theta = \int_\alpha^\beta \sqrt{r^2 + \left(\tfrac{dr}{d\theta} \right)^2} \, d\theta$.

29. $r = 2f(\theta)$, $\alpha \le \theta \le \beta \Rightarrow \tfrac{dr}{d\theta} = 2f'(\theta) \Rightarrow r^2 + \left(\tfrac{dr}{d\theta} \right)^2 = [2f(\theta)]^2 + [2f'(\theta)]^2 \Rightarrow$ Length $= \int_\alpha^\beta \sqrt{4[f(\theta)]^2 + 4[f'(\theta)]^2} \, d\theta$

$= 2 \int_\alpha^\beta \sqrt{[f(\theta)]^2 + [f'(\theta)]^2} \, d\theta$ which is twice the length of the curve $r = f(\theta)$ for $\alpha \le \theta \le \beta$.

8.4 CONICS IN POLAR COORDINATES

1. $16x^2 + 25y^2 = 400 \Rightarrow \tfrac{x^2}{25} + \tfrac{y^2}{16} = 1 \Rightarrow c = \sqrt{a^2 - b^2}$

$= \sqrt{25 - 16} = 3 \Rightarrow e = \tfrac{c}{a} = \tfrac{3}{5}$; $F(\pm 3, 0)$;

directrices are $x = 0 \pm \tfrac{a}{e} = \pm \tfrac{5}{\left(\tfrac{3}{5} \right)} = \pm \tfrac{25}{3}$

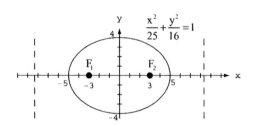

3. $2x^2 + y^2 = 2 \Rightarrow x^2 + \tfrac{y^2}{2} = 1 \Rightarrow c = \sqrt{a^2 - b^2}$

$= \sqrt{2 - 1} = 1 \Rightarrow e = \tfrac{c}{a} = \tfrac{1}{\sqrt{2}}$; $F(0, \pm 1)$;

directrices are $y = 0 \pm \tfrac{a}{e} = \pm \tfrac{\sqrt{2}}{\left(\tfrac{1}{\sqrt{2}} \right)} = \pm 2$

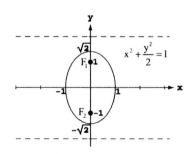

5. $3x^2 + 2y^2 = 6 \Rightarrow \frac{x^2}{2} + \frac{y^2}{3} = 1 \Rightarrow c = \sqrt{a^2 - b^2}$
    $= \sqrt{3-2} = 1 \Rightarrow e = \frac{c}{a} = \frac{1}{\sqrt{3}}$; $F(0, \pm 1)$;
    directrices are $y = 0 \pm \frac{a}{e} = \pm \frac{\sqrt{3}}{\left(\frac{1}{\sqrt{3}}\right)} = \pm 3$

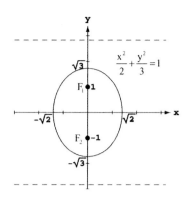

7. $6x^2 + 9y^2 = 54 \Rightarrow \frac{x^2}{9} + \frac{y^2}{6} = 1 \Rightarrow c = \sqrt{a^2 - b^2}$
    $= \sqrt{9-6} = \sqrt{3} \Rightarrow e = \frac{c}{a} = \frac{\sqrt{3}}{3}$; $F\left(\pm\sqrt{3}, 0\right)$;
    directrices are $x = 0 \pm \frac{a}{e} = \pm \frac{3}{\left(\frac{\sqrt{3}}{3}\right)} = \pm 3\sqrt{3}$

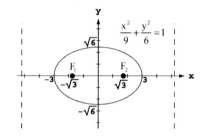

9. Foci: $(0, \pm 3)$, $e = 0.5 \Rightarrow c = 3$ and $a = \frac{c}{e} = \frac{3}{0.5} = 6 \Rightarrow b^2 = 36 - 9 = 27 \Rightarrow \frac{x^2}{27} + \frac{y^2}{36} = 1$

11. Vertices: $(0, \pm 70)$, $e = 0.1 \Rightarrow a = 70$ and $c = ae = 70(0.1) = 7 \Rightarrow b^2 = 4900 - 49 = 4851 \Rightarrow \frac{x^2}{4851} + \frac{y^2}{4900} = 1$

13. Focus: $\left(\sqrt{5}, 0\right)$, Directrix: $x = \frac{9}{\sqrt{5}} \Rightarrow c = ae = \sqrt{5}$ and $\frac{a}{e} = \frac{9}{\sqrt{5}} \Rightarrow \frac{ae}{e^2} = \frac{9}{\sqrt{5}} \Rightarrow \frac{\sqrt{5}}{e^2} = \frac{9}{\sqrt{5}} \Rightarrow e^2 = \frac{5}{9}$
    $\Rightarrow e = \frac{\sqrt{5}}{3}$. Then $PF = \frac{\sqrt{5}}{3} PD \Rightarrow \sqrt{\left(x - \sqrt{5}\right)^2 + (y-0)^2} = \frac{\sqrt{5}}{3}\left|x - \frac{9}{\sqrt{5}}\right| \Rightarrow \left(x - \sqrt{5}\right)^2 + y^2 = \frac{5}{9}\left(x - \frac{9}{\sqrt{5}}\right)^2$
    $\Rightarrow x^2 - 2\sqrt{5}x + 5 + y^2 = \frac{5}{9}\left(x^2 - \frac{18}{\sqrt{5}}x + \frac{81}{5}\right) \Rightarrow \frac{4}{9}x^2 + y^2 = 4 \Rightarrow \frac{x^2}{9} + \frac{y^2}{4} = 1$

15. Focus: $(-4, 0)$, Directrix: $x = -16 \Rightarrow c = ae = 4$ and $\frac{a}{e} = 16 \Rightarrow \frac{ae}{e^2} = 16 \Rightarrow \frac{4}{e^2} = 16 \Rightarrow e^2 = \frac{1}{4} \Rightarrow e = \frac{1}{2}$. Then
    $PF = \frac{1}{2}PD \Rightarrow \sqrt{(x+4)^2 + (y-0)^2} = \frac{1}{2}|x + 16| \Rightarrow (x+4)^2 + y^2 = \frac{1}{4}(x+16)^2 \Rightarrow x^2 + 8x + 16 + y^2$
    $= \frac{1}{4}(x^2 + 32x + 256) \Rightarrow \frac{3}{4}x^2 + y^2 = 48 \Rightarrow \frac{x^2}{64} + \frac{y^2}{48} = 1$

17. $x^2 - y^2 = 1 \Rightarrow c = \sqrt{a^2 + b^2} = \sqrt{1+1} = \sqrt{2} \Rightarrow e = \frac{c}{a}$
    $= \frac{\sqrt{2}}{1} = \sqrt{2}$; asymptotes are $y = \pm x$; $F\left(\pm\sqrt{2}, 0\right)$;
    directrices are $x = 0 \pm \frac{a}{e} = \pm \frac{1}{\sqrt{2}}$

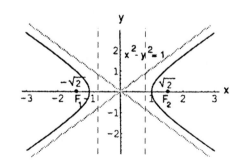

19. $y^2 - x^2 = 8 \Rightarrow \frac{y^2}{8} - \frac{x^2}{8} = 1 \Rightarrow c = \sqrt{a^2 + b^2}$
    $= \sqrt{8 + 8} = 4 \Rightarrow e = \frac{c}{a} = \frac{4}{\sqrt{8}} = \sqrt{2}$; asymptotes are
    $y = \pm x$; $F(0, \pm 4)$; directrices are $y = 0 \pm \frac{a}{e}$
    $= \pm \frac{\sqrt{8}}{\sqrt{2}} = \pm 2$

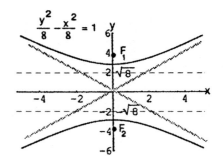

21. $8x^2 - 2y^2 = 16 \Rightarrow \frac{x^2}{2} - \frac{y^2}{8} = 1 \Rightarrow c = \sqrt{a^2 + b^2}$
    $= \sqrt{2 + 8} = \sqrt{10} \Rightarrow e = \frac{c}{a} = \frac{\sqrt{10}}{\sqrt{2}} = \sqrt{5}$; asymptotes
    are $y = \pm 2x$; $F\left(\pm\sqrt{10}, 0\right)$; directrices are $x = 0 \pm \frac{a}{e}$
    $= \pm \frac{\sqrt{2}}{\sqrt{5}} = \pm \frac{2}{\sqrt{10}}$

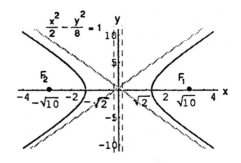

23. $8y^2 - 2x^2 = 16 \Rightarrow \frac{y^2}{2} - \frac{x^2}{8} = 1 \Rightarrow c = \sqrt{a^2 + b^2}$
    $= \sqrt{2 + 8} = \sqrt{10} \Rightarrow e = \frac{c}{a} = \frac{\sqrt{10}}{\sqrt{2}} = \sqrt{5}$; asymptotes
    are $y = \pm \frac{x}{2}$; $F\left(0, \pm\sqrt{10}\right)$; directrices are $y = 0 \pm \frac{a}{e}$
    $= \pm \frac{\sqrt{2}}{\sqrt{5}} = \pm \frac{2}{\sqrt{10}}$

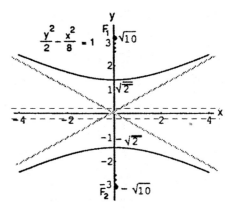

25. Vertices $(0, \pm 1)$ and $e = 3 \Rightarrow a = 1$ and $e = \frac{c}{a} = 3 \Rightarrow c = 3a = 3 \Rightarrow b^2 = c^2 - a^2 = 9 - 1 = 8 \Rightarrow y^2 - \frac{x^2}{8} = 1$

27. Foci $(\pm 3, 0)$ and $e = 3 \Rightarrow c = 3$ and $e = \frac{c}{a} = 3 \Rightarrow c = 3a \Rightarrow a = 1 \Rightarrow b^2 = c^2 - a^2 = 9 - 1 = 8 \Rightarrow x^2 - \frac{y^2}{8} = 1$

29. $e = 1, x = 2 \Rightarrow k = 2 \Rightarrow r = \frac{2(1)}{1 + (1)\cos\theta} = \frac{2}{1 + \cos\theta}$

31. $e = 5, y = -6 \Rightarrow k = 6 \Rightarrow r = \frac{6(5)}{1 - 5\sin\theta} = \frac{30}{1 - 5\sin\theta}$

33. $e = \frac{1}{2}, x = 1 \Rightarrow k = 1 \Rightarrow r = \frac{\left(\frac{1}{2}\right)(1)}{1 + \left(\frac{1}{2}\right)\cos\theta} = \frac{1}{2 + \cos\theta}$

35. $e = \frac{1}{5}, x = -10 \Rightarrow k = 10 \Rightarrow r = \frac{\left(\frac{1}{5}\right)(10)}{1 - \left(\frac{1}{5}\right)\sin\theta} = \frac{10}{5 - \sin\theta}$

37. $r = \frac{1}{1 + \cos\theta} \Rightarrow e = 1, k = 1 \Rightarrow x = 1$

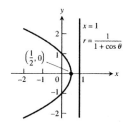

39. $r = \frac{25}{10 - 5\cos\theta} \Rightarrow r = \frac{\left(\frac{25}{10}\right)}{1 - \left(\frac{5}{10}\right)\cos\theta} = \frac{\left(\frac{5}{2}\right)}{1 - \left(\frac{1}{2}\right)\cos\theta}$

$\Rightarrow e = \frac{1}{2}, k = 5 \Rightarrow x = -5; a(1 - e^2) = ke$

$\Rightarrow a\left[1 - \left(\frac{1}{2}\right)^2\right] = \frac{5}{2} \Rightarrow \frac{3}{4}a = \frac{5}{2} \Rightarrow a = \frac{10}{3} \Rightarrow ea = \frac{5}{3}$

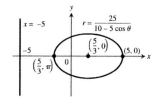

41. $r = \frac{400}{16 + 8\sin\theta} \Rightarrow r = \frac{\left(\frac{400}{16}\right)}{1 + \left(\frac{8}{16}\right)\sin\theta} \Rightarrow r = \frac{25}{1 + \left(\frac{1}{2}\right)\sin\theta}$

$e = \frac{1}{2}, k = 50 \Rightarrow y = 50; a(1 - e^2) = ke$

$\Rightarrow a\left[1 - \left(\frac{1}{2}\right)^2\right] = 25 \Rightarrow \frac{3}{4}a = 25 \Rightarrow a = \frac{100}{3}$

$\Rightarrow ea = \frac{50}{3}$

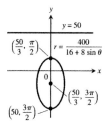

43. $r = \frac{8}{2 - 2\sin\theta} \Rightarrow r = \frac{4}{1 - \sin\theta} \Rightarrow e = 1,$

$k = 4 \Rightarrow y = -4$

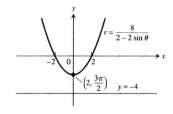

45. $r\cos\left(\theta - \frac{\pi}{4}\right) = \sqrt{2} \Rightarrow r\left(\cos\theta\cos\frac{\pi}{4} + \sin\theta\sin\frac{\pi}{4}\right)$

$= \sqrt{2} \Rightarrow \frac{1}{\sqrt{2}}r\cos\theta + \frac{1}{\sqrt{2}}r\sin\theta = \sqrt{2} \Rightarrow \frac{1}{\sqrt{2}}x + \frac{1}{\sqrt{2}}y$

$= \sqrt{2} \Rightarrow x + y = 2 \Rightarrow y = 2 - x$

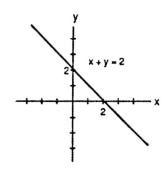

47. $r\cos\left(\theta - \frac{2\pi}{3}\right) = 3 \Rightarrow r\left(\cos\theta\cos\frac{2\pi}{3} + \sin\theta\sin\frac{2\pi}{3}\right) = 3$

$\Rightarrow -\frac{1}{2}r\cos\theta + \frac{\sqrt{3}}{2}r\sin\theta = 3 \Rightarrow -\frac{1}{2}x + \frac{\sqrt{3}}{2}y = 3$

$\Rightarrow -x + \sqrt{3}y = 6 \Rightarrow y = \frac{\sqrt{3}}{3}x + 2\sqrt{3}$

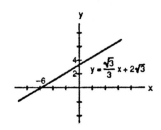

49. $\sqrt{2}x + \sqrt{2}y = 6 \Rightarrow \sqrt{2}r\cos\theta + \sqrt{2}r\sin\theta = 6 \Rightarrow r\left(\frac{\sqrt{2}}{2}\cos\theta + \frac{\sqrt{2}}{2}\sin\theta\right) = 3 \Rightarrow r\left(\cos\frac{\pi}{4}\cos\theta + \sin\frac{\pi}{4}\sin\theta\right)$

$= 3 \Rightarrow r\cos\left(\theta - \frac{\pi}{4}\right) = 3$

51. $y = -5 \Rightarrow r\sin\theta = -5 \Rightarrow -r\sin\theta = 5 \Rightarrow r\sin(-\theta) = 5 \Rightarrow r\cos\left(\frac{\pi}{2} - (-\theta)\right) = 5 \Rightarrow r\cos\left(\theta + \frac{\pi}{2}\right) = 5$

53.

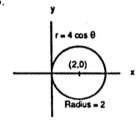

55.

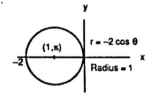

57. $(x - 6)^2 + y^2 = 36 \Rightarrow C = (6,0), a = 6$

$\Rightarrow r = 12\cos\theta$ is the polar equation

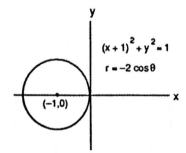

59. $x^2 + (y - 5)^2 = 25 \Rightarrow C = (0,5), a = 5$

$\Rightarrow r = 10\sin\theta$ is the polar equation

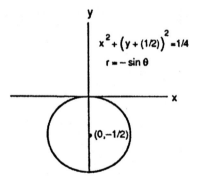

61. $x^2 + 2x + y^2 = 0 \Rightarrow (x + 1)^2 + y^2 = 1$

$\Rightarrow C = (-1,0), a = 1 \Rightarrow r = -2\cos\theta$ is the polar equation

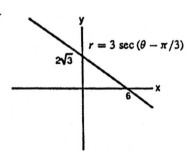

63. $x^2 + y^2 + y = 0 \Rightarrow x^2 + \left(y + \frac{1}{2}\right)^2 = \frac{1}{4}$

$\Rightarrow C = \left(0, -\frac{1}{2}\right), a = \frac{1}{2} \Rightarrow r = -\sin\theta$ is the polar equation

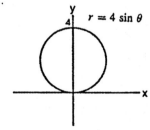

65.

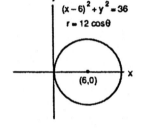

67.

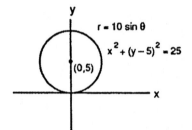

69.

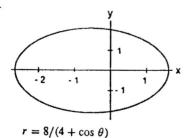

$$r = 8/(4 + \cos \theta)$$

71.

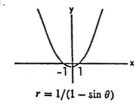

$$r = 1/(1 - \sin \theta)$$

73.

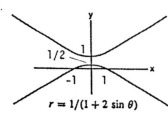

$$r = 1/(1 + 2 \sin \theta)$$

75. (a) Perihelion $= a - ae = a(1 - e)$, Aphelion $= ea + a = a(1 + e)$

    (b)

Planet	Perihelion	Aphelion
Mercury	0.3075 AU	0.4667 AU
Venus	0.7184 AU	0.7282 AU
Earth	0.9833 AU	1.0167 AU
Mars	1.3817 AU	1.6663 AU
Jupiter	4.9512 AU	5.4548 AU
Saturn	9.0210 AU	10.0570 AU
Uranus	18.2977 AU	20.0623 AU
Neptune	29.8135 AU	30.3065 AU
Pluto	29.6549 AU	49.2251 AU

77. $x^2 + y^2 - 2ay = 0 \Rightarrow (r \cos \theta)^2 + (r \sin \theta)^2 - 2ar \sin \theta = 0$
    $\Rightarrow r^2 \cos^2 \theta + r^2 \sin^2 \theta - 2ar \sin \theta = 0 \Rightarrow r^2 = 2ar \sin \theta$
    $\Rightarrow r = 2a \sin \theta$

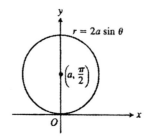

79. $x \cos \alpha + y \sin \alpha = p \Rightarrow r \cos \theta \cos \alpha + r \sin \theta \sin \alpha = p$
    $\Rightarrow r(\cos \theta \cos \alpha + \sin \theta \sin \alpha) = p \Rightarrow r \cos (\theta - \alpha) = p$

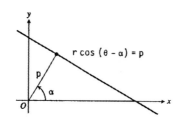

8.5 CONICS AND PARAMETRIC EQUATIONS; THE CYCLOID

1. $x = \cos t, y = \sin t, 0 \le t \le \pi$

    $\Rightarrow \cos^2 t + \sin^2 t = 1 \Rightarrow x^2 + y^2 = 1$

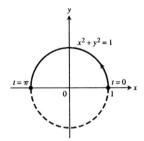

3. $x = 4 \cos t, y = 5 \sin t, 0 \le t \le \pi$

    $\Rightarrow \frac{16 \cos^2 t}{16} + \frac{25 \sin^2 t}{25} = 1 \Rightarrow \frac{x^2}{16} + \frac{y^2}{25} = 1$

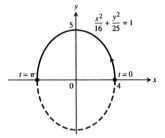

5. $x = t, y = \sqrt{t}, t \ge 0 \Rightarrow y = \sqrt{x}$

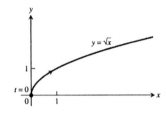

7. $x = -\sec t, y = \tan t, -\frac{\pi}{2} < t < \frac{\pi}{2}$

    $\Rightarrow \sec^2 t - \tan^2 t = 1 \Rightarrow x^2 - y^2 = 1$

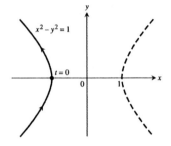

9. $x = t, y = \sqrt{4 - t^2}, 0 \le t \le 2$

    $\Rightarrow y = \sqrt{4 - x^2}$

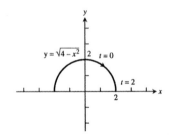

11. $x = -\cosh t, y = \sinh t, -\infty < 1 < \infty$

    $\Rightarrow \cosh^2 t - \sinh^2 t = 1 \Rightarrow x^2 - y^2 = 1$

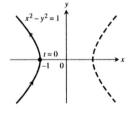

13. Arc PF = Arc AF since each is the distance rolled and

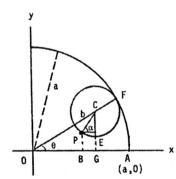

   $\frac{\text{Arc PF}}{b} = \angle FCP \Rightarrow \text{Arc PF} = b(\angle FCP); \frac{\text{Arc AF}}{a} = \theta$

   $\Rightarrow \text{Arc AF} = a\theta \Rightarrow a\theta = b(\angle FCP) \Rightarrow \angle FCP = \frac{a}{b}\theta;$

   $\angle OCG = \frac{\pi}{2} - \theta; \angle OCG = \angle OCP + \angle PCE$

   $= \angle OCP + \left(\frac{\pi}{2} - \alpha\right).$ Now $\angle OCP = \pi - \angle FCP$

   $= \pi - \frac{a}{b}\theta.$ Thus $\angle OCG = \pi - \frac{a}{b}\theta + \frac{\pi}{2} - \alpha \Rightarrow \frac{\pi}{2} - \theta$

   $= \pi - \frac{a}{b}\theta + \frac{\pi}{2} - \alpha \Rightarrow \alpha = \pi - \frac{a}{b}\theta + \theta = \pi - \left(\frac{a-b}{b}\theta\right).$

   Then $x = OG - BG = OG - PE = (a - b) \cos \theta - b \cos \alpha = (a - b) \cos \theta - b \cos \left(\pi - \frac{a-b}{b}\theta\right)$

   $= (a - b) \cos \theta + b \cos \left(\frac{a-b}{b}\theta\right).$ Also $y = EG = CG - CE = (a - b) \sin \theta - b \sin \alpha$

   $= (a - b) \sin \theta - b \sin \left(\pi - \frac{a-b}{b}\theta\right) = (a - b) \sin \theta - b \sin \left(\frac{a-b}{b}\theta\right).$ Therefore

   $x = (a - b) \cos \theta + b \cos \left(\frac{a-b}{b}\theta\right)$ and $y = (a - b) \sin \theta - b \sin \left(\frac{a-b}{b}\theta\right).$

If $b = \frac{a}{4}$, then $x = \left(a - \frac{a}{4}\right)\cos\theta + \frac{a}{4}\cos\left(\frac{a - \left(\frac{a}{4}\right)}{\left(\frac{a}{4}\right)}\theta\right)$

$= \frac{3a}{4}\cos\theta + \frac{a}{4}\cos 3\theta = \frac{3a}{4}\cos\theta + \frac{a}{4}(\cos\theta\cos 2\theta - \sin\theta\sin 2\theta)$

$= \frac{3a}{4}\cos\theta + \frac{a}{4}((\cos\theta)(\cos^2\theta - \sin^2\theta) - (\sin\theta)(2\sin\theta\cos\theta))$

$= \frac{3a}{4}\cos\theta + \frac{a}{4}\cos^3\theta - \frac{a}{4}\cos\theta\sin^2\theta - \frac{2a}{4}\sin^2\theta\cos\theta$

$= \frac{3a}{4}\cos\theta + \frac{a}{4}\cos^3\theta - \frac{3a}{4}(\cos\theta)(1 - \cos^2\theta) = a\cos^3\theta;$

$y = \left(a - \frac{a}{4}\right)\sin\theta - \frac{a}{4}\sin\left(\frac{a - \left(\frac{a}{4}\right)}{\left(\frac{a}{4}\right)}\theta\right) = \frac{3a}{4}\sin\theta - \frac{a}{4}\sin 3\theta = \frac{3a}{4}\sin\theta - \frac{a}{4}(\sin\theta\cos 2\theta + \cos\theta\sin 2\theta)$

$= \frac{3a}{4}\sin\theta - \frac{a}{4}((\sin\theta)(\cos^2\theta - \sin^2\theta) + (\cos\theta)(2\sin\theta\cos\theta))$

$= \frac{3a}{4}\sin\theta - \frac{a}{4}\sin\theta\cos^2\theta + \frac{a}{4}\sin^3\theta - \frac{2a}{4}\cos^2\theta\sin\theta$

$= \frac{3a}{4}\sin\theta - \frac{3a}{4}\sin\theta\cos^2\theta + \frac{a}{4}\sin^3\theta$

$= \frac{3a}{4}\sin\theta - \frac{3a}{4}(\sin\theta)(1 - \sin^2\theta) + \frac{a}{4}\sin^3\theta = a\sin^3\theta.$

15. Draw line AM in the figure and note that $\angle AMO$ is a right angle since it is an inscribed angle which spans the diameter of a circle. Then $AN^2 = MN^2 + AM^2$. Now, $OA = a$, $\frac{AN}{a} = \tan t$, and $\frac{AM}{a} = \sin t$. Next $MN = OP$

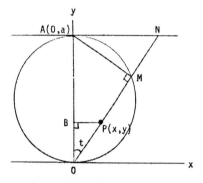

$\Rightarrow OP^2 = AN^2 - AM^2 = a^2\tan^2 t - a^2\sin^2 t$

$\Rightarrow OP = \sqrt{a^2\tan^2 t - a^2\sin^2 t}$

$= (a\sin t)\sqrt{\sec^2 t - 1} = \frac{a\sin^2 t}{\cos t}$. In triangle BPO,

$x = OP\sin t = \frac{a\sin^3 t}{\cos t} = a\sin^2 t\tan t$ and

$y = OP\cos t = a\sin^2 t \Rightarrow x = a\sin^2 t\tan t$ and $y = a\sin^2 t$.

17. $D = \sqrt{(x - 2)^2 + \left(y - \frac{1}{2}\right)^2} \Rightarrow D^2 = (x - 2)^2 + \left(y - \frac{1}{2}\right)^2 = (t - 2)^2 + \left(t^2 - \frac{1}{2}\right)^2 \Rightarrow D^2 = t^4 - 4t + \frac{17}{4}$

$\Rightarrow \frac{d(D^2)}{dt} = 4t^3 - 4 = 0 \Rightarrow t = 1$. The second derivative is always positive for $t \neq 0 \Rightarrow t = 1$ gives a local minimum for D^2 (and hence D) which is an absolute minimum since it is the only extremum $\Rightarrow$ the closest point on the parabola is $(1, 1)$.

19. (a)

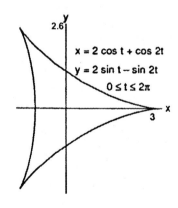

(b)

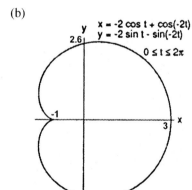

21.

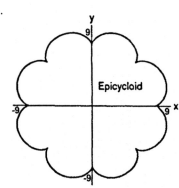

23.

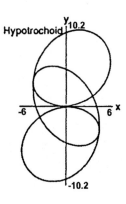

CHAPTER 8 PRACTICE AND ADDITIONAL EXERCISES

1. $r \cos\left(\theta + \frac{\pi}{3}\right) = 2\sqrt{3} \Rightarrow r\left(\cos\theta \cos\frac{\pi}{3} - \sin\theta \sin\frac{\pi}{3}\right)$

    $= 2\sqrt{3} \Rightarrow \frac{1}{2} r \cos\theta - \frac{\sqrt{3}}{2} r \sin\theta = 2\sqrt{3}$

    $\Rightarrow r \cos\theta - \sqrt{3} r \sin\theta = 4\sqrt{3} \Rightarrow x - \sqrt{3} y = 4\sqrt{3}$

    $\Rightarrow y = \frac{\sqrt{3}}{3} x - 4$

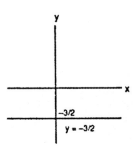

3. $r = 2\sec\theta \Rightarrow r = \frac{2}{\cos\theta} \Rightarrow r \cos\theta = 2 \Rightarrow x = 2$

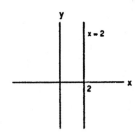

5. $r = -\frac{3}{2}\csc\theta \Rightarrow r \sin\theta = -\frac{3}{2} \Rightarrow y = -\frac{3}{2}$

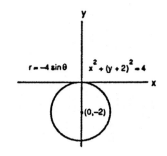

7. $r = -4\sin\theta \Rightarrow r^2 = -4r\sin\theta \Rightarrow x^2 + y^2 + 4y = 0$
    $\Rightarrow x^2 + (y+2)^2 = 4$; circle with center $(0, -2)$ and
    radius 2.

9. $r = 2\sqrt{2} \cos \theta \Rightarrow r^2 = 2\sqrt{2}\, r \cos \theta$

$\Rightarrow x^2 + y^2 - 2\sqrt{2}\, x = 0 \Rightarrow \left(x - \sqrt{2}\right)^2 + y^2 = 2;$

circle with center $\left(\sqrt{2}, 0\right)$ and radius $\sqrt{2}$

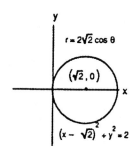

11. $x^2 + y^2 + 5y = 0 \Rightarrow x^2 + \left(y + \frac{5}{2}\right)^2 = \frac{25}{4} \Rightarrow C = \left(0, -\frac{5}{2}\right)$

and $a = \frac{5}{2}$; $r^2 + 5r \sin \theta = 0 \Rightarrow r = -5 \sin \theta$

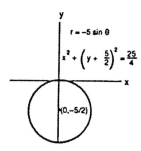

13. $x^2 + y^2 - 3x = 0 \Rightarrow \left(x - \frac{3}{2}\right)^2 + y^2 = \frac{9}{4} \Rightarrow C = \left(\frac{3}{2}, 0\right)$

and $a = \frac{3}{2}$; $r^2 - 3r \cos \theta = 0 \Rightarrow r = 3 \cos \theta$

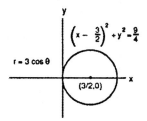

15.

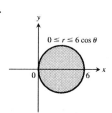

17. d   19. l   21. k   23. i

25. $A = 2\int_0^\pi \frac{1}{2} r^2 \, d\theta = \int_0^\pi (2 - \cos \theta)^2 \, d\theta = \int_0^\pi \left(4 - 4 \cos \theta + \cos^2 \theta\right) d\theta = \int_0^\pi \left(4 - 4 \cos \theta + \frac{1 + \cos 2\theta}{2}\right) d\theta$

$= \int_0^\pi \left(\frac{9}{2} - 4 \cos \theta + \frac{\cos 2\theta}{2}\right) d\theta = \left[\frac{9}{2} \theta - 4 \sin \theta + \frac{\sin 2\theta}{4}\right]_0^\pi = \frac{9}{2} \pi$

27. $r = 1 + \cos 2\theta$ and $r = 1 \Rightarrow 1 = 1 + \cos 2\theta \Rightarrow 0 = \cos 2\theta \Rightarrow 2\theta = \frac{\pi}{2} \Rightarrow \theta = \frac{\pi}{4}$; therefore

$A = 4\int_0^{\pi/4} \frac{1}{2} \left[(1 + \cos 2\theta)^2 - 1^2\right] d\theta = 2 \int_0^{\pi/4} \left(1 + 2 \cos 2\theta + \cos^2 2\theta - 1\right) d\theta$

$= 2\int_0^{\pi/4} \left(2 \cos 2\theta + \frac{1}{2} + \frac{\cos 4\theta}{2}\right) d\theta = 2 \left[\sin 2\theta + \frac{1}{2} \theta + \frac{\sin 4\theta}{8}\right]_0^{\pi/4} = 2 \left(1 + \frac{\pi}{8} + 0\right) = 2 + \frac{\pi}{4}$

29. $r = -1 + \cos \theta \Rightarrow \frac{dr}{d\theta} = - \sin \theta;$ Length $= \int_0^{2\pi} \sqrt{(-1 + \cos \theta)^2 + (- \sin \theta)^2} \, d\theta = \int_0^{2\pi} \sqrt{2 - 2 \cos \theta} \, d\theta$

$= \int_0^{2\pi} \sqrt{\frac{4(1 - \cos \theta)}{2}} \, d\theta = \int_0^{2\pi} 2 \sin \frac{\theta}{2} \, d\theta = \left[-4 \cos \frac{\theta}{2}\right]_0^{2\pi} = (-4)(-1) - (-4)(1) = 8$

31. $r = 8 \sin^3 \left(\frac{\theta}{3}\right), 0 \le \theta \le \frac{\pi}{4} \Rightarrow \frac{dr}{d\theta} = 8 \sin^2 \left(\frac{\theta}{3}\right) \cos \left(\frac{\theta}{3}\right); r^2 + \left(\frac{dr}{d\theta}\right)^2 = \left[8 \sin^3 \left(\frac{\theta}{3}\right)\right]^2 + \left[8 \sin^2 \left(\frac{\theta}{3}\right) \cos \left(\frac{\theta}{3}\right)\right]^2$

$= 64 \sin^4 \left(\frac{\theta}{3}\right) \Rightarrow L = \int_0^{\pi/4} \sqrt{64 \sin^4 \left(\frac{\theta}{3}\right)} \, d\theta = \int_0^{\pi/4} 8 \sin^2 \left(\frac{\theta}{3}\right) d\theta = \int_0^{\pi/4} 8 \left[\frac{1 - \cos \left(\frac{2\theta}{3}\right)}{2}\right] d\theta$

$= \int_0^{\pi/4} \left[4 - 4 \cos \left(\frac{2\theta}{3}\right)\right] d\theta = \left[4\theta - 6 \sin \left(\frac{2\theta}{3}\right)\right]_0^{\pi/4} = 4 \left(\frac{\pi}{4}\right) - 6 \sin \left(\frac{\pi}{6}\right) - 0 = \pi - 3$

33. $r = \frac{2}{1 + \cos \theta} \Rightarrow e = 1 \Rightarrow$ parabola with vertex at $(1, 0)$

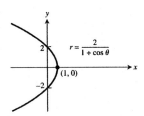

35. $r = \frac{6}{1 - 2 \cos \theta} \Rightarrow e = 2 \Rightarrow$ hyperbola; $ke = 6 \Rightarrow 2k = 6$

$\Rightarrow k = 3 \Rightarrow$ vertices are $(2, \pi)$ and $(6, \pi)$

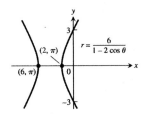

37. $e = 2$ and $r \cos \theta = 2 \Rightarrow x = 2$ is directrix $\Rightarrow k = 2$; the conic is a hyperbola; $r = \frac{ke}{1 + e \cos \theta} \Rightarrow r = \frac{(2)(2)}{1 + 2 \cos \theta}$

$\Rightarrow r = \frac{4}{1 + 2 \cos \theta}$

39. $e = \frac{1}{2}$ and $r \sin \theta = 2 \Rightarrow y = 2$ is directrix $\Rightarrow k = 2$; the conic is an ellipse; $r = \frac{ke}{1 + e \sin \theta} \Rightarrow r = \frac{(2)\left(\frac{1}{2}\right)}{1 + \left(\frac{1}{2}\right) \sin \theta}$

$\Rightarrow r = \frac{2}{2 + \sin \theta}$

41. (a) Around the x-axis: $9x^2 + 4y^2 = 36 \Rightarrow y^2 = 9 - \frac{9}{4} x^2 \Rightarrow y = \pm \sqrt{9 - \frac{9}{4} x^2}$ and we use the positive root:

$V = 2 \int_0^2 \pi \left(\sqrt{9 - \frac{9}{4} x^2}\right)^2 dx = 2 \int_0^2 \pi \left(9 - \frac{9}{4} x^2\right) dx = 2\pi \left[9x - \frac{3}{4} x^3\right]_0^2 = 24\pi$

(b) Around the y-axis: $9x^2 + 4y^2 = 36 \Rightarrow x^2 = 4 - \frac{4}{9} y^2 \Rightarrow x = \pm \sqrt{4 - \frac{4}{9} y^2}$ and we use the positive root:

$V = 2 \int_0^3 \pi \left(\sqrt{4 - \frac{4}{9} y^2}\right)^2 dy = 2 \int_0^3 \pi \left(4 - \frac{4}{9} y^2\right) dy = 2\pi \left[4y - \frac{4}{27} y^3\right]_0^3 = 16\pi$

43. (a) $r = \frac{k}{1 + e \cos \theta} \Rightarrow r + er \cos \theta = k \Rightarrow \sqrt{x^2 + y^2} + ex = k \Rightarrow \sqrt{x^2 + y^2} = k - ex \Rightarrow x^2 + y^2$

$= k^2 - 2kex + e^2 x^2 \Rightarrow x^2 - e^2 x^2 + y^2 + 2kex - k^2 = 0 \Rightarrow (1 - e^2) x^2 + y^2 + 2kex - k^2 = 0$

(b) $e = 0 \Rightarrow x^2 + y^2 - k^2 = 0 \Rightarrow x^2 + y^2 = k^2 \Rightarrow$ circle;

$0 < e < 1 \Rightarrow e^2 < 1 \Rightarrow e^2 - 1 < 0 \Rightarrow B^2 - 4AC = 0^2 - 4(1 - e^2)(1) = 4(e^2 - 1) < 0 \Rightarrow$ ellipse;

$e = 1 \Rightarrow B^2 - 4AC = 0^2 - 4(0)(1) = 0 \Rightarrow$ parabola;

$e > 1 \Rightarrow e^2 > 1 \Rightarrow B^2 - 4AC = 0^2 - 4(1 - e^2)(1) = 4e^2 - 4 > 0 \Rightarrow$ hyperbola